FUNGAL PATHOLOGY

Fungal Pathology

Edited by

J.W. Kronstad

KLUWER ACADEMIC PUBLISHERS
DORDRECHT / BOSTON / LONDON

Library of Congress Cataloging-in-Publication Data

Fungal pathology / edited by J.W. Kronstad.
    p. cm.
    ISBN 0-7923-6370-1 (alk. paper)
    1. Mycoses. 2. Medical mycology. I. Kronstad, James Warren.

    QR245 .F8535 2000
    579.5'165--dc21

                                                                    00-033061

ISBN 0-7923-6370-1

Published by Kluwer Academic Publishers,
P.O. Box 17, 3300 AH Dordrecht, The Netherlands.

Sold and distributed in North, Central and South America
by Kluwer Academic Publishers,
101 Philip Drive, Norwell, MA 02061, U.S.A.

In all other countries, sold and distributed
by Kluwer Academic Publishers,
P.O. Box 322, 3300 AH Dordrecht, The Netherlands.

Printed in the Netherlands.

TABLE OF CONTENTS

v

vi

PREFACE

This book brings together twelve chapters on fungal pathogens with the goal of presenting an overview of the current areas of activity and the common themes that pervade research on these important organisms. The timing of the book is appropriate because we have gained sufficient insight from molecular genetic analyses to begin to make some comparisons between different fungal pathogens and to discuss the key advances that have been made. The chapters provide a broad survey of the important topics in fungal pathogenesis including morphogenesis, virulence, avirulence, and signaling. The reader also will find clear discussions of parasitism, mutualism, symbiosis, evolution, phylogeny and ecology for those fungi where these issues are especially important. Finally, many of the chapters in this book illustrate the fact that we are on the verge of a revolution in our understanding of fungal pathogens because of the application of genomics to these organisms and their hosts.

The fungi included in this book represent many of the most intensively investigated fungal pathogens of plants; in this regard, a chapter is also included for pathogens in the *Phytophthora* group, even though these organisms are no longer classified as fungi. It is appropriate to include *Phytophthora* for historical reasons and, in addition, the insights in terms of pathogenesis and host-specific interactions are important to keep in mind when considering fungal pathogens. Chapters are also included on pathogens of insects and humans, as well as endophytic fungi. The endophytic fungi provide fascinating examples of the impact of fungi on human activities and food production including the influence on livestock. The chapters on the pathogens of insects and humans (*Aspergillus fumigatus*) are useful to provide comparisons with plant pathogens. It is clear that fungal pathogens share common suspected virulence properties (e.g., degradative enzymes) whether they attack plants or animals. These chapters also provide a perspective on how similar experimental questions and problems face investigators working on any fungal pathogen, no matter what the host. Although not included in this book, other fungal pathogens of animals have emerged as important models for understanding virulence. For example, information on *Candida albicans* and *Cryptococcus neoformans* is valuable for scientists working on plant pathogens to consider, particularly in the areas of attachment to host tissue and signal transduction. However, in this book we have limited ourselves to fungi that primarily display a filamentous growth morphology.

Space limitations and other constraints prevented the inclusion of chapters for other prominent plant pathogens such as the powdery and downy mildews, *Fusarium*, *Ophiostoma*, *Rhyncosporium*, and *Rhizoctonia*. These fungi will undoubtedly soon be included in other books on this subject. Fungicide resistance is also largely untouched in these chapters although some information on drug resistance is included for *A. fumigatus*. The reader is referred to an excellent web site for information on other plant pathogens and links to mycology resources to investigate these fungi (http:www.ifgb.uni-hannover.de/extern/ppigb/ppigb.htm)

Taken together, the chapters in this book present an excellent survey of the areas of investigation at the forefront of fungal pathogenesis. As the reader will note, many but not all of these chapters begin with a description of the interaction of the fungus with the host during the initiation of infection. Typically this involves the attachment or adhesion of a spore followed by germination, formation of infection structures or an infectious cell type and penetration of the host. The formation of infection structures such as appressoria have received considerable experimental attention. This is particularly true for *Magnaporthe grisea*, *Colletotrichum* spp., the insect pathogens and the rust fungi. In this regard, it is interesting to note the similarities between plant and insect pathogens (e.g., Chapters 7 and 8). The best view of appressorium formation has been developed for *M. grisea* and the genomic efforts with this fungus soon will undoubtedly reveal a detailed and complete view of the process in this fungus. The similarities between many of the fungal plant pathogens (and the insect pathogens) suggest that *M. grisea* will emerge as the paradigm for this process. It will be interesting to see whether the information from *M. grisea* will be applicable to fungi such as *Botrytis cinerea* where there is still debate about whether true infection structures are formed.

There are two other areas of note in the context of fungal differentiation during infection. These are the formation of the infectious dikaryon during the entry of *Ustilago maydis* (Chapter 12) into maize tissue and the formation of haustoria by rust fungi and other biotrophs (Chapter 10). The work on *U. maydis* provides interesting insight into the specialized role of mating-type loci in the disease process and the interconnections between pathogenesis, morphogenesis and sexual development. The chapter by Hahn (Chapter 10) provides an up to date view of haustoria in the rust fungi and the considerations for a molecular description of this process. This chapter provides an excellent description of the use of molecular approaches to identify fungal genes that are induced during the interaction of the rust with the host. A picture is emerging of the molecular details of nutrient transfer during the biotrophic interaction. Good comparisons can also be made with biotrophic growth for both *U. maydis* and *Cladosporium fulvum* (Chapter 3). Along with the work on rusts, these fungi also provide good examples of the efforts to identify and characterize fungal genes whose transcription is induced during the biotrophic interaction with the host. Similar information is also provided for *Phytophthora* spp. in Chapter 9. In addition, these chapters include information on the nutritional challenges that fungi face upon entry into the host. For example, Chapter 3 describes the results of a search for genes that are induced by starvation in *C. fulvum*. Biotrophic development also is described for *Colletotrichum* spp. in Chapter 5.

The identification and evaluation of virulence factors is a common theme for many of the chapters in this book. Restriction enzyme mediated insertion (REMI) has been a powerful tool for identifying virulence genes for a number of pathogens including *A. fumigatus*, *C. heterostrophus*, *U. maydis* and *Colletotrichum*. As noted in Chapter 3, the technique has not worked to date for *Cladosporium*. The best examples of virulence factors continue to be the host specific toxins defined for

Cochliobolus species, although toxins may also be important for *Botrytis cinerea*, *A. fumigatus* and for some of the insect pathogens. Other virulence factors include cell wall degrading enzymes and proteases as well as factors to avoid the host defense response (e.g., suppressins and antioxidant enzymes). Degradative enzymes are discussed in detail in the chapters on *A. fumigatus*, *B. cinerea*, the rust fungi, and the insect pathogens. The description of hypovirulence in *Cryphonectria parasitica* (Chapter 6) provides a fascinating example of how hypoviruses can attenuate fungal virulence. The detailed investigation of hypovirulence illustrates the importance of signal transduction in the interaction of fungi with their hosts (see below). An emerging and potentially novel story for virulence concerns the work on the path-1 mutant of *Colletotrichum magna* described in chapter 5. This mutant is able to colonize host plants but fails to cause disease. Interestingly, the mutant may stimulate a defense response and provide protection from infections by other pathogens. Other factors involved in virulence include the production of phytohormones (described for *B. cinerea* and *U. maydis*), the formation of melanin as a protective agent (*A. fumigatus*) or as part of infection structure formation (*M. grisea*, *Colletotrichum*), and the synthesis of hydrophobins by several pathogens.

Investigators are starting to accumulate avirulence genes from fungal pathogens and we will hopefully reach a level of understanding similar to that available for bacterial plant pathogens in the near future. Of course, the best studied system continues to be the *C. fulvum* - tomato interaction in which both avirulence gene products and the corresponding R genes in tomato have been characterized. Avirulence genes from *M. grisea* have also been characterized in detail and corresponding host R genes have been identified (Chapter 8). Information is also accumulating rapidly for avirulence elicitors in *Phytophthora* (Chapter 9). The chapters on rust fungi and *B. cinerea* also include information the interaction of the pathogens with the host defense response including elicitation and evasion of the host defense response. An interesting aspect of *B. cinerea* infections is the initiation of quiescent infections in which there is a delay in the formation of discernible symptoms. Host specialization, as describe in Chapter 11, also is an interesting aspect of the interactions between clavicipitaceous symbionts and grass species.

One area that has received tremendous attention in recent years is the role of signal transduction as a key aspect of fungal pathogenesis. Much of this work focuses on signaling and morphogenesis in *C. parasitica*, *M. grisea*, *U. maydis*, *Colletotrichum*, and the insect pathogens. This aspect of fungal pathogenesis has also proven to be very important in animal pathogens such as *Candida albicans* and *Cryptococcus neoformans*. As mentioned above, the story of hypovirulence for *C. parasitica* has provided an excellent illustration of the importance of signaling in virulence. Much of the work to date on fungal pathogens has involved identifying components of the signaling pathways (e.g., MAPK module genes, cAMP pathway genes). It will be particularly interesting to use our understanding of the signaling pathways to identify the upstream signals that will provide insight into the key features of the host that are perceived by pathogens. Also the detailed description of the downstream targets of the signaling pathways should provide a view of the connections between protein kinases (e.g., MAP kinases and PKA) and key factors involved in morphogenesis and virulence.

Finally, it is important to point out the common theme of genomics that pervades the chapters in this book. The chapters on *Phytophthora*, *C. heterostrophus* and *M. grisea*, in particular, include interesting information of genome structure. This is just the first glimmer of big things to come. Genomics projects are underway for many fungi and host species including most of the ones described in this book. For example, a review of the chapters will reveal discussions on genomics for *U. maydis*, *M. grisea* (and rice), *A. fumigatus*, *Phytophthora*, *Cochliobolus*, and the rust fungi. The application of genomics to fungal pathogenesis will open up a tremendous number of avenues of investigation and it is clear that we are on the verge of an amazing leap forward in our understanding of the pathogens and their interactions with hosts. In addition to the collection of EST and genomic sequence information, we will soon see microarray experiments with these fungi and their hosts.

J. W. Kronstad

Aspergillus fumigatus

C. D'ENFERT
Unité de Physiologie Cellulaire
Institut Pasteur, 25 rue du Docteur Roux, 75724 Paris Cedex 15,
France
Tel: 33 (0)1 40 61 32 57, Fax: 33 (0)1 45 68 87 90
E-mail: denfert@pasteur.fr

In the recent years, *Aspergillus fumigatus* has become the most prevalent airborne human fungal pathogen. Although *A. fumigatus* is a saprophyte that naturally develops in the soil, it is responsible for life-threatening diseases in cancer patients and organ transplant recipients in the hospital environment. The rise in the number of patients that are at risk of developing a fatal *A. fumigatus* infection together with the lack of appropriate therapies for the treatment of invasive aspergillosis has prompted a surge in the study of this opportunistic fungal pathogen. Recent reviews by Latgé (1999) and combined in a book edited by Brakhage *et al.* (1999) have extensively covered the various aspects of *A. fumigatus* pathogenicity, from clinical aspects to molecular approaches. This review will concentrate on the contribution that molecular biology has made to our understanding of *A. fumigatus* pathogenesis.

1. *Aspergillus fumigatus*: life cycle and clinical symptoms

A. fumigatus produces colonies on solid medium that are typical by their septated hyaline hyphae and by the structure of the conidiophores and colour of the conidia (Samson, 1999). Conidiophores are uniseriate with subclavate vesicles producing mostly compact columnar heads, with flask-shaped phialides and chains of gray-green to dark-blue green conidia. Conidia may be rough-walled to nearly smooth, globose to ellipsoidal and are 2.5 to 3 μm in diameter (see Fig. 2). *A. fumigatus* does not have a sexual stage and therefore only reproduces through the production of conidia. However, *A. fumigatus* is closely related both morphologically and genetically to several

1

J. W. Kronstad (ed.), Fungal Pathology, 1–32.
© 2000 *Kluwer Academic Publishers. Printed in the Netherlands.*

Neosartorya species that have a sexual stage but are not teleomorphs of *A. fumigatus*. A consequence of the lack of a sexual stage lies in the absence of classical genetics for *A. fumigatus* and this has somehow hampered the development of molecular tools for this species.

Several features of the *A. fumigatus* species might be relevant to its pathogenicity: small spore size that may facilitate the penetration of the conidia along the respiratory tract; spore hydrophobicity that probably facilitates the dispersion of the species in the environment and might contribute to some extent to the interaction with host proteins; thermophilicity that allows fast growth-rates at temperature of 37°C and above; ability to utilize a broad range of carbon and nitrogen sources as nutrients.

In immuno-competent individuals, *A. fumigatus* conidia are rapidly eliminated by innate immune mechanisms including phagocytosis of the conidia by alveolar macrophages and killing of the hyphae by polymorphonuclear leukocytes. However, allergic diseases can develop in individuals that are repeatedly exposed to *A. fumigatus* conidia. Three main pathologies are observed that are accompanied by mycelial development of the fungus (Denning, 1998; Latgé, 1999; Rüchel and Reichard, 1999 and references therein):

(1) *Allergic Bronchopulmonary Aspergillosis* (ABPA) occurs in patients suffering from atopic asthma or cystic fibrosis. ABPA is associated with a chronic inflammatory response of proximal bronchi that can lead to fibrosis and bronchiectasis.

(2) *Aspergilloma* occur in pre-existing pulmonary cavities resulting from previous lung diseases *e.g.* tuberculosis. Aspergilloma consist of a solid mass of *Aspergillus* mycelium (fungus ball) packed with a proteinaceous matrix. Aspergilloma are mostly asymptomatic although they can be associated with localized hemoptysis resulting from blood vessel disruption.

(3) *Invasive Aspergillosis* is a life-threatening disease occuring mostly in patients with acute leukemia or allogenic bone-marrow transplant but also in patients that have received cytotoxic treatments of blood diseases, autologous bone-marrow transplant or solid-organ transplant. These patients have a profound and prolonged neutropenia and frequently receive corticosteroid treatment. Invasive aspergillosis is always associated with the respiratory tract but can disseminate through blood vessels to other organs, in particular the brain. Current treatment of invasive aspergillosis is restricted to the use of amphotericin B or itraconazole that suffer from toxicity or lack of efficacy, respectively.

Because of the high mortality rate that is associated with systemic *A. fumigatus* infections in the hospital environment, a large number of studies are now focusing on an understanding of the physio-pathology of invasive aspergillosis. In this regard, animal models of invasive aspergillosis have been developed. Animals are subjected to a corticotherapy with or without treatment with an additional immunosuppressive drug (*e.g.* cyclophosphamide) and the inoculum is delivered via the intranasal route (Latgé,

1999). A major limitation of these models, however, is the lack of reliability and the need to use a high inoculum that differs substantially from what is observed in the human situation.

2. Molecular diagnosis and epidemiology

One of the applications of molecular biology techniques could be to improve the diagnosis of aspergillosis, in particular invasive aspergillosis for which diagnosis can not rely on the detection of a specific immune response directed towards antigens of the fungus but involves the detection of antigens produced by the fungus in the course of its development. In this regard, several PCR tests have been developed based on the sequence of *A. fumigatus* ribosomal RNA or mitochondrial DNA as well as genes encoding a ribotoxin (see section 7.1) and a secreted alkaline protease (see section 6.1) (Latgé, 1999 and references therein). Although application of these PCR tests yields false positives when broncho-alveolar lavages are assayed, very promising results have been obtained when they are applied to serum or plasma samples. Nevertheless detection of *A. fumigatus* is obtained at a later stage in the course of the disease than when fungal antigens are assayed. In this regard, detection in blood samples of galactomannnan - a major component of the *A. fumigatus* cell wall that is secreted in the course of the infection (Latgé *et al.*, 1994a) - appears the most sensitive and reliable diagnosis of invasive aspergillosis to date (Latgé, 1999).

In addition to the use of DNA technology in the diagnosis of aspergillosis, several genotypic methods have been developed in order to discriminate *A. fumigatus* isolates and to evaluate their genetic diversity and relatedness. Typing using RAPD (Rapid Amplification of Polymorphic DNA) has been used in several instances but is somewhat unreliable. In contrast, amplification and sequencing of microsatellites appear to be more reliable and allow efficient discrimination of a large number of isolates when the analysis of four microsatellites [$(CA)_9(GA)_{25}$, $(CA)_2C(CA)_{23}$, $(CA)_8$ and $(CA)_{21}$] is combined (Bart-Delabesse *et al.*, 1998).

An alternative to the PCR-based methods is the hybridization of genomic DNA with midly repetitive DNA sequences, a method that has been used successfully to type clinical and environmental isolates of *A. fumigatus* (Girardin *et al.*, 1994a; Girardin *et al.*, 1994b; Debeaupuis *et al.*, 1997; Neuvéglise *et al.*, 1997; Chazalet *et al.*, 1998). This method takes advantage of a DNA sequence identified by Girardin *et al.* (1993) that is specific for *A. fumigatus* and is occuring at least 10 times in the genome. This sequence can therefore be used to generate hybridization profiles that are specific of each *A. fumigatus* isolate. This repetitive element, referred to as *Afut1*, has been characterized in detail and corresponds to a defective retrotransposon of 6.9 kb that has a genetic organization typical of the *Drosophila* gypsy family of retrotranspons (Neuvéglise *et al.*, 1996). Inactivation of the two copies of *Afut1* that have been analyzed appears to have occured by a mechanism similar to RIP (repeated-induced point mutation) process of *Neurospora crassa* (Neuvéglise *et al.*, 1996; Selker, 1997). However, two basic phenomena that are associated with RIP, namely sexual reproduction and cytosine methylation, are absent in *A. fumigatus* suggesting that inactivation of *Afut1* may have arisen by a totally distinct process or prior to the loss of DNA methylation and sexual reproduction by *A. fumigatus*. Whether an active copy of *Afut1* or of other transposons exists in some *A. fumigatus* strains remains to be evaluated.

The *Afut1* sequence has been used to type a large number of *A. fumigatus* isolates and two important conclusions have been drawn from these experiments: 1) it is not possible to discriminate between environmental and clinical isolates suggesting that all *A. fumigatus* strains can colonize the host (Debeaupuis *et al.*, 1997); and 2) most patients with invasive aspergillosis are colonized by a single isolate that can be found in the hospital environment in relatively frequent instances, thus demonstrating the nosocomial origin of the disease (Chazalet *et al.*, 1998).

3. Genetic engineering of *Aspergillus fumigatus*

Analysis of *A. fumigatus* interaction with the host has been significantly improved since the development of transformation procedures for this fungus and the first identification of knock-out mutants in the early 1990's by the groups of Holden and co-workers (Tang *et al.*, 1992; Smith *et al.*, 1993) and Latgé and Monod and co-workers (Monod *et al.*, 1993b; Paris *et al.*, 1993). Indeed, the behavior in animal models of mutants that had been obtained through chemical or physical mutagenesis could not be taken as clear evidence of the involvement of a specific process in pathogenicity because of the lack of a sexual cycle in *A. fumigatus* that would allow back-crossings of the mutations. Transformation of *A. fumigatus* is now routinely used in several laboratories and several techniques have been adapted to allow the introduction of specific alleles in the genome of *A. fumigatus*. These techniques have been recently reviewed in detail by d'Enfert *et al.* (1999) and will only be summarized here.

3.1 TRANSFORMATION PROCEDURES

Entry of DNA into *A. fumigatus* has been obtained using standard protoplast fusion procedures (Paris, 1994; Smith, 1994) as well as electroporation of intact conidia (Brown *et al.*, 1998; Weidner *et al.*, 1998). While protoplast transformation yields lower efficiencies (2-100 transformants/µg DNA) than those obtained with electroporation (200-1000 transformants/µg DNA), it often results in higher frequencies of site-specific integration of the transforming DNA molecule. In this regard, protoplast transformation has been used preferentially to construct specific *A. fumigatus* mutant strains by homologous recombination while electroporation provides an efficient means to construct libraries of insertional mutants (Brown *et al.*, 1998).

3.2 TRANSFORMATION MARKERS

Selection of transformants requires the use of a transformation marker that will help to distinguish those cells that have integrated exogenous DNA from those that have not. *A. fumigatus*, like several other filamentous fungi, is sensitive to different antibiotics including hygromycin and phleomycin. Consequently, hygromycin-resistance and phleomycin-resistance transformation cassettes first developed for use in *A. niger* (Punt *et al.*, 1987; van den Hondel and Punt, 1991) and composed of a bacterial antibiotic resistance gene expressed under the control of the *A. nidulans gpdA* or *trpC* promotor and terminator, have been used successfully to transform *A. fumigatus* (Tang *et al.*, 1992; Monod *et al.*, 1993b; Jaton-Ogay *et al.*, 1994; Smith *et al.*, 1994; d'Enfert *et al.*, 1999). A major advantage of these dominant selective markers is that they can be applied to virtually all strains of *A. fumigatus*, including natural isolates.

Alternative transformation markers have been developed that rely on the complementation of the defect of an auxotrophic mutant. Although the lack of a sexual cycle would preclude the use of such mutants, strategies are available to positively select spontaneous auxotrophs that have defined recessive mutations: *pyrG* mutants defective for orotidine-5'-phosphate (OMP) decarboxylase activity and hence auxotrophic for uridine or uracil have been identified among strains that are resistant to 5-fluoro-orotic acid (5-FOA) (d'Enfert, 1996); *niaD* mutants lacking nitrate reductase activity and hence unable to grow on media containing nitrate as the nitrogen source have been identified among strains that are resistant to chlorate (Monod *et al.*, 1993b). Complementation of the *pyrG* defect using the *pyrG* gene from *A. fumigatus* (Weidner *et al.*, 1998) or from other species [*A. niger* (d'Enfert, 1996); *N. crassa*, C. d'Enfert, unpublished data] has been obtained and used to produce different types of mutants. In contrast, transformation of the *niaD* mutant with the *niaD* gene from *A. fumigatus* (Amaar and Moore, 1998) or from other species has not been reported. More recently, *niiA* mutants defective for the nitrite reductase, and hence unable to grow on media containing nitrite as the nitrogen source, have been constructed by genetic engineering and used as recipients for constructs containing the *A. nidulans* or *A. fumigatus niiA* gene (Y.G. Amaar and M.M. Moore, personal communication) thus providing an additional transformation marker for *A. fumigatus*. Similar approaches may be used with other cloned *A. fumigatus* nutritional markers (Borgia *et al.*, 1994; Gavrias *et al.*, 1997). However, the use of auxotrophic markers has several drawbacks: 1) they cannot be used to transform natural isolates of *A. fumigatus*; and 2) some auxotrophs of *A. fumigatus* [*e.g. pyrG* (d'Enfert *et al.*, 1996) and *pabaA* (Brown *et al.*, 1997)] have a reduced virulence in a mouse model of invasive aspergillosis and the integration of the transformation marker at a specific site in the genome other than the original locus may not be sufficient to complement this defect as observed in *Candida albicans* (Lay *et al.*, 1998).

3.3 GENETIC MANIPULATIONS

The avaibility of transformation procedures and transformation markers provides numerous approaches for the modification of the *A. fumigatus* genome and the study of gene function in this pathogenic fungus (d'Enfert *et al.*, 1999). Briefly, DNA constructs can integrate through either a single cross-over or a double cross-over at a locus of interest. Single cross-over integrations are used to convert a wild-type allele into two non-functional alleles (*gene disruption*), to integrate a gene fusion or a particular construct at the locus under study (Smith *et al.*, 1994) or at a specific locus in the genome *e.g. pyrG* (Weidner *et al.*, 1998), to integrate a mutant allele at the locus of interest and, when a counter-selectable marker is used, force the replacement of the wild-type allele by the mutant allele (*pop-in/pop-out*). Double cross-over integrations are used to introduce a null mutation at a locus of interest by *gene replacement*. The successive introduction of several null mutations by gene replacement at different loci can be achieved by using different transformation markers or a composite transformation marker referred to as a *pyrG*-blaster that contains the *A. niger pyrG* gene flanked by a direct repeat that encodes the neomycin phosphotransferase gene (*neo*) of transposon Tn*5* (d'Enfert, 1996). Recombination between the two elements of the direct repeat can be forced in the presence of 5-FOA, yielding a strain that has retained the null mutation because of the persistence of one element of the repeat at the locus and that can be transformed again with the *pyrG* gene.

The different methodologies highlighted in the previous section rely on two critical factors: 1) the unability of the transforming DNA molecule to replicate autonomously in the cell; and 2) its ability to recombine with homologous DNA sequences in the genome. However, autonomously replicating plasmids carrying the AMA1 plasmid replicator (Aleksenko and Clutterbuck, 1997) can be introduced into *A. fumigatus* with high transformation efficiencies and used to clone genes by functional complementation of *A. fumigatus* mutants (Langfelder *et al.*, 1998; d'Enfert *et al.*, 1999). Besides, DNA molecules that do not carry *A. fumigatus* sequences integrate randomly in the genome, a phenomenon that can be used to generate a collection of insertional mutants of *A. fumigatus* (Brown *et al.*, 1998; Weidner *et al.*, 1998). While electroporation probably represents the method of choice to generate such collections since it predominantly results in single-copy integration of the transforming DNA (Brown *et al.*, 1998; Weidner *et al.*, 1998), Brown *et al.* have shown that REMI [Restriction Enzyme –Mediated Insertion; (Schiestl and Petes, 1991)] can be applied to *A. fumigatus*: addition of some restriction enzymes (*Kpn*I, *Xho*I) in the transformation mixture can increase protoplast transformation efficiencies and promote single-copy integration of the transforming DNA at corresponding restriction enzyme cleavage sites in the genome.

4. Spore structure and interaction with the host

Conidia of *A. fumigatus* ensure the dispersal of the fungus in the air and its first encounter with the host. It is therefore likely that spore components are involved in an interaction with host cells, host matrix components including fibrinogen, laminin, collagen and circulating components of the innate immune response like complement. Specific interactions between *A. fumigatus* conidia and fibrinogen, human complement component C3 or laminin have been demonstrated and conidial proteins that are likely to mediate these interactions have been identified (Annaix *et al.*, 1992; Sturtevant and Latgé, 1992; Tronchin *et al.*, 1997). However, the precise role of these candidate receptors with respect to adhesion of the conidia to host components or virulence has not been tested yet.

In contrast, factors that mediate some of the distinctive features of *A. fumigatus* conidia, in particular their high hydrophobicity and green pigmentation, have been characterized at the molecular level and their role in virulence has been evaluated.

4.1 SPORE HYDROPHOBICITY AND THE ROLE OF THE RODLET LAYER

Like other asexual spores of many fungal species, *A. fumigatus* conidia have an outermost cell wall layer that is composed of 5-10 nm wide rodlets. Biochemical and molecular characterization of these rodlets in other species has shown that their structural components are small cysteine-rich proteins referred to as hydrophobins (reviewed in Kershaw and Talbot, 1998). Hydrophobins are capable of interfacial self-assembly, a process that is responsible for the hydrophobicity of aerial hyphae and spores. The *A. fumigatus rodA/hyp1* gene encoding a 159 amino acid hydrophobin has been characterized (Parta *et al.*, 1994; Thau *et al.*, 1994) and analysis of *rodA* mutants obtained by gene replacement has shown that the *rodA* gene product contributes to the hydrophobicity of *A. fumigatus* conidia and to the formation of the rodlet layer (Thau *et al.*, 1994). Analysis of this mutant did not reveal any significant defect in adhesion to pneumocytes, fibrinogen or laminin and in pathogenicity in a mouse model of invasive aspergillosis suggesting that conidial hydrophobicity is not an important factor in *A. fumigatus* entry into the host. However, it is important to stress that, in these experiments, *A. fumigatus* conidia were delivered to the lung of immuno-suppressed mice in a rehydrated state and this may partially mask the contribution of hydrophobicity to the infection process. In this regard, the reduced binding of the *rodA* mutant to collagen or BSA suggests that the RodA hydrophobin may mediate non-specific hydrophobic interactions with host components (Thau *et al.*, 1994). Whether these interactions contribute to some extent to the infection process is not known.

The *A. fumigatus* RodA/Hyp1 protein can functionally substitute the highly similar *A. nidulans* RodA protein (Parta *et al.*, 1994). In *A. nidulans*, an additional hydrophobin, DewA, has been identified and shown to contribute, although to a lesser extent than RodA, to spore hydrophobicity (Stringer and Timberlake, 1995). Biochemical characterization of the rodlet layer in A. *fumigatus* has revealed a second hydrophobic protein (S. Paris, personal communication). Its contribution to A. *fumigatus* interaction with host components and virulence remains to be investigated.

4.2 PIGMENT BIOSYNTHESIS

A wide variety of fungi produce pigments that are thought to contribute to their virulence. This is the case for melanin that has been proposed to protect human pathogenic fungi like *Cryptococcus neoformans* and *Exophiala (Wangiella) dermatitidis* from host defense mechanisms (Wang *et al.*, 1995; Scnitzler *et al.*, 1998) and to contribute to infection structure formation in plant pathogenic fungi like *Magnaporthe grisea* (Howard and Valent, 1996). Environmental isolates of A. *fumigatus* that are unable to produce the typical green pigment found in conidia display a reduced virulence in an animal model of invasive aspergillosis, although they are able to cause a fatal systemic aspergillosis (Latgé *et al.*, 1997; Aufauvre-Brown *et al.*, 1998; J. Sarfati and J.-P. Latgé, personal communication). Recently, two A. *fumigatus* genes have been cloned that are involved in melanin biosynthesis thus allowing a better understanding of the role that conidial pigments play with respect to the survival of the spore in the host.

arp1 encodes a protein with homology to scytalone dehydratase, an enzyme involved in 1,8-dihydroxynaphtalene-melanin synthesis (Fig. 1) in several phytopathogenic fungi. Inactivation of *arp1* results in colonies that produce reddish-pink conidia (Tsai *et al.*, 1997). Interestingly, conidia of the *arp1* mutant bind complement component C3 more avidly than wild-type conidia, suggesting that *in vivo* the green pigment may protect the conidia from complement opsonization and further killing by the immune system. However, the behavior of the *arp1* mutant in an animal model of invasive aspergillosis has not been reported. Whether the fact that a homologue of *arp1* is absent in A. *nidulans* (Tsai *et al.*, 1997) is relevant to the differential infectivity that is observed between A. *fumigatus* and A. *nidulans* remains to be evaluated.

alb1/pksP is located adjacent to *arp1* and encodes a polyketide synthase that is involved in the production of 1,3,6,8-tetrahydroxynaphtalene, a precursor of scytalone and melanin (Fig. 1; Langfelder *et al.*, 1998; Tsai *et al.*, 1998). *alb1/pksP* mutants produce white (albino) conidia that have lost their typical echinulate surfaces (Fig. 2; Jahn *et al.*, 1997; Langfelder *et al.*, 1998; Tsai *et al.*, 1998). Interesting additional phenotypes are observed: 1) increased susceptibility to damage by oxidants; 2) increased stimulation of reactive oxygen species production by PMN or monocytes challenged with conidia; 3) increased binding to complement component C3; 4) increased neutrophil-mediated phagocytosis; and 5) markedly reduced virulence in a model of aspergillosis that uses non-immunocompromized animals. Although this latterresult contrasts with those obtained when environmental albino mutants are used to

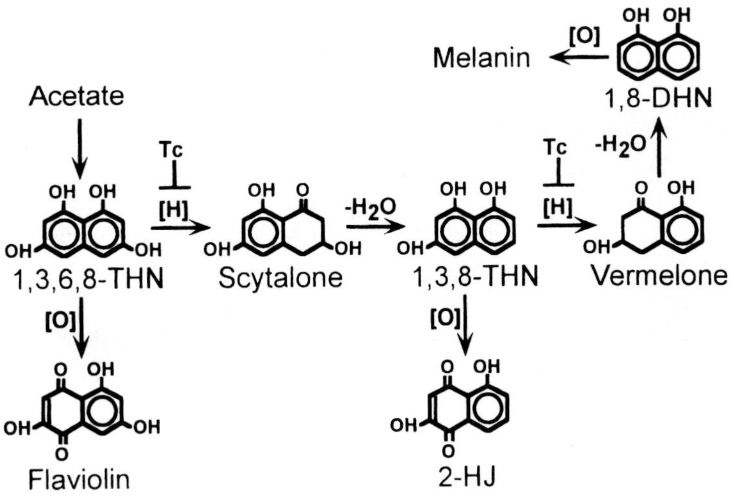

Figure 1 : Biosynthetic pathway of 1,8-dihydroxynaphtalene-melanin in brown and black fungi. Tc, tricyclazole. From Tsai *et al.* (1998) with permission.

infect immuno-suppressed mice through the air-ways (Latgé *et al.*, 1997; J. Sarfati and J.-P. Latgé, personal communication), they point to a role of the melanin-like pigment of *A. fumigatus* as a protectant against several host defense mechanisms, in particular complement deposition, phagocytosis and oxidative burst. It is not clear whether this protective role is served by the pigment itself or by the morphological modification that the conidia adopt upon active pigment biosynthesis and that may result in the masking of conidial proteins which have the potential to be recognized by the host immune system, *e.g.* complement component C3.

Transcription of the *arp1* and *alb1/pksP* genes and the consequent production of the melanin-like pigment has been shown to occur only during *A. fumigatus* conidiogenesis *in vitro*. While this process is likely to contribute to the protection of the mature conidia during the early stages of infection, there is no evidence that pigments are produced once spore germination is initiated and could protect invading hyphae from host defenses. Nevertheless, it has been shown in *A. nidulans* that genes which are normally induced during conidiogenesis can also be induced in response to different environmental stimuli (Skromne *et al.*, 1995; Hicks *et al.*, 1997). In this regard, analysis of the transcriptional status of the *arp1* and *alb1/pksP* genes during *A. fumigatus* development in the lung of immuno-suppressed animals might contribute to a better understanding of the role of pigments during aspergillosis.

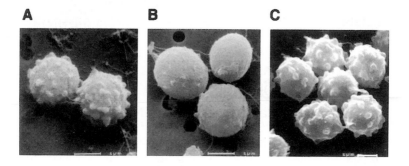

Figure 2: Scanning Electron Micrograph study of the surface of conidia of a wild-type *A. fumigatus* (A), *alb1* mutant (B) and *alb1*-complemented strain (C). Bars, 1 μm. From Tsai *et al.* (1998) with permission.

5. Biosynthetic pathways required for survival of *A. fumigatus* in the host

Establishment of an invasive aspergillosis requires that conidia are able to colonize specific ecological niches where some nutrient may not be readily accessible. An example of such a niche is the phagocytic compartment of alveolar macrophages where spore germination occurs when phago-lysosomal fusion is inhibited due to immuno-suppressive regimens. Analysis of auxotrophs of *A. fumigatus* or *A. nidulans* in murine model of invasive aspergillosis has revealed several biosynthetic pathways, the function of which is essential to the progression of the disease.

Early studies by Purnell (1973) and Sandhu *et al.* (1976) suggested a requirement for biosynthesis of the vitamin precursor *p*-aminobenzoic acid (PABA) for the establishment of aspergillosis. This has been recently confirmed using *A. nidulans* and *A. fumigatus* mutants. Tang *et al.* (1994) have shown that several well-defined *pabaA1* mutant strains of *A. nidulans* are strictly non-pathogenic in a mouse model of invasive pulmonary aspergillosis sensitive enough to assess the virulence of *A. nidulans*, a species that is not considered otherwise an opportunistic pathogen. Strikingly, *pabaA1* mutants regain virulence when PABA is supplied to the animals through their drinking water and colonization of the lung can then be stopped by withdrawal of the supplement, suggesting that PABA is probably not available in the lung. That PABA biosynthesis is essential for virulence of *A. fumigatus* in a mouse model of invasive aspergillosis has now been confirmed through the characterization of *A. fumigatus* strains with a null mutation in the *pabaA* gene encoding PABA synthase (Brown *et al.*, 1997). Interestingly, *pabaA* mutants were also recovered in a screen of insertional mutants of *A. fumigatus* that show reduced virulence (J. Brown and D. Holden, personal communication). Because biosynthesis of PABA does not occur in mammalian cells, it represents an interesting target for the development of novel antifungal compounds.

Similar results have been obtained by d'Enfert *et al.* (1996) using *pyrG* mutants of *A. fumigatus*. These mutants are auxotrophic for uridine and uracil because of a mutation in the *pyrG* gene encoding orotidine-5'-decarboxylase that catalyzes the penultimate step of *de novo* UMP biosynthesis. *A. fumigatus pyrG* mutants are entirely non-pathogenic in a murine model of invasive aspergillosis and mutant conidia remain ungerminated in alveolar macrophages (Fig. 3). Germination and virulence of the mutant can be restored by supplying uridine to the animals through the drinking water or by transformation with a functional *pyrG* gene, suggesting that uridine and uracil are limiting in the lung environment and more generally in mammalian tissues as suggested from the study of bacterial and fungal pathogens with defects in pyrimidine biosynthesis (Kirsch and Whitney, 1991; Mahan *et al.*, 1993; Cole *et al.*, 1995).

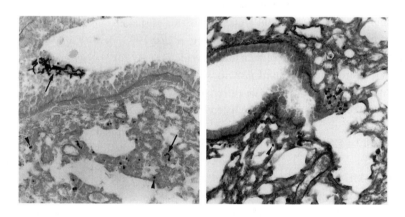

Figure 3: Paraffin section from the lungs of immunosuppressed mice inoculated with a wild-type strain (left panel) and a *pyrG* mutant strain (right panel) of *A. fumigatus* and sacrificed after 24 hours. Conidia and hyphal elements are indicated by arrowheads and arrows, respectively. Magnifcation, *ca.* x 50. Modified from d'Enfert*et al.* (1996).

Other biosynthetic pathways are likely to be required for the colonization of tissues by *A. fumigatus*. These include biosynthesis of adenine, pyridoxine and of the essential amino acid lysine (Purnell, 1973; Tang *et al.*, 1994). Analysis of bacterial pathogens has revealed additional biosynthetic pathways that are required for virulence or turned on during infection because of limiting avaibility of some nutrients in tissues (Groisman and Ochman, 1994; Valdivia and Falkow, 1997). Results obtained with the *A. fumigatus pyrG* mutant suggests that some of the information available for bacteria could be transposed to *A. fumigatus*. Analysis of the metabolic pathways in question and of their role in virulence may therefore help our understanding of *A. fumigatus* pathogenicity.

6. Is there a role for proteolysis during aspergillosis ?

Because proteins are a major constituent of the lung, the ability to degrade proteins, in particular elastin and collagen, has long been regarded as a potential virulence factor for *A. fumigatus*. This assumption was substantiated by different reports showing that, in particular, elastase production is correlated with the ability of *A. fumigatus* to cause invasive aspergillosis in mice (Khotary *et al.*, 1984) and that *Aspergillus* sp. strains isolated from patients produce elastase systematically in contrast to environmental isolates (Rhodes *et al.*, 1988). However, the fact that penetration of the hyphae in lung tissues of humans or experimental animals is associated with little or no alteration of the collagen and elastin matrix would argue against a significant role for proteolysis in the invasion process (Denning *et al.*, 1992; Tang *et al.*, 1993; Jaton-Ogay *et al.*, 1994). Since the development of molecular biology techniques for *A. fumigatus*, numerous studies have addressed the role of proteases with respect to *A. fumigatus* virulence (see Monod *et al.*, 1999 for review). The different proteases that have been characterized at the biochemical or molecular level are listed in Table 1.

6.1 EXTRACELLULAR ENDOPROTEASES

The ALP, MEP and PEP proteases are the three major proteases secreted when *A. fumigatus* is cultivated *in vitro* in the presence of proteins as the sole nitrogen source. Their expression is inhibited when peptides or free amino acids are available (Reichard *et al.*, 1990; Monod *et al.*, 1991; Monod *et al.*, 1993a; Markaryan *et al.*, 1994; Reichard *et al.*, 1994; Lee and Kolattukudy, 1995). Characterization of the genes encoding ALP, MEP and PEP has shown that these three proteases are synthesized as precursors with a signal peptide and a pro-peptide that is cleaved prior to the secretion of the enzymes in the external milieu (Jaton-Ogay *et al.*, 1992; Jaton-Ogay *et al.*, 1994; Ramesh *et al.*, 1994; Sirakova *et al.*, 1994; Lee and Kolattukudy, 1995; Reichard *et al.*, 1995). The mature ALP protease is a serine alkaline protease that belongs to the subtilisin family

TABLE 1 . Summary of proteases characterized at the biochemical or genetic level in *A. fumigatus*

Name	Type of activity	MW	Subcellular location	*A. fumigatus* mutant
ALP[1]	serine alkaline protease	33 kDa	Extracellular	+
ALP2[2]	serine alkaline protease	38 kDa	Cell wall	-
MEP[3]	metalloprotease	40 kDa	Extracellular	+
PEP[4]	aspartic protease	38 kDa	Extracellular	+
PEP2[5]	aspartic protease	36 kDa	Cell wall	-
MEP20[6]	neutral protease	20 kDa	Unknown	-
DPP-IV[7]	di-peptidyl-peptidase IV	94 kDa	Extracellular	-
DPP-V[8]	di-pepyidyl-peptidase V	88 kDa	Extracellular Spore wall	-
MepB[9]	Thymet oligopeptidase	82 kDa	Cytoplasmic	+

[1](Reichard *et al.*, 1990; Monod *et al.*, 1991; Jaton-Ogay *et al.*, 1992; Tang *et al.*, 1992; Monod *et al.*, 1993b; Moutaouakil *et al.*, 1993); [2](Reichard *et al.*, 1999a); [3](Monod *et al.*, 1993a; Jaton-Ogay *et al.*, 1994; Markaryan *et al.*, 1994; Sirakova *et al.*, 1994); [4](Reichard *et al.*, 1994; Lee and Kolattukudy, 1995; Reichard *et al.*, 1995; Reichard *et al.*, 1997); [5](Reichard *et al.*, 1997; Reichard *et al.*, 1999b); [6](Ramesh *et al.*, 1995); [7](Beauvais *et al.*, 1997c); [8](Moutaouakil *et al.*, 1993; Beauvais *et al.*, 1997b); [9](Ibrahim-Granet *et al.*, 1994; Ibrahim-Granet and d'Enfert, 1997)

and is able to cleave both collagen and elastin (Jaton-Ogay *et al.*, 1992; Ramesh *et al.*, 1994). MEP is a Zinc metalloprotease that cleaves collagen and has little elastinolytic activity (Jaton-Ogay *et al.*, 1994; Sirakova *et al.*, 1994). PEP is a member of the aspergillopepsin family of *Aspergillus* sp. aspartyl proteases and is active against collagen, laminin and elastin (Lee and Kolattukudy, 1995; Reichard *et al.*, 1995). Evidence for the production of these three proteases during the course of the infection has been established by immunodetection in infected tissues (Reichard *et al.*, 1990; Markaryan *et al.*, 1994; Reichard *et al.*, 1994; Lee and Kolattukudy, 1995) and in the case of the alkaline protease ALP, by detecting β-galactosidase produced in lungs of mice infected with a *A. fumigatus* strain carrying a gene fusion between the *ALP* gene and *lacZ* (Smith *et al.*, 1994). Furthermore, antibodies directed against ALP, MEP, and PEP are present in the sera of patients with an aspergilloma (Monod *et al.*, 1999).

However, the results obtained through the study of mutants that lack either one or two of the extracellular endo-proteases (*alp⁻, mep⁻, pep⁻, alp⁻ mep⁻*) suggest that these enzymes do not contribute significantly to the growth of the fungus in immuno-suppressed animals (Tang *et al.*, 1992; Monod *et al.*, 1993b; Jaton-Ogay *et al.*, 1994; Reichard *et al.*, 1997). Indeeed, no difference in pathogenicity could be observed between the single mutants, the double mutant and a wild-type strain in different animal models of invasive aspergillosis and when animals were co-infected with mutant and wild-type strains, an approach that can reveal rather subtle differences in pathogenicity (Tang *et al.*, 1993). Nevertheless, it can still be argued that proteolysis of lung matrix components has a role during invasive aspergillosis since the lack of one or two proteases could be compensated by an increased expression of other proteases. Such proteases could be ALP and/or MEP in the *pep⁻* mutant and PEP in the *alp⁻mep⁻* mutant although an increased expression of these enzymes has not been observed *in vitro* in the different mutants (Monod *et al.*, 1993b; Jaton-Ogay *et al.*, 1994; Reichard *et al.*, 1997). Other proteases could be the MEP20 neutral protease (Ramesh *et al.*, 1995), the PEP2 wall-associated aspartyl protease (Reichard *et al.*, 1997) and the ALP2 alkaline protease that has recently characterized (Reichard *et al.*, 1999a). In this regard, the construction of a A. *fumigatus* strain with multiple knock-outs in genes encoding extracellular proteases may help precise the role of proteolysis during systemic aspergillosis. Furthermore, while ALP, MEP and PEP are the three major proteases produced *in vitro*, this might not be the case during invasion of the lung tissues. A better knowledge of the respective levels of *in vivo* expression of the known genes encoding extracellular proteases would probably be quite informative.

6.2 *Af*AREA AND THE REGULATION OF NITROGEN UTILIZATION

The recent characterization of the *areA* gene of A. *fumigatus* and corresponding null mutants has shed additional light on the role of proteolysis and nitrogen utilization during invasive aspergillosis. In A. *nidulans*, the *areA* gene encodes a positive-acting transcriptional regulator that controls the expression of genes involved in the utilization of nitrogen sources other than ammonium and glutamine (Arst and Cove, 1973; Kudla *et al.*, 1990). Hensel *et al.* (1998) have cloned a A. *fumigatus* homolog of *areA* (*AfareA*) and constructed *AfareA⁻* strains by gene disruption or gene replacement. These mutant strains, like the *areA'* mutants of A. *nidulans*, are unable to use a broad range of nitrogen sources including proteins. This latter phenotype is likely to result from the observed defect in the synthesis of extracellular alkaline and acidic proteases. In addition, the *AfareA⁻* strains show a reduced virulence when compared to a wild-type strain in an animal model of invasive aspergillosis using single-strain infection as well as mixed infections. Furthermore, when reversion of the *AfareA⁻* is possible, because of the nature of the mutation, reversion to the wild-type genotype is observed at a high frequency *in vivo* suggesting that a functional AREA protein provides a selective advantage during invasion of lung tissues. Taken together, these data suggest that *Af*AREA contributes to, but is not essential for, virulence of A. *fumigatus* and this is probably because ammonium and/or glutamine are limiting in lung tissues.

Strains carrying extragenic suppressors of the *AfareA⁻* mutation have been selected for their ability to grow on some nitrogen sources that the *AfareA⁻* mutant cannot utilize (Hensel *et al.*, 1998). Interestingly, one of these strains, although it cannot use proteins as a nitrogen source, has a level of virulence that is higher than that of the *AfareA⁻* mutant but remains lower than that of a wild-type strain. A distinctive feature of this revertant is its ability to grow on different amino acids including alanine, arginine, aspartate, glycine and ornithine suggesting that utilization of one or several of these amino acids facilitates the growth of the fungus in the lung.

Although, it could here again be argued that some proteases are not subject to transcriptional control by *Af*AREA, the difference observed between the *AfareA⁻* strain and the suppressor strain that shows a correlation between the virulence level and the ability to utilize specific amino acids but not to degrade proteins, suggests that a key component of *A. fumigatus* pathogenicity lies in its ability to utilize a broad spectrum of nutrients to achieve fast growth rates. In this regard, proteolysis is probably not essential to the progression of the disease because of the availability in the lung of nitrogen sources other than proteins. A role for proteolysis might be found in the localized softening of the lung matrix through wall-associated proteases like the PEP2 and ALP2 proteases (Reichard *et al.*, 1997; Reichard *et al.*, 1999a). Whether such proteases are controlled by *Af*AREA and have a role during invasion remains to be investigated.

6.3 EXTRACELLULAR DIPEPTIDYL-PEPTIDASES

In addition to the extracellular endo-proteases described above, *A. fumigatus* secretes both *in vitro* and *in vivo* two dipeptidyl-peptidases (Beauvais *et al.*, 1997b; Beauvais *et al.*, 1997c). DPP-IV is a 95 kDa glycoprotein that belongs to the class IV of dipeptidyl-peptidases and removes X-Pro and to some extent X-Ala di-peptides from the N-termini of peptides (Beauvais *et al.*, 1997c). DPP-V is a 88 kDa glycoprotein that cleaves N-teminal X-Ala, His-Ser and Ser-Tyr dipeptides and therefore constitutes the unique member of a new class of dipeptidyl-petidases (Beauvais *et al.*, 1997b). A distinctive feature of DPP-IV and DPP-V is that they are secreted proteins unlike all the other known dipeptidyl-peptidases, suggesting that they may have a significant role during interactions between *A. fumigatus* and its host. Although this has not been addressed through the study of knock-out mutants yet, several lines of evidence are in agreement with such a role. First, DPP-IV can bind to and release a dipeptide from collagen and may participate in the localized softening of the lung matrix. Alternatively, this binding may result in the activation of T cells, a process that can be mediated in humans and animals by class IV dipeptidyl-peptidases like the CD26 surface differentiation marker (Naquet *et al.*, 1988; Dang *et al.*, 1990). Furthermore, class IV dipeptidyl-peptidases can inactivate the tumor necrosis factor alpha produced by the CD4⁺ Th1 population, a class of T cells whose inactivation has been shown to promote *A. fumigatus* infection (Bauvois *et al.*, 1992; Cenci *et al.*, 1997). On the other hand, DPP-V is one of the major antigens produced by *A. fumigatus* during the development of an aspergilloma in immunocompetent patients and has long been designated as the chymotryptic antigen although it does not have any chymotrypsin-like activity

(Kobayashi *et al.*, 1993; Beauvais *et al.*, 1997b). Interestingly, it has been shown that immuno-suppressed mice surviving an otherwise lethal infection have antibodies that monospecifically recognize DPP-V, suggesting that inactivation of DPP-V could protect the host against infection (Beauvais *et al.*, 1997b). Active DPP-V could therefore play a role similar to that proposed for DPP-IV in counteracting the action of the CD4+ Th1 population of T cells.

7. Fungal attack and resistance against host defense mechanisms

As proposed above for the dipeptidyl-peptidases, some of the proteins produced by *A. fumigatus* in the course of the infection may protect the fungus against host defense mechanisms or, alternatively, inactivate phagocytic cells that are mostly responsible for the protection against fungal infection. Several studies have attempted to determine the role of toxins produced by *A. fumigatus* as well as oxidative enzymes (catalase, superoxide dismutase) that could protect the fungus from the detrimental effect of reactive oxygen species produced by phagocytic cells.

7.1 TOXINS AND TOXIC PROTEINS

A. fumigatus produces several secondary metabolites that show toxicity to mammalian cells. These include gliotoxin, a potent immunosuppressant molecule that can inhibit macrophage phagocytosis, induce apoptotosis in macrophage and impair lymphocyte activation (Müllbacher and Eichner, 1984; Müllbacher *et al.*, 1985; Waring, 1990), and fumagillin, an inhibitor of angiogenesis whose target has been identified as methionine amino-peptidase 2 (Liu *et al.*, 1998). Despite the potential these molecules have as virulence factors during aspergillosis, relatively little is known about their direct implication in the progress of the disease. While it has been shown for instance that gliotoxin production may promote the establishment of the disease (Sutton *et al.*, 1996), gliotoxin-non producing strains of *A. fumigatus* are still able to cause an aspergillosis (Latgé, 1999). A better understanding of the role of gliotoxin and fumagillin as well as other toxic secondary metabolites produced by *A. fumigatus* (*e.g.* helvolic acid, fumigaclavine-C and aurasperone-C; (Turner and Aldridge, 1983; Mitchell *et al.*, 1997) will probably be facilitated by the identification and inactivation of the biochemical pathways that are required for their synthesis.

Restrictocin is a highly specific ribonuclease that inactivate eukaryotic ribosomes by cleavage of the 28S RNA of the large ribosomal subunit at a single phosphodiester bond, inhibiting protein synthesis (Fando *et al.*, 1985). The sequence around this cleavage site is a binding site for elongation factors, and is conserved in all cytoplasmic ribosomes (Lamy *et al.*, 1992). Restrictocin is the major antigen found in the urine of patients that have developed an invasive aspergillosis (Lamy *et al.*, 1991; Latgé *et al.*, 1991) and it has been detected within the kidney cells of mice infected with *A.*

fumigatus in regions of necrosis surrounding fungal colonies (Lamy *et al.*, 1991). Restrictocin is also known as one of the major allergens of *A. fumigatus,* AspfI, with *ca.* 85% of the patients with IgE directed against *A. fumigatus* that also have IgE antibodies to AspfI (Arruda *et al.*, 1990). Given its toxicity and its production *in vivo*, restrictocin has been postulated to participate in the virulence of *A. fumigatus*. However, evaluation of knock-out mutants in models of invasive aspergillosis has shown that restrictocin is not a necessary component of *A. fumigatus* pathogenicity (Paris *et al.*, 1993; Smith *et al.*, 1993; Smith *et al.*, 1994). This is in agreement with the occurence of restrictocin-like ribonucleases in non-pathogenic *Aspergillus* species and its absence in several pathogenic species (Latgé *et al.*, 1994b).

A. *fumigatus* also produces a hemolysin which has been detected in experimental aspergillosis and that could be involved in red blood cells lysis (Ebina *et al.*, 1984). Although the gene encoding this hemolysin has been cloned (Ebina *et al.*, 1994), no evidence has so far been provided that would demonstrate an active role of this potential toxin in the virulence of *A. fumigatus*.

7.2 ANTIOXIDANT ENZYMES

As mentioned in a previous section (4.2), the ability to interfere with some of the defense mechanisms of the host might be a component of *A. fumigatus* virulence. One of the defense mechanisms is the production of reactive oxigen species by phagocytic cells (macrophages, polymorphonuclear leukocytes) that contributes to the killing of the fungal cell. In this regard, anti-oxidant enzymes including catalases and superoxide dismutases that are active on hydrogen peroxide and superoxide ions, respectively, represent potential virulence factors. *A. fumigatus* Cat1 catalase, first identified as a major antigen recognized by sera of patients that have developed an aspergilloma (Hearn *et al.*, 1992; Lopez-Medrano *et al.*, 1996), was of particular interest in that it is encoded as a pre-pro-protein and is found both associated with the cell wall and in the culture medium (Calera *et al.*, 1997). However, disruption of the corresponding *cat1* gene does not impair *A. fumigatus* virulence nor increase *A. fumigatus* susceptibility to polymorphonuclear leukocytes killing and sensitivity to hydrogen peroxide (Calera *et al.*, 1997). Although this might be easily explained by functional redundancy since two other catalases have been identified in *A. fumigatus* (Hearn *et al.*, 1992; Calera *et al.*, 1997; Takasuka *et al.*, 1999), it appears that *A. fumigatus* strains that lack the Cat1 and Cat2 (CatA) catalases do not show any defect in virulence (Latgé, 1999). A Cu/Zn-superoxide dismutase and a Mn-superoxide dismutase have also been identified in *A. fumigatus* (Holdom *et al.*, 1995; Crameri *et al.*, 1996). The Cu/Zn-superoxide dismutase has been localized both in the cytoplasm and in the cell wall although secretion of the protein is not supported by the occurence of a signal peptide in the deduced amino-acid sequence of the corresponding gene (Hamilton *et al.*, 1996; Holdom and Monod, 1999). On the other hand, the Mn-superoxide dismutase is one of the major antigens recognized by IgE from *A. fumigatus*-allergic individuals (Crameri *et al.*, 1996). Whether these two enzymes in concert with the catalases contribute to the protection of *A. fmigatus* against host defenses remains to be evaluated through the construction of multiply mutated strains.

8. Cell wall biogenesis

Although the cell wall is not directly involved in the interplay between *A. fumigatus* and the host, it has attracted a lot of interest because it represents a target of choice for the development of novel antifungal molecules. Indeed, the cell wall is responsible for hyphal integrity and its alteration is likely to result in cell death or growth arrest. Furthermore, the major components of the cell wall, β-glucans and chitin, are not found in mammals and therefore, drugs that would inhibit glucan or chitin biosynthesis are likely to be devoid of adverse effects in the host.

Chemical analysis of the cell wall in *A. fumigatus* has revealed that it is composed predominantly of polysaccharides (β-1,3-glucans, chitin, α-1,3-glucans and galactomannan) and of glycoproteins (Fontaine *et al.*, 1997b). These various polymers form a network through inter-molecular cross-linking, a process that involves hydrolases and trans-glycosidases in addition to the synthases that are required for the biosynthesis of the linear polysaccharides (Fontaine *et al.*, 1997b).

Recently, several *A. fumigatus* enzymes that are responsible for the biosynthesis of the polymers and for their maturation have been identified through biochemical purification or identification of the corresponding gene. As summarized below, the picture that is emerging from the analysis of knock-out mutants confirms that cell wall biosynthesis is a rather complex and highly regulated process that is important for the normal morphology of the hyphae as well as the differentiation of the conidiophores.

8.1 β-GLUCAN BIOSYNTHESIS

Synthesis of linear β(1,3)-linked glucans in *A. fumigatus* is, like in *S. cerevisiae*, catalyzed by an enzyme complex, the β(1,3)-glucan synthase, that contains a plasma membrane catalytic subunit that is activated by a GTP-binding protein (Beauvais *et al.*, 1993; Mol, 1997). The *fks* gene has been identified based on its homology to the *S. cerevisiae FKS1* and *FKS2* genes and is likely to encode the catalytic subunit of *A. fumigatus* β-glucan synthase (A. Beauvais, personal communication; Beauvais *et al.*, 1997a). Similarly, two genes encoding GTP-binding proteins of the Rho-family have been characterized, of which Rho1 is a candidate regulatory subunit of the *A. fumigatus* β(1,3)-glucan synthase (Mol, 1997). Repeated attempts to inactivate these genes has failed suggesting that they may be essential for cell viability (A. Beauvais, P.C. Mol and J.-P. Latgé, personal communication). In this regard, lipopeptides of the echinochandin family that inhibit β(1,3)-glucan synthase activity are among the promising molecules for novel treatments of aspergillosis (Kurtz and Douglas, 1997).

As mentioned above, the linear polymers that are produced by the β(1,3)-glucan synthase are remodeled by hydrolases that cleave pre-existing bonds and trans-glycosidases that form new covalent bonds between glucan chains as well as between glucan chains and chitin (Fontaine *et al.*, 1997b). Two β(1,3)-glucanosyl transferases, two exo-β(1,3)-glucanases and an endo-β(1,3)-glucanase have been identified in the cell wall of *A. fumigatus* (Fontaine *et al.*, 1996; Hartland *et al.*, 1996; Fontaine *et al.*, 1997a; Mouyna *et al.*, 1998). One of the two β(1,3)-glucanosyl transferases, Bgt1, introduces intrachain β(1,6) linkages into β(1,3)-glucans (Mouyna *et al.*, 1998) while the second, Gel1, sequentially cleaves β(1,3)-glucans and links one of the resulting cleavage product to another glucan chain (Hartland *et al.*, 1996; I. Mouyna *et al.*, personal communication). Inactivation of the *bgt1* or *gel1* genes does not result in any growth phenotype, suggesting that the two corresponding enzymes are not essential for cell wall biogenesis (Mouyna *et al.*, 1998; I. Mouyna and J.-P. Latgé, personal communication). Nevertheless, Gel1 is a member of a family of glycophosphatidylinositol-anchored membrane proteins found in *S. cerevisiae* (e.g. Gas1; Ram *et al.*, 1998 and references therein) and *C. albicans* (Phr1 and Phr2; Saporito-Irwin *et al.*, 1995; Muhlschlegel and Fonzi, 1997). Each of these species contain several members of this family which are differentially regulated and may perform redundant functions. The situation appears to be similar in *A. fumigatus* where homologues of the *gel1* gene have been identified (I. Mouyna and J.-P. Latgé, personal communication). These proteins may compensate for the loss of function of Gel1 in the knock-out mutant and an understanding of the function of trans-glycosylation in the biogenesis of the cell wall in *A. fumigatus* will undoubtedly require the construction of multiply-disrupted strains.

8.2 CHITIN BIOSYNTHESIS

Chitin synthases catalyze the polymerization of N-acetyl-D-glucosamine through β(1,4) linkages. Fungal chitin synthases have been grouped into 5 classes based on structural features: Class I, II and III chitin synthases are zymogenic with a central catalytic domain and a carboxy-terminal hydrophobic domain that is likely to be involved in plasma membrane anchoring; class IV chitin synthases are non-zymogenic and have an organization similar to the zymogenic chitin synthases; and class V chitin synthases have a large amino-terminal domain sharing homology with myosins in addition to the central catalytic core and carboxy-terminal anchoring domain (Horiuchi and Takagi, 1999). In *A. fumigatus*, seven genes have been identified that encode chitin synthases: *chsA* encodes a class I chitin synthase; *chsB* encodes a class II chitin synthase; *chsC* and *chsG* encode class III chitin synthases; *chsF* encodes a class IV chitin synthase; *chsE* encodes a class V chitin synthase; and *chsD* encodes a chitin synthase that cannot be grouped within the five known classes and may define a novel class of chitin synthases (Mellado *et al.*, 1995; Mellado *et al.*, 1996a; Mellado *et al.*, 1996b).

Functional analysis of several *A. fumigatus* chitin synthases has been conducted through the construction of knock-out mutants. Inactivation of *chsD* although it results in lower mycelial chitin content does not yield any defect in mycelial growth or virulence of *A. fumigatus* (Mellado *et al.*, 1996b). Similarly, inactivation of *chsC* does not result in any particular phenotype (Mellado *et al.*, 1996a). In contrast, inactivation of the *chsE* or *chsG* genes results in impaired colony morphology with the *chsG* mutant producing highly branched hyphae and the *chsE* mutant displaying hyphae with balloon-shaped swellings in subapical regions and being unable to initiate conidiation except in the presence of an osmotic stabilizer (Aufauvre-Brown *et al.*, 1996; Mellado *et al.*, 1996a). The swollen regions observed on the hyphae of the *chsE* mutant show an intense calcofluor staining that may reflect disorganized chitin (Aufauvre-Brown *et al.*, 1996). This suggests that the different chitin synthases act in concert to allow the formation of a fully-organized chitin-containing cell wall. A better understanding of this process will probably need, in addition to the construction of multiply-disrupted strains, a detailed analysis of the spatio-temporal regulation of chitin synthases expression and localization.

Evaluation of the *chsE* and *chsG* mutant strains in models of invasive aspergillosis has shown that they retain their ability to cause disease despite their *in vitro* morphological defects (Aufauvre-Brown *et al.*, 1996; Mellado *et al.*, 1996a). However, the *chsG* mutant shows reduced mortality, delay in the onset of the disease and smaller fungal colonies in the lung that may reflect its *in vitro* reduced growth rate (Mellado *et al.*, 1996a). Interestingly, the hyphae that are formed by the *chsE* mutant in the lung of infected mice still display their balloon-shaped structures (Aufauvre-Brown *et al.*, 1996) suggesting that alteration of the hyphal wall, even if it results in a morphological defect, is not sufficient to prevent the development of *A. fumigatus in vivo*, unless it also has a significant impact on the growth rate of the fungus. In this regard, it is likely that the development of antifungals aimed at blocking chitin biosynthesis will need to target several of the identified chitin synthases.

9. Drug resistance

The two major antifungals that are curently in use for the treatment of invasive aspergillosis are the polyene amphotericin B and the triazole itraconazole, despite their respective toxicity and lack of efficacy which argue for the development of novel antifungal molecules (Latgé, 1999 and references therein). This need is also stengthened by the recent appearance in patients of *A. fumigatus* strains that display resistance to either amphotericin B or itraconazole (Denning *et al.*, 1997; Verweij *et al.*, 1998). The mechanisms that is involved in amphotericin B resistance – like its precise mode of action – has not been elucidated yet but might result from alterations in membrane lipids, especially ergosterol (van den Bossche *et al.*, 1998; White *et al.*, 1998). In contrast, analysis of itraconazole-resistant isolates of *A. fumigatus* shows that resistance can occur through either alteration of the properties of the sterol 14α–demethylase that is inhibited by itraconazole or increased efflux of the drug (Denning *et al.*, 1997). In this regard, Slaven *et al.* (1999) have recently identified a *A. fumigatus* gene, *ADR1*, encoding an ATP-binding cassette transporter of the multi-drug resistance family. These authors have shown that *ADR1* is over-expressed in an itraconazole-resistant *A. fumigatus* strain that showed reduced accumulation of itraconazole, suggesting that resistance in this strain is mediated by increased efflux of itraconazole through the Adr1 transporter (Slaven *et al.*, 1999). Two other ABC transporter of the MDR family have been identified in *A. fumigatus* (Tobin *et al.*, 1997). Expression of one of these transporters, *AfuMDR1*, in *S. cerevisiae* confers resistance to an echinochandin (Tobin *et al.*, 1997). However, the demonstration of an actual role of these transporters in antifungal resistance acquisition in *A. fumigatus* has not been achieved yet.

CONCLUSION

Introduction of molecular biology techniques in the study of *A. fumigatus* and its pathogenicity is relatively recent: the first cloning of a *A. fumigatus* gene was reported in 1991 (Lamy *et al.*, 1991) while the first construction of a *A. fumigatus* strain by DNA-mediated transformation was published in 1992 (Tang *et al.*, 1992). Since these early developments, a significant number of *A. fumigatus* genes have been characterized with three major aims : 1) identifying proteins that are essential to the growth of the fungus *in vitro* and/or *in vivo* and may therefore be used as targets for the development of novel antifungal molecules; 2) identifying proteins that are critical in the establishment and progression of the disease and in the interaction with the host components in order to decipher the underlying mechanisms of the diseases that are due to *A. fumigatus*; and 3) developing novel tools for the epidemiology of *A. fumigatus* infection. Although molecular probes have been developed that have brought new insights in the study of invasive aspergillosis, it appears that there has been little progress in our understanding of the molecular basis of *A. fumigatus* pathogenicity. As developed in this review, the only factors that have been shown to impair the pathogenicity of *A. fumigatus* are 1) biosynthetic pathways that are also needed for fast growth of the fungus in more or less specific *in vitro* conditions and may in some instances define the ecological niche that the fungus is colonizing (Tang *et al.*, 1994; d'Enfert *et al.*, 1996; Mellado *et al.*, 1996a; Hensel *et al.*, 1998) and 2) the melanin biosynthesis pathway that represents one of the mechanisms *A. fumigatus* is displaying to escape host defense mechanisms (Jahn *et al.*, 1997; Tsai *et al.*, 1997; Langfelder *et al.*, 1998; Tsai *et al.*, 1998). In this regard, several explanations can be brought up for these relatively discouraging results and could serve to define new research needs for the study of *A. fumigatus* pathogenicity:

(1) *A. fumigatus* is an opportunistic pathogen and its natural environment is not the immuno-compromised host. It is therefore very likely that there is no selective pressure in the *A. fumigatus* population for the development of virulence factors required to colonize mammalian hosts. This is supported by epidemiological studies that cannot discriminate environmental isolates from clinical isolates on the basis of RFLP profiles and ability to cause disease in animal models (Debeaupuis *et al.*, 1997). This latter trait is likely to be due to some of the characteristics that define the *A. fumigatus* species, *e.g.* small spore size and fast growth rate at relatively high temperature. Melanin biosynthesis, although it cannot be considered as a genuine virulence factor, may represent an example of species-specific pathways that contribute to *A. fumigatus* pathogenicity. In this regard, there is a need to better define the molecular basis for the differences between *A. fumigatus* and other non-pathogenic *Aspergillus* species.

(2) The models of invasive aspergillosis that have been used remain of limited value because they are based on acute infections induced by a large fungal inoculum and in this regard they do not mimic the situation that is observed in immuno-compromised humans. Furthermore, a precise evaluation of the growth rate of the fungus and of its interaction with host cells in the lung environment is difficult to achieve. While better and more reliable animal models of invasive aspergillosis are obviously needed, the development of assays that enable the study of interactions between host cells and the fungus will probably facilitate the evaluation of genetically-engineered mutants of *A. fumigatus* as well as the development of novel molecular strategies for the study of *A. fumigatus* pathogenicity.

(3) The ability to cause invasive aspergillosis might be multi-factorial, each factor contributing to some extent to the virulence of *A. fumigatus* but being not necessary because, in particular, of functional redundancy. This is reinforced by the results that have been obtained through the study of a large collection of insertional mutants using signature-tagged mutagenesis approach (Hensel *et al.*, 1995) and that have only resulted in the identification of a single attenuated strain impaired in *p*-aminobenzoic acid biosynthesis (J. Brown and D.W. Holden, personal communication). In this regard, the identification of novel antifungal targets will probably need to focus on genes that are essential for growth *in vitro* as well as *in vivo*. More importantly with respect to the study of *A. fumigatus* pathogenicity, novel approaches need to be developed in order to identify genes that are specifically/predominantly expressed in the course of the

infection and of the interaction with the host. The recent initiation of a program for the sequencing of the genome of *A. fumigatus* will undoubtedly contribute to the analysis of the *A. fumigatus* transcriptome in various conditions that are relevant to the infection process. Furthermore, genetic tools are now available to implement screens for insertional mutants that cannot successfully interact with host cells or to adapt to *A. fumigatus* strategies like IVET (*in vivo* expression technology; Mahan *et al.*, 1993) and DFI (differential fluorescence induction; Valvidia and Falkow, 1997) that have been succcessfully used in the study of bacterial pathogens.

ACKNOWLEDGEMENTS

This review would not have been possible without the long-lasting personal and scientific interactions that the author has had with Jean-Paul Latgé and all the members of the Aspergillus laboratory at Institut Pasteur. Thanks are due to Michel Monod, David Holden, Jeremy Brown, June Kwon-Chung, Axel Brakhage and Margo Moore for their respective contributions to this review.

REFERENCES

Aleksenko, A., and Clutterbuck, A. J. (1997) Autonomous plasmid replication in *Aspergillus nidulans*: AMA1 and MATE elements. *Fungal Genet Biol* **21**, 373-387.

Amaar, Y. G., and Moore, M. M. (1998) Mapping of the nitrate-assimilation gene cluster (crnA-niiA-niaD) and characterization of the nitrite reductase gene (niiA) in the opportunistic fungal pathogen *Aspergillus fumigatus*. *Curr Genet* **33**, 206-215.

Annaix, V., Bouchara, J.-P., Larcher, G., Chabasse, D., and Tronchin, G. (1992) Specific binding of human fibrinogen fragment D to *Aspergillus fumigatus* conidia. *Infect Immun* **60**, 1747-1755.

Arruda, L. K., Platts-Mills, T. A. E., Fox, J. W., and Chapman, M. D. (1990) *Aspergillus fumigatus* allergen I, a major IgE-binding protein, is a member of the mitogillin family of cytotoxins. *J Exp Med* **172**, 1529-1532.

Arst, H. N., Jr., and Cove, D. J. (1973) Nitrogen metabolite repression in *Aspergillus nidulans*. *Mol Gen Genet* **126**, 111-41.

Aufauvre-Brown, A., Brown, J. S., and Holden, D. W. (1998) Comparison of virulence between clinical and environmental isolates of *Aspergillus fumigatus*. *Eur J Clin Microbiol Infect Dis* **17**, 778-780.

Aufauvre-Brown, A., Mellado, E., Gow, N. A., and Holden, D. W. (1996) *Aspergillus fumigatus chsE*: a gene related to *CHS3* of *Saccharomyces cerevisiae* and important for hyphal growth and conidiophore development but not pathogenicity. *Fungal Genet Biol* **21**, 141-152.

Bart-Delabesse, E., Humbert, J. F., Delabesse, E., and Bretagne, S. (1998) Microsatellite markers for typing *Aspergillus fumigatus* isolates. *J Clin Microbiol* **36**, 2413-8.

Bauvois, B., Sancéau, J., and Wietzerbin, J. (1992) Human U937 cell surface peptidase activities : characterization and degradative effect on tumor necrosis factor-alpha. *Eur J Immunol* **22,** 923-930.

Beauvais, A., Chazalet, V., Ram, A. F. J., Klis, F. M., and Latgé, J.-P. (1997a) *Genbank Accession number U79728.*

Beauvais, A., Drake, R., Ng, K., Diaquin, M., and Latgé, J.-P. (1993) Characterization of the 1,3-β–glucan synthase of *Aspergillus fumigatus*. *J Gen Microbiol* **139,** 3071-3078.

Beauvais, A., Monod, M., Debeaupuis, J. P., Diaquin, M., Kobayashi, H., and Latgé, J. P. (1997b) Biochemical and antigenic characterization of a new dipeptidyl- peptidase isolated from *Aspergillus fumigatus*. *J Biol Chem* **272,** 6238-6244.

Beauvais, A., Monod, M., Wyniger, J., Debeaupuis, J.-P., Grouzmann, E., Brakch, N., Svab, J., Hovanessian, A. G., and Latgé, J. P. (1997c) Dipeptidyl-peptidase IV secreted by *Aspergillus fumigatus*, a fungus pathogenic to humans. *Infect Immun* **65,** 3042-3047.

Borgia, P. T., Dodge, C. L., Eagleton, L. E., and Adams, T. H. (1994) Bidirectional gene transfer between *Aspergillus fumigatus* and *Aspergillus nidulans*. *FEMS Microbiol Letters* **122,** 227-232.

Brakhage, A. A., Jahn, B., and Schmidt, A. (1999) *Aspergillus fumigatus. Biology, clinical aspects and molecular approaches to pathogenicity*, Karger, Basel.

Brown, J. S., Aufauvre-Brown, A., and Holden, D. W. (1998) Insertional mutagenesis of *Aspergillus fumigatus*. *Mol Gen Genet* **259,** 327-335.

Brown, J. S., Aufauvre-Brown, A., Tiffin, N., and Holden, D. W. (1997) Identification of virulence determinants of *A. fumigatus* using a REMI/STM approach. *Fungal Genet Newslett* **44A,** 27.

Calera, J.-A., Paris, S., Monod, M., Hamilton, A. J., Debeaupuis, J.-P., Diaquin, M., Lopez-Medrano, R., Leal, F., and Latgé, J.-P. (1997) Cloning and disruption of the antigenic catalase gene of *Aspergillus fumigatus*. *Infect Immun* **65,** 4718-4724.

Cenci, E., Perito, S., Enslle, K. H., Mosci, P., Latgé, J. P., Romani, L., and Bistoni, F. (1997) Th1 and Th2 cytokines in mice with invasive aspergillosis *Infect Immun* **65,** 564-570.

Chazalet, V., debeaupuis, J.-P., Sarfati, J., Lortholary, J., Ribaud, P., Shah, P., Cornet, M., Vu Thien, H., Gluckman, E., Brücker, G., and Latgé, J.-P. (1998) Molecular typing of environmantal and patient isolates of *Aspergillus fumigatus* from various hospital settings. *J Clin Microbiol* **36,** 1494-1500.

Cole, M. F., Bowen, W. H., Zhao, X.-J., and Cihlar, R. L. (1995) Avirulence of *Candida albicans* auxotrophic mutants in a rat model of oropharyngeal candidiasis. *FEMS Microbiol Lett* **126,** 177-180.

Crameri, R., Faith, A., Hemmann, S., Jaussi, R., Ismail, C., Menz, G., and Blazer, K. (1996) Humoral and cell-mediated auto-immunity in allergy to *Aspergillus fumigatus*. *J Exp Med* **184,** 265-270.

d'Enfert, C. (1996) Selection of multiple disruption events in *Aspergillus fumigatus* using the orotidine-5'-decarboxylase gene, *pyrG* as a unique transformation marker. *Current Genet* **30,** 76-82.

d'Enfert, C., Diaquin, M., Delit, A., Wuscher, N., Debeaupuis, J.-P., Huerre, M., and Latgé, J.-P. (1996) Attenuated virulence of uridine-uracil auxotrophs of *Aspergillus fumigatus*. *Infect Immun* **64,** 4401-4405.

d'Enfert, C., Weidner, G., and Brakhage, A. A. (1999) Transformation systems of *Aspergillus fumigatus*. In (Brakhage, A. A., Jahn, B., and Schmidt, A., eds.), *Aspergillus fumigatus: Biology, clinics and molecular approaches to pathogenicity*, S. Karger AG, Basel, pp. 149-166.

Dang, N. H., Torimoto, Y., Schlossman, S. F., and Morimoto, C. (1990) Human CD4 helper T cell activation : functional involvement of two distinct collagen receptors, IF7 and VLA integrin family. *J Exp Med* **172,** 649.

Debeaupuis, J. P., Sarfati, J., Chazalet, V., and Latgé, J.-P. (1997) Genetic diversity among clinical and environmental isolates of *Aspergillus fumigatus*. *Infect Immun* **65,** 3080-5.

Denning, D. W. (1998) Invasive aspergillosis. *Clin Infect Dis* **26,** 781-805.

Denning, D. W., Venkateswarlu, K., Oakley, K. L., Anderson, M. J., Manning, N. J., Stevens, D. A., Warnock, D. W., and Kelly, S. L. (1997) Itraconazole resistance in *A. fumigatus*. *Antimicrobiol Agents Chemother* **41,** 1364-1368.

Denning, D. W., Ward, P. N., Fenelon, L. E., and Benbow, E. W. (1992) Lack of vessel wall elastolysis in human invasive pulmonary aspergillosis. *Infect Immun* **60,** 5153-6.

Ebina, K., Ichinowatari, S., Yokota, K., and Sakaguchi. (1984) Studies on toxin of *Aspergillus fumigatus*. XIX. Biochemical alteration of sera after *Asp*-hemolysin inoculation or *Aspergillus* infection in mice. *Jpn J Med Mycol* **25,** 236-263.

Ebina, K., Sakagami, H., Yokota, K., and Kondo, H. (1994) Cloning and nucleotide sequence of a cDNA encoding Asp-hemolysin from *Aspergillus fumigatus*. *Biochim Biophys Acta* **1219,** 148-150.

Fando, J. L., Alaba, I., Escarmis, C., Fernandez-Luna, J. L., Mendez, E., and Salinas, M. (1985) The mode of action of restrictocin and mitogillin on eukaryotic ribosomes. *Eur J Biochem* **149,** 29-34.

Fontaine, T., Hartland, R. P., Beauvais, A., Diaquin, M., and P, L. J. (1996) Purification and characterization of an endo-ß-1,3-glucanase from *Aspergillus fumigatus*. *Eur J Biochem* **243,** 315-321.

Fontaine, T., Hartland, R. P., Diaquin, M., Simenel, C., and Latgé, J.-P. (1997a) Differential patterns of activity displayed by two exo-β−1,3-glucanases associated with the *Aspergillus fumigatus* cell wall. *J Bacteriol* **179,** 3154-3163.

Fontaine, T., Mouyna, I., Hartland, R. P., Paris, S., and Latgé, J.-P. (1997b) From the surface to the inner layer of the fungal cell wall. *Biochem Soc Transac* **25,** 194-199.

Gavrias, V., Thiede, N., Iartchouk, N., Saxton, J.-A., Hitchcock, C., Timberlake, W. E., and Koltin, Y. (1997) The construction of vectors and strains for efficient gene disruption in *Aspergillus fumigatus*. *Fungal Genet Newslet* **44A,** 24.

Girardin, H., Latgé, J.-P., Skirantha, T., Morrow, B., and Soll, D. (1993) Development of DNA probes for fingerprinting *Aspergillus fumigatus*. *J Clin Microbiol* **31,** 1547-1554.

Girardin, H., Sarfati, J., Kobayashi, H., Bouchara, J.-P., and Latgé, J.-P. (1994a) Use of DNA moderately repetitive sequence to type *Aspergillus fumigatus* isolates from aspergilloma patients. *J Infect Dis* **169,** 683-685.

Girardin, H., Sarfati, J., Traoré, F., Dupouy-Camet, J., Derouin, F., and Latgé, J.-P. (1994b) Molecular epidemiology of nosocomial invasive aspergillosis. *J Clin Microbiol* **32,** 684-690.

Groisman, E. A., and Ochman, H. (1994) How to become a pathogen. *Trends Microbiol* **2,** 289-294.

Hamilton, A. J., Holdom, M. D., and Jeavons, L. (1996) Expression of the Cu,Zn superoxide dismutase as determined by immunochemistry and immunoellectron microscopy. *FEMS Immunol Med Microbiol* **14,** 95-102.

Hartland, R. P., Fontaine, T., Debeaupuis, J. P., Simonel, C., Delepierre, M., and Latgé, J. P. (1996) A novel beta (1-3)-glucanosyltransferase from the cell wall of *Aspergillus fumigatus*. *J Biol Chem* **271**, 26843-26849.

Hearn, V. M., Wilson, E. V., and Mackenzie, D. W. R. (1992) Analysis of *Aspergillus fumigatus* catalases possessing antigenic activity. *J Med Microbiol* **36**, 61-67.

Hensel, M., Arst, H. N., Jr., Aufauvre-Brown, A., and Holden, D. W. (1998) The role of the *Aspergillus fumigatus areA* gene in invasive pulmonary aspergillosis. *Mol Gen Genet* **258**, 553-557.

Hensel, M., Shea, J. E., Gleeson, C., Jones, M. D., Dalton, E., and Holden, D. W. (1995) Simultaneous identification of bacterial virulence genes by negative selection. *Science* **269**, 400-403.

Hicks, J. K., Yu, J. H., Keller, N. P., and Adams, T. H. (1997) Aspergillus sporulation and mycotoxin production both require inactivation of the FadA G alpha protein-dependent signaling pathway. *Embo J* **16**, 4916-4923.

Holdom, M. D., Hay, R. J., and Hamilton, A. J. (1995) Purification, N-terminal amino acid sequence and partial characterization of a Cu,Zn superoxide dismutase from the pathogenic fungus *Aspergillus fumigatus*. *Free Radical Res* **22**, 519-531.

Holdom, M. D., and Monod, M. (1999) cDNA sequence of Cu,Zn superoxide dismutase from *Aspergillus fumigatus*. , Genbank AF128886.

Horiuchi, H., and Takagi, M. (1999) Chitin synthase genes of *Aspergillus* species. In (Brakhage, A. A., Jahn, B., and Schmidt, A., eds.), *Aspergillus fumigatus. Biology, clinical aspects and molecular approaches to pathogenicity*, Karger, Basel, pp. 193-204.

Howard, R. J., and Valent, B. (1996) Breaking and entering - host penetration by the fungal rice blast pathogen *Magnaporthe grisea*. *Ann Rev Microbiol* **50**, 491-512.

Ibrahim-Granet, O., Bertrand, O., Debeaupuis, J.-P., Planchenault, T., Diaquin, M., and Dupont, B. (1994) *Aspergillus fumigatus* metalloproteinase that hydrolyses native collagen : purification by dye-binding chromatography. *Prot Express Purif* **5**, 84-88.

Ibrahim-Granet, O., and d'Enfert, C. (1997) The *Aspergillus fumigatus mepB* gene encodes an 82 kDa intracellular metalloproteinase structurally related to mammalian thimet oligopeptidases. *Microbiology* **143**, 2247-2253.

Jahn, B., Koch, A., Schmidt, A., Wanner, G., Gehringer, H., Bhakdi, S., and Brakhage, A. A. (1997) Isolation and characterization of a pigmentless-conidium mutant of *Aspergillus fumigatus* with altered conidial surface and reduced virulence. *Infect Immun* **65**, 5110-5117.

Jaton-Ogay, K., Paris, S., Huerre, M., Quadroni, M., Falchetto, R., Togni, G., Latgé, J. P., and Monod, M. (1994) Cloning and disruption of the gene encoding an extracellular metalloprotease of *Aspergillus fumigatus*. *Mol Microbiol* **14**, 917-928.

Jaton-Ogay, K., Suter, M., Crameri, R., Falchetto, R., Fatih, A., and Monod, M. (1992) Nucleotide sequence of a genomic and a cDNA clone encoding an extracellular alkaline protease of *Aspergillus fumigatus*. *FEMS Microbiol* **92**, 163-168.

Kershaw, M. J., and Talbot, N. J. (1998) Hydrophobins and repellents: proteins with fundamental roles in fungal morphogenesis. *Fungal genet Biol* **23**, 18-33.

Khotary, M. H., Chase, T., and Macmillan, J. D. (1984) Correlation of elastase production by some strains of *Aspergillus fumigatus* with ability to cause pulmonary invasive aspergillosis in mice. *Infect Immun* **43**, 320-325.

Kirsch, D. R., and Whitney, R. R. (1991) Pathogenicity of *Candida albicans* auxotrophic mutants in experimental infections. *Infect Immun* **59**, 3297-3300.

Kobayashi, H., Debeaupuis, J. P., Bouchara, J. P., and Latgé, J. P. (1993) An 88-kilodalton antigen secreted by *Aspergillus fumigatus*. *Infect Immun* **61**, 4767-4771.

Kudla, B., Caddick, M. X., Langdon, T., Martinez-Rossi, N. M., Bennett, C. F., Sibley, S., Davies, R. W., and Arst, H. N., Jr. (1990) The regulatory gene *areA* mediating nitrogen metabolite repression in *Aspergillus nidulans*. Mutations affecting specificity of gene activation alter a loop residue of a putative zinc finger. *Embo J* **9**, 1355-1364.

Kurtz, M. B., and Douglas, C. M. (1997) Lipopeptide inhibitors of fungal glucan synthase. *J Med Vet Mycol* **25**, 79-86.

Lamy, B., Davies, J., and Schindler, D. (1992) The *Aspergillus* ribonucleolytic toxins (ribotoxins). In (Frankel, A. E., ed.), *Genetically Engineered Toxins*, Marcel Dekker, New York, pp. 237-257.

Lamy, B., Moutaouakil, M., Latgé, J. P., and Davies, J. (1991) Secretion of a potential virulence factor, a fungal ribonucleotoxin during human aspergillosis infection. *Mol Microbiol* **5**, 1811-1815.

Langfelder, K., Jahn, B., Gehringer, H., Schmidt, A., Wanner, G., and Brakhage, A. A. (1998) Identification of a polyketide synthase gene (*pksP*) of *Aspergillus fumigatus* involved in conidial pigment biosynthesis and virulence. *Med Microbiol Immunol* **187**, 79-89.

Latgé, J.-P. (1999) *Aspergillus fumigatus* and aspergillosis. *Clin Microbiol Rev* **12**, 310-350.

Latgé, J.-P., Paris, S., Sarfati, J., Debeaupuis, J.-P., Beauvais, A., Jaton-Ogay, K., and Monod, M. (1997) Infectivity of *Aspergillus fumigatus*. In (vanden Bossche, H., Stevens, D. A., and Odds, F., eds.), *Host-fungus interplay*, National Foundation for Infectious Diseases, Bethesda

Latgé, J. P., Kobayashi, H., Debeaupuis, J. P., Diaquin, M., Sarfati, J., Wieruszeski, J. M., Parra, E., Bouchara, J. P., and Fournet, B. (1994a) Chemical and immunological characterization of the extracellular galactomannan secreted by *Aspergillus fumigatus*. *Infect Immun* **62**, 5424-5433.

Latgé, J. P., Moutaouakil, M., Debeaupuis, J. P., Bouchara, J. P., Haynes, K., and Prévost, M. C. (1991) The 18-kilodalton antigen secreted by *Aspergillus fumigatus*. *InfectImmun* **59**, 2586-2594.

Latgé, J. P., Paris, S., Sarfati, J., Debeaupuis, J. P., and Monod, M. (1994b) Exoantigens of *Aspergillus fumigatus* : serodiagnosis and virulence. In (Powell, K. A., Renwick, A., and Peberdy, J., eds.), *The Genus Aspergillus*, Plenum Press, New York, pp. 321-339.

Lay, J., Henry, L. K., Clifford, J., Koltin, Y., Bulawa, C. E., and Becker, J. M. (1998) Altered expression of selectable marker *URA3* in gene-disrupted *Candida albicans* strains complicates interpretation of virulence studies. *Infect Immun* **66**, 5301-5306.

Lee, J. D., and Kolattukudy, P. E. (1995) Molecular cloning of the cDNA and gene for an elastinolytic aspartic proteinase from *Aspergillus fumigatus* and evidence of its secretion by the fungus during invasion of the host lung. *Infect Immun* **63**, 3796-3803.

Liu, S., Widom, J., Kemp, C. W., Crews, C. M., and Clardy, J. (1998) Structure of human methionine aminopeptidase-2 complexed with fumagillin. *Science* **282**, 1324-1327.

Lopez-Medrano, R., Ovejero, M. C., Calera, J. A., Puente, P., and Leal, F. (1996) Immunoblotting patterns in the serodiagnosis of aspergilloma: antibody response to the 90 kDA *Aspergillus fumigatus* antigen. *Eur J Clin Microbiol Infect Dis* **15,** 146-152.

Mahan, M. J., Slauch, J. M., and Mekalanos, J. J. (1993) Selection of bacterial virulence genes that are specifically induced in host tissues. *Science* **259,** 686-688.

Markaryan, A., Morozova, I., Yu, H., and Kolattukudy, P. E. (1994) Purification and characterization of an elastinolytic metalloprotease from *Aspergillus fumigatus* and immunoelectron microscopic evidence of secretion of this enzyme by the fungus invading the murine lung. *Infect Immun* **62,** 2149-2157.

Mellado, E., Aufauvre-Brown, A., Gow, N. A. R., and Holden, D. W. (1996a) The *Aspergillus fumigatus* *chsC* and *chsG* genes encode class III chitin synthases with different functions. *Mol Microbiol* **20,** 667-679.

Mellado, E., Aufauvre-Brown, A., Specht, C. A., Robbins, P. W., and Holden, D. W. (1995) A multigene family related to chitin synthase genes of yeast in the opportunistic pathogen *Aspergillus fumigatus*. *Mol Gen Genet* **246,** 353-359.

Mellado, E., Specht, C. A., Robbins, P. W., and Holden, D. W. (1996b) Cloning and characterization of *chsD*, a chitin synthase-like gene of *Aspergillus fumigatus*. *FEMS Microbiol Lett* **143,** 69-76.

Mitchell, C. G., Donaldson, K., and Slight, J. (1997) Diffusible component of the spore surface of *Aspergillus fumigatus* which inhibits the macrophage burst is distinct from gliotoxin and other hyphal toxins. *Thorax* **52,** 796-801.

Mol, P. C. (1997) Regulation of glucan synthase activity in *Aspergillus fumigatus*. *Fungal Genet Newslett* **44A,** 11.

Monod, M., Jaton-Ogay, K., and Reichard, U. (1999) *Aspergillus fumigatus* secreted proteases as antigenic molecules and virulence factors. In (Brakhage, A. A., Jahn, B., and Schmidt, A., eds.), *Aspergillus fumigatus: Biology, clinics and molecular approaches to pathogenicity*, S. Karger AG, Basel, pp. 182-192.

Monod, M., Paris, S., Sanglard, D., Janton-Ogay, K., Bille, J., and Latgé, J. P. (1993a) Isolation and characterisation of a secreted metallo-protease of *Aspergillus fumigatus*. *infect Immun* **61,** 4099-4104.

Monod, M., Paris, S., Sarfati, J., Jaton-Ogay, K., Ave, P., and Latgé, J.-P. (1993b) Virulence of alkaline protease-deficient mutants of *Aspergillus fumigatus*. *FEMS Microbiol Lett* **106,** 39-46.

Monod, M., Togni, G., Rahalison, L., and Frenk, E. (1991) Isolation and characterisation of an extracellular alkaline protease of *Aspergillus fumigatus*. *J Med Microbiol* **35,** 23-28.

Moutaouakil, M., Monod, M., Prévost, M. C., Bouchara, J. P., Paris, S., and Latgé, J. P. (1993) Identification of the 33-kDa alkaline protease of *Aspergillus fumigatus in vitro* and *in vivo*. *J Med Microbiol* **39,** 393-399.

Mouyna, I., Hartland, R. P., Fontaine, T., Diaquin, M., Simenel, C., Delepierre, M., Henrissat, B., and Latge, J. P. (1998) A 1,3-beta-glucanosyltransferase isolated from the cell wall of *Aspergillus fumigatus* is a homologue of the yeast Bgl2p *Microbiology* **144,** 3171-3180.

Muhlschlegel, F. A., and Fonzi, W. A. (1997) *PHR2* of *Candida albicans* encodes a functional homolog of the pH-regulated *PHR1* with an inverted pattern of pH-dependent expression. *Mol Cell Biol* **17,** 5960-5967.

Müllbacher, A., and Eichner, R. D. (1984) Immunosuppression *in vitro* by a metabolite of a human pathogenic fungus. *Proc Natl Acad Sci USA* **81**, 3835-3837.

Müllbacher, A., Waring, P., and Eichner, R. D. (1985) Identification of an agent in cultures of *Aspergillus fumigatus* displaying anti-phagocytic and immunomodulating activity. *J Gen Microbiol* **131**, 1251-1258.

Naquet, P., MacDonald, H. R., Brekelmans, P., Barbet, J., Marchetto, S., Van Ewijk, W., and Pierres, M. (1988) A novel T cell-activating molecule (Tham) highly expressed on CD4-CD8 murine thymocytes. *J Immunol* **141**, 4101-4109.

Neuvéglise, C., Sarfati, J., Debeaupuis, J.-P., Vu-Thien, H., Just, J., Tournier, G., and Latgé, J.-P. (1997) Longitudinal study of *Aspergillus fumigatus* strans isolated from cystic fibrosis patients. *Eur J Clin Microbiol Infect Dis* **16**, 747-750.

Neuvéglise, C., Sarfati, J., Latge, J. P., and Paris, S. (1996) Afut1, a retrotransposon-like element from *Aspergillus fumigatus*. *Nucleic Acids Res* **24**, 1428-1434.

Paris, S. (1994) Isolation of protease negative mutants of *Aspergillus fumigatus* by insertion of a disrupted gene. In (Maresca, B., and Kobayashi, G. S., eds.), *Molecular Biology of Pathogenic Fungi. A Laboratory Manual*, Telos Press, New York, pp. 49-55.

Paris, S., Monod, M., Diaquin, M., Lamy, B., Arruda, L. K., Punt, P. J., and Latgé, J. P. (1993) A transformant of *Aspergillus fumigatus* deficient in the antigenic cytotoxin ASPFI. *FEMS Microbiol Lett* **111**, 31-36.

Parta, M., Chang, Y., Rulong, S., Pinto-DaSilva, P., and Kwon-Chung, K. J. (1994) HYP1, a hydrophobin gene from *Aspergillus fumigatus*, complements the rodletless phenotype in *Aspergillus nidulans*. *Infect Immun* **62**, 4389-4395.

Punt, P. J., Oliver, R. P., Dingemanse, M. A., Pouwels, P. H., and van den Hondel, C. A. M. J. J. (1987) Transformation of *Aspergillus* based on the Hygromycin B resistance marker from *Escherichia coli*. *Gene* **56**, 117-124.

Purnell, D. M. (1973) The effects of specific auxotrophic mutations on the virulence of *Aspergillus nidulans* in mice. *Mycopath Mycol Appl* **50**, 195-203.

Ram, A. F. J., Kapteyn, J. C., Montijn, R. C., Caro, H. P., Douwes, J. E., Baginsky, W., Mazur, P., Van den Ende, H., and Klis, F. M. (1998) Loss of the plasma membrane-bound protein Gas1p in *Saccharomyces cerevisiae* results in the release of β(1-3) glucan into the medium and induces a compensation mechanism to ensure cell wall integrity. *J Bacteriol* **180**, 1418-1424.

Ramesh, M. V., Sirakova, T., and Kolattukudy, P. E. (1994) Isolation, characterization, and cloning of cDNA and the gene for an elastinolytic serine proteinase from *Aspergillus flavus*. *Infect Immun* **62**, 79-85.

Ramesh, M. V., Sirakova, T. D., and Kolattukudy, P. E. (1995) Cloning and characterization of the cDNAs and genes (*mep20*) encoding homologous metalloproteinases from *Aspergillus flavus* and *Aspergillus fumigatus*. *Gene* **165**, 121-125.

Reichard, U., Büttner, S., Eiffert, H., Staib, F., and Rüchel, R. (1990) Purification and characterisation of an extracellular serine protease from *Aspergillus fumigatus* and its detection in tissue. *J Med Microbiol* **33**, 243-251.

Reichard, U., Effert, H., and Rüchel, R. (1994) Purification and characterization of an extracellular aspartic proteinase from *Aspergillus fumigatus*. *J Med Vet Mycol* **32**, 427-436.

Reichard, U., Monod, M., Odds, F., and Ruchel, R. (1997) Virulence of an aspergillopepsin-deficient mutant of *Aspergillus fumigatus* and evidence for another aspartic proteinase linked to the fungal cell wall. *J Med Vet Mycol* **35**, 189-196.

Reichard, U., Monod, M., and Ruchel, R. (1995) Molecular cloning and sequencing of the gene encoding an extracellular aspartic proteinase from *Aspergillus fumigatus*. *FEMS Microbiol Lett* **130**, 69-74.

Reichard, U., Monod, M., and Rüchel, R. (1999a) Genbank Accession number AF243145. .

Reichard, U., Monod, M., and Rüchel, R. (1999b) Genbank Accession Number AJ132504. .

Rhodes, J. C., Bode, R. B., and McKuan-Kirsch, C. M. (1988) Elastase production in clinical isolates of *Aspergillus*. *Diagn Microbiol Inf Dis* **10**, 165-170.

Rüchel, R., and Reichard, U. (1999) Pathogenesis and clinical presentation of aspergillosis. In (Brakhage, A. A., Jahn, B., and Schmidt, A., eds.), *Aspergillus fumigatus. Biology, clinical aspects and molecular approaches to pathogenicity.*, Karger, Basel, pp. 21-43.

Samson, R. A. (1999) The genus *Aspergillus* with special regard to the *Aspergillus* group. In (Brakhage, A. A., Jahn, B., and Schmidt, A., eds.), *Aspergillus fumigatus. Biology, clinical aspects and molecular approaches to pathogenicity.*, Karger, Basel, pp. 5-20.

Sandhu, D. K., Sandhu, R. S., Khan, Z. U., and Damodaran, V. N. (1976) Conditional virulence of a *p*-aminobenzoic acid-requiring mutant of *Aspergillus fumigatus*. *Infect Immun* **13**, 527-532.

Saporito-Irwin, S. M., Birse, C. E., Sypherd, P. S., and Fonzi, W. A. (1995) *PHR1*, a pH regulated gene of *Candida albicans* is required for morphogenesis. *Mol Cell Biol* **15**, 601-613.

Schiestl, R. H., and Petes, T. D. (1991) Integration of DNA fragments by illegitimate recombination in *Saccharomyces cerevisiae. Proc Natl Acad Sci* **88**, 7585-7589.

Scnitzler, N., Peltroche-Llacsahuanga, H., Bestier, N., Zündorf, J., Lütticken, R., and Haase, G. (1998) Effect of melanin and carotenoids of *Exophiala (Wangellia) dermatitidis* on phagocytosis, oxidative burst, and killing by human neutrophils. *Infect Immun* **67**, 94-101.

Selker, E. U. (1997) Epigenetic phenomena in filamentous fungi: useful paradigms or repeat-induced confusion? *Trends Genet* **13**, 296-301.

Sirakova, T. D., Markaryan, A., and Kolattukudy, P. E. (1994) Molecular cloning and sequencing of the cDNA and gene for a novel elastinolytic metalloproteinase from *Aspergillus fumigatus* and its expression in *Escherichia coli. Infect Immun* **62**, 4208-4218.

Skromne, I., Sanchez, O., and Aguirre, J. (1995) Starvation stress modulates the expression of the *Aspergillus nidulans brlA* regulatory gene. *Microbiology* **141**, 21-28.

Slaven, J. W., Anderson, M. J., Sanglard, D., Dixon, G. K., Bille, J., Roberts, I. S., and Denning, D. W. (1999) Induced expression of a novel *Aspergillus fumigatus* putative drug efflux gene in response to itraconazole. *Fungal Genet Newslett* **66A**, 64.

Smith, J. M. (1994) Construction of mutants of *Aspergillus fumigatus* by gene disruption. In (Maresca, B., and Kobayashi, G. S., eds.), *Molecular Biology of Pathogenic Fungi. A Laboratory Manual*, Telos Press, New York, pp. 41-47.

Smith, J. M., Davies, J. E., and Holden, D. W. (1993) Construction and pathogenicity of *Aspergillus fumigatus* mutants that do not produce the ribotoxin restrictocin. *Mol Microbiol* **9**, 1071-1077.

Smith, J. M., Tang, C. M., Noorden, S. V., and Holden, D. W. (1994) Virulence of *Aspergillus fumigatus* double mutants lacking restrictocin and an alkaline protease in a low-dose model of invasive pulmonary aspergillosis. *Infect Immun* **62,** 5247-5254.

Stringer, M. A., and Timberlake, W. E. (1995) *dewA* encodes a fungal hydrophobin component of the *Aspergillus* spore wall. *Mol Microbiol* **16,** 33-44.

Sturtevant, J. E., and Latgé, J.-P. (1992) Interactions between Conidia of *Aspergillus fumigatus* and Human Complement Component C3. *Infect Immun* **60,** 1913-1918.

Sutton, P., Waring, P., and Müllbacher, A. (1996) Exacerbation of invasive aspergillosis by the immunosuppressive fungal metabolite, gliotoxin. *Immunol Cell Biol* **74,** 318-322.

Takasuka, T., Sayers, N. M., Anderson, M. J., Benbow, E. W., and Denning, D. W. (1999) *Aspergillus fumigatus* catalases: cloning of an *Aspergillus nidulans* catalase B homologue and evidence for at least three catalases. *FEMS Immunol Med MIcrobiol* **23,** 125-133.

Tang, C. M., Cohen, J., and Holden, D. W. (1992) An *Aspergillus fumigatus* alkaline protease mutant constructed by gene disruption is deficient in extracellular elastase activity. *Mol Microbiol* **6,** 1663-1671.

Tang, C. M., Cohen, J., Krausz, T., van Noorden, S., and Holden, D. W. (1993) The alkaline protease of *Aspergillus fumigatus* is not a virulence determinant in two murine models of invasive pulmonary aspergillosis. *Infect Immun* **61,** 1650-1656.

Tang, C. M., Smith, J. M., Arst, H. N., and Holden, D. W. (1994) Virulence studies of *Aspergillus nidulans* mutants requiring lysine or *p*-aminobenzoic acid in invasive pulmonary aspergillosis. *Infect Immun* **62,** 5255-5260.

Thau, N., Monod, M., Crestani, B., Rolland, C., Tronchin, G., Latgé, J. P., and Paris, S. (1994) *rodletless* mutants of *Aspergillus fumigatus*. *Infect Immun* **62,** 4380-4388.

Tobin, M. B., Peery, R. B., and Skatrud, P. L. (1997) Genes encoding multiple drug resistance-like proteins in *Aspergillus fumigatus* and *Aspergillus flavus*. *Gene* **200,** 11-23.

Tronchin, G., Esnault, K., Renier, G., Filmon, R., Chabasse, D., and Bouchara, J.-P. (1997) Expression and identification of laminin-binding protein in *Aspergillus fumigatus* conidia. *Infect Immun* **65,** 9-15.

Tsai, H. F., Chang, Y. C., Washburn, R. G., Wheeler, M. H., and Kwon-Chung, K. J. (1998) The developmentally regulated *alb1* gene of *Aspergillus fumigatus*: its role in modulation of conidial morphology and virulence. *J Bacteriol* **180,** 3031-3038.

Tsai, H. F., Washburn, R. G., Chang, Y. C., and Kwon-Chung, K. J. (1997) *Aspergillus fumigatus arp1* modulates conidial pigmentation and complement deposition. *Mol Microbiol* **26,** 175-183.

Turner, W. B., and Aldridge, D. C. (1983) *Fungal metabolites (II)*, Academic Press, Ltd, London, United Kingdom.

Valdivia, R. H., and Falkow, S. (1997) Probing bacterial gene expression within host cells. *Trends Microbiol* **5,** 360-363.

Valvidia, R. H., and Falkow, S. (1997) Fluorescence-based isolation of bacterial genes expressed within host cells. *Science* **277,** 2007-2011.

van den Bossche, H., Dromer, F., Improvisi, L., Lozano-Chiu, M., Rex, J. H., and Sanglard, D. (1998) Antifungal drug resistance in pathogenic fungi. *Medic Mycol* **36S1,** 119-128.

van den Hondel, C. A. M. J. J., and Punt, P. J. (1991) Gene transfer systems and vector development for filamentous fungi. In (Peberdy, J. F., Caten, C. E., Ogden, J. E., and Bennett, J. W., eds.), *Applied Molecular Genetics of Fungi*, Cambridge University Press, Cambridge, pp. 1-28.

Verweij, P. E., Oakley, K. L., Morrissey, J., Morrissey, G., and Denning, D. W. (1998) Efficacy of LY303366 against amphotericin B-susceptible and -resistant *Aspergillus fumigatus* in a murine model of invasive aspergillosis. *Antimicrobiol Agents Chemother* **42**, 873-878.

Wang, Y., Aisen, P., and Casadevall, A. (1995) *Cryptococcus neoformans* virulence and melanin: mechanism of action. *Infect Immun* **63**, 3131-3136.

Waring, P. (1990) DNA fragmentation induced in macrophages by gliotoxin does not require protein synthesis and is preceded by raised inositol triphosphate levels. *J Biol Chem* **265**, 14476-14480.

Weidner, G., d'Enfert, C., Koch, A., Mol, P. C., and Brakhage, A. A. (1998) Development of a homologous transformation system for the pathogenic fungus *Aspergillus fumigatus* based on the *pyrG* gene encoding orotidine monophosphate decarboxylase. *Current genetics* **33**, 378-385.

White, T. C., Marr, K. A., and Bowden, R. A. (1998) Clinical, cellular, and molecular factors that contribute to antifungal drug resistance. *Clin Microbiol Rev* **11**, 382-402.

INFECTION STRATEGIES OF *BOTRYTIS CINEREA* AND RELATED NECROTROPHIC PATHOGENS.

THEO W. PRINS[1], PAUL TUDZYNSKI[2], ANDREAS VON TIEDEMANN[3], BETTINA TUDZYNSKI[2], ARJEN TEN HAVE[1], MELANIE E. HANSEN[2], KLAUS TENBERGE[2] AND JAN A.L. VAN KAN[1,*]

(1) Wageningen University, Laboratory of Phytopathology, Binnenhaven 9, 6709 PD Wageningen, The Netherlands.
(2) Westfälische Wilhelms-Universität Münster, Institut für Botanik, Schloßgarten 3, 48149 Münster, Germany.
(3)University of Rostock, Faculty of Agriculture, Department of Phytomedicine, Satowerstraße 48, 18051 Rostock, Germany.
*, corresponding author.

1. Introduction

Botrytis cinerea Persoon: Fries (known as "grey mould fungus") causes serious pre- and post-harvest diseases in at least 235 plant species (Jarvis, 1977), including a range of agronomically important crops, such as grapevine, tomato, strawberry, cucumber, bulb flowers and ornamental plants. Graminaceous monocots are generally considered as poor hosts for grey mould. Disease control frequently relies on chemicals, although efforts to develop biological control strategies are increasingly successful (*e.g.* Köhl *et al.*, 1995; Elad, 1996) and biocontrol agents are marketed. The name of the asexual stage or anamorph, *Botrytis cinerea,* is preferred to the name of the teleomorph, *Botryotinia fuckeliana* (de Bary) Whetzel (XIth International Botrytis Symposium, 1996, Wageningen, The Netherlands). The teleomorph has rarely been detected in the field during the last century, but molecular population studies recently provided clear evidence that sexual reproduction occurs more frequently than previously anticipated (Giraud *et al.*, 1997). The pathogen is a typical necrotroph, inducing host cell death resulting in serious damage to plant tissues, culminating in rot of the plant or the harvested

33

J. W. Kronstad (ed.), Fungal Pathology, 33–64.
© 2000 *Kluwer Academic Publishers. Printed in the Netherlands.*

product. There are extensive descriptions of microscopic and biochemical studies on infection mechanisms (reviewed by Staples and Mayer, 1995). Comprehensive insight in the infection process, however, is hampered by the fact that various groups used different fungal strains and different host species for their studies.

It is relatively recently that molecular-genetic tools such as transformation (Hamada *et al.*, 1994), differential gene expression analysis (Benito *et al.*, 1996) and gene cloning (van der Vlugt-Bergmans *et al.*, 1997a,b) have come available to unravel the factors involved in pathogenicity of *B. cinerea*. Application of these tools has provided substantial new data in a short time and it is expected to yield even more information in the next decade, which will be used to design novel strategies for grey mould control. Here we present an overview of the current insights and hypotheses on the mechanisms employed by *B. cinerea* to infect its host plants. The role of different fungal compounds in consecutive stages of the infection process is described. The balance between induced plant defense responses and the evasion of these responses by the fungus plays an important role in determining the outcome of the attempted invasion. Finally, we briefly discuss the analogy between infection mechanisms utilized by *B. cinerea* and related necrotrophs, which are as yet less accessible for molecular-genetic research.

2. Disease cycle

For the purpose of this review, different stages are distinguished in the disease cycle of *B. cinerea* (Figure 1). It should be emphasized that, given the wide host range of the fungus, not all processes occur in every infection.

B. cinerea can produce conidia on every host plant. They are ubiquitous in the air and can be transported by wind over long distances before infecting the next host (Jarvis, 1977). Following attachment, the conidium germinates under favorable conditions and produces a germ tube that penetrates the host surface. Whether true infection structures are produced during this process is a matter of debate. After surface penetration the underlying cells are killed and the fungus establishes a primary lesion, in which necrosis and defense responses may occur. In some cases this is the onset of a period of quiescence of an undefined length, in which fungal outgrowth is negligible (reviewed by Prusky, 1996). At a certain stage the defense barriers are breached and the fungus starts a vigorous outgrowth, resulting in rapid maceration of plant tissue, on which the fungus finally sporulates to produce inoculum for the next infection. Under optimal conditions, one infection cycle may be completed in as little as 3-4 days, depending on the type of host tissue attacked. In the following paragraphs we will discuss the various stages in the disease cycle and the role that enzymes and metabolites (mainly of fungal origin) play in these stages.

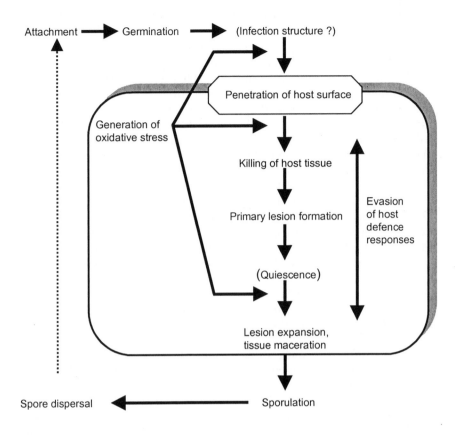

Figure 1: Different stages in the infection process of *B. cinerea*, to be discussed below. The shaded box represents the host tissue.

3. Attachment of conidia

3.1. DISPERSAL OF CONIDIA BY INSECTS

Before being able to penetrate host tissue the conidia of *B. cinerea* must land on and attach to the plant surface. The conidia are ubiquitous in the air and are capable of reaching any host by random chance. Specific trafficking by insects towards attachment sites of host tissue has also been described. The fruit fly *Drosophila melanogaster,* the New Zealand flower thrips *Thrips obscuratus*, as well as larvae of the grape berry moth *Lobesia botrana*, were shown to carry viable conidia of *B. cinerea* (Fermaud and Le Menn, 1992; Fermaud *et al.,* 1994; Fermaud and Gaunt, 1995; Louis *et al.,* 1996). Wounds inflicted on host tissue by the various insects increased disease development. The fruit fly not only carried conidia externally on its cuticle but also in the digestive tract, and microsclerotia were detected in the gut (Louis *et al.,* 1996).

3.2. SURFACE CHARACTERISTICS

A number of physical surface interactions is believed to play a role in the attachment of conidia to the plant epidermis. Conidia of *B. cinerea* have a hydrophobic surface which can usually only be wetted by solutions containing mild detergents. Williamson *et al.* (1995) described a granular structure on the surface of *B. cinerea* conidia, with a smooth area along the main axis. Doss *et al.* (1997) visualized numerous short (200-250 nm) protuberances on the rough surface of dry conidia of seven different *Botrytis* species. These protuberances disappeared upon hydration and redrying. The typical basket-weave pattern of hydrophobin rodlets present in *Magnaporthe grisea* (*MPG*1; Talbot *et al.*, 1996) was not observed on conidia of any of the *Botrytis* species tested (Doss *et al.*, 1997). Hydrophobins, which are involved in the interaction of *M. grisea* with the rice plant surface (Talbot *et al*, 1993; Beckerman and Ebbole, 1996; Talbot *et al.*, 1996), thus seem to be absent in the *Botrytis* species studied, unless *Botrytis* hydrophobins aggregate in a structure different from *M. grisea* and other fungi (Kershaw and Talbot, 1998). Molecular research is needed to clarify whether or not *B. cinerea* possesses hydrophobin genes.

Doss *et al.* (1993, 1995) distinguished two steps in the attachment to host tissue. The first stage, preceding the hydration of conidia, is characterized by relatively weak adhesive forces, presumably resulting mainly from hydrophobic interactions between the host and conidial surfaces (Doss *et al.*, 1993). A stronger binding occurs in the second stage of adhesion, several hours after inoculation, when it becomes increasingly difficult to wash off the conidia. It should be noted, however, that at this stage conidia have developed into germlings. At this stage a fibrillar-like matrix material is present on the tip of germ tubes (Cole *et al.*, 1996; see Figure 2). The attachment of germlings to either hydrophobic or hydrophilic substrata coincides with the production of a base-soluble compound, forming a fibrillate sheath around the germlings (Doss *et al.*, 1995), supposedly consisting of an extracellular β-(1,3)(1,6)-n-glucan (cinerean). Cinerean forms a slimy adhering capsule in liquid culture. Once glucose is exhausted from the medium, *B. cinerea* degrades the polymer by cinereanase into glucose and gentobiose, which is subsequently hydrolyzed to glucose (Stahmann *et al.*, 1992; Stahmann *et al.*, 1993; Monschau *et al.*, 1997). Besides serving as glue and nutrient storage, cinerean may have a role in protecting hyphae against dehydration and defense mechanisms of the host.

When inoculation was carried out using conidia in aqueous glucose, a sheath of fibrillar-like material (presumably cinerean) was detected on the germ tube surface, whereas in inoculations with dry conidia this sheath is absent (Cole *et al.*, 1996). Since dry-inoculated conidia were also firmly attached to the plant surface, it was concluded that the sheath is not essential for attachment of conidia to the host.

4. Germination

After landing on the host tissue, a number of factors influence the germination of a conidium. Free surface water or high relative humidity (>93% RH) is required to germinate and penetrate the host epidermis (Williamson *et al.*, 1995). In addition, moisture assists the pathogen in the uptake of nutrients residing on the host epidermis or pollen grains (Blakeman, 1980). When dry conidia are inoculated on plant surfaces and subsequently incubated in the absence of free surface water, the emerging germ tube usually remains shorter than the length of a conidium before it penetrates the surface (Salinas and Verhoeff, 1995; Williamson *et al.*, 1995; Cole *et al.*, 1996). Inoculation with conidia in an aqueous suspension, however, usually requires the addition of nutrients, which might mimic the situation in a wound on the plant epidermis, from which nutrients leach (Harper *et al.*, 1981; van den Heuvel, 1981). A highly efficient germination and synchronous infection of tomato leaves is obtained, when conidia are preincubated for 2-4 hours in liquid medium supplemented with phosphate and sugar (Benito *et al.*, 1998). The sugar is probably not only involved in stimulating the germination, but also in oxidative processes leading to host cell death (Edlich *et al.*, 1989; see paragraph 7).

Gaseous compounds might also be involved in the stimulation of germination. Elad and Volpin (1988) found a correlation between the level of ethylene production by flowers, petals and leaves of different rose cultivars, and the severity of grey mould symptoms. A stimulation of grey mould development by ethylene was also demonstrated in strawberry, tomato, cucumber and pepper (*e.g.* Elad and Volpin, 1988; McNicol *et al.*, 1989). This observation is usually ascribed to the weakening of the host tissue and senescence that coincides with ethylene production. The influence of ethylene on *B. cinerea* itself is only rarely investigated. Analogous to *Colletotrichum gloeosporioides* (Flaishman and Kolattukudy, 1994), germination of *B. cinerea* might be influenced by ethylene. Kepczynski and Kepczynska (1977) found that germination of *B. cinerea* conidia on a hydrophobic surface was stimulated by exogenous ethylene, but the germ tube length was unaffected. Application of 2,5-norbornadiene, a competitive inhibitor of ethylene perception in plants, inhibited germination in a reversible manner (Kepczynska, 1993). On the other hand, it was reported that in a hydrophilic environment, ethylene stimulated germ tube elongation without affecting the percentage of germination (Barkai-Golan *et al.*, 1989). It is tempting to speculate that ethylene produced by the plant during leaf senescence or fruit ripening might function as a signal for the conidia on the (hydrophobic) plant surface to germinate and initiate the infection. Subsequently the germ tube elongation might be stimulated by ethylene in the more hydrophilic environment of the invaded plant tissue. Thus, ethylene might favor grey mould development by weakening the host, as well as by stimulating germination of *B. cinerea* conidia and outgrowth of hyphae. Molecular and biochemical approaches are required to elucidate whether *B. cinerea* possesses ethylene receptors.

5. Differentiation of infection structures on the host surface

There is some debate whether *B. cinerea* forms appressoria during penetration. There is consensus that germ tubes do not differentiate into the highly organized appressoria that are typical for many plant pathogenic fungi (reviewed by Mendgen *et al.*, 1996). Several authors, however, reported an appressorium-like structure in *B. cinerea* but their nomenclature diverges. Van den Heuvel and Waterreus (1983) distinguished germ tube apices, *i.e.* tips of germ tubes appearing as an appressorium-like swelling, appressoria of different forms and infection cushions. Akutsu *et al.* (1981) grouped pre-penetration structures into primary, secondary or hyphal elongations. It is probable that these structures, which were all observed upon inoculation with conidia in an aqueous glucose suspension, correspond to the structure shown in Figure 2, a swollen hyphal tip with an "adhesive pad" (Cole *et al.*, 1996). Secretion of a phosphatase by *B. cinerea* grown *in vitro* occurs specifically from vesicles in the hyphal tip (Weber and Pitt, 1997), which is typical for protein secretion in filamentous fungi (Sietsma *et al.*, 1995). It remains to be determined whether fungal hydrolytic enzymes, involved in host surface penetration, are also secreted at the hyphal tip and whether these enzymes accumulate prior to the onset of penetration, in the indented area delimited by arrows in Figure 2.

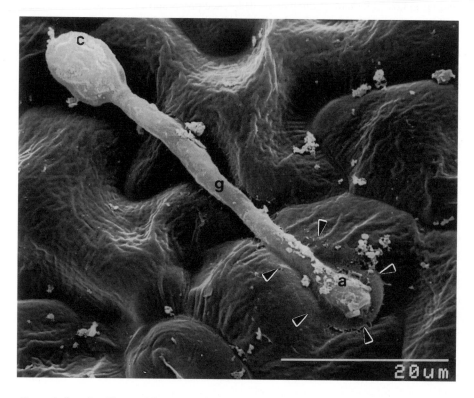

Figure 2: Scanning Electron Micrograph of an early stage of infection (5 h.p.i.) of tomato leaves. A conidium (c) has germinated and formed a germ tube (g), which terminates in a swollen tip (a) surrounded by matrix material, the adhesive pad. The host epidermal surface is partly indented around the site of contact with the adhesive pad (indicated by arrows). A stomatal cell is indicated (s).

The swelling of the hyphal tip may be the consequence of a rise in the osmotic value in the hyphal tip, resulting in water absorption. In the absence of a rigid layer in the outer wall, swelling can not result in an equally high turgor as is generated in appressoria of *M. grisea* (Howard *et al.*, 1991; de Jong *et al.*, 1997). The external sheath, presumably consisting of cinerean (see paragraph 3), may contribute to the swelling by retaining water, as cinerean is extremely hygroscopic. Whether the hydrophobic host surface or the release of volatile plant hormones provide signals to trigger the formation of hyphal swellings is purely speculative.

Swelling of hyphal tips is usually not observed when hosts are inoculated with dry conidia (Salinas and Verhoeff, 1995; Williamson *et al*. 1995; Cole *et al*., 1996). In the latter cases, the conidia produced short germ tubes which directly penetrated the epidermal cells, subsequently leading to cell death.

6. Penetration of the host surface

Invasion of host tissue can be achieved by active penetration or passive ingress. *B. cinerea* is a renowned opportunist that can initiate infection at wound sites, or at sites that have previously been infected by other pathogens. *B. cinerea* can also enter the substomatal cavity via an open stoma. Only direct penetration of the epidermal surface is discussed in this paragraph. For reasons of simplicity the penetration of dead or wounded tissue, as well as via stomata, is regarded as an expansion rather then a penetration process, and is dealt with in paragraph 10.

6.1. CUTINASES

When conidia land on aerial parts of a plant, the first barrier to overcome is the cuticle, which covers the epidermal cells. Its major structural component, cutin, is a polyester composed of hydroxylated and epoxidized C_{16}- and C_{18}-fatty acids (Martin and Juniper, 1970). Physical damage or mechanical penetration of the cuticle by *B. cinerea* is not usually observed (Williamson *et al*., 1995; Cole *et al*., 1996). Hence, cutinolytic activity is presumably required to penetrate this layer (Salinas and Verhoeff, 1995; van der Vlugt-Bergmans, 1997a). Salinas (1992) investigated whether a particular 18 kDa cutinase is important in this process and raised monoclonal antibodies against the enzyme. Application of the antibody to gerbera flowers prior to inoculation reduced lesion formation by 80%. A cutinase–deficient gene replacement mutant, however, did not have any discernible reduction in virulence on gerbera flowers nor on tomato fruits, as compared to the wild type (van Kan *et al.,* 1997). Although the observations of Salinas (1992) remain to be explained, it can be ruled out that this particular 18 kDa cutinase is essential in penetration.

It should be taken into account that the 18 kDa cutinase, like all other cutinases studied thus far in plant pathogenic fungi, is most likely an exo-hydrolase. Enzyme activity of cutinases is usually defined by their ability to release soluble fatty acid monomers from the water-insoluble substrate (Purdy and Kolattukudy, 1973). It would be much more efficient for a pathogen to produce an endo-hydrolase, in order to create openings for penetrating a polymer by fungal hyphae. Such endo-cutinase, however, will not release water-soluble products from the insoluble cutin, since cleavage products most likely remain attached in the network. Hence, endo-cutinase activity is difficult to detect in biochemical assays.

One candidate for an enzyme with such activity is a 60 kDa lipase that is induced upon growth in liquid medium with apple cutin as the sole carbon source (Comménil *et al.,* 1998). This lipase possesses low but significant cutinolytic activity and it has clearly distinct kinetic properties than the 'typical' cutinases mentioned above. When polyclonal antibodies raised against this lipase were applied prior to inoculation with *B. cinerea* conidia, germ tubes were no longer able to penetrate the cuticle. The antibodies did not affect germination (Comménil *et al.,* 1998). Whether the lipase plays an essential role in host tissue penetration should be assessed by cloning the corresponding gene, constructing a targeted lipase-deficient mutant and determining its virulence.

6.2. PECTINASES

Microscopic studies have shown that after penetration of the cuticle, hyphae of *B. cinerea* frequently invade the anticlinal wall between two epidermal cells. The concomitant swelling of the epidermal cell wall (Mansfield and Richardson, 1981) is indicative for the degradation of the pectin in the matrix of the epidermal wall, presumably as a result of water absorption, as will be discussed below in paragraph 7. Biochemical evidence suggested that pectinases might be involved in primary infection. At least one (basic) endopolygalacturonase (endoPG) is expressed constitutively and it was therefore proposed to be involved in early stages of the infection process (Van der Cruyssen *et al.*, 1994). Gene cloning revealed that *B. cinerea* contains an endoPG gene family, consisting of six members encoding basic as well as acidic isozymes (ten Have *et al.,* 1998; Wubben *et al.*, 1999a). Targeted deletion mutants were made in both genes encoding the basic endoPGs (*Bcpg*1 and *Bcpg*2) by gene replacement. Both types of mutants were still able to cause primary necrotic lesions on non-wounded tomato and bean leaves (ten Have *et al.*, 1998; ten Have *et al.*, 1999), excluding an essential role for BcPG1 and BcPG2 in host surface penetration.

6.3. PROTEASES

Movahedi and Heale (1990a) detected extracellular aspartic protease (AP) activity in ungerminated conidia as well as during germination, prior to the appearance of pectinase activity. Application of the specific AP inhibitor pepstatin during inoculation markedly reduced infection of carrot slices, suggesting an important role for AP during primary infection (Movahedi and Heale, 1990b). Recently, a gene was cloned encoding an aspartic protease, *BcAP*1, and targeted mutants were made to study its involvement in the infection of detached tomato leaf tissue (Prins *et al.*, 1999a). No discernable loss of virulence was observed for the BcAP1-deficient mutant, indicating that this protease is not essential for virulence. Since *B. cinerea* probably contains at least one additional AP gene, the importance of aspartic proteases in pathogenesis can not yet be excluded.

7. Killing the host

B. cinerea kills host cells before they are invaded by hyphae (Clark and Lorbeer, 1976). Recent studies have demonstrated that invasion of plant tissue by *B. cinerea* triggers nuclear condensation and plant membrane damage, two indicators for programmed cell death, in a ring of cells around the hyphae (Govrin and Levine, unpublished results). These results imply that diffusible factors have a direct or indirect phytotoxic activity. Several phytotoxic compounds that have been proposed to play a role in killing host cells are discussed below.

7.1. TOXINS

Culture filtrates of *B. cinerea* may induce toxic effects when applied to plant tissue (Rebordinos *et al.*, 1996), but there is still little evidence for the existence of a causal fungal toxin *sensu strictu*. A *B. cinerea* isolate excreted in liquid culture a highly substituted lactone, botcinolide, which inhibited wheat coleoptile elongation and induced necrosis or chlorosis on bean, corn and tobacco plants (Cutler *et al.*, 1993). Homobotcinolide, a natural botcinolide homologue, did not exhibit phytotoxic effects on bean but it was more potent in inhibiting wheat coleoptile elongation (Cutler *et al.*, 1996). Four additional derivatives of botcinolides have been isolated and structurally characterized without elucidation of their biological activity (Collado *et al.*, 1996). The tricyclic sesquiterpenes, botrydial and dehydrobotrydial, were purified from liquid culture and were phytotoxic on tobacco leaf discs. However, both metabolites were secreted by *B. cinerea* only when the medium contained a high glucose level (>3%). Moreover the damage inflicted by the purified metabolites only occurred at concentrations significantly above the natural concentrations found in liquid media (Rebordinos *et al.*, 1996). Before one can envisage a significant role of phytotoxins of *B. cinerea* in pathogenesis, more information is needed on the timing of secretion, the occurrence of toxic metabolites *in situ* and their correlation with the aggressiveness of individual fungal strains.

7.2. OXALIC ACID

Secretion of oxalic acid (OA) is a widespread property of fungi from various taxonomic classes. Its occurrence and ecological function was recently reviewed by Dutton and Evans (1996). A key role for OA in pathogenesis has been postulated for several plant pathogens like *Sclerotinia sclerotiorum* (Godoy *et al.*, 1990), *Sclerotium rolfsii* (Kritzman *et al.*, 1977), *Mycena citricolor* (Rao and Tewari, 1987), and *Sclerotium cepivorum* (Stone and Armentrout, 1985). The wide range of host species, tissue types and plant growth stages that are parasitized by some of these pathogens, has been ascribed to the phytotoxicity of OA. The sensitivity of bean cultivars to OA was closely correlated with the severity of infection by *Sclerotinia* white mould (Tu, 1985; Tu, 1989). Mutants of *S. sclerotiorum*, which are deficient in OA production, were unable to infect *Arabidopsis* plants (Dickman and Mitra, 1992) and the deficiency could be restored by supplementing the inoculum with OA (Godoy *et al.*, 1993). These studies do, however, not prove that OA acts solely and directly as a phytotoxin. It may simply be a component in a cascade of events leading to host cell death. This view is supported by other studies in which the damage induced by *Sclerotinia* could not be explained solely by the toxicity of OA (Callahan and Rowe, 1991; Rowe, 1993).

   B. cinerea produces OA both *in vitro* (Gentile, 1954; Schroeder, 1972; Donèche *et al.*, 1985) and *in planta* (Verhoeff *et al.*, 1988). The sizes of lesions induced by several strains of *B. cinerea* on grapevine and bean leaves strongly correlated with the amount of OA secreted *in vitro* by these strains (Germeier *et al.*, 1994). However, in a time course study with inoculated bean leaf discs the amounts of OA, secreted at 14 hours post inoculation, when the first host cells collapsed, were insufficient to justify the tissue damage observed (Tiedemann *et al.*, unpublished). It is thus likely that OA is an important co-determinant of pathogenicity but not the primary phytotoxic agent. In fact, the role of OA might be in a synergistic action with endoPGs during macerating plant tissue, as has been described in various plant pathogens (Bateman and Beer, 1965; Amadioha, 1993; Punja *et al.*, 1985; Stone and Armentrout, 1985). Many fungal endoPGs have an activity optimum at low pH (Rombouts and Pilnik, 1980) and are therefore stimulated by the simultaneous secretion of OA. In addition, OA is believed to enhance the pectin degradation resulting from endoPG activity, by sequestering the Ca^{2+} ions from (intact or partially hydrolyzed) Ca-pectates in the cell walls. The resulting calcium oxalate complex is insoluble and crystallizes in the infected leaf tissue (Figure 3). The removal of Ca^{2+} ions disturbs intermolecular interactions between pectic polymers and disrupts the integrity of the pectic backbone structure. Consequently, the pectic structure absorbs water and swells, as was described by Mansfield and Richardson (1981).

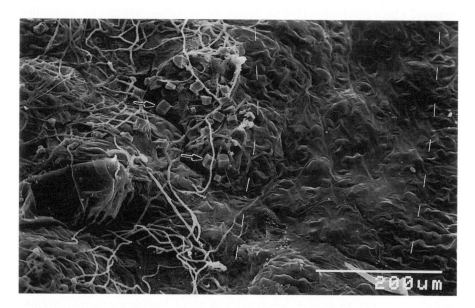

Figure 3: Formation of crystals (indicated by arrows) in tomato leaves infected by *B. cinerea*, visualized by Scanning Electron Microscopy. The white dashed line in the center of the picture indicates the border of the lesion between the external (ectotrophic) mycelium surrounded by crystals (on the left-hand side), and a concentric zone of collapsed epidermal cells (on the right-hand side). Both leaf zones are colonized by mycelium, growing below the epidermis. The white dashed line at the right-hand side of the picture represents the border between this endotrophically colonized area and the non-invaded leaf tissue.

7.3. INDUCTION OF ACTIVE OXYGEN SPECIES

Evidence that Active Oxygen Species (AOS) are involved in the attack by a pathogen, rather then in triggering plant defense, has been presented for several fungi, such as *Drechslera siccans, D. avenae* (Gönner and Schlösser, 1993), *Phytophthora infestans* (Jordan and DeVay, 1990) and *Cercospora beticola* (Daub and Hangarter, 1983). Studies on AOS production in relation to *B. cinerea* pathogenicity are more recent. Various antioxidants were found to reduce grey mould disease development (Elad, 1992; Tiedemann, 1997). In a screening of a number of bean (*Phaseolus vulgaris*) genotypes, a correlation was found between the sensitivity of a particular genotype to oxidative stress and its susceptibility to *B. cinerea* (Tiedemann, 1997). Toxic AOS were detected *in situ* in bean leaf tissue infected with an aggressive strain of *B. cinerea,* and their appearance coincided with the occurrence of host tissue damage (Tiedemann, 1997). Transgenic tobacco plants, expressing an alfalfa ferritin, were more tolerant to oxidative reagents such as paraquat, and they were more resistant to infection by *B. cinerea* than the non-transformed control (Deák *et al.*, 1999).

One of the AOS produced at the host-fungal interface is hydrogen peroxide (H_2O_2, see Figure 4). The level of H_2O_2, released from bean leaf discs inoculated with different *B. cinerea* isolates, correlated with the aggressiveness of the isolate on such leaf tissue (Tiedemann, 1997). In *B. cinerea*-infected *Arabidopsis thaliana* leaves, H_2O_2 was detected in the apoplastic space as much as 5-10 cell layers away from the fungal hyphae (Govrin and Levine, unpublished results).

The mechanism by which *B. cinerea* induces (presumably toxic levels of) AOS in its host is still unclear. Fungal sugar oxidases might provide the source (Edlich *et al.*, 1989). It was proposed that the oxidative burst is a consequence of lipid oxidation by the invading fungus (Weigend and Lyr, 1996). Exogenous application of glucose oxidase or H_2O_2, released during glucose oxidation, mimicked this phenomenon. The conclusion of Weigend and Lyr (1996) that a fungal glucose oxidase is responsible for the H_2O_2 accumulation at the host-fungus interface should be validated by molecular-genetic studies. Liu *et al.* (1998) purified and characterized a (presumably intracellular) glucose oxidase from mycelium of *B. cinerea*. More recently, a gene was cloned that encodes a different glucose oxidase, which contains a signal peptide and is therefore probably extracellular (Liu and Tudzynski, unpublished results). This gene will be used for gene replacement studies in order to study the role of extracellular glucose oxidase in pathogenicity.

Besides the production of AOS by fungal enzymes, it is probable that also the host contributes to a large extent to the generation of AOS, because of their extremely short half-life time. A prime source for generation of AOS is the plant 'oxidative burst' system, which consists of a plasma membrane-bound NADPH oxidase, inducible by fungal elicitors and requiring extracellular Ca^{2+} in the range of 1.5 mM (Schwacke and Hager, 1992). Infiltration of the specific NADPH oxidase inhibitor DPI into leaves of *Arabidopsis thaliana* prior to inoculation with *B. cinerea* resulted in a reduction of AOS production and a slower colonization of host tissue by the fungus (Govrin and Levine, unpublished results). Oligogalacturonides released from the plant cell wall by pectinases of *B. cinerea* are potential elicitors of an oxidative burst (Legendre *et al.*, 1993). This was demonstrated earlier in cotton leaf tissue, generating H_2O_2 in response to treatment with endoPG, in the presence of divalent cations (Mussell, 1973). Furthermore, OA may serve as a precursor of AOS in different ways, either through its enzymatic degradation by oxalate oxidase or by the chelation of ferric iron (Dutton and Evans, 1996). Although it is well established that oxidative damage is evoked by *B. cinerea* in the initial stages of attack, it remains to be clarified which factors determine the induction, generation and regulation of its phytotoxic components and how crucial this process is for penetration and colonization.

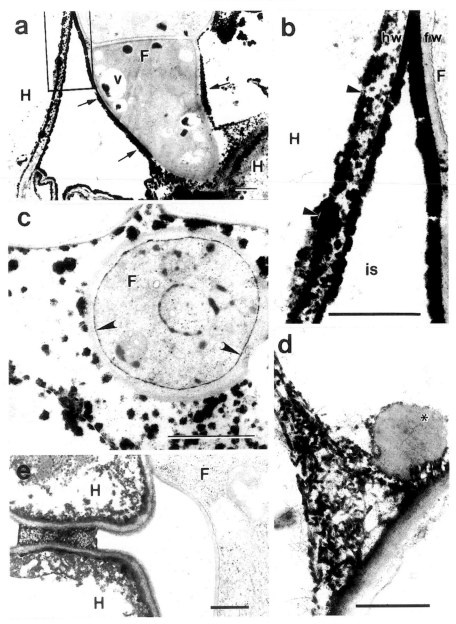

Figure 4: Localization of H_2O_2 in infected tomato leaf with cerium chloride (a-d) and visualization by Transmission EM. An electron-dense precipitate of cerium perhydroxide (indicated by arrows) is formed in the presence of H_2O_2. (a), overview picture of the interface between host cells (H) and *B. cinerea* hyphae (F). A fungal vacuole is indicated by (v). (b), close-up of a section of (a), H_2O_2 is produced in the host cell (H) at the plasma membrane and diffuses through the host cell wall (hw) into the intercellular space (is). Electron-dense precipitate is observed in the periplasmic space (indicated by arrow-heads) and at the outside of the host cell wall (hw), as well as at the outside of the fungal wall (fw). The host cell is electron translucent and appears to be dead. (c), H_2O_2 production in hyphae. Electron-dense precipitate is observed inside a fungal cell that appears vital. (d), control, illustrating H_2O_2 production in lignifying xylem tissue. The asterisk indicates a cell wall thickening. (e), control of infected tissue that was not treated with cerium chloride. Panels a, b, c and e were stained 48 h.p.i, panel d was stained 24 h.p.i.

8. Formation of primary lesions, defense responses in the host

One potent defense mechanism of plants against invasion by pathogens is the deposition of lignin, occurring during hypersensitive response. A more limited response involves the deposition of callose at the site of penetration. Brett and Waldron (1996) proposed that disturbance of the plant plasma membrane, which maintains a calcium gradient across the membrane, results in a net calcium influx, which in turn triggers callose deposition by the calcium-dependent $\beta(1-3)$ glucan synthase complex. Both lignin and callose formation may put up a barricade that prevents further penetration of the pathogen into the protoplast.

The host surface penetration and the subsequent rupture of the plant cell walls by cell wall degrading enzymes (CWDEs) of *B. cinerea* triggers a cascade of processes in the fungus as well as the host. This paragraph deals with molecules, generated at the host-fungus interface, which trigger plant defense responses and with the impact of such responses on the progress of the infection.

8.1 ELICITORS

Among the plant cell wall breakdown products, cutin monomers may act as endogenous elicitors. Dihydroxypalmitic acid, the major component of potato cutin, is the active monomer that stimulates ethylene production, alkalinization, membrane depolarization, and transcriptional activation of defense-related genes (Schweizer *et al.*, 1996). Such elicitors may be released by the action of the 18 kDa cutinase discussed in paragraph 6. Fragments of host-derived pectic polysaccharides released by fungal pectinases may act as elicitors of different defense responses (Legendre *et al.*, 1993; Brett and Waldron, 1996).

In addition it was shown that several components released from fungal hyphae may serve as exogenous elicitors of defense responses in *in vitro* assays with cell suspensions like ergosterol (Granado *et al.*, 1995), chitin fragments (Felix *et al.*, 1993) and glycopeptides (Basse and Boller, 1992; Fath and Boller, 1996). The host may respond by transient alkalinization and membrane depolarization, coinciding with an oxidative burst.

8.2. INDUCTION OF NECROSIS

The initial establishment of primary necrotic lesions coincides with (and is in fact the result of) host defense activation in the neighboring tissue in response to the death of an invaded cell. It is as yet unclear whether cell death caused by a necrotroph, such as *B. cinerea*, is equivalent to cell death during a hypersensitive response (HR) to a biotrophic pathogen (reviewed by Lamb and Dixon, 1997). Recent studies by Govrin and Levine (unpublished results) have shown that an oxidative burst occurs in plant tissue several cell layers away from the fungal hyphae. Cytological staining provided evidence for rapid nuclear condensation and irreversible membrane damage, indicative of a programmed cell death process (Govrin and Levine, unpublished results).

Largely the same defense responses are activated during an infection by *B. cinerea* as during HR to avirulent races of a biotrophic pathogen: lignification (Maule and Ride, 1976; Heale and Sharman, 1977), biosynthesis of phytoalexins (*e.g.* Bennett *et al.*, 1994), and PR proteins (*e.g.* Benito *et al.*, 1998). The total spectrum of defense responses results in a primary necrotic lesion in which the fungus is effectively restricted. Depending on the type of host tissue and yet unidentified physiological aspects of the host, the lesions enter a lag phase in which they become darker but do not expand. Only a proportion of the primary lesions is eventually able to develop into aggressive, expanding lesions that rapidly colonize the entire tissue (van den Heuvel, 1981; De Meyer and Höfte, 1997; Benito *et al.*, 1998). Even when only a small proportion of the primary lesions expands, the entire tissue can be destroyed within 24-48 hours. In the non-expanding lesions the fungus is not killed, since viable fungal mycelium could be recovered from all lesions (Benito, unpublished results).

Thus, an active defense contributes to restricting the fungus within the primary lesions. This was further emphasized by the work of Benito *et al.* (1998) who incubated inoculated tomato leaves at low temperature (4°C) and in darkness. Tomato is a rather cold-sensitive plant and its induced defense responses are limited and ineffective at 4°C. *B. cinerea*, however, is renowned for its ability to infect plant tissues, even when they are stored cold. Under these conditions the plant serves as a living albeit rather inert substrate, whereas *B. cinerea* can readily grow and has an advantage over the host. The primary penetration sites become water-soaked instead of necrotic, and all primary lesions develop into expanding lesions, often at fairly even growth rate (Benito *et al.*, 1998; ten Have *et al.*, 1998).

8.3. SYSTEMIC ACQUIRED RESISTANCE

It has recently been reported that an effective defense to *B. cinerea* may be pre-activated by a mechanism resembling Systemic Acquired Resistance (SAR, reviewed by Ryals *et al.*, 1996). De Meyer and Höfte (1997) demonstrated that bacterization of bean roots with a *Pseudomonas aeruginosa* strain resulted in a reduced number of expanding *B. cinerea* lesions on the leaves. The production of salicylic acid by the bacterium was required for triggering the SAR to *B. cinerea* (De Meyer and Höfte, 1997). Transgenic plants expressing a bacterial salicylate hydroxylase gene (NahG, Gaffney *et al.*, 1993) may be used to validate whether lesion restriction is mediated by salicylate signaling, providing further evidence for the importance of SAR in restricting *B. cinerea* lesion expansion. The potent grey mould biocontrol agent *Trichoderma harzianum* T39 (Elad *et al.*, 1994; O'Neill *et al.*, 1996) appears to act in part by conferring SAR (De Meyer *et al.*, 1998).

More recent studies have indicated that effective lesion restriction occurs in non-induced wild type *Arabidopsis thaliana* ecotypes. The ability to restrict *B. cinerea* lesion outgrowth was abolished in mutant *A. thaliana* genotypes carrying the *coi*1 mutation (deficient in the jasmonate signaling pathway; Thomma *et al.*, 1998) and the *lsd*1 mutation (involved in HR in gene-for-gene interactions; Govrin and Levine, unpublished results).

Even when lesion restriction occurs, the defense responses are eventually unable to stop *B. cinerea* colonization completely. As a necrotroph, *B. cinerea* is well adapted to surviving in a hostile environment in which toxic plant metabolites and antifungal proteins accumulate. Various screenings were performed *in vitro* to identify plant proteins that are capable of reducing growth of *B. cinerea*. Many plant defense proteins appeared to be potent growth inhibitors of *B. cinerea* like chitinase, β-1,3-glucanase, and a class of low molecular weight proteins known as defensins (*e.g.* Broekaert *et al.*, 1989; Terras *et al.*, 1993; Chung *et al.*, 1997; Tabei *et al.*, 1997; Salzmann *et al.*, 1998). Several of the corresponding genes have been or will be used in molecular resistance breeding, but no major success has yet been reported.

8.4. QUIESCENCE

In non-green tissues, *B. cinerea* often causes quiescent infections (reviewed by Prusky, 1996), in which no symptoms are discernible at first. Most prominent examples are soft fruits such as strawberry, raspberry and grape. In these hosts, *B. cinerea* frequently infects during the flowering stage and resides in the developing fruit tissue. No fungal growth occurs and the mycelium degenerates into fragments of irregular shaped cells. Fungal growth only resumes at the onset of fruit ripening. It has often been considered that high levels of fungitoxic or fungistatic compounds in immature fruits contribute to grey mould quiescence, since the level of these compounds decreases during the ripening process. Therefore, attempts have been undertaken to increase the levels of compounds of interest, or to prevent their degradation during ripening. The level of the stilbene phytoalexin resveratrol in grapes was shown to be correlated with grey mould resistance (Langcake, 1981; Bavaresco *et al.*, 1997). The effect of over-expressing stilbene synthase genes from *Vitis* in transgenic plants on the level of resistance towards *B. cinerea* was evaluated. A significant, partial resistance was obtained in tobacco (Hain *et al.*, 1993) but not in tomato (Thomzik *et al.*, 1997).

Besides phytoalexins, fruits usually contain high levels of proteinaceous inhibitors of fungal cell wall degrading enzymes, the PolyGalacturonase Inhibiting Proteins (PGIPs) and their level decreases during ripening. In view of this correlation and the significant role that polygalacturonases play in the infection (see paragraph 10), efforts to produce transgenic plants overexpressing PGIPs have been undertaken to obtain resistance towards *B. cinerea* (Graham *et al.*, 1996). This strategy has thus far met with limited success if any. One of the problems in this strategy is that PGIPs have a differential activity towards individual fungal endoPGs (Desiderio *et al.*, 1997). This makes it relevant to utilize PGIPs that are most potent against the *B. cinerea* endoPG isozymes that are important in virulence (see paragraph 10).

9. Evasion of chemical defense

Plant pathogenic fungi have developed various mechanisms to overcome the deleterious effect of preformed (phytoanticipins) or induced (phytoalexins) chemical defense agents. The major strategy is an enzymatic detoxification of these compounds (reviewed by VanEtten *et al.*, 1995; Osbourn, 1996; Osbourn *et al.*, 1998). Although the ability to detoxify these chemicals is widespread among phytopathogens, an essential role for the successful colonization of a host has so far only been proven in a few cases (reviewed by Tudzynski and Tudzynski, 1998).

Because of its broad host range, *B. cinerea* encounters a wide spectrum of antimicrobial compounds synthesized by the various host plants. The ability of *B.*

cinerea to degrade or detoxify phytoalexins was intensively studied already over two decades ago (see review of the "older" literature by Mansfield, 1980). In some cases this ability obviously contributed to the virulence, *e.g.* in *Capsicum frutescens* (capsidiol, Stoessl *et al.*, 1972) and *Phaseolus vulgaris* (phaseollin, van den Heuvel, 1976). *B. cinerea* is also able to degrade the phytoalexins rishitin, medicarpin and maackiain, but this ability did not directly influence its ability to infect the respective host plants (Mansfield, 1980).

The best studied example for phytoalexin detoxification by *B. cinerea* that correlates with virulence, is the detoxification of the *Vitis* phytoalexins pterostilbene and resveratrol. The ability of fungal isolates to detoxify these phytoalexins was correlated to their virulence (Sbaghi *et al.*, 1996). Conversely, the resistance level of *Vitis* genotypes against grey mould is correlated to their phytoalexin content (Langcake, 1981; Jeandet *et al.*, 1992). *B. cinerea* produces a substrate-specific laccase (stilbene oxidase) that is able to oxidize both compounds to non-toxic derivatives (Pezet *et al.*, 1991). The enzyme has been purified and characterized (Pezet, 1998) and the mechanism of stilbene degradation has been elucidated in detail (Breuil *et al.*, 1998). The elicitation of stilbene biosynthesis in *Vitis* plants by a soil bacterium can be used as a biological control mechanism against *B. cinerea* infection of grape (Bernard *et al.*, 1998). Thus, in spite of the capability of *B. cinerea* isolates to degrade the *Vitis* phytoalexins, these compounds remain effective chemical defense compounds. The exact role of stilbene oxidase in pathogenicity of *B. cinerea* remains to be validated by a molecular-genetic approach, by cloning and deletion of the stilbene oxidase gene.

B. cinerea is also able to detoxify preformed antimicrobial compounds. Verhoeff and Liem (1975) showed that *B. cinerea* can degrade the tomato saponin α-tomatine, and suggested that this ability is correlated with the resistance against this phytoanticipin and with latency on tomato. Recently, Quidde *et al.* (1998) analyzed the degradation of α-tomatine by a field isolate of *B. cinerea* in detail. They showed that the saponin is not completely deglycosylated (as suggested by Verhoeff and Liem, 1975), but that only the terminal xylose is removed, yielding β_1-tomatine. β_1-tomatine appeared to be far less toxic than α-tomatine, confirming that this deglycosylation step represents a detoxification. The corresponding enzyme, a "tomatinase", was purified and characterized biochemically. A field survey showed that most *B. cinerea* isolates tested (from various host plants and geographic origin) possessed tomatinase activity, with only one exception (Quidde *et al.*, 1998). Interestingly, the strain lacking tomatinase activity was highly sensitive to α-tomatine, completely non-pathogenic on tomato, but highly aggressive on *Phaseolus*, strongly suggesting that the ability to detoxify α-tomatine is correlated with the virulence of *B. cinerea* on tomato. However, this strain might have further defects that cause the specific loss of virulence on tomato; a disruption of the tomatinase gene will be necessary for a final proof.

B. cinerea is also able to detoxify saponins other than α-tomatine. Quidde *et al.* (1999) demonstrated that the fungus can deglycosylate digitonin, avenacin, and avenacosides. They purified an "avenacinase" and showed that it is highly specific to avenacin and has no side activity, *e.g.* against avenacosides. Deletion of a putative "saponinase" gene (*sap1*) led to loss of avenacinase activity only. The role of an "avenacinase" for virulence of *B. cinerea* is open, since avenacin has so far only been detected in *Avena* roots, which are not infected by *B. cinerea*. Probably comparable saponins occur in other plants that are potential hosts for the fungus. Taken together, the available data demonstrate that *B. cinerea* synthesizes at least three different saponinases: two glucosidases (avenacinase/avenacosidase) and a xylosidase (tomatinase, digitoninase). Since such analyses are limited by the availability of substrates, it is very likely that *B. cinerea* possesses an even larger set of enzymes for the effective detoxification of phytoanticipins.

In addition to substrate-specific detoxifying enzymes, a less specific enzyme might be involved in counteracting the effect of plant defense compounds that inhibit the growth of *B. cinerea*, namely glutathione S-transferase (GST). This enzyme has been studied especially in mammals and in plants in the past few years (Marrs, 1996). GST is able to conjugate glutathione to toxic compounds that accumulate in the cytoplasm. The resulting conjugate is subsequently transported to the vacuole or secreted. Many organisms contain a set of GST isozymes, which accept a variety of toxic substrates, including xenobiotics, phenolics, heavy metals and H_2O_2. A GST gene was cloned from *B. cinerea* (Prins *et al.*, 1999b), which is present in a single copy in the genome. A targeted gene disruption was performed and the resulting mutant was tested for a possible loss of virulence on tomato leaves. The mutant was still pathogenic, indicating that the GST gene is not essential for infecting this host tissue (Prins *et al.*, 1999b). In view of the wide substrate range of GSTs, it is difficult to predict against which toxic compounds the *B. cinerea* GST is supposedly conferring protection, and whether this has consequences for the infection on particular host plants.

Besides detoxifying antifungal chemicals that accumulate in plant tissue, a second mechanism has been proposed for fungi to overcome the growth inhibiting effects of these compounds. Energy-dependent secretion by ABC-transporters was postulated to provide plant pathogenic fungi with a tool to prevent the fungistatic or fungitoxic effects of such compounds (de Waard, 1997). ABC transporters have been cloned from *B. cinerea* and their role in resistance towards plant defense compounds and virulence on various hosts is being studied (Del Sorbo, Schoonbeek and de Waard, unpublished results). Some ABC-transporters are located in the vacuolar membrane and are involved in pumping complexes of glutathione and toxic compounds, generated by GST, into the vacuole (Ishikawa *et al.*, 1997).

The oxidative burst that occurs at the host-pathogen interface (see Figure 4 and paragraphs 7 and 8) imposes an oxidative stress on the host as well as the pathogen. During the infection on tomato leaves, the host catalase mRNA level increased from the moment of necrotic lesion appearance (van der Vlugt-Bergmans *et al.*, 1997b). The pathogen needs to be able to cope with external oxidative stress in order to survive in the necrotic tissue. Successful detoxification of extracellular AOS by *B. cinerea* was indicated by the observation that an intracellular catalase of *B. cinerea* is not expressed *in planta* in necrotic tissue, in spite of the fact that it can be induced *in vitro* in liquid culture by exogenously applied hydrogen peroxide (van der Vlugt-Bergmans *et al.*, 1997b). The removal of potentially toxic external AOS by *B. cinerea* is likely to be mediated by extracellular enzymes, such as Superoxide Dismutase (SOD) and catalase. Genes encoding SOD (Quidde, Weltring and Tudzynski, unpublished results) and extracellular catalase (Schouten and van Kan, unpublished results) have been cloned, and are currently characterized. If SOD and extracellular catalase are involved in self-protection of the pathogen against the oxidative stress that it imposes on the host, targeted replacement of the corresponding genes in *B. cinerea* might yield a mutant that is killed by the oxidative burst occurring within the necrotic lesion.

Taken together, the biochemical and preliminary molecular-genetic data demonstrate the enormous metabolic versatility of *B. cinerea*, enabling the fungus to overcome a wide spectrum of antifungal plant defense mechanisms. Whether this metabolic capability contributes to the broad host-range of the fungus is under investigation in various laboratories.

10. Disease expansion and tissue maceration

Besides needing to cope with host defense responses, as described in paragraph 9, *B. cinerea* must be able to macerate plant tissue and utilize it for its own growth. The initial step in expansion of primary lesions is presumably the killing of neighboring cells by mechanisms similar to the ones described in paragraph 7.

In order to grow out of the primary lesion, the plant tissue must be actively degraded. First, the lignified barrier surrounding the primary lesion may need to be breached. It has not yet been studied how this process is achieved, but it is likely to involve peroxidases and other enzymes with lignin degrading capacity. It was proposed that laccase activity secreted by *B. cinerea* can interfere with plant defense responses by oxidizing phenolic compounds that serve as lignin precursors or as phytoalexins (Viterbo *et al.*, 1992). Moreover, some fungal laccases are capable of degrading lignin (Evans and Betts, 1991). Cucurbitacins, a group of compounds from Cucurbitaceae, inhibited the production of extracellular laccase activity by *B. cinerea* and conferred resistance to the pathogen (Bar Nun and Mayer, 1990).

Once through the lignin barrier, access to the surrounding tissue is obtained and this tissue is rapidly colonized. Cell wall degradation facilitates the entry of the pathogen and it provides nutrients for growth. For this purpose, *B. cinerea* possesses a set of cell wall degrading enzymes (CWDEs). The enzymes are produced during all stages of infection. Among the CWDEs of *B. cinerea* described by various groups are one or more pectin lyases (Movahedi and Heale, 1990b), pectin methylesterase (Reignault *et al.*, 1994), endopolygalacturonase (Johnston and Williamson, 1992), exopolygalacturonase (Johnston and Williamson, 1992) and cellulase (Barkai-Golan *et al.*, 1988).

It has been reported by several groups that treatment of plants with calcium fertilizer contributes to inhibition of grey mould development (Elad and Volpin, 1988; Wisniewski *et al.*, 1995; Chardonnet *et al.*, 1997; Klein *et al.*, 1997), although the effect may be tissue dependent. In cucumber, calcium treatment was able to reduce disease, by reduction of enzymatic pectin degradation on fruits but not on leaves (Chardonnet and Donèche, 1995). It is most likely that the protective effect of calcium is caused by the reinforcement of Ca-pectate complexes into egg-box structures, which are less accessible to CWDEs.

Recently, a number of *B. cinerea* genes have been cloned that encode CWDEs: pectin lyase (Mulder, unpublished), rhamnogalacturonan-hydrolase (Chen *et al.*, 1997) and six genes encoding endoPGs (ten Have *et al.*, 1998, Wubben *et al.*, 1999a). The endoPG genes, denoted *Bcpg*1-6, constitute a well studied and, most probably, complete gene family. The expression patterns of the individual endoPG genes *in planta* depend on the host tissue that is infected, on the progress of infection, as well as on the external conditions at which infection occurs (ten Have *et al.*, 1999). Expression patterns *in planta* are in accordance with the expression data obtained in liquid cultures (Wubben *et al.*, 1999b) and they can largely be explained by combinations of inducing and repressing conditions in the host tissue. *Bcpg*1 is expressed constitutively in all tissues and on all carbon sources tested. Other genes (*Bcpg*4 and 6) are induced by the end product of PG activity, galacturonic acid, while the *Bcpg*3 gene is induced at low pH and the *Bcpg*5 gene is induced by growing on apple pectin. These studies suggest that *B. cinerea* can degrade pectate under different environmental conditions. It may be argued that there is a relation between the host range and the need to possess such a large gene family. The related plant pathogenic fungus *Sclerotinia sclerotiorum* also has a substantial endoPG gene family, as well as a broad host range (Fraissinet-Tachet *et al.*, 1995). However, Southern analysis showed that homologues of all members of the *B. cinerea* endoPG gene family are present in *Botrytis* species that can only infect a single host plant species (Wubben *et al.*, 1999a). This observation makes it unlikely that the mere presence of a family of endoPG genes is sufficient to explain the broad host range of *B. cinerea*, as compared to other *Botrytis* species.

Two genes encoding endoPGs (*Bcpg*1 and *Bcpg*2), were eliminated by gene replacement and both mutants were reduced in virulence on tomato as well as on broad bean leaves (ten Have *et al.*, 1998; ten Have *et al.*, 1999). The effect of the deletion was most pronounced for the *Bcpg*1 gene. This gene is expressed constitutively and could therefore be important in facilitating intercellular growth at the outer edge of the diseased tissue. Another explanation for the reduction in virulence may be purely nutritional. The absence of one enzyme will hamper the release of pectin degradation products that serve as nutrients, consequently resulting in slower growth of the fungus through the tissue. These studies clearly show that *B. cinerea* CWDEs can be involved in the invasion of host tissue surrounding the primary infected area. In various other plant pathogenic fungi, targeted deletion of CWDE genes resulted in strains with no discernable loss of virulence (Schaeffer *et al.*, 1994; Bowen *et al.*, 1995; Sposato *et al.*, 1995; Apel-Birkhold and Walton, 1996; Gao *et al.*, 1996; Görlach *et al.*, 1998; Scott-Craig *et al.*, 1998). The only other fungus in which a significant role of an endoPG in pathogenesis has been reported, is the saprophyte *Aspergillus flavus*, where endoPG deletion resulted in a reduced ability to invade cotton bolls (Shieh *et al.*, 1997).

We hypothesize that *B. cinerea* should be regarded as a "pectolytic" fungus. The host range of *B. cinerea* is confined to dicotyledons and non-graminaceous monocotyledons. All these hosts have a rather high content of pectin in the cell walls, when compared to the graminaceous non-host plants (Carpita and Gibeaut, 1993). In view of the role of endoPGs in virulence of *B. cinerea*, a suitable control strategy might be based on the use of PGIPs, which can inhibit endoPGs from *B. cinerea* (Johnston *et al.*, 1994). Therefore, PGIP genes may be used in molecular resistance breeding (Graham *et al.*, 1996), taking into account the differential activity of PGIPs towards individual fungal endoPGs (Desiderio *et al.*, 1997).

11. Phytohormone production by *B. cinerea*

Changes in the level of one individual plant hormone, or in the balance between different hormones, cause developmental changes in the plant which may result in symptoms such as stunting, overgrowth, leaf curls, and can even lead to plant death. Many of these symptoms are similar to those caused by plant pathogenic fungi. Interestingly, some plant pathogenic fungi are able to synthesize phytohormones in culture. Infected plant tissue often contains higher levels of certain phytohormones than uninfected plants (Isaac, 1992). However, it is unclear whether the hormone increase originates from the plant or from the fungus, whether pathogens influence the plant-directed hormone synthesis, or whether external application of hormones influences the disease. These knowledge gaps hamper a clear evaluation of the role of plant hormones in disease etiology.

B. *cinerea* is often regarded as a pathogen that mainly attacks weakened or senescing tissues. Hence, it was logical to determine the effects of applying exogenous hormones that stimulate plant senescence or ripening. Increased susceptibility to grey mould was shown to be associated with factors that enhance aging of the host tissues, such as abscisic acid (ABA) and ethylene (Elad, 1988; McNicol et al., 1989; Elad and Evensen, 1995). ABA was also found to be a strong inhibitor of the accumulation of the phytoalexins rishitin and lubimin in potato tuber. The disease development on rose flowers was promoted by application of ABA, although germination and germ-tube elongation of *B. cinerea* were unaffected by ABA (Shaul et al., 1996).

B. *cinerea* is able to produce several plant hormones (see Tudzynski, 1997). ABA is one of the naturally occurring phytohormones produced by *B. cinerea* (Marumo et al., 1982; Kettner and Dörffling, 1995; Wu and Shi, 1998) and other phytophathogenic fungi such as Cercospora rosicola (Assante et al., 1977), Ceratocystis, Fusarium (Dörffling and Petersen, 1984; Tuomi et al., 1995) and Alternaria brassicae (Dahiya et al., 1988).

Among 95 strains of *B. cinerea* tested, 39 produced ABA in axenic culture at a yield up to several mg/l (Hirai et al., 1986). Kettner and Dörffling (1995) determined the levels of ABA in tomato leaves infected with ABA producing or non-producing *B. cinerea* strains and concluded that at least four processes control the level of ABA: 1. Release of ABA or its precursor by the pathogen; 2. Stimulation of ABA biosynthesis in the host plant; 3. Inhibition of plant ABA metabolism; 4. Promotion of biosynthesis of fungal ABA by the host tissue.

The ABA level in tomato leaves increased after infection, both with an ABA-producing and a non-producing *B. cinerea* isolate, albeit much more in response to the ABA-producing strain (Kettner and Dörffling, 1995). This indicates that ABA favors infection, but the ability of the fungus to produce ABA is not a prerequisite for infection of tomato leaves. In studies on the interaction of ABA-producing *B. cinerea* isolates with *Salix* leaves, a positive correlation was found between the frequency of infection and the amount of ABA produced in the infected tissue (Tuomi et al., 1993). However, the ABA quantities detected in the tissue were too high to be secreted by the fungus, and presumably resulted mainly from increased ABA production by the plant.

Several *B. cinerea* strains isolated from infected fruits are able to produce ethylene in liquid culture. The ethylene levels are physiologically significant, suggesting that ethylene produced by *B. cinerea* itself may play a role in the interaction between host and pathogen (Qadir *et al.*, 1997).

Cytokinin-like substances with cytokinin activity were detected in culture fluids of pathogenic *B. cinerea* isolates and other pathogens (Talieva and Filimonova, 1992). Since exogenous application of cytokinin causes abnormalities in plants such as tumor formation, overgrowth, galls and rusts, a possible involvement of fungal cytokinins in development of plant infection was suggested.

A systematic study of phytohormone production by *B. cinerea* and its role in pathogenicity is only in its early stages. It is difficult to determine whether the observed effects are due to growth regulators produced by the invading fungus itself, or by the plant in response to the stress evoked by the pathogen. Cloning of the biosynthetic genes of *B. cinerea* phytohormone pathways, followed by gene expression and gene replacement experiments are needed to resolve the role of hormone production by the fungus in pathogenesis.

12. Other *Botrytis* species and other necrotrophs

12.1. *BOTRYTIS* SPECIES WITH A NARROW HOST RANGE

In contrast to *Botrytis cinerea* that can infect at least 235 plant species, virtually all other (± 20) pathogenic *Botrytis/Botryotinia* species have a narrow host range, which is restricted to one or a few species of the corroliferous monocotyledons, the *Ranunculaceae* or the *Leguminosae* (Jarvis, 1977). The symptoms and development of diseases caused by the other *Botrytis* species on their host are often similar to those caused by *B. cinerea*, *i.e.* primary necrotic lesions of which a proportion subsequently expands into healthy tissue, thereby destroying large parts of the plant. Most of the infection mechanisms of narrow host range *Botrytis* species are probably similar to the mechanisms described above for *B. cinerea*.

Epidemics of *B. elliptica* in lilies and *B. tulipae* in tulips can occur simultaneously in regions in The Netherlands where tulips and lilies are co-cultivated. When *B. elliptica* is inoculated onto the non-host tulip, it is able to cause primary necrotic lesions that are barely distinguishable from those caused by *B. tulipae*. The lesions of *B. elliptica*, however, do not expand on tulip whereas those of *B. tulipae* do. The reciprocal is true for *B. tulipae* on the non-host lily. *B. cinerea* frequently occurs as a secondary opportunistic pathogen on tulip and lily leaves previously invaded by either of the other *Botrytis* species.

The determinants of the limited host range of *Botrytis* species, other than *B. cinerea,* are yet unclear. Specificity is apparently not determined by recognition of the host or signal exchange needed to establish an infection. In many cases, host-specific *Botrytis* species are able to cause primary lesions on a non-host, but these primary lesions fail to expand. The decisive step whether a plant is regarded as a host or a non-host thus lies in the lesion expansion. The effective restriction of the pathogen within the primary lesion needs to be breached, by neutralizing the fungitoxic and fungistatic (phenolic) metabolites that have accumulated within the lesion and by breaking through the fortified lignin barriers surrounding the lesion.

The broad host range of *B. cinerea* might thus be a consequence of its ability to neutralize defense mechanisms of a large number of host species (as discussed in paragraph 9), whereas other *Botrytis* species are possibly only able to neutralize the defense mechanisms of one particular plant species, *i.e.* their specific host. Only very few studies have been reported on the ability of host-specific *Botrytis* species to neutralize fungitoxic compounds from their respective hosts.

As discussed in paragraph 9, Quidde *et al.* (1998) characterized a tomatinase-deficient *B. cinerea* field isolate that was non-pathogenic on tomato, yet retained the ability to infect bean as any other isolate. This suggests that host specialization or preference may occur even within the *B. cinerea* population. There is recent evidence that genetically isolated, sexually reproducing, sympatric populations (denominated *transposa* and *vacuma*) co-exist within the species *Botryotinia fuckeliana* in the Champagne region (Giraud *et al.*, 1997). One of the populations appeared to be more stable over time than the other and it was proposed that this represented a resident population, which was better adapted to the host plant(s) in the area (Giraud *et al.*, 1997). It will be highly interesting to investigate whether separate fungal populations have a preference for certain (single or groups of) host plant species. If such subgroups are indeed specifically adapted to subsets of host plants, the statement that *B. cinerea* is able to infect at least 235 host species (Jarvis, 1977) may represent an oversimplification.

12.2. *SCLEROTINIA SCLEROTIORUM*

Sclerotinia sclerotiorum is a wide host range pathogen with many similarities to *B. cinerea*. Both fungi belong to the family Sclerotiniaceae. The major difference is that *S. sclerotiorum* ("white mould") does not normally produce asexual spores. In the field it usually infects the base of the stem by ascospores released from apothecia that develop in the top layer of the soil in the spring. There are probably many things in common among the infection mechanism utilized by the two pathogens. Alike *B. cinerea* (see paragraph 10), *S. sclerotiorum* secretes a spectrum of CWDEs with differential regulation patterns (Fraissinet-Tachet *et al.*, 1995; Fraissinet-Tachet and Fèvre, 1996; Martel *et al.*, 1996). Whether these enzymes play a role in the infection process could thus far not be validated, since gene replacement has not yet been achieved in *S. sclerotiorum*. Mutants of *S. sclerotiorum* that are deficient in producing oxalic acid (OA) are significantly less virulent than the wild type and the loss of virulence can be restored by adding exogenous OA in the inoculum (Godoy *et al.*, 1993). The genes responsible for the OA deficiency in *S. sclerotiorum* have not been identified. It is likely that OA has an analogous role in pathogenesis of *B. cinerea* (see paragraph 7), but final validation of this hypothesis should come from the construction of an OA-deficient *B. cinerea* mutant.

13. Summary and concluding remarks

Tools have come available to start validating the numerous microscopic and biochemical observations that have been made on the infection strategies of *B. cinerea*. Figure 5 recapitulates our current view on the separate stages of the infection process and summarizes the involvement of various fungal enzymes and metabolites in these stages, as discussed in this chapter. Some of the corresponding genes have been analyzed in recent years, others will soon follow.

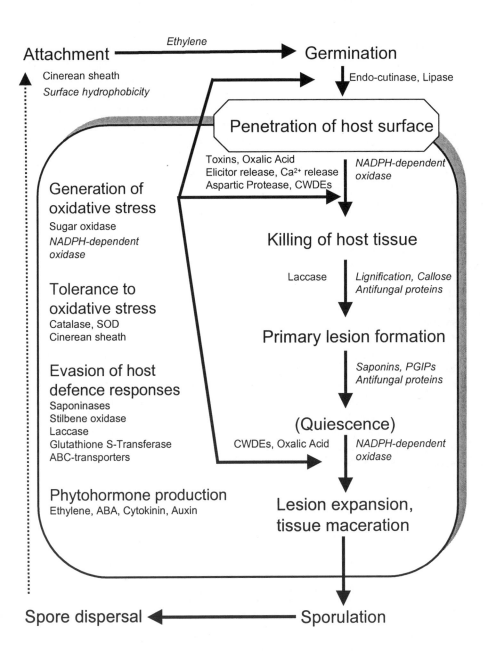

Figure 5: Schematic model of the involvement of various enzymes and metabolites in different stages of the infection process of *B. cinerea*. The shaded box represents the host tissue. Compounds produced by the pathogen are indicated in normal font, compounds originating from the host plant are indicated in italics.

The major challenge is in integrating the various factors in a comprehensive scheme of events. Such knowledge will be useful for developing new, rational disease control strategies.

Botrytis cinerea is a complex pathogen that can probably not be controlled by simple measures. This review demonstrates that the pathogen is versatile and uses a combination of factors during pathogenesis. In spite of the fact that necrotrophs appear to avoid the (usually subtle) interactions with their host that are typical for biotrophs, there seems to be a delicate balance between the attack mechanisms of the fungus and the defense of the host. Rather than combating the fungus vigorously by one single approach, it should be attempted to alter this balance in favor of the plant. Such a strategy will likely consist of a sophisticated combination of biological control agents, appropriate (partially) resistant plant genotypes and chemicals that either enhance the plant defense response or interfere with crucial steps in the infection process.

14. Acknowledgements

The authors acknowledge all the colleagues who have agreed to incorporate unpublished results in this chapter: Thomas Quidde, Klaus Weltring, Songjie Liu, Eri Govrin, Alex Levine, Sander Schouten, Giovanni Del Sorbo, Henkjan Schoonbeek and Maarten de Waard. We are grateful to Sander Schouten and Ernesto P. Benito for critical reading of the manuscript.

The research of T.W. Prins was supported by the Life Sciences Foundation (SLW), which is subsidized by The Netherlands Organization for Scientific Research (NWO), grant SLW805.45.003. The research of A. ten Have was supported by the Technology Foundation (STW), grant WBI33.3046. The research in all three contributing laboratories is partly supported by the European Commission, in the framework of an EU-FAIR project entitled "Oxidative attack by necrotrophic pathogens – New approaches for an innovative and non-biocidal control of plant diseases".

15. References

Akutsu, K., Kobayashi, Y., Matsuzawa, Y., Watanabe, T., Ko, K. and Misato, T. (1981) Morphological studies on infection process of cucumber leaves by conidia of *Botrytis cinerea* stimulated with various purine-related compounds, *Annals of the Phytopathological Society of Japan* **47**, 234-243.

Amadioha, A.C. (1993) A synergism between oxalic acid and polygalacturonases in the depolymerization of potato tuber tissue, *World J. of Microbiology and Biotechnology* **9**, 599-600.

Apel-Birkhold, P.C. and Walton, J.D. (1996) Cloning, disruption, and expression of two endo-beta-1,4-xylanase genes, XYL2 and XYL3, from *Cochliobolus carbonum*, *Applied and Environmental Microbiology* **62**, 4129-4135.

Assante, G., Merlini, L., and Nasini, G. (1977) (+)-Abscisic acid, a metabolite of the fungus *Cercospora rosicola*, *Experientia* **33**, 1556-1557.

Bar Nun, N. and Mayer, A.M. (1990) Cucurbitacins protect cucumber tissue against infection by *Botrytis cinerea*, *Phytochemistry* **29**, 787-792.

Barkai-Golan, R., Lavy Meir, G., and Kopeliovitch, E. (1988) Pectolytic and cellulolytic activity of *Botrytis cinerea* Pers. related to infection of non-ripening tomato mutants, *J. of Phytopathology* **123**, 174-183.

Barkai-Golan, R., Lavy Meir, G., and Kopeliovitch, E. (1989) Effects of ethylene on the susceptibility to *Botrytis cinerea* infection of different tomato genotypes, *Annals of Applied Biology* **114**, 391-396.

Basse, C.W. and Boller, T. (1992) Glycopeptide elicitors of stress responses in tomato cells. *N*-linked glycans are essential for activity but act as suppressors of the same activity when released from the glycopeptides, *Plant Physiology* **98**, 1239-1247.

Bateman, D.F. and Beer, S.V. (1965) Simultaneous production and synergistic action of oxalic acid and polygalacturonase during pathogenesis by *Sclerotium rolfsii, Phytopathology* **55**, 04-211.

Bavaresco, L., Petegolli, D., Cantu, E., Fregoni, M., Chiusa, G., and Trevisan, M. (1997) Elicitation and accumulation of stilbene phytoalexins in grapevine berries infected by *Botrytis cinerea, Vitis* **36**, 77-83.

Beckerman, J.L. and Ebbole, D.J. (1996) MPG1, a gene encoding a fungal hydrophobin of *Magnaporthe grisea*, is involved in surface recognition, *Molecular Plant-Microbe Interactions* **9**, 450-456.

Benito, E.P., Prins, T.W., and Kan, J.A.L., van. (1996) Application of differential display RT-PCR to the analysis of gene expression in a plant fungus interaction, *Plant Molecular Biology* **32**, 947-957.

Benito, E.P., ten Have, A., van 't Klooster, J.W., and van Kan, J.A.L. (1998) Fungal and plant gene expression during synchronized infection of tomato leaves by *Botrytis cinerea, European J. of Plant Pathology* **104**, 207-220.

Bennett, M.H., Gallagher, M.D.S., Bestwick, C.S., Rossiter, J.T., and Mansfield, J.W. (1994) The phytoalexin response of lettuce to challenge by *Botrytis cinerea, Bremia lactucae* and *Pseudomonas syringae* pv. phaseolicola, *Physiological and Molecular Plant Pathology* **44**, 321-333.

Bernard, P., Chereyathmanjiyil, A., Masih, I., Chapuis, L., and Benoit, A. (1998) Biological control of *Botrytis cinerea* causing grey mould disease of grapevine and elicitation of stilbene phytoalexin (resveratrol) by a soil bacterium, *FEMS Microbiology Letters* **165**, 65-70.

Blakeman, J.P. (1980) Behavior of conidia on aerial plant surfaces, in J.R. Coley-Smith, K. Verhoeff and W.R. Jarvis (eds), *The biology of Botrytis,* Academic Press, London, pp. 115-151.

Bowen, J.K., Templeton, M.D., Sharrock, K.R., Crowhurst, R.N., and Rikkerink, E.H.A. (1995) Gene inactivation in the plant pathogen *Glomerella cingulata*: three strategies for the disruption of the pectin lyase gene pnlA, *Molecular and General Genetics* **246**, 196-205.

Brett, C. and Waldron, K. (1996) The cell wall and interactions with other organisms, in M. Black, B. Charwood (eds), *Physiology and biochemistry of plant cell walls,* Chapman & Hall, London, pp. 173-191.

Breuil, A.C., Adrian, M., Pirio, N., Meunier, P., Bessis, R., and Jeandet, P. (1998) Metabolism of stilbene phytoalexins by *Botrytis cinerea*: I. Characterization of a resveratrol dehydrodimer, *Tetrahedron Letters* **39**, 537-540.

Broekaert, W.F., Van Parijs, J., Leyns, F., Joos, H., and Peumans, W.J. (1989) A chitin-binding lectin from stinging nettle rhizomes with antifungal properties, *Science* **245**, 1100-1102.

Callahan, F.E. and Rowe, D.E. (1991) Use of a host-pathogen interaction system to test whether oxalic acid is the sole pathogenic determinant in the exudate of *Sclerotinia trifoliorum, Phytopathology* **81**, 1546-1550.

Carpita, N.C. and Gibeaut, D.M. (1993) Structural models of primary cell walls in flowering plants: consistency of molecular structure with the physical properties of the walls during growth, *Plant J.* **3**, 1-30.

Chardonnet, C. and Donèche, B. (1995) Influence of calcium pre-treatment of pectic substance evolution in cucumber fruit (*Cucumis sativus*) during *Botrytis cinerea* infection, *Phytoparasitica* **23**, 335-344.

Chardonnet, C., L'Hyvernay, A., and Donèche, B. (1997) Effect of calcium treatment prior to *Botrytis cinerea* infection on the changes in pectic composition of grape berry, *Physiological and Molecular Plant Pathology* **50**, 213-218.

Chen, H.J., Smith, D.L., Starrett, D.A., Zhou, D.B., Tucker, M.L., Solomos, T., and Gross, K.C. (1997) Cloning and characterisation of a rhamnogalacturonan hydrolase gene from *Botrytis cinerea, Biochemistry and Molecular Biology International* **43**, 823-838.

Chung, R.P.T., Neumann, G.M., and Polya, G.M. (1997) Purification and characterisation of basic proteins with *in vitro* antifungal activity from seeds of cotton, *Gossypium hirsutum, Plant Science* **127**, 1-16.

Clark, C.A. and Lorbeer, J.W. (1976) Comparative histopathology of *Botrytis squamosa* and *B. cinerea* on onion leaves, *Phytopathology* **66**, 1279-1289.

Cole, L., Dewey, F.M., and Hawes, C.R. (1996) Infection mechanisms of *Botrytis* species: Pre-penetration and pre-infection processes of dry and wet conidia, *Mycological Research* **100**, 277-286.

Collado, I.G., Aleu, J., Hernandez-Galan, R., and Hanson, J.R. (1996) Some metabolites of *Botrytis cinerea* related to botcinolide, *Phytochemistry* **42**, 1621-1624.

Comménil, P., Belingheri, L., and Dehorter, B. (1998) Antilipase antibodies prevent infection of tomato leaves by *Botrytis cinerea, Physiological and Molecular Plant Pathology* **52**, 1-14.

Cutler, H.G., Jacyno, J.M., Harwood, J.S., Dulik, D., Goodrich, P.D., and Roberts, R.G. (1993) Botcinolide: a biologically active natural product from *Botrytis cinerea, Bioscience,* Biotechnology *and Biochemistry* **57**, 980-1982.

Cutler, H.G., Parker, S.R., Ross, S.A., Crumley, F.G., and Schreiner, P.R. (1996) Homobotcinolide: a biologically active homologue of botcinolide from *Botrytis cinerea, Bioscience, Biotechnology and Biochemistry* **60**, 656-658.

Dahiya, J.S., Tewari, J.P., and Woods, D.L. (1988) Abscisic acid from *Alternaria brassicae, Phytochemistry* **27**, 2983-2984.

Daub, M.E. and Hangarter, R.P. (1983) Light-induced production of singlet oxygen and superoxide by fungal toxin, cercosporin, *Plant Physiology* **73**, 55-857.

de Jong, J.C., McCormack, B.J., Smirnoff, N., and Talbot, N.J. (1997) Glycerol generates turgor in rice blast, *Nature* **389**, 244-245.

Deák, M., Horváth, G.V., Davletova, S., Török, K., Sass, L., Vass, I., Barna, B., Király, Z., and Dudits, D. (1999) Plants ectopically expressing the iron-binding protein, ferritin, are tolerant to oxidative damage and pathogens, *Nature Biotechnology* **17**, 192-196.

De Meyer, G. and Höfte, M. (1997) Salicylic acid produced by the rhizobacterium *Pseudomonas aeruginosa* 7NSK2 induces resistance to leaf infection by *Botrytis cinerea* on bean, *Phytopathology* **87**, 588-593.

De Meyer, G., Bigirimana, J., Elad, Y., and Höfte, M. (1998) Induced systemic resistance in *Trichoderma harzianum* T39 biocontrol of *Botrytis cinerea, European J. of Plant Pathology* **104**, 279-286.

Desiderio, A., Aracri, B., Leckie, F., Mattei, B., Salvi, G., Tigelaar, H., Van Roekel, J.S.C., Baulcombe, D.C., Melchers, L.S., De Lorenzo, G., *et al.* (1997) Polygalacturonase-inhibiting proteins (PGIPs) with different specificities are expressed in *Phaseolus vulgaris, Molecular Plant-Microbe Interactions* **10**, 852-860.

de Waard, M.A. (1997) Significance of ABC transporters in fungicide sensitivity and resistance, *Pesticide Science* **51**, 271-275.

Dickman, M.B. and Mitra, A. (1992) *Arabidopsis thaliana* as a model for studying *Sclerotinia sclerotiorum* pathogenesis, *Physiological and Molecular Plant Pathology* **41**, 255-263.

Donèche, B., Roux, F., and Ribereau-Gayon, P. (1985) Degradation de l'acide malique *par Botrytis cinerea, Canadian J. Botany* **63**, 1820-1824.

Dörffling, K. and Petersen, W. (1984) Abscisic acid in fungi of the genera *Botrytis, Ceratocystis* and *Rhizoctonia, Z. Naturforsch.* **39c**, 683-684.

Doss, R.P., Potter, S.W., Chastagner, G.A., and Christian, J.K. (1993) Adhesion of nongerminated *Botrytis cinerea* conidia to several substrata, *Applied and Environmental Microbiology* **59**, 1786-1791.

Doss, R.P., Potter, S.W., Soeldner, A.H., Christian, J.K., and Fukunaga, L.E. (1995) Adhesion of germlings of *Botrytis cinerea, Applied and Environmental Microbiology* **61**, 260-265.

Doss, R.P., Potter, S.W., Christian, J.K., Soeldner, A.H., and Chastagner, G.A. (1997) The conidial surface of *Botrytis cinerea* and several other *Botrytis* species, *Canadian J. of Botany* **75**, 612-617.

Dutton, M.V. and Evans, C.S. (1996). Oxalate production by fungi: its role in pathogenicity and ecology in the soil environment, *Canadian J. Microbiology* **42**, 881-895.

Edlich, W., Lorenz, G., Lyr, H., Nega, E., and Pommer, E.H. (1989) New aspects on the infection mechanism of *Botrytis cinerea* Pers, *Netherlands J. of Plant Pathology* **95**, 53-62.

Elad, Y. (1988) Involvement of ethylene in the disease caused by *Botrytis cinerea* on rose and carnation flowers and the possibility of control, *Annals of Applied Biology* **113**, 589-598.

Elad, Y. (1992) The use of antioxidants (free radical scavengers) to control grey mould (*Botrytis cinerea*) and white mould (*Sclerotinia sclerotiorum*) in various crops, *Plant Pathology* **41**, 417-426.

Elad, Y. (1996) Mechanisms involved in the biological control of *Botrytis cinerea* incited diseases, *European J. of Plant Pathology* **102**, 719-732.

Elad, Y. and Evensen, K. (1995) Physiological aspects of resistance to *Botrytis cinerea, Phytopathology* **85**, 637-643.

Elad, Y. and Volpin, H. (1988) The involvement of ethylene, and calcium in gray mold of pelargonium, Ruscus, and rose plants, *Phytoparasitica* **16**, 119-132.

Elad, Y., Köhl, J., and Fokkema, N.J. (1994) Control of infection and sporulation of *Botrytis cinerea* on bean and tomato by saprophytic bacteria and fungi, *European J. of Plant Pathology* **100**, 315-336.

Evans, C.S. and Betts, W.B. (1991) Enzymes of lignin degradation, *Biodegradation: natural and synthetic materials* 175-184.

Fath, A. and Boller, T. (1996) Solubilization, partial purification, and characterization of a binding site for a glycopeptide elicitor from microsomal membranes of tomato cells, *Plant Physiology* **112**, 1659-1668.

Felix, G., Regenass, M., and Boller, T. (1993) Specific perception of subnanomolar concentrations of chitin fragments by tomato cells: induction of extracellular alkalinization, changes in protein phosphorylation, and establishment of a refactory state, *Plant J.* **4**, 307-316.

Fermaud, M. and Gaunt, R.E. (1995) *Thrips obscuratus* as a potential vector of *Botrytis cinerea* in kiwifruit, *Mycological Research* **99**, 267-273.

Fermaud, M. and Le Menn, R. (1992) Transmission of *Botrytis cinerea* to grapes by grape berry moth larvae, *Phytopathology* **82**, 1393-1398.

Fermaud, M., Gaunt, R.E., and Elmer, P.A.G. (1994) The influence of *Thrips obscuratus* on infection and contamination of kiwifruit by *Botrytis cinerea*, *Plant Pathology* **43**, 953-960.

Flaishman, M.A. and Kolattukudy, P.E. (1994) Timing of fungal invasion using host's ripening hormone as a signal, *Proceedings of the National Academy of Sciences USA* **91**, 6579-6583.

Fraissinet-Tachet, L. and Fèvre, M. (1996) Regulation by galacturonic acid of pectinolytic enzyme production by *Sclerotinia sclerotiorum*, *Current Microbiology* **33**, 49-53.

Fraissinet-Tachet, L., Reymond Cotton, P., and Fèvre, M. (1995) Characterisation of a multigene family encoding an endopolygalacturonase in *Sclerotinia sclerotiorum*, *Current Genetics* **29**, 96-99.

Gaffney, T., Friedrich, L., Vernooij, B., Negrotto, D., Nye, G., Uknes, S., Ward, E., Kessmann, H., and Ryals, J. (1993) Requirement of salicylic acid for the induction of systemic acquired resistance, *Science* **261**, 754-756.

Gao, S.J., Choi, G.H., Shain, L., and Nuss, D.L. (1996) Cloning and targeted disruption of enpg-1, encoding the major in vitro extracellular endopolygalacturonase of the chestnut blight fungus, *Cryphonectria parasitica*, *Applied and Environmental Microbiology* **62**, 1984-1990.

Gentile, A.C. (1954) Carbohydrate metabolism and oxalic acid synthesis by *Botrytis cinerea*, *Plant Physiology* **29**, 257-261.

Germeier, C., Hedke, K., and Tiedemann, A., von (1994) The use of pH-indicators in diagnostic media for acid-producing plant pathogens, *Zeitschrift Pflanzenkrankheiten und Pflanzenschutz* **101**, 498-507.

Giraud, T., Fortini, D., Levis, C., Leroux, P., and Brygoo, Y. (1997) RFLP markers show genetic recombination in *Botryotinia fuckeliana* (*Botrytis cinerea*) and transposable elements reveal two sympatric species, *Molecular Biology and Evolution* **14**, 1177-1185.

Godoy, G., Steadman, J.R., Dickman, M.B., and Dam, R. (1990) Use of mutants to demonstrate the role of oxalic acid in pathogenicity of *Sclerotinia sclerotiorum* on *Phaseolus vulgaris*, *Physiological and Molecular Plant Pathology* **37**, 179-191.

Gönner, M. von and Schlösser, E. (1993) Oxidative stress in interactions between *Avena sativa* L. and *Drechslera* spp., *Physiological and Molecular Plant Pathology* **42**, 221-234.

Görlach, J.M., Knaap, E. van der, and Walton, J.D. (1998) Cloning and targeted disruption of MLG1, a gene encoding two of three extracellular mixed-linked glucanases of *Cochliobolus carbonum*, *Applied and Environmental Microbiology* **64**, 385-391.

Graham, J., Gordon, S.C., and Williamson, B. (1996) Progress towards the use of transgenic plants as an aid to control soft fruit pests and diseases, *Brighton Crop Protection Conference: Pests & Diseases*: Proceedings of an International Conference, Brighton, UK, 18-21 November 1996, **3**, 777-782.

Granado, J., Felix, G., and Boller, T. (1995) Perception of fungal sterols in plants. Subnanomolar concentrations of ergosterol elicit extracellular alkalinization in tomato cells, *Plant Physiology* **107**, 485-490.

Hain, R., Reif, H.J., Krause, E., Langebartels, R., Kindl, H., Vornam, B., Wiese, W., Schmelzer, E., and Schreier, P.H. (1993) Disease resistance results from foreign phytoalexin expression in a novel plant, *Nature* **361**, 153-156.

Hamada, W., Reignault, P., Bompeix, G., and Boccara, M. (1994) Transformation of *Botrytis cinerea* with the hygromycin B resistance gene, *hph*, *Current Genetics* **26**, 251-255.

Harper, A.M., Strange, R.N., and Langcake, P. (1981) Characterisation of the nutrients required by *Botrytis cinerea* to infect broad bean leaves, *Physiological Plant Pathology* **19**, 153-167.

Heale, J.B. and Sharman, S. (1977) Induced resistance to *Botrytis cinerea* in root slices and tissue cultures of carrot (*Daucus carota* L.), *Physiological Plant Pathology* **10**, 51-61.

Hirai, N., Okamoto, M., and Koshimizu, K. (1986) The 1',4'-trans-diol of abscisic acid, a possible precursor of abscisic acid in *Botrytis cinerea*, *Phytochemistry* **25**, 1865-1868.

Howard, R.J., Ferrari, M.A., Roach, D.H., and Money, N.P. (1991) Penetration of hard substrates by a fungus employing enormous turgor pressures, *Proceedings of the National Academy of Sciences of the United States of America* **88**, 11281-11284.

Isaac, S. (1992) Fungal-plant interactions, Chapman & Hall, London, pp. 252-265.

Ishikawa, T., Li, Z.S., Lu, Y.P., and Rea, P.A. (1997) The GS-X pump in plant, yeast and animal cells: structure, function and gene expression, *Bioscience Reports* **17**, 189-207.

Jarvis, W.R. (1977) *Botryotinia* and *Botrytis* species - Taxonomy, physiology and pathogenicity. A guide to the literature, Monograph no. 14, Ottawa, Research Branch, Canada Department of Agriculture.

Jeandet, P., Sbaghi, M., and Bessis, R. (1992) The use of phytoalexin induction and of *in vitro* methods as a tool for screening grapevines for resistance to *Botrytis cinerea*, in K. Verhoeff, N.E. Malathrakis and B. Williamson (eds), *Recent advances in Botrytis research*, Pudoc Scientific Publishers, Wageningen, pp. 109-118.

Johnston, D.J. and Williamson, B. (1992) Purification and characterisation of four polygalacturonases from *Botrytis cinerea*, *Mycological Research* **96**, 343-349.

Johnston, D.J., Williamson, B. and McMillan, G.P. (1994) The interaction *in planta* of polygalacturonases from *Botrytis cinerea* with a cell wall-bound polygalacturonase-inhibition protein (PGIP) in raspberry fruits, *J. Experimental Botany* **45**, 1837-1843.

Jordan, C.M. and DeVay, J.E. (1990) Lysosome disruption associated with hypersensitive reaction in the potato-*Phytophthora infestans* host-parasite interaction, *Physiological and Molecular Plant Pathology* **36**, 221-236.

Kepczynska, E. (1993) Involvement of ethylene in the regulation of growth and development of the fungus *Botrytis cinerea* Pers. ex. Fr, *Plant Growth Regulation* **13**, 65-69.

Kepczynski, J. and Kepczynska, E. (1977) Effect of ethylene on germination of fungal spores causing fruit rot, *Fruit Science reports* **15**, 31-35.

Kershaw, M.J. and Talbot, N.J. (1998) Hydrophobins and repellents: proteins with fundamental roles in fungal morphogenesis, *Fungal Genetics and Biology* **23**, 18-33.

Kettner, J. and Dörffling, K. (1995) Biosynthesis and metabolism of abscisic acid in tomato leaves infected with *Botrytis cinerea*, *Planta* **196**, 627-634.

Klein, J.D., Conway, W.S., Whitaker, B.D., and Sams, C.E. (1997) *Botrytis cinerea* decay in apples is inhibited by post-harvest heat and calcium treatments, *J. of the American Society for Horticultural Science* **122**, 91-94.

Köhl, J., Molhoek, W.M.L., Van Der Plas, C.H., and Fokkema, N.J. (1995) Effect of *Ulocladium atrum* and other antagonist on sporulation of *Botrytis cinerea* on dead lily leaves exposed to field conditions, *Phytopathology* **85**, 393-401.

Kritzman, G., Chet, I., and Henis, Y. (1977) The role of oxalic acid in the pathogenic behaviour of *Sclerotium rolfsii* Sacc., *Experimental Mycology* **1**, 280-285.

Lamb, C.J. and Dixon, R.A. (1997) The oxidative burst in plant disease resistance, *Annual Review of Plant Physiology and Plant Molecular Biology* **48**, 251-275.

Langcake, P. (1981) Disease resistance of *Vitis* spp. and the production of the stress metabolites resveratrol, epsilon -viniferin, alpha - viniferin and pterostilbene, *Physiological Plant Pathology* **18**, 213-226.

Legendre, L., Rueter, S., Heinstein, P.F., and Low, P.S. (1993) Characterisation of the oligogalacturonide-induced oxidative burst in cultured soybean (*Glycine max*) cells, *Plant Physiology* **102**, 233-240.

Liu, S., Oeljeklaus, S., Gerhardt, B., and Tudzynski, B. (1998) Purification and characterisation of glucose oxidase of *Botrytis cinerea*, *Physiological and Molecular Plant Pathology* **53**, 123-132.

Louis, C., Girard, M., Kuhl, G., and Lopez Ferber, M. (1996) Persistence of *Botrytis cinerea* in its vector *Drosophila melanogaster*, *Phytopathology* **86**, 934-939.

Mansfield, J.W. (1980) Mechanisms of resistance to *Botrytis*, in J.R. Coley-Smith, K. Verhoeff and W.R. Jarvis (eds), *The biology of Botrytis*, Academic Press, New York, pp. 181-218.

Mansfield, J.W. and Richardson, A. (1981) The ultrastructure of interactions between *Botrytis* species and broad bean leaves, *Physiological Plant Pathology* **19**, 41-48.

Marrs, K.A. (1996) The functions and regulation of Glutathione S-transferases in plants, *Annual Review of Plant Physiology and Plant Molecular Biology* **47**, 127-158.

Martel, M.B., Letoublon, R., and Fèvre, M. (1996) Purification of endopolygalacturonases from *Sclerotinia sclerotiorum*: Multiplicity of the complex enzyme system, *Current Microbiology* **33**, 243-248.

Martin, J.T. and Juniper, B.E. (1970) *The cuticles of plants*, St. Martin's, New York.

Marumo, S., Katayama, M., Komori, E., Ozaki, Y., Natsume, M., and Kondo, S. (1982) Microbial production of abscisic acid by *Botrytis cinerea*, *Agric. Biol. Chem.* **46**, 1967-1968.

Maule, A.J. and Ride, J.P. (1976) Ammonia-lyase and O-methyl transferase activities related to lignification in wheat leaves infected with *Botrytis*, *Phytochemistry* **15**, 1661-1664.

McNicol, R.J., Williamson, B., and Young, K. (1989) Ethylene production by black currant flowers infected by *Botrytis cinerea*, *Acta Horticulturae* **262**, 209-215.

Mendgen, K., Hahn, M., and Deising, H. (1996) Morphogenesis and mechanisms of penetration by plant pathogenic fungi, *Annual Review of Phytopathology* **34**, 367-386.

Monschau, N., Stahmann, K.P., Pielken, P., and Sahm, H. (1997) In vitro synthesis of beta-(1-3)-glucan with a membrane fraction of *Botrytis cinerea*, *Mycological Research* **101**, 97-101.

Movahedi, S. and Heale, J.B. (1990a) Purification and characterisation of an aspartic proteinase secreted by *Botrytis cinerea* Pers ex. Pers in culture and in infected carrots, *Physiological and Molecular Plant Pathology* **36**, 289-302.

Movahedi, S. and Heale, J.B. (1990b) The roles of aspartic proteinase and endo-pectin lyase enzymes in the primary stages of infection and pathogenesis of various host tissues by different isolates of *Botrytis cinerea* Pers ex. Pers, *Physiological and Molecular Plant Pathology* **36**, 303-324.

Mussell, H.W. (1973) Endopolygalacturonase: evidence for involvement in Verticillium wilt of cotton, *Phytopathology* **63**, 62-70.

O'Neill, T.M., Niv, A., Elad, Y., and Shtienberg, D. (1996) Biological control of *Botrytis cinerea* on tomato stem wounds with *Trichoderma harzianum*, *European J. of Plant Pathology* **102**, 635-643.

Osbourn, A.E. (1996) Preformed antimicrobial compounds and plant defence against fungal attack, *Plant Cell* **8**, 1821-1831.

Osbourn, A.E., Melton, R.E., Wubben, J.P., Flegg, L.M., Oliver, R.P., and Daniels, M.J. (1998) Saponin detoxification and fungal pathogenesis, in K. Kohmoto, O.C. Yoder (eds), *Molecular genetics of host-specific toxins in plant diseases*, Kluwer Academic Publishers, Dordrecht, pp. 309-315.

Pezet, R. (1998) Purification and characterisation of a 32-kda laccase-like stilbene oxidase produced by *Botrytis cinerea* Pers. :Fr, *FEMS Microbiology Letters* **167**, 203-208.

Pezet, R., Pont, V., and Hoang Van, K. (1991) Evidence for oxidative detoxification of pterostilbene and resveratrol by a laccase-like stilbene oxidase produced by *Botrytis cinerea*, *Physiological and Molecular Plant Pathology* **39**, 441-450.

Prins, T.W., Wagemakers, C.A.M., and van Kan, J.A.L. (1999a) Aspartic protease *BcAP1* is not a virulence factor for *Botrytis cinerea*, manuscript in preparation.

Prins, T.W., Wagemakers, C.A.M., and van Kan, J.A.L. (1999b) Cloning and characterization of a Glutathione S-Transferase homologue from the plant pathogenic fungus *Botrytis cinerea*, submitted.

Prusky, D. (1996) Pathogen quiescence in post-harvest diseases, *Annual Review of Phytopathology* **34**, 413-434.

Punja, Z.K., Huang, J.-S., and Jenkins, S.F. (1985) Relationship of mycelial growth and production of oxalic acid and cell wall degrading enzymes to virulence in *Sclerotium rolfsii*, *Canadian J. Plant Pathology* **7**, 109-117.

Purdy, R.E. and Kolattukudy, P.E. (1973) Depolymerization of a hydroxy fatty acid biopolymer, cutin, by an extracellular enzyme from *Fusarium solani* f. *pisi*: isolation and some properties of the enzyme, *Archives of Biochemistry and Physics* **159**, 61-69.

Qadir, A., Hewett, E.W., and Long, P.G. (1997) Ethylene production by *Botrytis cinerea*, *Postharvest Biology and Technology* **11**, 85-91.

Quidde, T., Osbourn, A.E., and Tudzynski, P. (1998) Detoxification of alpha-tomatine by *Botrytis cinerea*, *Physiological and Molecular Plant Pathology* **52**, 151-165.

Quidde, T., Büttner, P., and Tudzynski, P. (1999) Evidence for three different specific saponin-detoxifying activities in *Botrytis cinerea* and cloning and functional analysis of a gene coding for a putative avenacinase, *European J. of Plant Pathology* in press.

Rao, D.V. and Tewari, J.P. (1987) Production of oxalic acid by *Mycena citricolor*, causal agent of the American leaf spot of coffee, *Phytopathology* **77**, 780-785.

Rebordinos, L., Cantoral, J.M., Prieto, M.V., Hanson, J.R., and Collado, I.G. (1996) The phytotoxic activity of some metabolites of *Botrytis cinerea*, *Phytochemistry* **42**, 383-387.

Reignault, P., Mercier, M., Bompeix, G., and Boccara, M. (1994) Pectin methylesterase from *Botrytis cinerea*: Physiological, biochemical and immunochemical studies, *Microbiology* **140**, 3249-3255.

Rombouts, F.M. and Pilnik, W. (1980) Pectic enzymes, in A.H. Rose (ed.), *Economic Microbiology*, Academic Press, New York, pp. 228-282.

Rowe, D.E. (1993) Oxalic acid in exudates of *Sclerotinia trifoliorum* and *S. sclerotiorum* and potential use in selection, *Crop Science* **33**, 1146-1149.

Ryals, J.A., Neuenschwander, U.H., Willits, M.G., Molina, A., Steiner, H.Y., and Hunt, M.D. (1996) Systemic acquired resistance, *Plant Cell* **8**, 1809-1819.

Salinas, J. and Verhoeff, K. (1995) Microscopical studies of the infection of gerbera flowers by *Botrytis cinerea*, *European J. of Plant Pathology* **101**, 377-386.

Salinas, J. Calvete (1992) *Function of cutinolytic enzymes in the infection of gerbera flowers by* Botrytis cinerea, Ph.D.-thesis, University of Utrecht, The Netherlands.

Salzman, R.A., Tikhonova, I., Bordelon, B.P., Hasegawa, P.M., and Bressan, R.A. (1998) Coordinate accumulation of antifungal proteins and hexoses constitutes a developmentally controlled defence response during fruit ripening in grape, *Plant Physiology* **117**, 465-472.

Sbaghi, M., Jeandet, P., Bessis, R., and Leroux, P. (1996) Degradation of stilbene-type phytoalexins in relation to the pathogenicity of *Botrytis cinerea* to grapevines, *Plant Pathology* **45**, 139-144.

Schaeffer, H.J., Leykam, J., and Walton, J.D. (1994) Cloning and targeted gene disruption of EXG1, encoding exo-beta-1,3-glucanase, in the phytopathogenic fungus *Cochliobolus carbonum*, *Applied and Environmental Microbiology* **60**, 594-598.

Schroeder, C. (1972) Untersuchungen zum Wirt-Parasit-Verhältnis von Tulpe und *Botrytis* spp., *J. Plant Diseases and Protection* **79**, 94-104.

Schwacke, R. and Hager, A. (1992) Fungal elicitors induce a transient release of active oxygen species from cultured spruce cells that is dependent on Ca^{2+} and protein-kinase activity, *Planta* **187**, 136-141.

Schweizer, P., Felix, G., Buchala, A., Muller, C., and Metraux, J.P. (1996) Perception of free cutin monomers by plant cells, *Plant J.* **10**, 331-341.

Scott-Craig, J.S., Cheng, Y.Q., Cervone, F., Lorenzo, G. de, Pitkin, J.W., and Walton, J.D. (1998) Targeted mutants of *Cochliobolus carbonum* lacking the two major extracellular polygalacturonases, *Applied and Environmental Microbiology* **64**, 1497-1503.

Shaul, O., Elad, Y., and Zieslin, N. (1996) Suppression of *Botrytis* blight in cut rose flowers with gibberellic acid. Effects of exogenous application of abscisic acid and paclobutrazol, *Postharvest Biology and Technology* **7**, 145-150.

Shieh, M.T., Brown, R.L., Whitehead, M.P., Cary, J.W., Cotty, P.J., Cleveland, T.E., and Dean, R.A. (1997) Molecular genetic evidence for the involvement of a specific polygalacturonase, P2c, in the invasion and spread of *Aspergillus flavus* in cotton bolls, *Applied and Environmental Microbiology* **63**, 3548-3552.

Sietsma, J.H., Wösten, H.A.B., and Wessles, J.G.H. (1995) Cell wall growth and protein secretion in fungi, *Canadian J. of Botany* **73**, Supplement 1, S388-S395.

Sposato, P., Ahn, J.H., and Walton, J.D. (1995) Characterisation and disruption of a gene in the maize pathogen Cochliobolus carbonum encoding a cellulase lacking a cellulose binding domain and hinge region, *Molecular Plant-Microbe Interactions* **8**, 602-609.

Stahmann, K.P., Pielken, P., Schimz, K.L., and Sahm, H. (1992) Degradation of extracellular beta-(1,3)-(1,6)-D-glucan by *Botrytis cinerea*, *Applied and Environmental Microbiology* **58**, 3347-3354.

Stahmann, K.P., Schimz, K.L., and Sahm, H. (1993) Purification and characterisation of four extracellular 1,3-beta-glucanases of *Botrytis cinerea*, *J. General Microbiology* **139**, 2833-2840.

Staples, R.C. and Mayer, A.M. (1995) Putative virulence factors of *Botrytis cinerea* acting as a wound pathogen, *FEMS Microbiology Letters* **134**, 1-7.

Stoessl, A., Unwin, C.H., and Ward, E.W.B. (1972) Postinfectional inhibitors from plants I. Capsidiol, an antifungal compound from *Capsicum frutescens*, *Phytopathologische Zeitschrift* **74**, 141-152.

Stone, H.E. and Armentrout, V.N. (1985) Production of oxalic acid by *Sclerotium cepivorum* during infection of onion, *Mycologia* **77**, 526-530.

Tabei, Y., Kitade, S., Nishizawa, Y., Kikuchi, N., Kayano, T., Hibi, T., and Akutsu, K. (1997) Transgenic cucumber plants harbouring a rice chitinase gene exhibit enhanced resistance to gray mold (*Botrytis cinerea*), *Plant Cell Reports* **17**, 159-164.

Talbot, N.J., Ebbole, D.J., and Hamer, J.E. (1993) Identification and characterisation of MPG1, a gene involved in pathogenicity from the rice blast fungus Magnaporthe grisea, *Plant Cell* **5**, 1575-1590.

Talbot, N.J., Kershaw, M.J., Wakley, G.E., Vries, O.M.H., de, Wessels, J.G.H., Hamer, J.E., and De-Vries, O.M.H. (1996) MPG1 encodes a fungal hydrophobin involved in surface interactions during infection-related development of *Magnaporthe grisea*, *Plant Cell* **8**, 985-999.

Talieva, M.N. and Filimonova, M.V. (1992) On parasitic specialisation of *Botrytis* species in the light of new experimental data, *Zhurnal Obshchei Biologii* **53**, 225-231.

ten Have, A., Mulder, W., Visser, J., and van Kan, J.A.L. (1998) The endopolygalacturonase gene *Bcpg*1 is required for full virulence of *Botrytis cinerea*, *Molecular Plant-Microbe Interactions* **11**, 1009-1016.

ten Have, A., Wubben, J.P., Oude-Breuil, W., Mulder, W., Visser, J., and van Kan, J.A.L. (1999) *Botrytis cinerea* endopolygalacturonase genes are differentially expressed *in planta* and contribute to virulence. Manuscript in preparation.

Terras, F.R.G., Schoofs, H.M.E., Thevissen, K., Osborn, R.W., Vanderleyden, J., Cammue, B.P.A., and Broekaert, W.F. (1993) Synergistic enhancement of the antifungal activity of wheat and barley thionins by radish and oilseed rape albumins and by barley trypsin inhibitors, *Plant Physiology* **103**, 1311-1319.

Thomma, B.P.H.J., Eggermont, K., Penninckx, I.A.M.A., Mauch-Mani, B., Vogelsang, R., Cammue, B.P.A., and Broekaert, W.F. (1998) Separate jasmonate-dependent and salicylate-dependent defence response pathways in Arabidopsis are essential for resistance to distinct microbial pathogens, *Proceedings of the National Academy of Sciences USA* **95**, 15107-15111.

Thomzik, J.E., Stenzel, K., Stocker, R., Schreier, P.H., Hain, R., and Stahl, D.J. (1997) Synthesis of a grapevine phytoalexin in transgenic tomatoes (*Lycopersicon esculentum* mill.) conditions resistance against *Phytophthora infestans, Physiological and Molecular Plant Pathology* **51**, 265-278.

Tiedemann, A. von (1997) Evidence for a primary role of active oxygen species in induction of host cell death during infection of leaves with *Botrytis cinerea, Physiological and Molecular Plant Pathology* **50**, 151-166.

Tu, J.C. (1985) Tolerance of white bean (*Phaseolus vulgaris*) to white mould *(Sclerotinia sclerotiorum)* associated with tolerance to oxalic acid, *Physiological and Molecular Plant Pathology* **26**, 111-117.

Tu, J.C. (1989) Oxalic acid induced cytological alterations differ in beans tolerant or susceptible to white mould, *New Phytologist* **112**, 519-525.

Tudzynski, B. (1997) Fungal phytohormones in pathogenic and mutualistic associations, in G.C. Carrol, P. Tudzynski (eds), *The Mycota V Part A: Plant relationships,* Springer Verlag, Berlin, pp. 167-184.

Tudzynski, P. and Tudzynski, B. (1998) Genetics of plant pathogenic fungi, *Progress in Botany* **59**, 169-193.

Tuomi, T., Ilvesoksa, J., Laakso, S., and Rosenqvist, H. (1993) Interaction of abscisic acid and indole-3-acetic acid-producing fungi with *Salix* leaves, *J. of Plant Growth Regulation* **12**, 149-156.

Tuomi, T., Laakso, S., and Rosenqvist, H. (1995) Plant hormones in fungi and bacteria from malting barley, *J. of the Institute of Brewing* **101**, 351-357.

van den Heuvel, J. (1976) Sensitivity to, and metabolism of, phaseollin in relation to the pathogenicity of different isolates of *Botrytis cinerea* to bean (*Phaseolus vulgaris*), *Netherlands J. of Plant Pathology* **82**, 153-160.

van den Heuvel, J. (1981) Effect of inoculum composition on infection of French bean leaves by conidia of *Botrytis cinerea, Netherlands J. of Plant Pathology* **87**, 55-64.

van den Heuvel, J. and Waterreus, L.P. (1983) Conidial concentration as an important factor determining the type of infection structures formed by *Botrytis cinerea* on leaves of French bean (*Phaseolus vulgaris*), *Plant Pathology* **32**, 263-272.

Van der Cruyssen, G., De Meester, E., and Kamoen, O. (1994) Expression of polygalacturonases of *Botrytis cinerea in vitro and in vivo, Mededelingen Faculteit Landbouwkundige en Toegepaste Biologische Wetenschappen Universiteit Gent* **59**, 895-905.

van der Vlugt-Bergmans, C.J.B., Wagemakers, C.A.M., and van Kan, J.A.L. (1997a) Cloning and expression of the cutinase A gene of *Botrytis cinerea, Molecular Plant-Microbe Interactions* **10**, 21-29.

van der Vlugt-Bergmans, C.J.B., Wagemakers, C.A.M., Dees, D.C.T., and Van Kan, J.A.L. (1997b) Catalase A from *Botrytis cinerea* is not expressed during infection on tomato leaves, *Physiological and Molecular Plant Pathology* **50**, 1-15.

van Kan, J.A.L., van 't Klooster, J.W., Wagemakers, C.A.M., Dees, D.C.T., and van der Vlugt-Bergmans, C.J.B. (1997) Cutinase A of *Botrytis cinerea* is expressed, but not essential, during penetration of gerbera and tomato, *Molecular Plant-Microbe Interactions* **10**, 30-38.

VanEtten, H.D., Sandrock, R.W., Wasmann, C.C., Soby, S.D., McCluskey, K., and Wang, P. (1995) Detoxification of phytoanticipins and phytoalexins by phytopathogenic fungi, *Canadian J. of Botany* **73**, S518-S525.

Verhoeff, K. and Liem, J.I. (1975) Toxicity of tomatine to *Botrytis cinerea*, in relation to latency, *Phytopathologische Zeitschrift* **82**, 333-338.

Verhoeff, K., Leeman, M., van Peer, R., Posthuma, L., Schot, N., and van Eijk, G.W. (1988) Changes in pH and the production of organic acids during colonisation of tomato petioles by *Botrytis cinerea, J. of Phytopathology* **122**, 327-336.

Viterbo, A., Bar Nun, N., Mayer, A.M., (1992) The function of laccase from *Botrytis cinerea* in host infection. in K. Verhoeff, N.E. Malathrakis and B. Williamson (eds), *Recent advances in Botrytis research*, Heraklion, Greece, 5-10 April 1992.

Weber, R.W.S. and Pitt, D. (1997) Purification, characterisation and exit routes of two acid phosphatases secreted by *Botrytis cinerea, Mycological Research* **101**, 1431-1439.

Weigend, M. and Lyr, H. (1996) The involvement of oxidative stress in the pathogenesis of *Botrytis cinerea* on *Vicia faba* leaves, *Zeitschrift für Pflanzenkrankheiten und Pflanzenschutz* **103**, 310-320.

Williamson, B., Duncan, G.H., Harrison, J.G., Harding, L.A., Elad, Y., and Zimand, G. (1995) Effect of humidity on infection of rose petals by dry-inoculated conidia of *Botrytis cinerea, Mycological Research* **99**, 1303-1310.

Wisniewski, M., Droby, S., Chalutz, E., and Eilam, Y. (1995) Effects of Ca^{2+} and Mg^{2+} on *Botrytis cinerea* and *Penicillium expansum* in vitro and on the biocontrol activity of *Candida oleophila, Plant Pathology* **44**, 1016-1024.

Wu, J. and Shi, Z.X. (1998) Isolation of abscisic acid producing strains of phytopathogens, *Phytochemistry* **49**, 89-90.

T.W. Prins et al.

Wubben, J.P., Mulder, W., ten Have, A., van Kan, J.A.L., and Visser, J. (1999a) Cloning and partial characterization of the endopolygalacturonase gene family from B*otrytis cinerea*, *Applied and Environmental Microbiology* **65**, 1596-1602.

Wubben, J.P., ten Have, A., van Kan, J.A.L., and Visser, J. (1999b) Regulation of endopolygalacturonase gene expression in *Botrytis cinerea* by galacturonic acid, ambient pH and carbon catabolite repression, *Current Genetics*, in press.

CLADOSPORIUM FULVUM, CAUSE OF LEAF MOULD OF TOMATO

R.P. OLIVER, B. HENRICOT, G. SEGERS
Department of Physiology
Carlsberg Laboratory
Gamle Carlsberg Vej 10
DK-2500 Copenhagen Valby
Denmark

Cladosporium fulvum is a potentially serious fungal pathogen of tomato which has attracted attention as a model system in fungal phytopathology. *C. fulvum* is described, somewhat controversially, as an example of a biotrophic pathogen on the grounds that it causes little or no obvious damage to the plant. However, it is fully capable of axenic growth, unlike any of the classical biotrophs. It is also unique, at least amongst well studied fungi, in remaining entirely within the intercellular spaces in the spongy mesophyll of the sole host tomato. This has had the happy consequence that it has been possible to isolate from so-called apoplastic fluid, a number of significant fungal proteins including the first two avirulence gene products to be cloned from fungi (Van Kan *et al.*, 1991; Joosten *et al.*, 1994). *C. fulvum* was also the first biotroph whose complementary resistance genes were cloned (Jones *et al.*, 1994).

1. History

The disease was first described in 1883 by M.C. Cooke in England on leaf samples sent from South Carolina (Cooke, 1883). Tomato had been brought to Europe from Central America in the 16th Century but the disease was unknown or at least undescribed for the next 300 years. It is likely that the requirement for high humidity prevented development of the disease in more temperate climates in Europe and N. America until intensive greenhouse cultivation became widespread in the early part of this century. The disease was by 1908 present in all tomato cultivating areas (Makemson, 1918).

   From 1908, a number of pathologists noted that resistant lines of tomato could be found but that these were resistant to one but not all races of the pathogen (Lind, 1909; Norton, 1914). Thus the pathogen exhibits "physiologic specialization" and exists in what we loosely term races (Day, 1956). Furthermore, resistant lines of tomato broke down with depressing rapidity, at least until the introduction of Cf9, a resistance gene which has remained effective to this day.

   Some 25 resistance genes have been described with greater or lesser certainty, including 4 that have been cloned (Jones *et al.*, 1993, 1994; Dixon *et al.*, 1996, 1998). Resistance genes have been bred in by interspecific crosses involving *L. hirsutum*, *L. peruvianum*, *L. pimpinellifolium* and *L. glandosum*.

J. W. Kronstad (ed.), Fungal Pathology, 65–91.
© 2000 *Kluwer Academic Publishers. Printed in the Netherlands.*

1.1. RACES OF *C. FULVUM*

The nomenclature of the races refers entirely to the observed interaction with cultivars of the tomato. This has led to a stable nomenclature system as shown in Table 1. The interaction between races and cultivar is fully consistent with the gene-for-gene hypothesis. Thus the races are defined by the expression of a number of avirulence genes (*vide infra*). The definition of race is not entirely satisfactory as (1) it is based solely on the expression of a small number of genes and (2) a race may be divided if a further differential resistance gene is used. This nomenclature tells us little about the evolutionary or phylogenetic relationships of individual isolates. It is still unclear whether each race evolved from pre-existing races or was already present in the existing population. No suitable molecular markers have yet been used to examine populations of this fungus to answer such questions.

Table 1. Interaction of *Cladosporium fulvum* races with tomato cultivars

Race	Cf0	Cf2	Cf4	Cf2.4	Cf5	Cf9	Cf11	Presumed fungal genotype[b]				
0	S	R	R	R	R	R	R	A2,	A4,	A5,	A9,	A11
2	S	S	R	R	R	R	R	a2,	A4,	A5,	A9,	A11
4	S	R	S	R	R	R	R	A2,	a4,	A5,	A9,	A11
2,4	S	S	S	S	R	R	R	a2,	a4,	A5,	A9,	A11
5	S	R	R	R	S	R	R	A2,	A4,	a5,	A9,	A11
2,4,5	S	S	S	S	S	R	R	a2,	a4,	a5,	A9,	A11
2,4,9,11	S	S	S	S	R	S	S	a2,	a4,	A5,	a9,	a11
2,4,5,9,11	S	S	S	S	S	S	S	a2,	a4,	a5,	a9,	a11

Tomato cultivar[a] (column header spanning Cf0–Cf11)

[a] Indicates specific *C. fulvum* resistance genes (*Cf* genes) carried by cultivar.
   S, susceptible; R, resistant.
[b] A2 indicates avirulence to Cf2, a2 virulence to Cf2, etc.

1.2. TAXONOMY AND PHYLOGENETICS

Cooke described the pathogen and placed it in the large, diverse genus Cladosporium (Cooke, 1883). Cladosporium is an imperfect form genus within the Deuteromycotina comprising mainly saprotrophic but also some pathogenic species. *C. fulvum* has subsequently been placed in the Fulvia genus (Ciferri, 1952) (indeed *Fulvia fulva* is the recognized name) and also in a novel genus Mycovellosiela (von Arx, 1983). However, neither change was greeted with any confidence and *C. fulvum* is the name used by breeders and growers as well as pathologists.

The revolution wrought on fungal taxonomy by DNA sequence comparisions has rendered the Deuteromycotina obsolete. An initial study using rDNA sequences of *C. fulvum*, a sample of Cladosporium species and the available outgroup sequences had three conclusions (Curtis *et al.*, 1994). Firstly, all the *C. fulvum* isolates were identical in the region sequenced, implying a recent radiation of the races. Secondly, all the Cladosporium sequences were reasonably similar and were each other's closest neighbours, consistent with all belonging in the same genus. Thirdly, all the

Cladosporia could be placed confidently in the Ascomycotina but no close affiliation to any well-established lower taxon could be made.

Many more sequences have been made available since then and many established taxa within the Ascomycotina have been confirmed, whilst others have been discarded (Taylor, 1995). Amongst the discarded was the Loculoascomycetes and amongst the retained was the Dothideales. Recently Ueng *et al.* (1998) noted that *C. fulvum* was closely related to *Septoria tritici* (alias *Mycosphaerella graminicola*). Further analysis of this finding is underway (Wirsel, Oliver, Mendgen, unpubl.) but it is already apparent that a reclassification of both species into the same family, if not the same genus, within the Dothideales, will be justified. This comparison is intriguing because both pathogens have striking similarities such as stomatal penetration as well as profound differences, in their mode of infection. For example both penetrate their host through stomata and both have narrow host ranges.

2. Mode of infection – compatible interactions

Systematic and sustained study of *C. fulvum* dates from the work of de Wit (1977) and Higgins (Lazarowitz and Higgins, 1976a,b). Infection starts with bipolar germination of conidia on the surfaces of tomato leaves. Runner hyphae extend considerable distances until an open stomata is encountered (Fig.1). The requirement for open stomata explains the requirement for high humidity during infection. There is, however, no evidence for directional growth of the runner hyphae. After 1-4 days, hyphae penetrate the stomata and enter first the substomatal space and then the spongy mesophyll. Penetration is accompanied by a thickening of the hyphae. The hyphae ramify throughout the apoplast, in an apparently random manner, although there is some indication that the vascular tissue is more heavily colonized. No specialized infection structures have been observed. Instead, the fungus merely grows over, around and between the host cells.

After 10-12 days post-infection, the mesophyll is heavily colonized. At this stage, there are still no macroscopically obvious symptoms. The fungus then emerges from the stomata in a coordinated bundle of up to 20 hyphae. It is likely that this severely disrupts the functioning of the stomata. The emergent hyphae then extend to form the visible, superficial, whitish mycelium which then turns yellow or brown as the conidiophores produce conidia (Fig. 2). The whole process takes 12-17 days under ideal conditions. During the later stages of infection, some chlorosis can occur and even necrosis, but in general, tomato tolerates the pathogen as well as, for example, barley tolerates powdery mildew.

2.1. MODE OF INFECTION – INCOMPATIBLE INTERACTIONS

Isolates (races) of the pathogen carrying avirulence genes corresponding to the resistance (Cf) genes in the host fail to infect (Table 1). The early stages are identical but at different times after penetration, fungal development ceases. Different avr/Cf gene combinations have varying effectiveness and most, if not all of the R genes, appear to be partially dominant. The weakest R-genes are Cf11 and Cf3 and these permit extensive hyphal growth and even some weak sporulation. Cf2, 5, 9 and 4 are much more effective with 2 the strongest and 4 the weakest. Homozygous Cf2 effectively stops the fungus in the sub-stomatal space, whereas heterozygous Cf4 plants

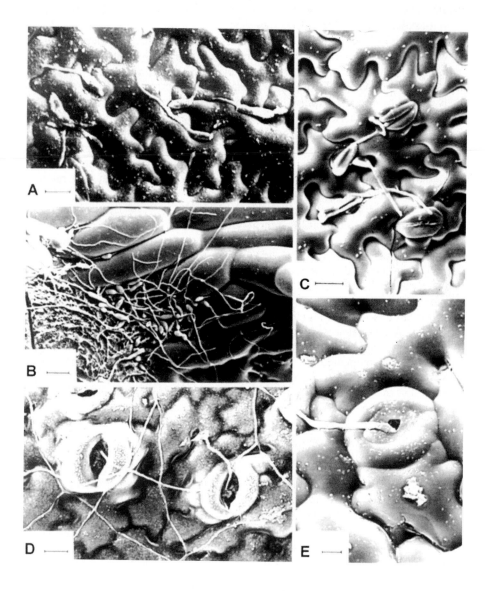

Figure 1. Scanning electron micrographs of stages in a compatible interaction.

(A) Germination of spores on the leaf surface after 24 h (Bar = 15 μm)
(B) Growth of runner hyphae on the leaf surface after 48 h (Bar = 30 μm)
(C) Penetration of a stomata after 48 h (Bar = 15 μm)
(D) Growth of runner hyphae over the leaf surface and penetration of a stomata after 72 h (Bar = 15 μm)
(E) Penetration of a stomata after 72 h (Bar = 5 μm)

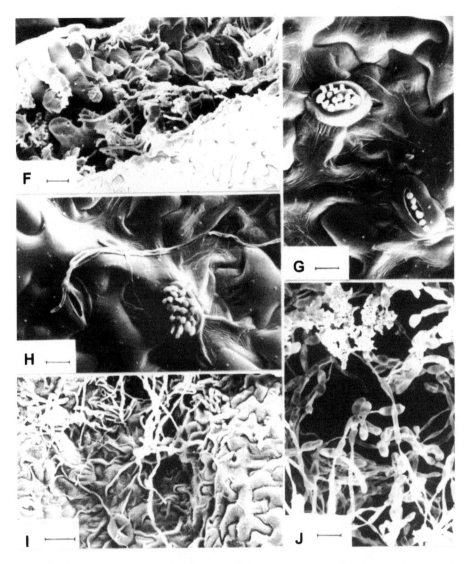

Figure 1. Scanning electron micrographs of stages in a compatible interaction (cont.)

(F) Freeze-fracture of a cotyledon showing the intercellular growth of hyphae after 9 days (Bar = 30 μm)

(G) Aggregation of hyphae in a stomatal aperture prior to growth of conidiophores 9 days after inoculation (Bar = 10 μm)

(H) Penetration of a stomata by an infecting runner hyphae and the emergence of conidiophores from a neighbouring stomata prior to sporulation 9 days after inoculation (Bar = 10 μm)

(I) Growth of conidiophores from a stomata 11 days after inoculation (Bar = 30 μm)

(J) Conidiophores bearing conidia 14 days after inoculation of the cotyledon (Bar = 15 μm)

Figure 2. Underside of tomato leaf showing heavy sporulation of *F. fulva* two weeks after inoculation.

permitted growth through an average of five cell layers (Hammond-Kosack and Jones, 1994). The mechanism of resistance and pathogenicity are considered below.

2.2. A DIGRESSION ON BIOTROPHY

Fungal pathogens have long been divided into two broad classes, necrotrophs and biotrophs. The definition of the two classes is based on the mode of nutrition. Lewis (1973) defined biotrophs as "organisms entirely dependent on another living organism as a source of nutrients". In contrast, necrotrophs are defined as feeding off dead cells. Thus necrotrophs typically produce cell wall degrading enzymes and toxins, whereas biotrophs would not. As discussed recently by Lucas (1998), biotrophs typically have narrow host ranges and are incapable of growth *in vitro*.

The definition of *C. fulvum* as a biotroph still causes dispute. However, there can be no doubt that, in its pathology on tomato, it is as biotrophic as the rusts or mildews. Unlike the mildews and the rusts, it is capable of *in vitro* growth, indeed on simple, defined media. However, the rusts would still be defined as biotrophic even if a suitable axenic media could be found. Furthermore, whilst biotrophs typically have narrow host ranges, this is not true of (e.g.) the polyphagous powdery mildews such as *Erysiphe cichoracearum* which can infect at least 14 species of plants from 3 families (Adam *et al.*, 1991), whereas *C. fulvum* is restricted to tomato and close relatives in the Lycopersicon genus (Bond, 1926).

Thus *C. fulvum* fulfils the criteria of belonging to biotrophs, but can be grown in culture and has therefore been described, by this and other authors, as a model system for biotrophy (Oliver, 1992). However, there are clear differences between *C. fulvum* and the classical biotrophs such as rusts and mildews of which the most prominent is the lack of specialized infection structures such as haustoria or appressoria. Furthermore, it is increasingly apparent that the division between necrotrophs and biotrophs is artificial and that a continuum exists. What, then, is the justification for studying this organism ? The key feature which attracts us to *C. fulvum* is the simplicity of its infection mode. We speculate that *C. fulvum* represents a primitive pathogen possessing only basic pathogenicity factors. Other, more sophisticated pathogens may have evolved from species similar to *C. fulvum* at the time of the angiosperm radiation. *C. fulvum* may thus be regarded as a baseline pathogen and unravelling its mode of infection may tell us about other, more developed pathogens, both necrotrophs and biotrophs, and may suggest general methods to control fungal disease.

3. Methods

3.1. FUNGAL ISOLATES AND CULTURE CONDITIONS
Isolates of *Cladosporium fulvum* are routinely subcultured on V8 juice agar (20% v/v Campbells V8 juice, 2% w/v agar, pH6) containing an alternating antibiotic selection of ampicillin (50 µg/ml) or tetracycline (10 µg/ml). The fungal plates are incubated at 23°C in the dark. For long term storage, conidia are stored in 10% glycerol at -80°C (Harling *et al.*, 1988). The fungus is subcultured no more than three times from the glycerol stock to avoid the development of pale green or white aconidial cultures.

For culturing in liquid media, 100 ml of Gamborg's B5 medium (Gamborg *et al.*, 1968) is inoculated with 10^5-10^6 conidia/ml, and the culture grown at 23°C shaking at 120 rpm. The mycelium is harvested by filtration, washed, frozen in liquid N_2 and freeze-dried overnight.

3.2. PLANT INFECTIONS
Plant infections in growth chambers are carried out essentially as described by Jones *et al.* (1993). Tomato seedlings are treated, ten days after sowing, with a solution of 10^{-5} M paclobutrazol (ICI Agrochemicals) with 50 ml per 500 ml of compost. Paclobutrazol is an inhibitor of gibberellin synthesis, and stops etiolation under conditions of high humidity (Jones *et al.*, 1993). Approximately 21 days after sowing, when the first pair of leaves are expanded, the plants are infected by submerging the leaves in a conidia suspension of approximately 10^5 conidia per ml. The plants are kept in closed propagators at 100% humidity for three days to facilitate germination of the conidia. After this period, the propagator lids are opened during the day and closed at night. To stimulate stomatal opening, the plants are grown in a growth chamber, under conditions of 60-70% humidity at 24°C, where a light intensity of 100 $\mu E.m^{-2}.s^{-2}$ and photoperiod of 16h is provided by 400W fluorescent metal halide lamps.

Plant infections in the greenhouse are carried out essentially as decribed by de Wit (1977) and de Wit and Flach (1979). High humidity is maintained for 72 hours and at night thereafter as above.

3.3. NUCLEIC ACID PREPARATION
Genomic DNA of *C. fulvum* is efficiently isolated according to Raeder and Broda (1985). RNA is extracted from fungal and plant tissue essentially as described in Coleman *et al.* (1997) or by using the Hybaid Recovery RNA purification kit (Hybaid) according to the manufacturer's instructions. In cases of high RNase activity in the tissue, 0.01% β-mercaptoethanol is added to the extraction buffer.

3.4. PREPARATION OF PROTOPLASTS
Protoplasts are prepared essentially as described by Harling *et al.* (1988). Flasks with B5 medium are inoculated with 10^7-10^8 conidia per 100 ml medium. Flasks are grown

with 120-130 rpm shaking for 48 h. Mycelium is then harvested on membrane filters (5 μm pore size), washed with 100 ml of sterile distilled water, then washed with 100 ml of MM solution (20 mM 2-(N-morpholino)(ethanol-sulfonic acid, pH 5.8, 1 M MgSO4).

A solution of 15 mg/ml of Glucanex (Novo Nordisk, Bio Industri, U.K.) in MM is prepared, and sterilised by filtration through a 0.2 μm filter. The mycelium is digested in 10 ml of Glucanex solution for 3 to 4 hours in a shaking water bath (100 rpm) at 28°C. The mixture is then filtered through a funnel containing glasswool to remove undigested mycelium and centrifuged in 50 ml conical tubes, at 3500 rpm for 10 min at room temperature. The supernatant is transferred to a new tube, and 2 volumes of NM buffer (20 mM MES, pH 5.8, 1 M NaCl) added. The protoplasts are pelleted by centrifugation at 4500 rpm for 10 min at room temperature and resuspended in 1 ml of MTC buffer (1 M MgSO4, 10 mM CaCl2, 10 mM Tris-HCl, pH 7.5), using a 1 ml wide bore pipette tip.

3.5. TRANSFORMATION

Three selectable markers have been developed for *C. fulvum*. These are hygromycin (Oliver, 1987), phleomycin (Mattern, 1988) and pyrimidine auxotrophy (Marmeise, 1993). Use of the latter is complicated by the consideration that the host strain is non-pathogenic. Complementation with orotidine-5'-phosphate decarboxylase genes restores pathogenicity, but only partially (Henricot, 1998). Transformation of protoplasts is carried out essentially as described by Oliver *et al.* (1987). Protoplasts are diluted with MTC to give at least 5×10^6 protoplasts per ml for a transformation. DNA (5-10 μg) is added to 0.5 ml of protoplasts, mixed and incubated for 30 min on ice. Two volumes of PTC (20% [v/v] polyethylene glycol 4000, 10 mM CaCl₂, 10 mM Tris-HCl, pH 7.5) are added and mixed gently. After incubation at room temperature for 10 min, 200-300 μl of the mixture is added to 10 ml of molten OCM media (Harling *et al.*, 1988) + 1% agar, cooled to 48°C, and poured into a petri dish. The plates are left overnight at room temperature, after which they are overlayed with 10 ml OCM (1% agar) containing 100 μg/ml of hygromycin B (Boehringer Mannheim) or 30 μg/ml phleomycin. Transformants are obtained after incubation for three to four weeks of incubation at 24°C.

For the transformation of the pyr⁻ mutant, the above protocol is followed with the following modifications. Instead of 2 volumes of PTC, 3 volumes of this buffer are added to the mixture of protoplasts/DNA and incubated at room temperature for 15 min. To the transformed DNA, 5 ml of MCD (NaNO₃ 0.5 g/l, KCl 0.5 g/l, MgSO₄ 0.5 g/l, KH₂PO₄ 1 g/l, FeSO₄ 30 mg/l, sucrose 273.8 g/l, pH 6, 1% agar) is mixed and plated out on 10 ml MCD (2% agar) plates.

3.5.1. *Reporter genes*

Reporter gene systems based on GUS (Roberts *et al.*, 1989) and GFP (Henricot, 1998) have been developed. The GUS system has been used to facilitate biomass determination (Oliver *et al.*, 1993) and to measure promoter activity *in vivo* (Van den Ackerveken *et al.*, 1994). The GFP system has the advantage that living tissue can be examined.

Table 2. Analysis of gene disruption experiments

Locus	Length of homology (kb) (a)	R/I (b)	linear/ circular (c)	P/H (d)	% homologous recombination (e)	Authors
ecp1	2.5 (1.5+1)	R	circular	N/A	2% (2/100)	Lauge *et al.* (1997)
ecp2	4.3 (2.3+2)	R	linear 3 cuts	P	2% (2/90)	Marmeisse *et al.* (1994)
ecp1+ecp2	2.5 (1.5+1)	R	circular	N/A	0.3% (1/350)	Lauge *et al.* (1997)
avr9	5.2 (2.5+2.7)	R	linear 1 cut	P	3% (3/95)	Marmeisse *et al.* (1993)
adh (pSI-10)	1	I	circular	N/A	0% (0/70)	Bradshaw (1999)
aldh (pSI-9)	1.4	I	circular	N/A	0% (0/50)	Bradshaw (1999)
aldh (pSI-9)	4.1 (1.9 + 2.2)	R	linear 1 cut	H	1.3% (1/75)	Bradshaw (1999)
aox (pSI-47)	5.5 (2.5 + 3)	R	linear 1 cut	P	12% (13/105)	Segers (1998)
pSI-1	0.653	I	circular	N/A	0% (0/75)	Henricot (1998)
pSI-1	0.8 (0.4 + 0.4)	R	linear 2 cuts	P	0% (0/120)	Henricot (1998)
HCf-1	3.5 (1.4 + 2.1)	R	nd	nd	1.1% (3/271)	Spanu (1998)

(a) length of homology (kb) of transgene with chromosomal copy. Numbers in brackets indicate length of continuous homology.
(b) R indicates a replacement event of endogenous gene copy by transgene; I indicates integration of transgene within region of homology.
(c) type of construct transformed: linear and number of linearization sites of circular.
(d) region where linearization occurred: P = polylinker, H = region of homology.
(e) percentage of homologous recombination. Numbers in brackets indicate the number of targeted integrations out of transformants screened.

N/A: non applicable, nd: not determined.

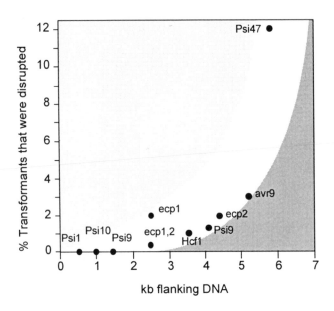

Figure 3. Relationship between the length of homologous DNA in a disruption vector and the % of transformants with the desired gene replacement.

3.5.2. *Gene disruption*
The ability to specifically ablate gene expression has been crucial in our developing understanding of the pathogenicity of *C. fulvum*. The technique that has been found to be successful is a double crossover method, involving replacement of the open-reading frame with the hygromycin or phleomycin selectable marker. This technique has been applied to ten genes (Table 2) and trends are becoming apparent. Firstly, the technique is still inefficient, with disruption ratios of more than 3% being only found on one occasion. Secondly, there is a reasonable correlation between the extend of the flanking sequences and the frequency of disruption (Fig. 3). A minimum of 2.5 kb seems essential whereas 5 kb seems a safe figure. Smaller flanking sequences could be used if the knock-outs can be screened phenotypically *ex-planta* (Spanu, 1998). Apart from the length of the homologous sequence, no other factor appears to have a significant effect on the efficiency of gene replacement.

3.5.3. *Co-suppression*
Gene disruption is a valuable but laborious technique. It suffers from the disadvantage that multi-gene families and genes essential for growth *in vitro* are not applicable to this technique. A method to partially down-regulate gene expression would also have value in these circumstances. This is particularly the case when validating potential fungicide targets. Complete ablation of gene functions is rarely, if ever, achieved by a fungicide. Therefore, to be a target, a gene product must be required at or near full expression for full pathogenicity.

Antisense technology has failed to produce useful reductions in gene expression of *Adh* in *C. fulvum* (Bradshaw, 1999; Henricot, 1998). Co-suppression, however, shows some promise. Expression of sense copies of two *C. fulvum* genes (*HCf-1* (Hamada & Spanu, 1998) and *Adh* (*Psi10*) (Bradshaw, 1999; Henricot, 1998) see below) have resulted in a high frequency (13-30%) of strains which show marked reductions in gene expression (50 to 100%). The phenotypes were unstable but still may be useful in assessing the genes' role in pathogenicity. The mechanism is unclear, but may involve RNA-RNA interactions (Hamada and Spanu, 1988) and may effect other members of the gene family (Segers *et al.*, 1999).

3.5.4. *Screening methods*

Transformants can be screened for the presence or absence of particular DNA sequences using a colony lift hybridisation method (Spanu, pers. comm.; Arganoza and Akins, 1995). Plates with V8 medium containing the appropriate antibiotic are inoculated with conidia of the transformants, at about 1 cm distance. A sterile nylon filter (Hybond N, Amersham) is placed on top of the agar, and the mycelium is allowed to grow for 7 days, after which the hyphae have penetrated into the membrane. The filters are then lifted off the plates, incubated for 1-2 h in 20% SDS, placed colony side up on top of two layers of SDS-soaked 3MM paper (Whatman) and treated in a microwave oven for 2-3 min on high power (900W). The filters are then incubated for at least 5 min in each of the following reagents: denaturing buffer (1.5 M NaOH, 0.5M NaCl), neutralising buffer (1.5 M Tris-HCl pH 7.5, 0.5 M NaCl) twice. The filters are then rinsed in 2x SSC (SSC is 0.15 M NaCl, 0.015 M Na citrate, pH 7.0). The DNA is crosslinked to the filters by UV treatment. The filters are hybridized with a labeled probe fragment prepared from the gene being replaced. Absence of hybridization indicates successful gene knock-out; presence indicates successful transformation.

3.6. PATHOGENICITY ASSAYS

Several qualitative and quantitative pathogenicity assays have been developed. Microscopic examination with a fungal specific fluorescent dye is a useful, quick qualitative assay. Mycelium growing on the leaf epidermis can be visualised using the lipophilic fluorochrome 3,3'-dihexyloxacarbocyanine iodide ($DiOC_6$) (Duckett and Read, 1991). Using this rapid technique, it is possible to distinguish the thin runner hyphae that emerge from conidia from the thicker conidiogenic hyphae that emerge from stomata. Trypan blue (Keogh *et al.*, 1980; Melton *et al.*, 1998) can be used quantitatively and is particularly valuable when assessing partial resistant interactions. For compatible interactions, assays of mannitol dehydrogenase (MtDH) (Joosten *et al.*, 1990; Noeldner *et al.*, 1994) can be used. This is based on an assumption that MtDH expression is constitutive. This seems a reasonable assumption, but proof requires comparison with another marker. A monoclonal antibody-based ELISA has also been developed (Karpovich-Tate, 1998) and used. Finally, the most biologically relevant measure of pathogenicity is undoubtedly the quantitation of the fecundity of the fungus. To do this, spores are collected from infected leaves and counted (Lauge *et al.*, 1997).

3.6.1. $DiOC_6$

A stock solution is prepared (5 mg/ml in ethanol). Tomato leaves are submerged in a 10x aqueous dilution for 1 min, and then rinsed with water. Samples are then examined using e.g. a Zeiss Axioplan microscope, fitted with epifluorescence optics with a 395-440 nm excitation filter.

3.6.2. *Trypan Blue*

Tissue is immersed in boiling trypan blue/lactophenol (10 ml of lactic acid, 10 g of phenol, 10 ml of glycerol, 10 ml of water, 10 mg of trypan blue mixed 1:1 with ethanol) for 1 minute and destained over several days (depending on tissue thickness amongst other factors) in 2.5 g/ml chloral hydrate. Microscopic bright-field examination reveals intercellular hyphae but individual conidia are lost. This method is slow but visualizes apoplastic hyphae within the tomato leaf.

3.6.3. *Mannitol dehydrogenase assay*

Three leaflets from two leaves are ground in liquid N_2, and the powder transferred to an Eppendorf containing 1 ml of 50 mM Tris-HCl (pH 7.5) with 500 μM phenyl-methyl-sulfonyl fluoride (PMSF). Mannitol dehydrogenase (MtDH) activity is assayed by monitoring oxidation of NADPH at 340 nm with fructose as substrate (Morton *et al.*, 1985). The reaction mixture consisted of 0.25 mM NADPH, 0.8 M fructose and 25 μl of crude protein extract in 50 mM Tris-HCl (pH 7.5), in a total volume of 1 ml. Fructose is added to start the reaction.

3.6.4. *Sporewashes*

0.5 g of infected plant material, taken from 3 leaflets, is transferred to a 15 ml Falcon tube containing 5 ml H_2O and 0.01% Tween-20. The tubes are shaken vigorously for 1 h. A 1 ml sample is taken and transferred to an Eppendorf tube, and centrifuged for 30 sec to concentrate the spores. 900 μl of the supernatant is removed, and the spores in the remaining 100 μl are counted using a haemocytometer.

3.7. INTERCELLULAR FLUID

The preparation of intercellular fluid and the demonstration that it contain fungal (and plant) proteins of crucial importance, has been a major factor in the development of our understanding of this pathosystem (de Wit and Spikman, 1982). The original method is simple, does not require special equipment (the packaging of 50 ml syringes can be used) and produces large quantities of material. However, there are two major disadvantages. Firstly, it is not possible to accurately determine the degree to which the intercellular fluid is diluted by the external buffer during preparation. Hence it is not possible to measure concentrations of solutes in the apoplast. Secondly, it has long been suspected that infiltrating with a hypotonic solution would both draw solutes out of plant and fungal cells and burst cells, releasing their contents. The presence of virus-like particles (McHale, 1992) and the NADPH-linked enzyme mannitol dehydrogenase (Joosten *et al.*, 1990) in such intercellular fluid preparations were indicative of the release of fungal cytoplasmic contents. For these reasons, we have adapted an isotonic method of intercellular fluid preparation developed originally for brassica (Husted and Schjoerring, 1995; Hansen and Solomon, unpubl.). This is proving an ideal method to study the low-molecular weight components of intercellular fluid.

3.8. GENETICS

No sexual stage has been observed for *C. fulvum* either in the laboratory or in the greenhouse. This represents the single most important disadvantage in the use of *C. fulvum*. Considerable effort was made to develop a parasexual system (Talbot *et al.*, 1988; Arnau and Oliver, 1993; Arnau *et al.*, 1994) but no workable system emerged.

This has been ascribed to karyotype instability during both protoplast regeneration and after protoplast fusion.

In the absence of a (a)sexual crossing system, the use of chemical or radiation induced mutagenesis was always likely to be problematic (Kenyon *et al.*, 1993).

DNA-mediated transformation offered an alternative mutagenesis method, partially obviating the need for recombination to locate and map mutations. However, extensive trials (Horskins, 1994 and Spanu, unpubl.) involving several thousand transformants have proved it to be unreliable. The problems may be related to genetic instability – control cultures lose pathogenicity spontaneously and integrated DNA undergoes frequent rearrangements (Arnau and Oliver, 1993; Horskins, 1994). A small scale experiment using REMI (e.g. as used by Bolker *et al.*, 1995) was also unsuccessful in that the insertion frequency was not enhanced and insertions were not in intact restriction sites (Coleman, unpubl.). Transposon tagging may thus represent the best way forward to obtain mutants.

3.8.1. *Genomics*
The genome of *C. fulvum* has been estimated to be 44.5 Mbp (Talbot *et al.*, 1991). Nine bands were observed in pulse-field gel experiments ranging in size from 1.9 to 5.4 Mbp. Telomeres have been cloned and analysed (Coleman *et al.*, 1993). The repeat unit is TTAGGG. A retrotransposon family named CfT-1 has been extensively characterized (McHale, 1992). It is a 7 kb element, present in about 25 copies. No evidence that it is active was found.

Polymorphisms between the "races" (*vide supra*) have been found using PFGE (Talbot *et al.*, 1981) telomere RFLP (Coleman *et al.*, 1993), RFLPs using cDNAs (Arnau, unpubl.), RAPDs (Arnau *et al.*, 1994) and AFLPs (Majer *et al.*, 1996).

3.8.2. *Gene structure*
A small but growing number of genes have been cloned and characterized. This now permits some analysis of gene structure in this fungus (Table 3). The gene structure of 8 genomic sequences were analysed and compared to the 5', 3' non coding and intervening sequences of other filamentous fungi. In higher eukaryotes, the TATA and CAAT are widespread and thought to play an important role in transcription (Latchman, 1990). However, the occurrence and significance of these motifs in filamentous fungi is not well defined (Ballance, 1986; Gurr *et al.*, 1987). In the studied plant-induced genes, only *Psi-5* had a TATA motif, but the CAAT motif was present in 3 of the sequences analysed. TATA and CAAT motifs are usually found 30 bp and between 70 to 90 bp upstream the major transcriptional start site, respectively. Because the transcriptional start sites of the plant-induced (Psi) genes were not experimentally determined, it was not possible to tell if the locations of these putative motifs were consistent with other filamentous fungi. In most fungal genes, pyrimidine-rich sequences represented by a CT tract are preceding the transcriptional start site(s) and are usually present in genes where no TATA and CAAT motifs are observed (Gurr *et al.*, 1987). CT tracts of lengths varying from 8 to 16 nucleotides were found in the 5 promoters analysed but their significance as promoter elements, if any, is not known.

CREA protein mediates carbon catabolite repression in *A. nidulans* (Kulmburg *et al.*, 1993). The presence of several consensus CREA binding sequences in promoter region in *Psi1*, *Psi5* and *Psi7* suggests that a similar carbon catabolite repression

Table 3. Structure of genomic sequences of *Psi1*, *Psi5*, *Psi7*, *Psi18* and the genes *ecp1*, *ecp2*, *avr9*, *HCf-1* (Van der Ackerveken et al., 1992, 1993[a], Spanu, 1997) compared to filamentous fungal genes

Intron	Filamentous fungi	Psi1	Psi5	Psi7	Psi18 Intron 1	Intron 2	Intron 3	Intron 4
size	Average <100 bp	67 bp	64 bp	82 bp	134 bp	109 bp	53 bp	59 bp
5' splice site	GT	GT	GT	GT	GT	GT	GT	GT
splicing signal	RCTRAC	GCTGAC	ACTGAC	(T)CTGAC	GCTAAC	GCTAAC	ACCCGT	ACTGAC
3' splice site	AG	AG	AG	AG	AG	AG	AG	AG
distance of splicing signal from 3' splice site	0-22 bp	11 bp	20 bp	10 bp	14 bp	16 bp	8 bp	15 bp

Regulatory sequences in the 5' non coding region

	Filamentous fungi	Psi1	Psi5	Psi7	Psi18
AUG environment	CCRCCATGC	ACAAGATGC	TCAAAATGG	TGGCAATG A	CCGCAATGG
TATA box	TATWAW	-	TATATA (-121)	-	-
CAAT	CAAT	-	CAAT (-39)	CAAT (-854,-778)	CAAT (-13)
Pyrimidine rich sequence	CT tracts (length and position not defined)	-	8CT^a (-110,-272)	8CT^a (-214), 10CT^a (-29)	12CT^a (-26), 16CT^a (-95)
CREA motif	SYGGRG	CTGGAG (-95) GTGGAG (-199)	GTGGGG (-671) GTGGAG (-665)	CCGGGG (-733) CTGGAG (-873)	
GATA motif	GATA	-		GATA (-292, -335, -796)	GATA (-18)
Metal binding motif	HTHNNGCTGD	-	CTCGAGCTGG (-323) TTCCGGCTGA (-723)	-	-

Regulatory sequences in the 3' non coding region

	Filamentous fungi	Psi1	Psi5	Psi7	Psi18
Polyadenylation signal	AAUAAA	-	-	-	-

^{a}number of consecutive cytosine and thymine. Numbers into brackets represent the position of the motif in relation to the translational start.
W = A or T, S = G or C; R = A or G; Y = C or T, H = A, C or T, D = A, G or T, N = A, G, C or T

Table 3.

Intron	ecp1		ecp2	avr9	HCf-1	
	Intron 1	Intron 2			Intron 1	Intron 2
size	86 bp	69 bp	58 bp	59 bp	57 bp	56 bp
5' splice site	GT	GT	GT	GT	GT	GT
splicing signal	ACTAAC	-	ACTGAC	ACTAAC	A(T)TGAC	ACTGAC
3' splice site	AG	AG	AG	AG	AG	AG
distance of splicing signal from 3' splice site	20 bp	-	14 bp	13 bp	10 bp	15 bp

Regulatory sequences in the 5' non coding region	ecp1	ecp2	avr9	HCf-1
AUG environment	CCAAGATGC	TCACGATGC	CTATCATGA	TCAAAATGC
TATA box	TATAAA (-37)	TATAGT (-36)	TATAAGT (-128)	TATAA (-35)
CAAT	CAAT (-175)	CAAT (-80)	-	CAAT (-114, -127)
Pyrimidine rich sequence	9 CT^a (+37)	11 CT^a (+51)	-	-
CREA motif	GCGGAG (-156)	-	-	-
GATA motif	GATA (-605)	-	GATA (-270, -257, -198, -171, -101, -98, -54)	-

Regulatory sequences in the 3' non coding region	ecp1	ecp2	avr9	HCf-1
Polyadenylation signal	-	-	-	-

[a] number of consecutive cytosine and thymine.

Numbers into brackets represent the position of the motif in relation to the transcriptional start.

regulatory system exists in *C. fulvum*. Putative GATA motifs were also present in *Psi7* and *Psi18* promoter sequences suggesting that these genes are under nitrogen regulation. The GATA motifs are known to bind the proteins NIT2, AREA and NRE of *N. crassa*, *A. nidulans* and *P. chrysogenum*, respectively, which mediate nitrogen metabolite repression in these organisms (Fu and Marzluf, 1990; Kudla *et al.*, 1990; Haas *et al.*, 1995). One way to determine if these motifs are involved in the regulation of the plant-induced genes would be to study the expression of their promoters in *areA* or *creA* mutants of *A. nidulans* (S. Snoeijers, pers. comm.; Screen *et al.*, 1997).

In higher eukaryotes, selection of the correct AUG depends on its sequence context and its position within the mRNA (Kozak, 1986). Initiation of translation occurs in 90% of cases at the proximal AUG triplet if the context of this one corresponds to the sequence CC(A/G)CCAUGG, where a purine in position –3 is required (Kozak, 1986). If the context of the first AUG is not optimal, it will not be selected by the 40S ribosomal unit to start translation. This accounts for the 10% of eukaryotic mRNAs that have AUG triplet(s) upstream the translation start (Kozak, 1986). This observation might be particularly important where a choice has to be made between two open reading frames. For example, *Psi5* had two small ORFs but only the proximal one had the ATG in the optimal sequence context.

It was found that the translation start environment of *Psi1, 5, 7, 18, 23, 25, 26* and *28* had a translational start context sequence that fulfils the requirements of the Kozak consensus. Indeed, the position at –3 by an A or G was conserved in all the cDNAs analysed. In 7 of the 9 cDNAs, the –3 position was occupied by an A rather than a G. This observation was consistent with the observations in higher of eukaryotes, where 75% of the mRNAs analysed had an A in that position (Kozak, 1986). The polyadenylation signal of higher eukaryotes AATAAA does not appear in many sequences of fungal genes (Gurr *et al.*, 1987) and was not found in any of the *C. fulvum* Psi genes. This reinforces the impression that this sequence does not seem to be a necessary feature of fungal genes.

A high proportion of the genes (about two thirds) isolated from filamentous fungi have introns in contrast to yeast genes where only 10% of the genes are interrupted (Gurr *et al.*, 1987; Ballance, 1986; Bruchez *et al.*, 1993). The role of the fungal introns is so far poorly understood although it is speculated that they might provide an additional site of gene regulation (Gurr *et al.*, 1987; Ballance, 1986). All the tested genes possess at least one intron, each having the short size characteristic of filamentous fungi. The 5' and 3' splice sites of all introns identified are represented by the nucleotide pairs GT and AG, respectively, which is characteristic not only in filamentous fungi but also in yeast, plants and vertebrates (Gurr *et al.*, 1987). The lariat formation sequence (splicing signal) is, however, less stringent in filamentous fungi, usually fits the consensus (A/G)CT(A/G)AC and is located near the 3' splice site. It was found that only the intron of *Psi7* did not fully fulfil this consensus sequence and all putative splicing signals were located within 20 bp from the 3' splice site. The high conservation of these particular motifs and their position in the intron sequence suggests that these elements be recognized by specific elements involved in the splicing process.

3.8.3. *Codon usage*

An analysis of the codon usage in a sample of 2,960 codons of starvation induced genes indicates a clear bias which can prove useful when designing PCR primers (Table 4). The main themes are a preference for C in the wobble position in all possible cases except Pro (where CCA is favoured). Conversely G in the wobble position is disfavoured in most cases. The exceptions are Gln, Glu, Lys, Leu, Ser, Val where A or T (Ser) are disfavoured.

Table 4. Percentage of codons in codon usage table

Ala	GCG	15.4	Gln	CAG	65.5	Leu	TTG	65.8	Ser	AGT	4.4
	GCA	18.6		CAA	34.5		TTA	2.8		AGC	26.3
	GCT	23.1	Glu	GAG	69.7		CTG	15.6		TCG	18.0
	GCC	42.9		GAA	30.3		CTA	7.3		TCA	11.8
Arg	AGG	10.2	Gly	GGG	6.0		CTT	18.3		TCT	15.4
	AGA	10.8		GGA	17.2		CTC	37.2		TCC	24.1
	CGG	9.6		GGT	31.0	Phe	TTT	25.5	Trp	TGG	100.0
	CGA	15.7		GGC	45.7		TTC	74.5	Tyr	TAT	19.8
	CGT	19.9	His	CAT	29.7	Pro	CCG	12.9		TAC	80.2
	CGC	33.7		CAC	70.3		CCA	49.7	Val	GTG	18.5
Asn	AAT	13.0	Ile	ATA	5.7		CCT	19.0		GTA	3.6
	AAC	87.0		ATT	20.7		CCC	18.4		GTT	21.0
Asp	GAT	35.3		ATC	73.6	Thr	ACG	13.8		GTC	56.9
	GAC	64.7	Lys	AAG	81.3		ACA	23.0	End	TGA	63.6
Cys	TGT	24.1		AAA	18.7		ACT	20.9		TAG	18.2
	TGC	75.9	Met	ATG	100.0		ACC	42.3		TAA	18.2

4. Mechanisms of pathogenicity

Pathogenicity is defined as the ability to cause disease on a susceptible host. Molecular analysis of pathogenicity, therefore, involves the study of genes, mRNAs, proteins and other components expressed or produced during growth on the plant. This definition does not exclude components produced during growth on (any one of many different) artificial media. Components produced specifically during growth on the plant are particularly interesting, but are not the only targets for our investigation.

Our ultimate goal in understanding pathogenicity is to develop strategies to improve, in various ways, the efficacy of crop protection. With that in mind, the importance of all factors required for pathogenicity, whether or not they are also expressed in axenic culture, comes into focus.

Three general strategies (Oliver and Osbourn, 1995) for the study of pathogenicity have been employed, with sharply varying degrees of success on *C. fulvum*. The first is mutagenesis. This has the advantages of being unbiased and directing attention directly at genes only needed for pathogenicity. However, such studies have proved difficult as described above (3.8).

The second strategy is to isolate plant-induced genes (mRNAs or cDNAs) or proteins. The isolation of plant induced genes has been a major preoccupation of this laboratory for the past few years. The particular strategy adopted has been to isolate *in*

R.P. Oliver *et al.*

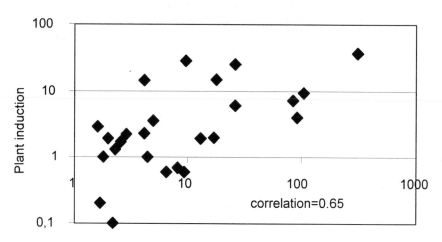

Figure 4. The plant-induction ratios of genes isolated as carbon and nitrogen starvation induced in relation to the degree of starvation induction.

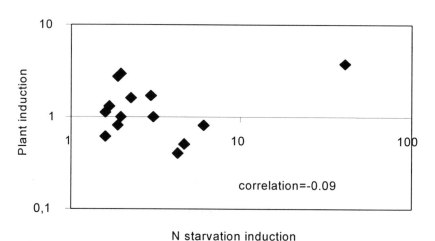

Figure 5. The plant-induction ratios of genes isolated as nitrogen-starvation induced is not correlated with starvation induction.

vitro starvation-induced mRNAs. This strategy was based on the technical advantage of avoiding contamination with plant mRNA – hence the use of *in vitro* grown material – and the observation that many so-called pathogenicity genes also were found to be starvation-induced *in vitro* (detailed in Coleman *et al.*, 1997). The strategy was developed as a rapid route to isolate a moderate number of interesting genes. Validation of the strategy has involved the isolation of genes representing 61 unique genes (Coleman *et al.*, 1997; Faber, unpubl.; Henricot, 1998; Segers, 1998) (Table 5). The overall correlation between the degree of plant-induction and starvation-induction of genes isolated from fully starved mycelium was 0.65 (Fig. 4). This indicates that the isolation of starvation-induced clones does represent an efficient method to isolate at least a sub-set of plant-induced clones. Surprisingly, shifting the focus specifically to nitrogen starvation, did not give a positive correlation with plant induction (*pS152 et seg.*) (Fig. 5).

The identity of some of the genes is indicated by similarity to genes in databases (Table 5). Apart from an unsurprising focus on carbon metabolism, the number of genes is still too small to make general conclusions about metabolism during infection. This approach, and indeed any involving differential screening of libraries, has now largely been superceded by technological developments in DNA sequencing and data handling which combine to make EST approaches more meaningful. Nonetheless analysis of these genes has and will continue to provide useful tools and some insight into pathogenicity.

Table 5. Characterization of Genes Isolated as Starvation Induced

	Starvation Induction	Plant Induction	Similarity
pS 1-1	26	25	-
5	18	15	Metallothionein
7	5	3.6	YCR 068
9	26	6	aldehyde dehydrogenase
10	92	4	alcohol dehydrogenase
15	0.8	0.38	-
16	1.6	-	-
18	1.3	11	ADP/ATP translocator
19	2.5	-	-
21	1.7	0.8	-
22	3	-	-
23	0.6	4.6	Ribosomal protein L41
24	2.2	-	-
25	0.7	3.7	Ribosomal protein S18
26	4.2	14.4	-
28	2.6	1.7	YPL 004
30	0.7	60	NIP1 (weak) (secreted?)
31	1.2	11	methyl transferase
32	1.8	1	-

	Starvation Induction	Plant Induction	Similarity
33	9.8	28	-
34	1	3.1	HCf-1 hydrophobin
35	4.5	1	-
36	1	0.2	-
37	6.6	0.6	-
38	104	9.4	(secreted?)
39	1.2	1.3	Zn-finger protein
40	13.3	1.9	-
41	4.2	2.3	-
42	2.0	1.9	-
43	2.3	1.3	-
44	1.6	2.9	Aspergillus ORF G4PO6
45	0.4	0.3	-
46	2.9	22	Flavoprotein
47	308	37	Alcohol oxidase
48	9.4	0.6	Hexose transporter
49	17	2.0	Chitinase
50	84	7.2	-
51	8.3	0.7	-
52	1.9	2.7	-
53	4.7	0.5	-
54	0.2	0.2	-
55	1.7	1.3	Adenylosuccinate synthase
56	41	3.7	-
57	0.6	1	Ribosomal protein L32
58	0.2	0.2	HCf-2 hydrophobin
59	1	1.9	RNA binding protein
60	2.0	2.9	-
61	4.3	0.4	scytalone dehydratase
62	2.3	-	
63	104	-	Hydrophobin HCf-5
64	3.1	1.1	-
65	1.6	0.7	-
66	6.1	0.8	p450
67	1.6	1.1	-
68	1.9	1.3	-
69	3	2.1	-
70	1.3	1.1	Acyl-carrier protein
71	1.1	0.4	Hydrophobin HCf-3
72	1.4	0.7	Hydrophobin HCf-4
73	2.0	1.0	CON7 Zn-finger
74	2.3	1.8	-

4.1. CHARACTERIZATION OF INDIVIDUAL GENES

Functional characterization of these can best be achieved by assessing the phenotype of knock-out mutants of the genes. This has been achieved for two; *Psi9* (aldehyde dehydrogenase) which seemed unaffected in pathogenicity (Bradshaw, 1999) and *Psi47* (alcohol oxidase) which significantly decreased pathogenicity (Segers, 1998).

4.1.2 *Hydrophobins*

An interesting class of proteins found in many filamentous fungi, is the hydrophobins. Five have been isolated in *C. fulvum* (Spanu; Segers *et al.*, 1999), the largest number of any fungus (Fig. 6.). All possess canonical signal sequences for secretion. They include both class I (HCf 1-4) and class II (HCf-5) proteins. No other fungus has so far been

```
HCf-1    1   MQFTSFAILAISAVASARVTRR..................
HCf-2    1   MQFTTIVMTLAAAVAVTAYPGS..................
HCf-3    1   MQF.IASILAVAAVAYAVAIP...................
HCf-4    1   MQFTTFALLAVAAATASAQAPQAYYGQGAKSAQVHTFETR
HCf-5    1   MQFLVLALASLAAAAPSIKLR...................
Cryp     1   MQFSIIAISFLASLAMASPAK...................

HCf-4   41   KAVPTRVAEVYGEHEQERVTKTKVYHALVTEEAQHHGEEH

HCf-4   81   KAAPYKAYKVYSVASSYSAQPRATHAAEHYGEGKKADHYA

HCf-1   23   ..................DDSSATGAD....KGGT....C
HCf-2   23   ..SSAFGVGQDEHKHHSSDDHSATGAS....KGAT....C
HCf-3   21   ................DDNSATGAS....KGST....C
HCf-4  121   EPAKAVHADPHHVDPVKAR.PTMAATEMKQPEKEAPSTVC
HCf-5   22   ......................RAPS...............DVC
Cryp    22   ...GGGGGGSGS....GSGSGSGSGSG....GGSTTYTAC

HCf-1   37   A..VGSQISCCTTNSSGSDILGNVLGGSCLLDNVSLISSL
HCf-2   53   A..VGSQVSCCTTDSSGSDVLGNVLGGSCLVDNLSLISIL
HCf-3   35   A..TGAQVACCTTNSSNSDLLGNVVGGSCLLDNLSLLSSL
HCf-4  160   AK..GSEISCCTTDSSNSGALGNVLGGSCLLQNLSLLSSL
HCf-5   29   P..ALDTPLCCQADVLGVLDLTCE.....APSDDTSVSNF
Cryp    51   SSTLYSEAQCCATDVLGVADLDCET....VPETPTSASSF

HCf-1   75   NSNCPAGN..TFCCPSNQDGTLNINVSCIPVSA
HCf-2   91   NSQCPGAN..TFCCPSNQDGTLNIHAACIPVAL
HCf-3   73   NSNCPAGN..TFCCPSNSDGTLNINAQCIPISA
HCf-4  198   NSNCAAAN..TFCCPTTQEGTLNINLSCIPISL
HCf-5   62   EAACATTGLTARCCTLPLLGEALLCTTP
Cryp    87   ESICATSGRDAKCCTIPLLGQQALLCQDPVGL
```

Figure 6. Protein sequence alignment. Alignment of deduced aminoacid sequences of HCf-1, -2, -3, -4 and -5 and cryparin. The conserved cysteine residues are shaded. The predicted cleavage sites are marked by ∇ and the N-termini of the mature proteins as deduced by sequencing are marked by ↓ (HCf-1 and –2 only).

shown to have both classes of hydrophobins. A further class II gene has also been serendipitously isolated (Nielsen, unpubl.). The single copy genes show differential expression patterns in relation both to nutrition and sporulation. The role these genes and their encoded proteins play is still unclear. *C. fulvum* is unusually resistant to cell wall degrading enzymes (Joosten *et al.*, 1995) and it may be speculated that the hydrophobin diversity may have a role here. Nonetheless, disruption of *HCf-1* had no obvious effect on pathogenicity (Spanu, 1998).

4.1.3. *Plant-induced proteins*

Using conventionally-produced intercellular fluid as a starting material, de Wit's group has isolated a number of proteins which appear in infected plant material but are absent in uninfected plants. These proteins can be arbitrarily divided into pathogenicity proteins and avirulence proteins, a division that has now become blurred (*vide infra*). The pathogenicity proteins are named ECP for extracellular proteins and two of these have been characterized in some detail. ECP1 is a 36 amino peptide derived from a 96 amino acid primary translation product via both N- and C-terminal processing. ECP2 is processed from a 165 amino acid primary product via N-terminal cleaving only. Both ECP1 and ECP2 are strongly induced during growth on the plant, though weak, *ex planta* expression was observed constitutively for ECP2. No clues to their functions could be deduced from the sequence. Initial experiments (Marmeisse *et al.*, 1994) indicated that ECP2 had no role in pathogenicity either. However, a more realistic pathogenicity assay has revealed that both affected the degree of infection. ECP1-deficient strains were weakly and ECP2-deficient strains were strongly affected in pathogenic ability. Double knock-outs were as poorly infecting as δ-ECP2 strains (Lauge *et al.*, 1997). It is not clear whether the strains are inherently less pathogenic, or if they induce a stronger or faster resistant response.

ECP1 and ECP2 are present in all known races of *C. fulvum*. Therefore it was very surprising to learn that ECP2 acted like an avirulence product. Using PVX as an expression vector, 21 cultivars of tomato were tested for an HR response against ECP2. Four lines were found which responded (Lauge *et al.*, 1998). These lines contain a gene which confers resistance to *C. fulvum*, which is dependent on ECP2 expression. Thus ECP2 may be a virulence factor, whose loss is so damaging to pathogenicity that strains lacking it cannot survive. Such factors and plant genes recognizing them hold out promise as durable resistance factors (Lauge and de Wit, 1998).

5. Avirulence and resistance

Resistance in the tomato-*C. fulvum* is clearly dependent on the expression of fungal avirulence genes. Two *avr* genes have been characterized. Avr9 is a 28 amino acid cystine knot peptide (Vervoort *et al.*, 1997) and the identity of the amino acids inducing necrosis has been determined (Kooman-Gersmann *et al.*, 1997, 1998). Virulence against the cognate resistance gene is associated with complete absence of the gene (Van Kan *et al.*, 1991). No evidence of direct interactions between the Avr9 peptide and Cf9 has been obtained (Kooman-Gersmann *et al.*, 1998). Perhaps the Avr9 binds to an ubiquitous protein and the complex interacts with Cf9 to induce the defence response.

A significant feature of *Avr9* is that its expression can be induced *in-vitro* by starvation of nitrogen (Van den Ackerveken *et al.*, 1994). This induction is correlated with the presence, in the avr9 promoter, of sequences to which the *Aspergillus nidulans* protein AREA binds. *AreA* controls the global induction of nitrogen source utilization genes in Aspergillus. To follow this up, a search has been initiated for a *C. fulvum* homologue of AreA (Snoeijers *et al.*, pers. commun.).

The product of Avr4 has also been extensively characterized. The mature peptide is 87 amino acids and is derived from a 135 amino acid preproprotein by both N- and C-terminal processing. Avr4 expression is sufficient to explain the avirulence of races expressing it in Cf4 plants. Cf4-virulent isolates contain various alleles which in most cases are altered at cysteine residues, which are replaced by tyrosine or by substitutions of Thr66 to Ile or Tyr67 to His (Joosten *et al.*, 1994). A third class of virulent allele encodes a truncated, frame-shifted peptide. This latter product suggests the locus does not produce an essential protein that can only be marginally interfered with. Hence the "intrinsic" role of both *Avr4* and *9* genes is still mysterious. The virulent Avr4 peptides cannot be detected *in-vivo*, suggesting they are unstable (Joosten *et al.*, 1997).

The resistance of tomato to *C. fulvum* is very largely accounted for by the avr-induced hypersensitive response. The resistance genes corresponding to the *Avr4* and *Avr9* genes have been cloned (Thomas *et al.*, 1997; Jones *et al.*, 1994) as well as *Cf2* and *Cf5* (Dixon *et al.*, 1996, 1998). *Cf4* and *Cf9* are components of one complex locus and Cf2 and Cf5 components of a second (Jones, 1993). The structure of the loci is strongly suggestive of an intricate mechanism to preserve allelic variation in the putative recognition domains (leucine rich repeats) of the encoded proteins (Parniske *et al.*, 1997). Furthermore, two more plant genes *Rcr1* and *Rcr2* appear to be required for full expression of resistance (Hammond-Kosack *et al.* 1994).

Surprising, the detailed mechanisms by which the pathogen is halted and killed in the HR are poorly understood (Hammond-Kosack and Jones, 1996). Reactive oxygen species, lipid peroxidation, and glutathione oxidation have all been observed in plants undergoing resistance responses (May *et al.*, 1996) but it has not been possible to dissect the mechanisms of pathogen limitation so far. A number of PR proteins have been detected in apoplastic fluid. These proteins have powerful antifungal hydrolytic activities which could contribute to resistance (Joosten and de Wit, 1989) but *C. fulvum* is unusually resistant to such enzymes (Joosten *et al.*, 1995). The possible role of phytoalexins in the interaction has been investigated. Tomatine is a steroidal glycoalkaloid found in tomato leaves. Tomatine is a glucosidase found in many necrotrophic tomato pathogens, that degrades tomatine to the less toxic tomatidine. Expression of a *Septoria lycopersici* tomatinase gene in *C. fulvum*, lead to enhanced growth in both compatible and incompatible interactions (Melton *et al.*, 1998). This was surprising because *C. fulvum* is very sensitive to the phytoalexin tomatine and it was therefore assumed that the fungus was not exposed to it during infection. The result indicates, but does not prove, that tomatine release is a component of the defence response of tomato. However, the fact that the tomatinase expressing strains were still, albeit belatedly, halted in incompatible interactions makes it clear that factors other than tomatine are responsible for fungal arrest.

Overall it seems likely that a range of factors – peroxide, hydrolytic enzymes, tomatine, water stress, nutrient limitations to name but a selection – are involved in pathogen killing. This multiplicity of weapons implies that the fungus is deploying a corresponding range of counter defensive strategies.

Perspectives

It will be apparent that only sketchy outlines of pathogenicity and resistance are yet clear in the *C. fulvum*-tomato interaction. In both the plant and the pathogen, genetic methodologies need to be developed and applied on a systematic level to achieve further knowledge. An understanding of this very rudimentary pathosystem is likely to be of widespread significance in the plant pathogen field, leading, hopefully, to general strategies for crop protection.

Acknowledgements

Thanks are due to Pierre de Wit and Pietro Spanu for supplying preprints, to Inge Sommer and Nina Rasmussen for preparing the manuscript.

References

Adam, L. and Somerville, S. (1996) Genetic characterization of 5 powdery mildew disease resistance loci in *Arabidopsis thaliana*. *Plant J.* **9**, 341-356.

Arganoza, M.T. and Akins, R.A. (1995) A modified colony-filter-hybridization protocol for filamentous fungi. *Trends Genet.* **11**, 381-382.

Arnau, J. and Oliver, R. (1993) Inheritance and alteration of transforming DNA during an induced parasexual cycle in the imperfect fungus *Cladosporium fulvum*. *Curr. Genet.* **23**, 508-511.

Arnau, J., Housego, A., and Oliver, R. (1994) The use of RAPD markers in the genetic, analysis of the plant pathogenic fungus *Cladosporium fulvum*. *Curr. Genet.* **25**, 438-444.

Ballance, D.J. (1986) Sequences important for gene expression in filamentous fungi. *Yeast* **2**, 229-236.

Bolker, M., Bohnert, H.U., Braun, K.H., Gorl, J., and Kahmann, R. (1995) Tagging pathogenicity genes in Ustilago maydis by restriction enzyme-mediated integration (REMI). *Mol. Gen. Genet.* **248**, 547-552.

Bond, T. (1938) Infection experiments with *Cladosporium fulvum* (Cooke) and related species. *Ann. Appl. Biol.* **25**, 277-305.

Bradshaw, N. (1999) Alcohol metabolism in filamentous fungi in relation to toxigenicity and phytopathogenicity. Ph.D. thesis. University of East Anglia, Norwich, UK.

Bruchez, J.J.P., Eberle, J., and Russo, V.E.A. (1993) Regulatory sequences in the transcription of *Neurospora crassa* genes: CAAT box, TATA box, Introns, Poly(A) tail formation sequences. *Fungal Genetics Newsletter* **40**, 89-97.

Ciferri, R. (1952) A few critical Italian fungi. *Atti* **10**, 237-247.

Coleman, M., McHale, M., Arnau, J., Watson, A., and Oliver, R. (1993) Cloning and characterisation of telomeric DNA from *Cladosporium fulvum*. *Gene* **132**, 67-73.

Coleman, M., Henricot, B., Arnau, J., and Oliver, R.P. (1997) Starvation-induced genes of the tomato pathogen *Cladosporium fulvum* are also induced during growth in planta. *Mol. Plant-Microbe Interact.* **10**, 1106-1109.

Cooke, M. (1883) New American fungi. *Grevillea*. XII:32.

Curtis, M.D., Gore, J., and Oliver, R.P. (1994) The phylogeny of the tomato leaf mould fungus *Cladosporium fulvum* syn. *Fulvia fulva* by analysis of rDNA sequences. *Curr. Genet.* **25**, 318-322.

Day, P. (1956) Race names of *Cladosporium fulvum*. *Tomato Genetics Cooperative*. **6**, 13-15.

de Wit, P.J. (1977) A light and scanning-electron microscopic study of infection of tomato plants by virulent and avirulent races of *Cladosporium fulvum*. *Neth. J. Plant Path.* **83**,109-122.

de Wit, P. and Flach, W. (1979) Differential accumulation of phytoalexins in tomato leaves but not in fruits after inoculation with virulent and avirulent races of *Cladosporium fulvum*. *Phys. Plant Path.* **15**, 257-267.

de Wit, P. and Spikman, G. (1982) Evidence for the occurrence of race and cultivar-specific elicitors of necrosis in the intercellular fluids of compatible interactions of *Cladosporium fulvum* and tomato. *Phys. Plant Path.* **21**, 1-11.

Dixon, M., Ones, D., Keddie, J., Thomas, C., Harrison, K., and Jones, J. (1996) The tomato *Cf-2* disease resistance locus comprises 2 functional genes encoding leucine-rich repeat proteins. *Cell* **84**, 451-459.

Dixon, M.S., Hatzixanthis, K., Jones, D.A., Harrison, K., and Jones, J.D.G. (1998) The tomato *Cf-5* disease resistance gene and six homologs show pronounced allelic variation in leucine-rich repeat copy number. *Plant Cell* 10, 1915-1925.

Duckett, J. and Read, D. (1991) The use of the fluorescent dye, 3,3'-dihexyloxacarbocyanin iodide, for selective staining of ascomycete fungi associated with liverwort rhizoids and ericoid mycorrhizal roots. *New Phytol.* 118, 259-272.

Fu, Y.H. and Marzluf, G.A. (1990) *nit-2*, the major nitrogen regulatory gene of *Neurospora crassa*, encodes a protein with a putative zinc-finger DNA-binding domain. *Mol. Cell Biol.* 10, 1056-1065.

Gamborg, O.L., Miller, R.A., and Ojima, K. (1968) Nutrient requirements of suspension cultures of soybean root cells. *Exp. Cell Res.* 50, 151-158.

Gurr, S., Unkles, S., and Kinghorn, J. (1987) The structure and organisation of nuclear genes of filamentous fungi, in J.R. Kinghorn (ed), *Gene structure in eukaryotic microbes*, IRL Press, pp. 117-127.

Haas, H., Bauer, B., Redl, B., Stöffler, G., and Marzluf, G.A. (1995) Molecular cloning and analysis of *nre*, the major nitrogen regulatory gene of *Penicillium chrysogenum*. *Curr. Genet.* 27, 150-158.

Hamada, W. and Spanu, P.D. (1998) Co-suppression of the hydrophobin gene *HCf-1* is correlated with antisense RNA biosynthesis in *Cladosporium fulvum*. *Mol. Gen. Genet.* 259, 630-638.

Hammond-Kosack, K.E. and Jones, J.D.G. (1994) Incomplete dominance of tomato *Cf* genes for resistance to *Cladosporium fulvum*. *Mol. Plant-Microbe Interact.* 7, 58-70.

Hammond-Kosack, K.E., Jones, D.A., and Jones, J.D.G. (1994) Identification of two genes required in tomato for full *Cf-9*-dependent resistance to *Cladosporium fulvum*. *Plant Cell* 6, 361-374.

Hammond-Kosack, K.E. and Jones, J.D.G. (1996) Resistance gene-dependent plant defense responses. *Plant Cell* 8, 1773-1791.

Harling, R., Kenyon, L., Lewis, B., Oliver, R., Turner, J., and Coddington, A. (1988) Conditions for efficient isolation and regeneration of protoplasts from *Fulvia fulva*. *J. Phytopathol.* 122, 143-146.

Henricot, B. (1998) A study of the starvation-induced genes of *Cladosporium fulvum* (Cooke). Ph.D. thesis, University of East Anglia, Norwich, UK.

Horskins, A. (1994) Molecular analysis of the pathogenicity of *Cladosporium fulvum* (Cooke). Ph.D. thesis, University of East Anglia, Norwich, UK.

Husted, S. and Schjoerring, J.K. (1995) Apoplastic pH and ammonium concentration in leaves of *Brassica napus* L. *Plant Physiol.* 109, 1453-1460.

Jones, D., Dickinson, M., Balint-Kurti, P., Dixon, M., and Jones, J. (1993) Two complex resistance loci revealed in tomato by classical and RFLP mapping of the *Cf-2, Cf-4, Cf-5*, and *Cf-9* genes for resistance to *Cladosporium fulvum*. *Mol. Plant-Microbe Interact.* 6, 348-357.

Jones, D., Thomas, C., Hammond-Kosack, K., Balintkurti, P., and Jones, J. (1994) Isolation of the tomato *Cf-9* gene for resistance to *Cladosporium fulvum* by transposon tagging. *Science* 266, 789-793.

Joosten, M. and de Wit, P. (1989) Identification of several Pathogenesis-Related proteins in tomato leaves inoculated with *Cladosporium fulvum* syn *Fulvia fulva* as 13-β-glucanases and chitinases. *Plant Physiol.* 89, 945-951.

Joosten, M., Hendrickx, L.M., and de Wit, P. (1990) Carbohydrate composition of apoplastic fluids isolated from tomato leaves inoculated with virulent or avirulent races of *Cladosporium fulvum* syn *Fulvia fulva*. *Neth. J. Plant Pathol.* 96, 103-112.

Joosten, M., Cozijnsen, T., and de Wit, P. (1994) Host-resistance to a fungal tomato pathogen lost by a single base-pair change in an avirulence gene. *Nature* 367, 384-386.

Joosten, M.H.A.J., Verbakel, H.M., Nettekoven, M.E., Van Leeuwen, J., Van der Vossen, R.T.M., and de Wit, P.J.G.M. (1995) The phytopathogenic fungus *Cladosporium fulvum* is not sensitive to the chitinase and β-1,3-glucanase defence proteins of its host, tomato. *Physiol. Mol. Plant Pathol.* 46, 45-59.

Joosten, M.H.A.J., Vogelsang, R., Cozijnsen, T.J., Verberne, M.C., and de Wit, P.J.G.M. (1997) The biotrophic fungus *Cladosporium fulvum* circumvents Cf-4-mediated resistance by producing unstable AVR4 elicitors. *Plant Cell* 9, 367-379.

KarpovichTate, N., Spanu, P., and Dewey, F.M. (1998) Use of monoclonal antibodies to determine biomass of *Cladosporium fulvum* in infected tomato leaves. *Mol. Plant-Microbe Interact.* 11, 710-716.

Kenyon, L., Lewis, B., Coddington, A., Harling, R., and Turner, J. (1993) Pathogenicity mutants of the tomato leaf mould fungus *Fulvia fulva* (cooke) Ciferri (syn *Cladosporium fulvum* cooke). *Physiol. Mol. Plant Pathol.* 43, 173-191.

Keogh, R.C., Deverall, B.J., and Mcleod, S. (1980) Comparison of histological and physiological responses to *Phakopsora pachyrhizi* in resistant and susceptible soybean. *Trans. Br. Mycol. Soc.* 74, 329-333.

Kooman-Gersmann, M., Vogelsang, R., Hoogendijk, E.C.M., and de Wit, P.J.G.M. (1997) Assignment of amino acid residues of the AVR9 peptide of *Cladosporium fulvum* that determine elicitor activity. *Mol. Plant-Microbe Interac.* **10**, 821-829.

Kooman-Gersmann, M., Vogelsang, R., Vossen, P., Van den Hooven, H.W., Mahe, E., Honee, G., and de Wit, P.J.G.M. (1998) Correlation between binding affinity and necrosis-inducing activity of mutant AVR9 peptide elicitors. *Plant Physiol.* **117**, 609-618.

Kozak, M. (1986) Point mutations define a sequence flanking the AUG initiator codon that modulates translation by eukaryotic ribosomes. *Cell* **44**, 283-292.

Kudla, B., Caddick, M.X., Langdon, T., Martinez-Rossi, N.M., Bennet, C.F., Sibley, S., Davies, R.W., and Arst, H.N. Jr. (1990) The regulatory gene *areA* mediating nitrogen repression in *Aspergillus nidulans*. Mutations affecting specificity of gene activation after a loop residue of a putative zinc finger. *Embo J.* **9**, 1355-1364.

Kulmburg, P., Mathieu, M., Dowzer, C., Kelly, J., and Felenbok, B. (1993) Specific binding-sites in the *alcR* and *alcA* promoters of the ethanol regulon for the *creA* repressor mediating carbon catabolite repression in *Aspergillus nidulans*. *Mol. Microbiol.* **7**, 847-857.

Latchman, D.S. (1990) *Gene Regulation: A Eukaryotic Perspective*, Unwin Hyman, London.

Lauge, R., Joosten, M.H.A.J., Van den Ackerveken, G.F.J.M., Van den Broek, H.W.J., and de Wit, P.J.G.M. (1997) The *in planta*-produced extracellular proteins ECP1 and ECP2 of *Cladosporium fulvum* are virulence factors. *Mol. Plant-Microbe Interact.* **10**, 725-734.

Lauge, R. and de Wit P.J. (1998) Fungal avirulence genes: structure and possible functions. *Fungal Genet. Biol.* **24**, 285-297.

Lauge, R., Joosten, M.H.A.J., Haanstra, J.P.W., Goodwin, P.H., Lindhout, P., and de Wit, P.J.G.M. (1998) Successful search for a resistance gene in tomato targeted against a virulence factor of a fungal pathogen. *Proc. Natl. Acad. Sci. USA* **95**, 9014-9018.

Lazarovitz, G. and Higgins, V. (1976) Histological comparison of *Cladosporium fulvum* race 1 on immune resistant and susceptible varieties. *Can. J. Bot.* **54**, 224-234.

Lazarovitz, G. and Higgins, V. (1976) Ultrastructure of susceptible resistant and immune reactions of tomato to races of *Cladosporium fulvum*. *Can. J. Bot.* **54**, 235-249.

Lewis, D.H. (1973) Concepts in fungal nutrition and the origin of biotrophy. *Biol. Rev.* **48**, 261-278.

Lind, J. (1909) En tomatsort der ikke angribes af sygdom tomatbladenes fløjlsplet. *Gartner Tidende* xxv:201.

Lucas, J.A. (1998) *Plant Pathology and Plant Pathogens*, Blackwell, Oxford.

Majer, D., Lewis, B.G., Devos, P., Mithen, R.F., and Oliver, R.P. (1996) Use of AFLPs to detect and measure variation in fungal populations. *Mycol. Res.* **100**, 1107-1111.

Makemson (1918) The leaf mould of tomatoes caused by *Cladosporium fulvum*. *Michigan Acad. Sci. Ann. Rept.* **20**, 309-348.

Marmeisse, R., Van den Ackerveken, G., Goosen, T., de Wit, P., and Van den Broek, H. (1993) Disruption of the avirulence gene AVR9 in 2 races of the tomato pathogen *Cladosporium fulvum* cause virulence on tomato genotypes with the complementary resistance gene CF9. *Mol. Plant-Microbe Interact.* **6**, 412-417.

Marmeisse, R., Van den Ackerveken, G.F.J.M., Goosen, T., de Wit, P.J.G.M., and Van den Broek, H.W.J. (1994) The *in-planta* induced *ecp2* gene of the tomato pathogen *Cladosporium fulvum* is not essential for pathogenicity. *Curr. Genet.* **26**, 245-250.

Mattern, I., Punt, P., and Van den Hondel, C. (1988) A vector for Aspergillus transformation conferring phleomycin resistance. *Fungal Genetics Newsletter* **35**, 25.

May, M.J., Hammond-Kosack, K.E., and Jones, J.D.G. (1996) Involvement of reactive oxygen species, glutathione metabolism, and lipid peroxidation in the *Cf*-gene-dependent defense response of tomato cotyledons induced by race-specific elicitors of *Cladosporium fulvum*. *Plant Physiol.* **110**, 1367-1379.

McHale, M.T., Roberts, I.N., Noble, S.M., Beaumont, C., Whitehead, M.P., Seth, D., and Oliver, R.P. (1992) CfT-I: an LTR-retrotransposon in *Cladosporium fulvum*, a fungal pathogen of tomato. *Mol. Gen. Genet.* **233**, 337-347.

Melton, R.E., Flegg, L.M., Brown, J.K.M., Oliver, R.P., Daniels, M.J., and Osbourn, A.E. (1998) Heterologous expression of *Septoria lycopersici* tomatinase in *Cladosporium fulvum*: Effects on compatible and incompatible interactions with tomato seedlings. *Mol. Plant-Microbe Interact.* **11**, 228-236.

Morton, N., Dickerson, A., and Hammond, J. (1985) Mannitol metabolism in *Agaricus bisporus*: Purification and properties of mannitol dehydrogenase. *J. Gen. Microbiol.* **131**, 2885-2890.

Noeldner, P.KM., Coleman, M.J., Faulks, R., and Oliver, R.P. (1994) Purification and characterization of mannitol dehydrogenase from the fungal tomato pathogen *Cladosporium fulvum* (syn. *Fulvia fulva*). *Physiol. Mol. Plant Pathol.* **45**, 281-289.

Norton, J. (1914) Resistance to *Cladosporium fulvum* in tomato varieties. *Phytopath.* **4**, 398.

Oliver, R., Roberts, I., Harling, R., Kenyon, L., Punt, P., Dingemanse, M., and Van den Hondel, C. (1987) Transformation of *Fulvia fulva*, a fungal pathogen of tomato to hygromycin B resistance. *Curr. Genet.* **12**, 231-233.

Oliver, R. (1992) A model system for the study of plant-fungal interactions: Tomato leaf mold caused by *Cladosporium fulvum*, in D.P.Verma (ed) *Molecular Signals in Plant-Microbe Communications*, Boca Raton, CRC Press, pp. 97-108.

Oliver, R., Farman, M., Jones, J., and Hammond-Kosack, K. (1993) Use of fungal transformants expressing beta-glucuronidase activity to detect infection and measure hyphal biomass in infected-plant tissues. *Mol. Plant-Microbe Interact.* **6**, 521-525.

Oliver, R. and Osbourn, A. (1995) Molecular dissection of fungal phytopathogenicity. *Microbiol.* **141**, 1-9.

Parniske, M., Hammond-Kosack, K.E., Golstein, C., Thomas, C.M., Jones, D.A., Harrison, K., Wulff, B.B.H., and Jones, J.D.G. (1997) Novel disease resistance specificities result from sequence exchange between tandemly repeated genes at the *Cf-4/9* locus of tomato. *Cell* **91**, 821-832.

Raeder, U. and Broda, P. (1985) Rapid preparation of DNA from filamentous fungi. *Lett. Appl. Microbiol.* **1**, 17-20.

Roberts, I., Oliver, R., Punt, P., and Van den Hondel, C. (1989) Expression of the *Escherichia coli* β-glucuronidase gene in industrial and phytopathogenic filamentous fungi. *Curr. Genet.* **15**, 177-180.

Screen, S., Bailey, B., Charnley, K., Cooper, R., and Clarkson, J. (1997) Carbon regulation of the cuticle-degrading enzyme PR1 from *Metarhizium anisopliae* may involve a trans-acting DNA-binding protein CRR1, a functional equivalent of the *Aspergillus nidulans* CREA protein. *Curr. Genet.* **31**, 511-518.

Segers, G.C. (1998) Isolation and characterisation of fungal genes, expressed during the infection of tomato by *Cladosporium fulvum*. Ph.D. thesis, University of East Anglia, Norwich, UK.

Segers, G.C., Hamada, W., Oliver, R.P., and Spanu, P.D. (1999) Isolation and characterisation of five different hydrophobins of the fungal tomato pathogen *Cladosporium fulvum*. *Mol. Gen. Genet.* (in press).

Spanu, P. (1998) Deletion of *HCf-1*, a hydrophobin gene of *Cladosporium fulvum*, does not affect pathogenicity in tomato. *Physiol. Mol. Plant Pathol.* **52**, 323-334.

Talbot, N., Coddington, A., Roberts, I., and Oliver, R. (1988) Diploid construction by protoplast fusion in *Fulvia fulva* syn. *Cladosporium fulvum*; genetic analysis of an imperfect fungal plant pathogen. *Curr. Genet.* **14**, 567-572.

Talbot, N., Coddington, A., and Oliver, R. (1991) Pulsed field gel electrophoresis of reveals chromosome length polymorphisms between strains of *Cladosporium fulvum* (syn. *Fulvia fulva*). *Mol. Gen. Genet.* **229**, 267-272.

Taylor, J.W. (1995) Making the Deuteromycota redundant: a practical integration of mitosporic and meiosporic fungi. *Can. J. Bot.* **73**, (Suppl.) S754-S759.

Thomas, C.M., Jones, D.A., Parniske, M., Harrison, K., Balint-Kurti, P.J., Hatzixanthis, K., and Jones, J.D.G. (1997) Characterization of the tomato *Cf-4* gene for resistance to *Cladosporium fulvum* identifies sequences that determine recognitional specificity in *Cf-4* and *Cf-9*. *Plant Cell* **9**, 2209-2224.

Ueng, P.P., Subramaniam, K., Chen, W., Arseniuk, E., Wang, L.X., Cheung, A.M., Hoffmann, G.M., and Bergstrom, G.C. (1998) Intraspecific genetic variation of Stagonospora avenae and its differentiation from S-nodorum. *Mycol. Res.* **102**, 607-614.

Van den Ackerveken, G., Dunn, R., Cozijnsen, A., Vossen, J., Van den Broek, H., and de Wit, P. (1994) Nitrogen limitation induces expression of the avirulence gene *Avr9* in the tomato pathogen *Cladosporium fulvum*. *Mol. Gen. Genet.* **243**, 277-285.

Van Kan, J., Van den Ackerveken, G., and de Wit, P. (1991) Cloning and characterisation of a cDNA of avirulence gene *Avr9* of the fungal pathogen *C. fulvum* causal agent of tomato leaf mold. *Mol. Plant-Microbe Interact.* **4**, 52-60.

Vervoort, J., Van den Hooven, H.W., Berg, A., Vossen, P., Vogelsang, R., Joosten, M.H., and de Wit, P.J. (1997) The race-specific elicitor AVR9 of the tomato pathogen *Cladosporium fulvum*: a cystine knot protein. Sequence-specific 1H NMR assignments, secondary structure and global fold of the protein. *FEBS Lett.* **404**, 153-158.

von Arx, J. (1983) Mycosphaerella and its anamorphs. *Proceedings Koninklijke Nederlandse Akademie van Wetenschappen.* **86**, 47-48.

EVOLUTION OF HOST SPECIFIC VIRULENCE IN *COCHLIOBOLUS HETEROSTROPHUS*

B. GILLIAN TURGEON
SHUN-WEN LU
Department of Plant Pathology, Cornell University
Ithaca, New York, 14853, USA

1. Introduction

Cochliobolus heterostrophus is the cause of Southern Corn Leaf Blight. Two races of the fungus are known; both are pathogenic. Race O was formally described as a corn pathogen in 1925 (Drechsler 1925). Race T, which produces a host-specific virulence factor called T-toxin, appeared suddenly in the US in 1969 and caused an epidemic on Texas male sterile cytoplasm (T-cms) corn in 1970 (Hooker 1974). Concurrently, a new disease, Yellow Leaf Blight (Arny and Nelson 1971), caused by *Mycosphaerella zeae-maydis* (Mukunya and Boothroyd 1973) was recognized. *M. zeae-maydis* is also specific to T-corn and produces PM-toxin, a polyketide with structural and functional similarity to T-toxin (Danko et al. 1984; Yoder 1973). The sudden appearance of these virulent strains suggests an identifiable genetic step in the evolution of a new race in each of these fungal species. Indeed, thirty years later, we know that the essential difference between *C. heterostrophus* races T and O is the *Tox1* locus which resides on a large segment of DNA that is not found in race O. Since there is evidence that race T arose from race O (Alcorn 1975; Kodama et al. 1999; Leonard 1977; Yang et al. 1996), a molecular understanding of the *Tox1* locus in race T and O should advance understanding of how new pathogenic races of fungi evolve.

Although the ability to produce T-toxin segregates as a single genetic locus in crosses between naturally occurring races T and O, *Tox1* is actually two loci (*Tox1A* and *Tox1B*), located at or near translocation breakpoints on two different chromosomes. This was shown physically by examination of electrophoretic karyotypes and genetically when toxin minus mutants of *Tox1*⁺ strains were created and crossed to each other. In certain crosses, twenty five percent of the progeny were Tox⁺, indicating independent segregation of two loci (designated *Tox1A* and *Tox1B*). *Tox1A* and *Tox1B* encode a polyketide synthase (PKS1) and a decarboxylase (DEC1), respectively; both are required for T-toxin biosynthesis. DNA gel blots show these genes are lacking in race O and in all other *Cochliobolus* spp. examined to date. This suggests that race T may have evolved from race O by acquiring DNA horizontally from an unknown organism rather than by inheriting it from an ancestor (Rose 1996; Yang et al. 1996).

Additional insight into the genetic events that gave rise to these virulent races may be gained by comparison of genes responsible for production of T-toxin and PM-toxin since both were first identified during the Southern Corn Leaf Blight epidemic and since the producing fungi are taxonomically distinct, even though they share pathological characteristics. Sequencing of the *MzTox1* locus that controls production of PM-toxin

J. W. Kronstad (ed.), Fungal Pathology, 93–126.

by *M. zeae-maydis* has revealed a gene cluster including *MzPKS1*, encoding a polyketide synthase, and two ketoreductase-encoding genes, *RED1* and *RED2*. All are necessary for PM-toxin biosynthesis. Since the organization of the *Tox* loci differs in *C. heterostrophus* and *M. zeae-maydis* and since the central gene in both gene clusters, the polyketide synthase, is only 60% identical at the amino acid level, we know that these two fungi did not acquire the genes for toxin production from the same organism at the time of the 1970 epidemic, nor were genes transferred from one to the other at that time. ·What then, is the source of these genes and when were they acquired?

Although T-toxin and PM-toxin determine specificity of race T and *M. zeae-maydis* for T-corn, it is clear that other factors are necessary for successful pathogenesis, e.g., race O is pathogenic but Tox⁻. Analysis of one mutant exhibiting reduced virulence/pathogenicity has led to the cloning of a gene (*CPS1*) encoding a putative peptide synthetase, that is part of a large gene cluster. Targeted disruption of *CPS1* and other genes in the cluster causes loss of virulence in both race T and race O. This gene is present not only in *C. heterostrophus* races O and T but also in many other fungi. To determine if *CPS1* is also needed for general pathogenicity in another fungus with different host specificity, its homologue was disrupted in *C. victoriae,* a pathogen of oats. *C. victoriae* produces the cyclic peptide victorin, a host-specific toxin that affects only oats with the *Vb* allele; victorin is required for pathogenicity (Yoder et al. 1997). *CPS1⁻* mutants of *C. victoriae* are unaltered in victorin production, but suffer drastic loss of virulence, indicating that victorin alone is insufficient for normal pathogenesis and suggesting that the hypothetical CPS1 peptide is a general factor in fungal pathogenesis, an idea supported by the observation that *CPS1* has homologues in a wide array of fungal species.

Discovery of a possible general virulence factor prompts us to consider the larger set of fungal molecules required for pathogenesis, how they work together in disease development, and the genetic events leading to the evolution of new pathogenic races with either wide or narrow host ranges. The path to understanding evolution of pathogenicity and the role of secondary metabolism in it, is the subject of our review.

This review is divided into three major sections. The first (the past) describes early discoveries and the biology of *Cochliobolus*. The second (the middle years) documents genetic analyses that suggested a single locus controlled production of the virulence factor, T-toxin, and is followed by a summary of the molecular and physical analyses which showed that the genetic evidence was confounded by a chromosomal rearrangement; ability to produce toxin is actually controlled by two loci (two gene clusters) on two different chromosomes. This section also includes a discussion of what molecules, in addition to toxins, are required for pathogenicity. The third section (the future) addresses how questions of this sort might be tackled in the genomics era.

2. The Past: *Cochliobolus* and 'Host Specific' Toxins

2.1. *COCHLIOBOLUS HETEROSTROPHUS* AND T-TOXIN

2.1.1. *The Fungus*
Early in 1925, Drechsler reported a leafspot of maize in Florida and the Philippine Islands that was distinct from leaf blights caused by several known species of fungi (Drechsler 1925). He first identified this pathogen as the fungus *Ophiobolus*

heterostrophus based on its teleomorph (Drechsler 1925; 1927). About ten years later, he chose this fungus as the type species of a newly-erected genus *Cochliobolus* in the Family Pleosporaceae (Class Ascomycetes, Order Pleosporales) which included those *Helminthosporium* (a Deuteromycete classification) species that have a sexual stage with ascospores arranged in a helicoid pattern in the ascus (Drechsler 1934). However, the proper teleomorph designation, *Cochliobolus heterostrophus,* was not widely used by mycologists and plant pathologists until nomenclature conventions for plant pathogenic fungi were proposed (Yoder et al. 1986). In most of the older literature, including some publications from the late 1980s, the fungus was more frequently referred to by one of its anamorphic designations, *Helminthosporium maydis* Nisikado and Miyake, *Drechslera maydis* (Nisikado and Miyake) Subramanian and Jain or *Bipolaris maydis* (Nisikado and Miyake) Shoemaker.

C. heterostrophus is the most widely distributed species in the genus *Cochliobolus.* As a natural pathogen of corn, it can be found in many tropical and subtropical areas of the world (Drechsler 1925; 1934; Orillo 1952; Yu 1933). In the United States, the fungus is usually found in the warmer southern states, thus, the disease is commonly known as Southern Corn Leaf Blight (Hooker 1974).

The life cycle of the fungus includes both asexual and sexual stages (Fig. 1). Asexual reproduction can occur repeatedly in a single growing season; conidia infect corn leaves predominantly by direct penetration (Wheeler 1977) and cause small, tan lesions which, in epidemic conditions, cover the entire leaf, killing the plant (Miller et al. 1970; Smith et al. 1970). Although appressoria are usually associated with penetration sites, they may not be necessary for infection (Horwitz et al. 1999). The fungus overwinters as conidia or mycelia on debris of dead corn plants. The sexual stage has rarely, if ever, been observed in the field, but can be induced easily in the laboratory. Sexual structures are properly termed pseudothecia (Sivanesan 1984). The fungus is heterothallic: only strains of opposite mating type can cross to each other and undergo sexual reproduction (Nelson 1957; 1959; Sivanesan 1984; Turgeon et al. 1993).

2.1.2. *The Epidemic of Southern Corn Leaf Blight*

Prior to 1970, *C. heterostrophus* was known as an endemic pathogen that was of minor economic importance in the United States. However, in 1970, a severe epidemic swept all corn growing areas of the world. In a short period (May to September) of that year, the disease spread from southern states, such as Florida, north to Maine and finally into southern Canada (Moore 1970, Fig. 2). More than thirty states reported serious damage to corn (ranging from 50% reduced yields to complete loss) valued at more than one billion dollars. The disease was prominent in 1971 but the overall severity was much reduced because of cooler weather, unfavorable for the fungus (Ullstrup 1972). The two years of epidemic had an unprecedented impact on the US corn production industry and made Southern Corn Leaf Blight one of the most widely known crop diseases in the US. It also sparked an extensive investigation of the disease that resulted in the discovery of a new race of the fungus.

2.1.3. *The Discovery of Race T*

Although increased susceptibility of corn with Texas male sterile cytoplasm (T-cytoplasm) to certain isolates of *C. heterostrophus* was reported in the Philippines as early as 1961 (Mercado and Lantican 1961) and also in the United States (southern Iowa and Illinois) in 1968 (Scheifele et al. 1970; Ullstrup 1970; 1972), the presence of

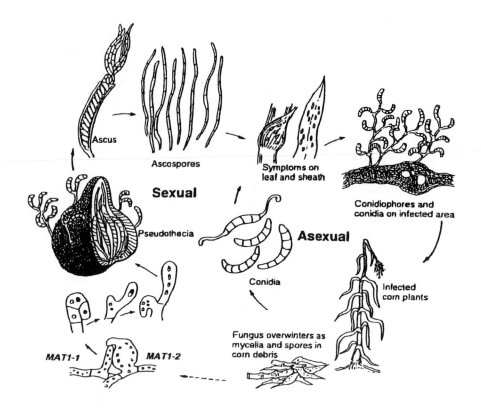

Figure 1. Life and disease cycles of *Cochliobolus heterostrophus.* Mycelia and both asexual and sexual spores are multicellular; each cell is multinucleate. The sexual cycle, which involves mating of two individuals of opposite mating type, is completed in lab conditions in about 18 days. The disease cycle can occur in as little as a week (Yoder, 1988). All stages are haploid except for karyogamy.

a new race of the fungus was not seriously considered until the epidemic in 1970. Shortly after summer of that year, Smith *et al* (Smith et al. 1970) and (Hooker et al. 1970; Lim and Hooker 1971; 1972) published their studies on a *C. heterstrophus* isolate obtained from severely diseased corn leaves collected in central Illinois in 1969. They designated this isolate "Race T" for its high virulence on T-cytoplasm corn and for its ability to produce a substance called T-toxin that specifically affects T-corn. In a previous report, toxic preparations from *C. heterostrophus* were not evaluated by comparing activity against both N- (normal) and T-cytoplasm corn; thus the activity described in that report was not T-toxin, which by definition is T-cytoplasm-specific (Smedegard-Peterson and Nelson 1969). The "old race" of the fungus that had been known since 1925 was called "race O". Race O does not produce T-toxin and is mildly virulent on both T- and N-cytoplasm (Hooker et al. 1970; Smith et al. 1970). The concept of two races of *C. heterostrophus* was further formalized by subsequent genetic and biochemical analyses (Lim and Hooker 1971; Yoder 1976; 1980; Yoder and Gracen 1975). A detailed history and commentary on the origins of race T can be found in Wise et al. (Wise et al. 1999).

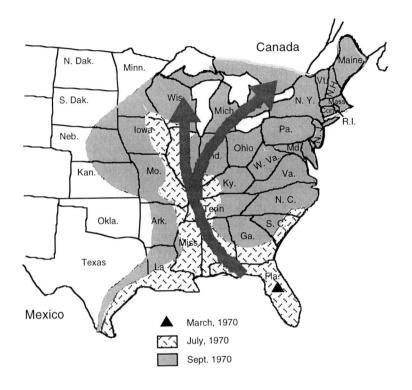

Figure 2. The 1970 outbreak of Southern Corn Leaf Blight in the United States (adapted from Moore, 1970). The epidemic was first reported in Belle Glade, Florida (black triangle) then spread northward (arrows) during the summer to more than thirty states and Canada (shaded areas).

2.1.4. The *Nature of T-Toxin*

Host specificity of T-toxin was first studied using culture filtrates produced by the fungus and inbred corn lines with T- or N-cytoplasm. Seedling assays for inhibition of root growth and leaf injection tests confirmed T-toxin caused the same symptoms on T-cytoplasm corn as did the fungus itself (Gracen et al. 1971; Lim and Hooker 1971; Yoder 1976). Ten years later, it was reported that T-toxin is actually a mixture of several linear polyketols (Fig. 3A) ranging in length from C_{35} to C_{41} with the C_{39} and C_{41} components predominating (60%-90%) in the native toxin (Kono and Daly 1979; Kono et al. 1981). Each of these components has the same specific toxicity against T-cytoplasm corn. Chemically synthesized T-toxin analogs, available in the early 1980s (Suzuki et al. 1982; 1983) displayed the same activity as fungal preparations in a bioassay of inhibition of dark CO_2 fixation by susceptible corn leaves (Yoder and Gracen 1977) thereby validating the structure.

2.1.5. *T-cytoplasm Corn and the URF-13 Protein*

Just one year after the discovery of race T and T-toxin it was reported that the T-toxin target site is T-cytoplasm (Gengenbach et al. 1972; Miller and Koeppe 1971). Subsequent studies further demonstrated that T-toxin binds to a 13 kDa inner

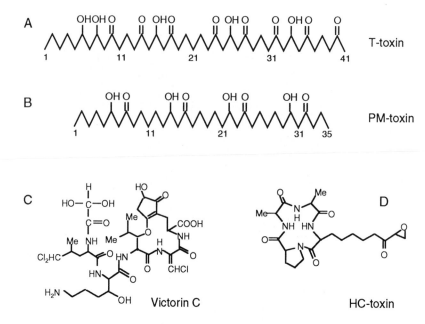

Figure 3. Structures of host specific toxins discussed in this chapter (see text for details). All molecules shown represent the most abundant member of each toxin family found in culture filtrate.

mitochondrial membrane protein, which is the product of the mitochondrial gene, *T-urf13* (Forde et al. 1978). Binding of T-toxin causes pores in the URF-13 oligomeric complex; small molecules required for normal mitochondrial function leak out (Levings 1990; Levings and Siedow 1992; Levings et al. 1995), resulting in the cessation of ATP synthesis and subsequent cell death (Fig. 4). The specific interaction between T-toxin and the URF-13 protein was also proven by heterologous expression of *T-urf13* in other organisms, including *E. coli* (Dewey et al. 1988; Huang et al. 1990), tobacco (Vonallmen et al. 1991), insects (Korth and Levings 1993) and *C. heterostrophus* itself (Yang et al. 1994). All transformants of these organisms that express *T-urf13* become sensitive to T-toxin (Fig. 4). Since T-toxin blocks energy production in host cells, the plant is unable to activate host defense mechanisms, thus allowing fungal colonization (Yoder and Turgeon 1996). The URF-13 protein confers a second phenotype-cytoplasmic male sterility (cms). The history of studies on T-cms in maize and the mechanisms of host susceptibility to T-toxin were extensively reviewed recently (Wise 1999).

In terms of the history of corn production, T-cytoplasm corn had been widely used for hybrid seed production and breeding since the 1950s to avoid hand or mechanical emasculation. By 1970, 85% of hybrid corn produced in the US was T-cytoplasm corn. This monoculture of the *C. heterostrophus* host resulted in the epidemic of Southern Corn Leaf Blight in the US, either because race T originated at that time or because it had been in the field population at low levels all along but had no advantage over race O until a susceptible host was available.

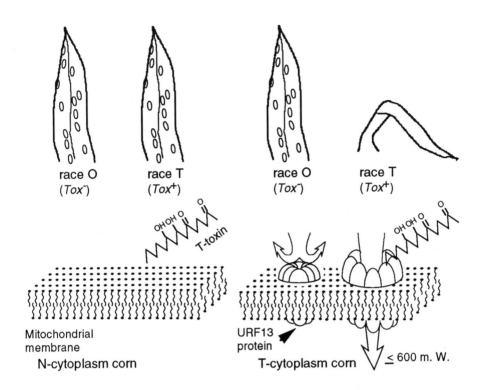

Figure 4. Diagrammatic representation of disease phenotypes resulting from either specific interaction or no interaction between T-toxin and the URF-13 protein. Left: Race O (*Tox⁻*) and race T (*Tox⁺*) are weakly virulent on N-cytoplasm corn because the target site for T-toxin is absent. Right: Race T is highly virulent (collapsed leaves) to T-cytoplasm corn because binding of T-toxin to the URF-13 oligomeric complex causes pores in the mitochondrial membrane through which small molecules, required for normal mitochondrial function, leak. This results in cell death. The structure of URF-13 protein is based on Levings and Siedow, 1992.

2.2. *COCHLIOBOLUS VICTORIAE* AND VICTORIN, *COCHLIOBOLUS CARBONUM* AND HC-TOXIN

Studies on two other *Cochliobolus* species, *C. victoriae* (Nelson) {anamorph = *Bipolaris victoriae (Shoemaker), Helminthosporium victoriae (Meehan and Murphy)*} and *C. carbonum* (Nelson) {anamorph = *Bipolaris zeicola (*Shoemaker), *Helminthosporium carbonum* (Ullstrup)} have interesting parallels with *C. heterostrophus.* This reflects not only their close taxonomic relationship but also the extraordinary abilities of taxa in this group to produce secondary metabolites involved in pathogenesis.

2.2.1. *C. victoriae*
Like the prevalence of T-cytoplasm corn in the 1960s, genetically uniform oat varieties possessing the Victoria type of resistance to crown rust disease were widely planted during the 1940s, such that, by the end of 1945, 75% of US oats were Victoria (now called *Vb* oats). *Vb* was introduced into commercial oat varieties because it is highly

associated with resistance to Crown Rust caused by *Puccinia coronata*. A severe blight became epidemic in 1946 in the US, spreading to almost twenty states, from Texas to New York and from Florida to Idaho (Murphy and Meehan 1946). The pathogen that caused the disease was identified as a new fungal species, *Helminthosporium victoriae* based on its asexual stage and the fact that it specifically attacked only Victoria oats (Meehan and Murphy 1946; Murphy and Meehan 1946).

Subsequent studies revealed the presence of a toxin (victorin) in culture filtrates that showed the same specific toxicity as the fungus when tested on oat seedlings (Litzenberger 1949). This was the first report of a host specific toxin produced by plant pathogenic fungi. Victorin was later characterized as a chlorinated cyclized peptide. The most abundant component in culture filtrates is victorin C, which consists of six residues: glyoxylic acid, 5,5-dichloroleucine, erythro-β-hydroxyleucine, 2-alanyl-3,5-dihydroxy-D^2-cyclopentenone-1 or victalanine (victala), threo-β-hydroxylysine, and α-amino-β-chloro-acrylic acid (aclaa) (Fig. 3D, (Macko et al. 1985; Wolpert et al. 1985).

Victorin is the most potent toxin known. It inhibits growth of susceptible oat roots half-maximally at 100 pM (Walton and Earle 1984). Studies on the target site of victorin have focused on the characterization of victorin-binding proteins (VBPs), components of glycine decarboxylase, a multiple enzyme complex essential for mitochondrial function. It has been proposed that victorin kills oat plants by inhibiting the activity of this enzyme complex in susceptible oats (Navarre and Wolpert 1995; Navarre and Wolpert 1999a; 1999b; Wolpert et al. 1994; 1998).

2.2.2. *C. carbonum*

In contrast to T-toxin and victorin, both of which were found concurrently with characterization of a new race or a newly recognized fungus, HC-toxin produced by *C. carbonum* was discovered more than twenty years after the fungus was identified. The original *C. carbonum* isolate, discovered in 1938, was called *Helminthosporium maydis* (Ullstrup 1944). When it was renamed *H. carbonum* in 1944, two races (race 1 and race 2) were distinguished based on differences in disease symptoms (Ullstrup 1944). Unlike T-toxin and victorin, HC-toxin toxin (named after the former anamorph) has never been linked to an epidemic of the disease it causes, Northern Leaf Spot, probably because the susceptible host (corn with the genotype *hm1/hm1*) has never been widely planted, there being no agronomic reason to do so.

The ability of *C. carbonum* race 1 to produce HC-toxin was not demonstrated until 1965 when culture filtrates of the fungus were used to test susceptibility of certain inbred corn lines (Scheffer and Ullstrup 1965). HC-toxin was characterized as a cyclic peptide, cyclo-(D-Pro-L-Ala-D-Ala-L-[2-amino-9,10-epoxi-8-oxodecanoic acid]) (Fig. 3C, (Gross et al. 1982; Pope et al. 1983; Walton et al. 1982). Three closely related HC-toxins (HC-toxin I, II and III) are produced by the fungus, but HC-toxin I is most abundant in culture filtrate.

HC-toxin and *hm1/hm1* corn represents an unique host-parasite specificity. Host susceptibility is due to the lack of production of a detoxification enzyme called HC-toxin reductase (HCTR), encoded by *Hm1*, the first plant resistance gene cloned (Johal and Briggs 1992), which appears to be present in all plants except for corn with the *hm1/hm1* genotype (Meeley et al. 1992; Meeley and Walton 1991). A mutation in *Hm1* is thought to have occurred in a maize line in Kansas under little disease pressure (Briggs and Johal 1994). When this germplasm was moved to the cornbelt in the 1930s, disease resulted. The mode of action of HC-toxin, assuming it is not first detoxified by HCTR, has been proposed to be the inhibition of histone deacetylase, which results in suppression of the host defense response (Brosch et al. 1995; Ransom and Walton 1997; Walton 1996).

Three host specific toxins had been identified from *Cochliobolus* species by the end of 1970. All are secondary metabolites and all are associated with a new race or species and all cause a unique disease. How do these closely related fungi produce structurally diverse secondary metabolites? Are these potent toxins alone enough to kill the host plants? How do these toxins determine the outcome of the plant-fungus interaction? Does the acquisition of ability to produce these secondary metabolites convert a non pathogen to a pathogen?

2.3. *MYCOSPHAERELLA ZEAE-MAYDIS* AND PM-TOXIN

M. zeae-maydis Mukunya & Boothroyd (Mukunya and Boothroyd 1973), the causal agent of the Yellow Leaf Blight, appeared suddenly in the field in the late 1960s, early 1970s, as did race T of *C. heterostrophus*. When first identified, it was frequently described by its anamorph, *Phyllosticta maydis* (Arny and Nelson 1971), later renamed *Phoma zeae-maydis* (Punithalingam 1990). It is, in fact, a homothallic ascomycete in the Order Dothideales and is not closely related to *C. heterostrophus,* in the Order Pleosporales. However, *M. zeae-maydis* has the same biological specificity as *C. heterostrophus,* showing high virulence to T-cytoplasm corn (Arny and Nelson 1971; Arny et al. 1970; Scheifele and Nelson 1969). The discovery of T-toxin prompted researchers to screen *M. zeae-maydis* for a toxin that had the same specificity as T-toxin. Two years later, this was identified in culture filtrates and named PM-toxin based on the anamorph (Comstock et al. 1972; 1973; Yoder and Mukunya 1972; Yoder 1973). Later, structural studies confirmed that PM-toxin is also a family of linear polyketides, similar, although not identical in structure, to T-toxin. The major difference is in chain length, C_{33} to C_{35} for PM-toxin *versus* C_{35} to C_{41} for T-toxin (Danko et al. 1984; Kono et al. 1983). Fig. 3B.). Unlike *C. heterostrophus* which occurs in two forms, race T which produces T-toxin and race O which does not, all known isolates of *M. zeae-maydis* produce PM-toxin. The discovery of PM-toxin from *M. zeae-maydis*, with the same structure and biological activity as T-toxin, provided an additional comparative route to tracing the evolutionary origin of these toxins. Are the genes encoding the ability to produce toxin the same in the two fungi, did one fungus acquire them from the other, or did they both get them from independent sources?

3. The Middle Ages: Genetic and Molecular Analyses

The 1970's epidemic led to the genetic domestication of *C. heterostrophus* and to its development as a model eukaryotic plant pathogen for rigorous analysis of fungal pathogenesis to plants by both conventional genetic and molecular genetic approaches (Kohmoto and Yoder 1997). For the former, a line of inbred near-isogenic lab strains called C- (Leach et al. 1982; Tegtmeier et al. 1982) and K (Klittich and Bronson 1986) strains are available. In this section, we first review the early genetic analysis of characters responsible for T-toxin production, then the molecular discoveries that refined the genetic conclusions and identified the genes involved.

3.1. GENETIC ANALYSES

3.1.1. *Tox1 is Defined as a Single Genetic Locus*

Heterothallism, haploidy and ease of mating the fungus in the laboratory allowed early investigators to quickly demonstrate that the ability to produce T-toxin is genetically inseparable from the high virulence of the fungus on T-cytoplasm corn (Fig. 5). In a cross between races O and T, all progeny producing T-toxin are highly virulent and all progeny not producing T-toxin are weakly virulent on T-corn; both races are pathogenic on N-cytoplasm corn (Lim and Hooker 1971). This 1:1 segregation of parental phenotypes was observed in most crosses conducted in studies using either field isolates or inbred laboratory strains (C-strains) (but see exceptions Section 3.1.2) (Yoder 1976; Yoder and Gracen 1975), thus defining a single genetic locus, designated *Tox1* (Leach et al. 1982; Lim and Hooker 1971; Tegtmeier et al. 1982) that controls T-toxin production and high virulence on T-cytoplasm corn of *C. heterostrophus* race T.

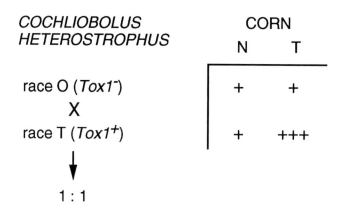

Figure 5. Tox1 appears to be a single genetic locus. When the two naturally occuring races of *C. heterostrophus* are crossed to each other, progeny segregate 1:1 for parental types only (see text for details), suggesting a single Mendelian element that determines the essential difference between the two races. N = N-cytoplasm; T = T-cytoplasm; "+" = mildly virulent; "+++" = highly virulent (as diagrammed in *Fig. 4*)

3.1.2. *Odd ratios*

Exceptions to 1:1 segregation of parental phenotypes (Tox$^+$:Tox$^-$) in crosses between race O and race T strains were described (Yoder and Gracen 1975). Certain crosses between field isolates yielded an excess of Tox- progeny, which might be explained if more than one locus were responsible for toxin production or, alternatively, if non-random ascospore abortion were occurring. The former explanation was ruled out by crossing field isolates to genetically defined laboratory strains and laboratory strains to each other (Bronson et al. 1990). Non random ascospore abortion was demonstrated to be due to the presence of a spore killer gene in about 50% of race O field isolates (Taga et al. 1985). Progeny of crosses between a race O strain carrying this gene and a race T strain which does not, are non viable if they lack the spore killer gene. Progeny of crosses between defined race O and T laboratory strains, lacking the spore killer factor, always segregate 1:1 for ability to produce toxin and for high virulence.

3.1.3. One Difference Between Race O and T is a Reciprocal Translocation
Although early studies suggested a simple inheritance pattern, *Tox1* was subsequently found to be genetically inseparable from a reciprocal translocation breakpoint (Fig. 7A). This possibility was first proposed based on the comparison of nonviable ascospore frequencies in crosses homozygous *vs.* heterozygous at *Tox1* (Bronson 1988). The pattern of ascospore abortion suggested that race T and race O differ by a reciprocal translocation and that *Tox1* is at or near the breakpoint. Note that there were, in fact, two genetic complexities confounding *Tox1* analysis- the spore killer gene, which, once recognized, could be eliminated by choosing strains which lacked it, and the reciprocal translocation which could not be eliminated until it was possible to make Tox⁻ mutants of a Tox⁺ strain (Section 3.2.1.3, 3.2.1.4).

3.2. MOLECULAR ANALYSES

3.2.1. Tools
All molecular techniques commonly used for manipulation of DNA (*in vivo* and *in vitro*) can be used routinely for *C. heterostrophus*. The following is a brief summary of molecular technological milestones in the development of this particular fungal pathosystem.

3.2.1.1. Transformation. *C. heterostrophus* was the first phytopathogenic fungus to be transformed. The technique used, $CaCl_2$-PEG-mediated protoplast transformation, (Turgeon et al. 1985; Yoder 1988), remains the procedure of choice for this fungus. Stable integration of transforming DNA into the chromosome by homologous recombination occurs at high frequency (Keller et al. 1990; Turgeon et al. 1987; Wirsel et al. 1996). At least five selectable markers can be used for construction of transforming vectors, including the *A. nidulans amdS* gene, for growth on acetamide as the sole carbon source (Turgeon et al. 1985), the *E. coli hygB* gene, for resistance to hygromycin B (Turgeon et al. 1987), the *C. heterostrophus TRP1* gene (Mullin et al. 1993; Turgeon et al. 1986), the *Streptomyces hygroscopicus bar* for resistance to bialaphos (Straubinger et al. 1992), and the *A. terreus BSD* gene for resistance to blasticidin S (Kimura et al. 1994). This versatile system makes it possible to clone genes and manipulate fungal genomic DNA by targeted gene disruption or gene replacement.

3.2.1.2. Microbial bioassay for T-toxin Production. Early evaluation of T-toxin action relied on plant assays; T-toxin activity was indicated by a typical race T symptom on susceptible corn plants, or by inhibition of root growth, dark CO_2 fixation in leaf discs, or death of protopasts from the susceptible host (Bhullar et al. 1975; Lim and Hooker 1971; Yoder et al. 1976; Yoder et al. 1977). Although sensitive enough, these assays are laborious. In addition, certain chemical methods, such as a colorimetric assay were developed (Karr et al. 1975) but found unreliable (Yoder and Gracen 1977).

The introduction of the *T-urf13* gene from mitochondrial DNA of T-cytoplasm corn (Dewey et al. 1988) into *E. coli* cells provided a highly specific and efficient way to detect T-toxin (Fig. 6) (Ciuffetti et al. 1992). *E. coli* cells expressing the *T-urf13* gene (conferring sensitivity to T-toxin) are evenly spread on LB medium containing ampicillin and plates air dried. Agar plugs bearing fungal mycelia are inoculated onto the *E. coli* cell lawn and the plates incubated. T-toxin-producing strains of the fungus will inhibit growth of the *E. coli* cells and produce halos. *Tox⁻* mutants can be distinguished from wild type by failure to produce a halo and thus appear like wild type race O, or by production of halos smaller (leaky) or larger (overproducing) than wild type (Fig. 6). Without this microbial assay, the characterization of *Tox⁻* mutants described below would have been very difficult.

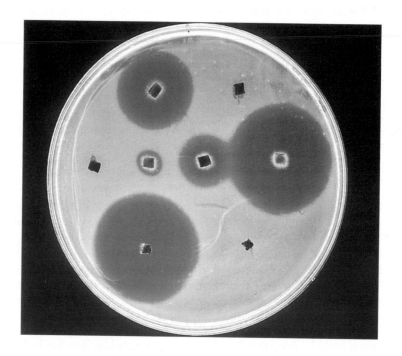

Figure 6. T-toxin and PM-toxin production assay plate. The plate was first overlayed with T-toxin-sensitive *E. coli* cells, then inoculated with agar plugs bearing mycelia and spores of: Top: *C. heterostrophus* race T (left), race O (right); Middle: four REMI-induced, race T derived "*Tox⁻*" mutants (left to right): R.C4.2326 (tight, no halo), R.C4.1771 and R.C4.1434 (leaky, halos smaller than wild type), R.C4.1957 (toxin-overproducing mutant, halo larger than wild type) (Lu, 1998); Bottom: wild type *M. zeae-maydis* (left) and PT2, a REMI-induced PM-toxin deficient mutant (right) (Yun et al., 1997). Other *C. heterostrophus Tox⁻* mutants discussed in this chapter i.e., R.C4.186 and R.C4.350L (Lu et al., 1994), C4.PKS.13 (Rose, 1996) and ctm45 (Yang et al., 1996) have tight Tox- phenotypes like that of R.C4.2326.

3.2.1.3. *Enrichment for Tox⁻ Mutants.* Chemical mutagenesis of *Cochliobolus* was used in the early 1980s to develop a large collection of auxotrophic and morphological mutants for genetic analyses. For this, conidia were treated with chemical mutagens such as ethyl methanesulphonate (EMS) and *N*-methyl-*N*'-nitro-*N*-nitrosoguanidine (NTG) (Leach et al. 1982). In the mid 1990s, this procedure was combined with transformation technology in an attempt to enrich for *Tox⁻* mutants. In this case, conidia of a *Tox⁺* strain that had been transformed with a version of the *T-urf13* gene (conditionally expressed and shown to confer T-toxin sensitivity to the fungus itself) were first mutagenized with EMS, then grown on a medium that induced expression of the lethal *T-urf13* gene. The assumption was that survivors, if any, should include those that sustained mutations in the gene(s) for T-toxin production. Several leaky and one tight mutant (ctm45) were recovered; the latter completely failed to produce T-toxin as determined by the microbial assay (Yang et al. 1996). This was the first, laboratory-produced, *Tox⁻* mutant of a *Tox⁺* strain available for mutational analysis of the *Tox1* locus. Although valuable, chemically induced mutants are not ideal in terms of molecular cloning because the mutation sites are not tagged.

3.2.1.4. *Insertional Mutagenesis by REMI*. *C. heterostrophus* was also the first plant pathogenic fungus for which the Restriction Enzyme-Mediated Integration (REMI) procedure to generate insertional mutations was reported (Lu et al. 1994). REMI was originally used to study transformation efficiency in *Saccharomyces cerevisiae* (Schiestl and Petes 1991) and then pioneered for gene tagging in *Dictyostelium discoideum* (Kuspa and Loomis 1992). Like transposon tagging in prokaryotes, REMI has had a profound impact on molecular investigation of pathogenesis in many fungal species (Akamatsu et al. 1997; Boelker et al. 1995; Lu et al. 1994; Sweigard 1996; Yun et al. 1998).

A REMI insertional mutant library has been generated (Lu 1998). Genetic analysis of 40 mutants with a clear phenotype indicates that on average, most (>90%) carry a single gene mutation and 60% of these are tagged. These tagged mutations represent different genetic loci controlling diverse phenotypes, including ability to produce T-toxin, pathogenicity, mating ability, auxotrophy, pigmentation, ability to conidiate and growth habit. This suggests that most, if not all, genes of *C. heterostrophus* can be mutated and simultaneously tagged by this procedure.

Tagging efficiency in *Cochliobolus* varies from as low as 25% to as high as 83%, depending on the restriction enzyme used and the phenotype screened for. For example, 70% of mutations generated by *Kpn*I were tagged, in contrast to 25% with *Sac*I. Mutations in the T-toxin pathway were tagged more frequently (83%) than mutations with other phenotypes, perhaps because of the size of the target (Lu 1998). Molecular characterization suggests that the vector (single copy) usually integrates into the same site as that recognized by the restriction enzyme used for REMI; mutations are stable and chromosome rearrangements occur at low frequency (1 out of 6). Because the genomic DNA flanking the insertion sites can be recovered efficiently by plasmid rescue (Yang et al. 1996), this procedure has greatly facilitated cloning genes involved in pathogenesis.

3.2.1.5. *Electrophoretic Karyotype Analysis*
Plant pathogenic fungi have small genomes compared to their hosts, allowing separation of whole chromosomes by gel electrophoresis. The *Cochliobolus* genome consists of 15 easily resolved small chromosomes, ranging in size from about 1.3 to 3.7 Mb (Chang and Bronson 1996; Kodama et al. 1999; Tasma and Bronson 1998; Tzeng et al. 1992). Electrophoretic separation of these chromosomes, combined with subsequent blotting and probing of gels with markers or genes of interest was a key experimental approach in unraveling and documenting the fact that *Tox1* is on two chromosomes in race T, translocated with respect to a pair in race O (Chang and Bronson 1996; Kodama et al. 1999; Tzeng et al. 1992).

3.2.1.6. *Restriction Fragment Length Polymorphism (RFLP) Mapping.* Although over 50 loci had been identified in early studies, *C. heterostrophus* genetic linkage groups were not established until 1992 when a combined genetic and RFLP map was developed (Tzeng et al. 1992). This map was based on the segregation pattern of known phenotypic or RFLP markers among progeny of a cross between a *Tox⁻ MAT-1* field isolate, Hm540, and a *Tox, MAT-2* lab strain, B30.A3.R.45, combined with physical placement of these markers by probing electrophoretically separated chromosomes of both parents. Fifteen chromosomes (numbered sequentially, largest to smallest based on their sizes in Hm540) and a dispensable chromosome were identified and the chromosomal location of 125 markers were determined; the total map length was estimated to be 1501 cM and the total genome size ~35 Mb (kb/cM=~23) (Tzeng et al. 1992). The map has been updated to include additional markers and telomeres and the map length recalculated as 1598 cM (Tasma and Bronson 1998). This map confirmed the earlier genetic evidence (Section 3.1.3) that the *Tox1* locus is tightly linked to the breakpoints of a reciprocal translocation (Fig. 7).

3.2.2. *Resolving the Genetic Puzzle of the Tox1 Locus*

3.2.2.1. *Tox1 Appears to be a Single Locus.* As noted above (Section 3.1.1., Fig. 5), in a cross between race O and race T, all progeny producing T-toxin are highly virulent and all progeny not producing T-toxin are weakly virulent on T-corn (Lim and Hooker 1971; Yoder 1976; Yoder and Gracen 1975). This 1:1 segregation of parental phenotypes identifies a single genetic locus designated *Tox1* (Leach et al. 1982) that controls T-toxin production and high virulence.

3.2.2.2. *Tox1 is Actually Two Unlinked Genetic Loci.* The 1988 discovery that *Tox1* is genetically inseparable from a reciprocal translocation breakpoint (Section 3.1.3) was supported by the RFLP map (Tzeng et al. 1992) which revealed a four-armed linkage group, diagnostic of a reciprocal translocation, with *Tox1* located at the intersection (Fig. 7A). The linkage group consists of a pair of race O chromosomes (6 and 12) and a pair of race T chromosomes (6;12 and 12;6) that are reciprocally translocated with respect to the race O pair (Tzeng et al. 1992) (Fig. 7B). These findings demonstrated that *Tox1* may not be a simple, single, Mendelian character and that the conclusions from the earlier genetic analyses were misleading. Although RFLP mapping placed *Tox1* at the intersection of the four-armed linkage group, it did not reveal its exact chromosomal location. Which translocated chromosome was it on, or was it on both?

The answer to this question was not clear until it became possible to do genetic analysis of toxin production with tagged *Tox⁻* mutants of a progenitor *Tox1⁺* strain (Section 3.2.1.3, 3.2.1.4). Until this time, all genetic analyses relied on crosses between naturally occurring race O and race T strains, which as noted above, are distinguished by a reciprocal translocation intimately associated with toxin production. The translocation complicates genetic analysis since the pair of chromosomes carrying *Tox1* is heterozygous with respect to the pair in race O. Crosses between the induced, tagged, *Tox⁻* mutants and a wild type race O tester (the translocated chromosomes are heterozygous) confirmed that all induced *Tox⁻* mutations mapped at the previously defined *Tox1* locus since all progeny from these crosses were *Tox⁻*. Crosses between the *Tox⁻* mutants and a wild type race T tester are between isogenic strains that differ only by the mutation at *Tox1* and, by necessity, at *MAT*. This type of cross confirmed

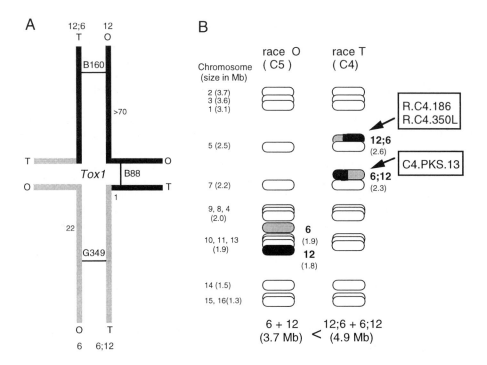

Figure 7. The *Tox1* locus of *C. heterostrophus* is two genetic elements on two reciprocally translocated chromosomes and is associated with 'extra DNA'. (A) The four-armed linkage group identified in the RFLP map (Tzeng et al. 1992). Linkage data show that *Tox1* is inseparable from the translocation breakpoints, but do not reveal whether *Tox1* is positioned at one, the other, or both breakpoints. In crosses heterozygous at *Tox1* (represented here), all markers that are linked to the breakpoints on any of the four chromosomes will appear linked to each other. Positions of RFLP markers B160, B88, and G349 are shown; each uniquely identifies one translocated and one normal sequence chromosome. Numbers near lines indicate map distances in cM (Tzeng et al. 1992; Tasma and Bronson 1998). T = race T, O = race O. (B) Diagram of electrophoretically separated *C. heterostrophus* chromosomes, adapted from Kodama et al. 1998. The karyotype of a race O field isolate Hm540 (not shown here) was chosen (Tzeng et al. 1992) as the reference for numbering the chromosomes. C4 and C5 are lab strains bred for near-isogenicity except for heterozygosity at *Tox1* and *MAT* (mating type) and therefore have similar karyotypes; Note that i) C5 chromosomes 6 and 12 (shaded) are missing in C4; C4 chromosomes 12;6 and 6;12 (shaded) are missing in C5 and are translocated with respect to their counterparts in C5; CHEF gel blot analysis (Kodama et al. 1998) located two *Tox−* mutations on chromosome 12;6, and one on chromosome 6;12, confirming that *Tox1* is two loci, on different chromosomes. ii) The sum of the sizes of chromosomes 12;6 and 6;12 is approximately 1.2 Mb more than that of chromosomes 6 and 12, indicating 'extra' DNA in race T that maps at *Tox1* (Chang and Bronson, 1996).

that each mutation was at a single site, since progeny segregated 1:1 for toxin production. Up to this point, these data led to the same conclusion as race O by race T crosses and one might conclude that *Tox1* is indeed a single locus. However, when mutants were crossed to each other, (these crosses are also between isogenic strains differing only by the mutation at *Tox1* and at *MAT*) the major hypothesis, that *Tox1* is a single locus, collapsed. Progeny of a cross between one particular mutant (Section 3.2.3.2, 3.2.3.3) and any of the other induced mutants were 25% Tox$^+$, indicating that the mutation in this mutant is not linked to the others. Furthermore, heterokaryons between an auxotroph of the unique mutant and different auxotrophs of any of the other induced mutants were all found to produce halos (produce T-toxin) about the same size as that of a control formed by two wild type race T strains, indicating that the defects in the induced mutants can be complemented by the nucleus of the unique mutant (Kodama et al. 1999).

Physical mapping of gel separated chromosomes of the tagged Tox^- mutants confirmed that $Tox1$ is not a single locus. The majority of the mutations are located on translocated chromosome 12;6, however, the unique mutation is located on chromosome 6;12, the other chromosome involved in the (Kodama et al. 1999). Thus, after about twenty five years of data which suggested that $Tox1$ is a single locus, it was demonstrated that $Tox1$ is on two different chromosomes and that the two loci map to the chromosomes (Kodama et al. 1999) that are reciprocally translocated in race T (i.e., chromosomes 6;12 and 12;6) with respect to the race O counterparts (chromosomes 6 and 12) (Bronson 1988; Tzeng et al. 1992). $Tox1$ behaves like a single Mendelian element because these two chromosomes must co- segregate during meiosis whenever a cross is heterozygous at $Tox1$ (i.e., race O x race T) (Kodama et al. 1999; Turgeon et al. 1995). When they do not co-segregate, ascospore progeny are non-viable due to duplications and deletions. When crosses are homozygous, (i.e., Tox^- mutant of a $Tox1^+$ strain x another such mutant) the complexities of reciprocal translocation genetics are eliminated and the two translocated chromosomes segregate independently. To reflect the unique genetic organization of $Tox1$, the two loci have been designated $Tox1A$ and $Tox1B$ (Fig. 7B) (Kodama et al. 1999). Details regarding the genes encoded at these two loci are given below.

3.2.3. *Cloning of Genes at Tox1*

3.2.3.1. *A Polyketide Synthase at Tox1A.* The fact that T-toxin is a polyketide suggested a simple hypothesis (Yoder et al. 1993) that $Tox1$ is a polyketide synthase (PKS)-encoding gene since biosynthesis of polyketides generally depends on PKS activity (Hopwood and Sherman 1990) Indeed, the first gene identified at $Tox1$ was found to encode this type of enzyme. This gene (designated Ch*PKS1*) was cloned from the REMI tagged Tox^- mutant R.C4.350L (which maps at $Tox1A$), using the plasmid rescue procedure (Yang et al. 1996). Ch*PKS1* is a 7.6 kb open reading frame (ORF) after splicing of four introns. The putative multifunctional protein has six enzymatic domains, including β-ketoacyl synthase (KS), acyltransferase (AT), dehydratase (DH), enoyl reductase (ER), β-ketoacyl reductase (KR), and acyl carrier protein (ACP, the REMI vector insertion site), all identified by conserved motifs found in known type I PKSs (Fig. 8) (Yang et al. 1996). Among known fungal PKSs, ChPKS1 is one of only two that has all of the six possible domains for polyketide chain extension and carbonyl group reduction (Proctor et al. 1999; Yang et al. 1996). It lacks a chain-terminating thioesterase domain (TE) found at the C-terminus of some PKSs.

Ch*PKS1* has several features that have helped in unraveling of the origin of $Tox1$. 1) When a gel blot of genomic DNAs from race T and race O, other *Cochliobolus* spp. and related genera was probed with Ch*PKS1*, a single copy was detected only in race T; Ch*PKS1* is absent from the race O genome and from the genome of any other *Cochliobolus* species or related genus (Yang et al. 1996). 2) The Ch*PKS1* G+C content is only 50%, which is lower than that of most *Cochliobolus* genes (54 to 62%) identified to date and of PKS-encoding bacterial genes (62 to 74%). 3) Unlike most

C. heterostrophus

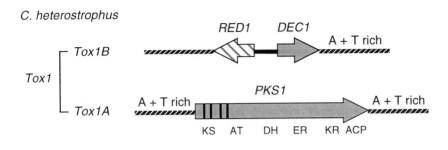

M. zeae-maydis

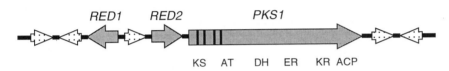

Figure 8. Organization of genes for biosynthesis of T-toxin and PM-toxin (see text for details). The arrows represent open reading frames (ORFs) and transcription directions (the embedded vertical bars indicate introns). Stippled and hatched ORFs represent genes that are required or not required for toxin production, respectively. Dotted ORFs in the *MzTox* gene cluster are putative transposons. *PKS1*: polyketide synthase; *RED1* and *RED2*: reductases; *DEC1*: decarboxylase. Note that *Tox* genes in *C. heterostrophus* are located at two unlinked loci (*Tox1A* and *Tox1B*) and are surrounded by A+T rich repetitive DNA; in *M. zeae-maydis*, *Tox* genes are found in a single cluster that contains a number of transposon-like elements. *ChRED1* has no similarity to *MzRED1* or *MzRED2* and *MzRED1* and *MzRED2* are not similar to each other except at conserved motifs. *ChPKS1* and *MzPKS1* are identical in length, domain structure and intron position, but they encode proteins that have only 60% identity.

Cochliobolus genes and genes from other fungi, such as *Aspergillus, Fusarium, Magnaporthe* and *Penicillium, ChPKS1* has no preference for a C residue in the third position of codons for any of the eight amino acids that have the option of all four bases at this position. The third position bias is also different from all PKS-encoding bacterial genes, except for *Mycobacterium leprae pksE*, which encodes part of a polyketide synthase that has high similarity to Ch*PKS1*. 4) DNA on both flanks of the Ch*PKS1* open reading frame is non-coding and A+T rich (~70%).

3.2.3.2. *A Decarboxylase and a Reductase at Tox1B.* In contrast to the relatively straightforward cloning of Ch*PKS1* from *Tox1A*, the cloning of genes at *Tox1B* was complicated. A plasmid tagged, large deletion mutation mapping at *Tox1B*, which eliminated T-toxin production and high virulence on T-cytoplasm corn, was identified in a race T strain. Comparisons among this mutant, wild type race T and race O DNA revealed that a diagnostic *Not*I fragment associated with wild type race T (not found in race O), was missing from the mutant and replaced by a smaller fragment. Screening a wild type cDNA library with probes made from the polymorphic *Not*I restriction fragment from the deletion mutant and its wild type *Tox1B* counterpart identified two single copy genes, a decarboxylase (*DEC1*) and a reductase (*RED1*), present in the wild type but absent in the mutant. Disruption of *DEC1* eliminates T-toxin production and high virulence on T-cytoplasm corn. *DEC1* is similar to the acetoacetate decarboxylase gene of *Clostridium acetobutylicum*. The phenotype of mutants carrying *RED1* mutations is still in question; preliminary results suggest that *RED1* is not required for T-toxin production. *RED1* is similar to members of the medium-chain dehydrogenase/reductase gene superfamily. *DEC1* (46% G+C) and *RED1* (44% G+C) are adjacent in the *C. heterostrophus* genome and divergently transcribed. DNA gel blot analysis indicates that they, like *PKS1*, are unique to race T. Like *PKS1*, *DEC1* is flanked by A+T-rich (72%), highly repetitive, non-coding DNA (Fig. 8) (Rose 1996; Rose et al. 1996).

We favor the idea that the function of the decarboxylase may be to remove the terminal carboxyl group from the completed T-toxin chain. This would explain why all members of the T-toxin family have an odd number of carbons, and why each lacks a terminal carboxyl group (Fig. 3). Alternatively, the decarboxylase could provide precursors for polyketide biosynthesis (Bisang et al. 1999).

3.2.3.3. *Tox1A and Tox1B Appear to be Gene Clusters.* Both genetic and physical mapping have confirmed that *PKS1* is at *Tox1A* on chromosome 12;6 and *DEC1/RED1* are at *Tox1B* on chromosome 6;12. *PKS1* and *DEC1* are required for T-toxin production, but are they sufficient to make active T-toxin molecules? To answer this question, *PKS1* was introduced into *Tox⁻* mutant R.C4.186, which sustained a deletion of ~ 700 kb at *Tox1A* (Kodama et al. 1999). Transformants remained Tox⁻ (Zhu 1999). Similarly, *DEC1* was introduced into *Tox⁻* mutant C4.PKS.13, which sustained a deletion of ~ 100 kb at *Tox1B* (Kodama et al. 1999; Rose 1996). Transformants also remained Tox⁻ (Zhu 1999). The failure of these single genes at *Tox1A* and *Tox1B* respectively, to restore ability to produce T-toxin suggests that there are as yet undiscovered genes at both *Tox1A* and *Tox1B* that are necessary for T-toxin biosynthesis.

3.2.3.4. *Tox1 is Associated with an Insertion of DNA.* RFLP mapping and karyotype analysis demonstrated that the *Tox1* locus is complex. In addition to the translocation, *Tox1* is associated with a large insertion and highly repeated DNA (Chang and Bronson 1996; Tzeng et al. 1992) based on three lines of evidence: 1) The sizes of translocated chromosomes 6;12 and 12;6 in race T sum to about 1.2 Mb more than the sum of the sizes of chromosomes 6 and 12 in near-isogenic race O (Fig. 7). Since the strains used for this analysis are highly inbred, but heterozygous at *Tox1*, the inserted DNA must be located at or near *Tox1*. 2) Half of the RFLP probes mapping within 4 cM of *Tox1* are repetitive, in contrast to only ~4% repetitive probes in the remainder of race T genome (Chang and Bronson 1996; Turgeon et al. 1995; Tzeng et al. 1992; Yoder et al. 1994). Cloning of genes from the two *Tox1* loci has confirmed that the *Tox1* genes are indeed embedded in highly repeated DNA (A+T-rich). 3) The genes are completely missing in race O and the codon bias and G+C content are atypical of most fungal genes (Rose 1996; Yang et al. 1996). These results suggest that *Tox1* DNA may have been inserted *via* horizontal transfer, the source of which is unknown. The insertion, perhaps, caused the reciprocal translocation of chromosomes 6 and 12 in race O that gave rise to race T (Fig. 9).

3.2.4. *Are Additional Tox Loci Involved in T-Toxin Production?*
The working hypothesis, based on the fact that reconstruction experiments failed to restore toxin producing ability (Section 3.2.3.3), is that there are two gene clusters, one at *Tox1A* and one at *Tox1B*, each carrying genes beyond *PKS1* and *DEC1* required for the T-toxin biosynthesis and that these genes include those involved in biosynthesis of precursors, regulation, secretion etc. Furthermore, since the ability to produce T-toxin segregates with *Tox1*, these two clusters should contain all of the genes necessary for toxin production; genes at loci other than *Tox1* should not be required. In earlier studies, evidence for non 1:1 segregation of *Tox⁺:Tox⁻* progeny of crosses involving certain race O field isolates was presented (Section 3.1.2). These data, however, were subsequently attributable to the presence of a *Tox1*-linked spore killer gene and not to the requirement for products of genes at loci, other than *Tox1*, controlling T-toxin production (Section 3.1.2, 3.1.3). This conclusion was strengthened by a survey of the field population of the fungus; no locus other than *Tox1* was found to affect T-toxin production (Bronson et al. 1990). Mutational analyses of T-toxin production also support this conclusion: twelve T-toxin production-deficient mutants generated by random mutagenesis (Section 3.2.1.3, 3.2.1.4) all map to the known *Tox1* locus. In addition, race O lacks the *PKS1* and *DEC1* genes required for T-toxin production implying that race O does not carry DNA associated with this trait. Therefore, we were surprised when a REMI screen identified four loci, unlinked to *Tox1*, and present in race O, which when mutated, in race T, affect toxin production.

3.2.4.1. *Phenotypic Characterization of New Tox Loci.* Of the four mutations affecting ability to produce toxin, three are tagged and one is not. One of these mutants shows a complete loss (tight, no halo produced in the microbial assay), two show a reduction (leaky, small halos) and one shows an increase (overproducing, halos larger than wild type) in ability to produce T-toxin (Fig. 6). When these mutants were crossed to a race T tester, progeny segregated 1:1 (Tox⁺:Tox⁻), indicating a single gene mutation in each mutant. When crossed to a race O tester, 25% of the progeny were Tox⁺, indicating that none of these mutations is linked to the known *Tox1* locus. When the mutants were crossed to each other in all possible combinations, 25% of the progeny segregated Tox⁺, indicating that none of these mutations is linked to any of the others. Based on these genetic analyses, four new loci that control toxin production in the genus *Cochliobolus* are proposed.

3.2.4.2. *Molecular Analysis of New Tox Genes.* Genomic DNA flanking the vector insertion sites of the three tagged mutants (R.C4.1434, R.C4.1771 and R.C4.1957) were recovered. In R.C4.1434 (*Tox⁻* leaky), the REMI vector insertion site was in an open reading frame that encodes a protein with high similarity to the yeast proteins, Hmd1 and Nab3, both of which have been shown to be RNA-binding proteins required for pre-mRNA processing (Sugimoto et al. 1995; Wilson et al. 1994). Amino acid alignment suggests that the N-terminal end of this ORF contains a RNA-binding motif highly conserved in known RNA/ssDNA binding proteins (Kim and Baker 1993). In mutant R.C4.1771 (*Tox⁻* leaky), the REMI vector disrupted an ORF encoding a protein with high similarity to the yeast RLR1 protein, a putative transcription regulator (Mallet et al. 1995). In R.C4.1957 (T-toxin overproducer), the vector inserted in an ORF encoding a protein with high similarity (50% identity and 71% similarity) to the human transcription factor ZFM1 (Toda et al. 1994), which is a putative tumor suppressor (Wrehlke et al. 1997).

Site-specific disruption of these three putative regulatory genes in race T genome restored the original mutant phenotypes, confirming that these genes are involved in T-toxin production. Sequencing of these genes has not been completed and how they affect the T-toxin biosynthetic pathway is yet to be determined.

3.2.4.3. *New Tox Loci are Present in the Genome of Both Race O and Race T.*

Previous studies demonstrated that *Tox1* is represented only in the genome of race T. In contrast, all of the new *Tox* loci are present in the genomes of both races. Segregation analysis of crosses between the new *Tox⁻* mutants and wild type race O testers, yields 25% wild type Tox^+ recombinants, suggesting, as noted above, that none of these mutations maps at the known *Tox1* locus AND that race O has genes that can complement these *Tox⁻* mutations. In addition, both race O and race T DNA hybridize to genomic DNA fragments recovered from the insertion sites and both show the same hybridization patterns. This indicates that the new *Tox* loci are not T-toxin production specific, rather, they could be involved in both T-toxin biosynthesis and in other common, but dispensable, pathways. Indeed, linkage analyses revealed that two of the new Tox loci (R.C4.1771 and R.C4.1957) also contribute to a second mutant phenotype (reduced pigmentation) and a double mutation at these two loci is lethal.

3.2.5. *Cloning the PM-toxin Gene Cluster*

Because PM-toxin is structurally similar to T-toxin (both are polyketides), a PCR based strategy was used to clone *PKS* genes from *M. zeae-maydis*. Using primers corresponding to the conserved KS domain of known PKSs, a predicted 290 bp product was amplified and cloned from genomic DNA of wild type. The same PCR fragment yielded four different clones. One of them (*MzPKS1*) had 82% identity to the KS domain of *C. heterostrophus PKS1*, suggesting that it was a homologue.

Using the *MzKS1* PCR product as a starting point, the entire *MzPKS1* gene and its flanking sequences (totaling 23 kb, Fig. 8) were cloned by a combination of TAIL-PCR (Liu and Whittier 1995) and plasmid rescue (Yang et al. 1996). Sequencing revealed that *MzPKS1* has 60% identity to *ChPKS1* over its entire length; identity in some enzymatic domain signature motifs is as high as 90%. The two genes have identical organization, each has four introns in conserved positions, and each is 7.6 kb after intron splicing (Yun 1998).

In contrast to *ChPKS1*, which is flanked by A+T-rich, repeated, noncoding DNA, *MzPKS1* is flanked by ORFs, including those with similarity to transposases. Two of the ORFs, designated *MzRED1* and *MzRED2* have high similarity to ketoreductases involved in polyketide and fatty acid biosynthesis but are not similar to *RED1* (Section 3.2.3.2) in *C. heterostrophus*. When *MzRED1*, *MzRED2*, or *MzPKS1* is disrupted, PM-toxin production is lost, indicating they are indispensable for PM-toxin biosynthesis and that a gene cluster is involved (Yun 1998).

3.2.6. *Toxins as Virulence/Pathogenicity Factors*

3.2.6.1. *T-Toxin is a Virulence Factor.*

T-toxin is one of the few fungal secondary metabolites that has been thoroughly tested for a role in pathogenesis (Yoder 1980;

1997). This evaluation is based on three kinds of analyses. 1) Conventional genetic analysis: as indicated above (Section 3.1.1), this analysis first demonstrated that the ability to produce T-toxin is inseparable from high virulence on T-cytoplasm corn. 2) Mutational analysis: this analysis confirmed that all induced mutants, with a tight Tox⁻ phenotype, cause disease symptoms that are indistinguishable from those caused by race O, when tested on both T and N-cytoplasm corn; mutants with a leaky Tox⁻ phenotype cause weak race T symptoms. 3) Site-specific disruption of *PKS1* and *DEC1*: these experiments provided the strongest evidence that the consequence of loss of ability to produce T-toxin is the loss of high virulence to T-cytoplasm corn. Thus, the conclusion that T-toxin is a host specific virulence factor is difficult to avoid.

Studies on the role of T-toxin in pathogenesis have contributed fundamental knowledge about how host specific toxins play key roles in determining outcomes of fungal pathogen/plant interactions. The fact that T-toxin acts as a virulence factor by blocking production of energy in host cells is one of the best examples of the way in which a pathogen can suppress host defenses and colonize more aggressively (Yoder and Turgeon 1996). These studies have also shown that pathogenesis by *C. heterostrophus* must involve factors in addition to those required for T-toxin production because race O, which does not produce T-toxin and race T-derived *Tox⁻* mutants are still effective pathogens on corn.

3.2.6.2. *PM-toxin is a Pathogenicity Factor.* As mentioned above (Section 2.3), all isolates of *M. zeae-maydis* produce PM-toxin. The lack of naturally-occurring *Tox⁻* strains and the fact that the fungus is homothallic made it impossible, in the past, to evaluate the role of PM-toxin in pathogenesis by conventional genetic analysis as used for *C. heterostrophus*. This problem was not alleviated until PM-toxin-deficient mutants were generated by REMI (Yun et al. 1998). *Tox⁻* mutants generated by site-specific disruption of PM-toxin genes in the wild type genome fail to cause the disease. Thus it is now clear that all *Tox⁻* mutants that have lost ability to produce PM-toxin as indicated by the microbial assay (Section ,3.2.1.2, which works for PM-toxin as well as T-toxin, Fig. 6) are nonpathogenic on T-cytoplasm corn. *M. zeae-maydis* does not generally cause disease on corn without T-cytoplasm. Thus PM-toxin appears to be required for pathogenicity itself, rather than contributing to virulence as does T-toxin (Yun et al. 1998).

3.2.6.3. *HC-toxin and Victorin are Pathogenicity Factors.* HC-toxin and victorin, along with T-toxin, were the first toxins determined to be correlated with fungal plant pathogenesis, based on co-segregation of the ability to produce toxin and the host specificity of the fungus.

For T-toxin, this was readily demonstrated by segregation analysis of progeny of crosses between race O and race T. For HC-toxin and victorin, interspecific crosses between a HC-toxin-producing isolate of *C. carbonum* and a victorin-producing isolate of *C. victoriae* illustrate the point (Nelson and Hebert 1960; Scheffer et al. 1967). Initially, the choice of parents in the cross was made largely due to the inability to mate *C. victoriae* isolates; it has been shown subsequently that all *C. victoriae* isolates in extant collections are *MAT-2* and all appear to be female sterile (Christiansen et al. 1998). Progeny of these *C. carbonum* by *C. victoriae* interspecific crosses segregate 1:1:1:1 for ability to produce HC-toxin only, victorin only, both toxins or neither toxin. All progeny producing HC-toxin are pathogenic to susceptible corn only (*hm1hm1*), all progeny producing victorin are pathogenic to susceptible oats (*Vb*) only, progeny producing both toxins are pathogenic to both hosts, and nonproducing progeny are nonpathogenic (Scheffer et al. 1967). These data indicate that the production of each toxin is under the control of a single, unlinked locus. The loci controlling HC-toxin and victorin production have been formally designated *Tox2* and *Tox3*, respectively (Yoder et al. 1986; 1989).

Although molecular evaluation of *C. victoriae Tox3* is not possible because attempts to clone *TOX3* genes have been unsuccessful, three lines of evidence indicate that victorin is a pathogenicity factor. The first is the genetic evidence described above: all victorin-producing progeny of *C. victoriae* specifically attack oat plants with the *Vb* allele; all victorin-nonproducing progeny are nonpathogenic to oats (Yoder et al. 1997). The second is that spontaneous Tox⁻ mutants are nonpathogenic. The third line of evidence comes from analysis of victorin-deficient mutants generated using REMI; victorin minus mutants are nonpathogenic to oats and leaky mutants kill oats at a much lower rate than does wild (Churchill et al. 1995).

The organization of the *C. carbonum Tox2* locus, like *C. heterostrophus Tox1,* is complex. Molecular analyses have revealed that: 1) *TOX2* is actually a cluster of genes distributed over a large region (540 kb, on a 2.2 or 3.5 Mb chromosome in different isolates) (Ahn and Walton 1996). This cluster includes at least six genes: *HTS1* (15.7 kb) encoding a cyclic peptide synthetase (Scott-Craig et al. 1992), *TOXA* encoding a HC-toxin pump (Walton et al. 1994), *TOXC* encoding a fatty acid synthase for biosynthesis of Aeo, one of the amino acid constituents of HC-toxin (Ahn and Walton 1997), *TOXE* encoding a DNA binding protein (Ahn and Walton 1998), *TOXF* encoding a cytochrome P450, and *TOXD* of unknown function (Scott-Craig et al. 1995). 2) Most *TOX2* genes are present as multiple copies (Ahn and Walton 1996; 1998; Scott-Craig et al. 1995). 3) All *TOX2* genes are present only in HC-toxin-producing isolates of the fungus.

HC-toxin must be considered a pathogenicity factor because in addition to the genetic evidence mentioned above, all HC-toxin deficient mutants generated by targeted disruption of genes in the *TOX2* cluster (all copies of genes occurring as multiple copies must be deleted or disrupted) are nonpathogenic on susceptible corn plants (Ahn and Walton 1997; Panaccione et al. 1992).

3.2.7. *Cps1 Acts as a General Pathogenicity Determinant*

To explore molecular mechanisms underlying pathogenesis by *C. heterostrophus* more broadly, the REMI mutant library (Section 3.2.1.4) was screened for mutations associated with pathogenicity. The cloning and characterization of DNA from one such mutation site identified a peptide synthetase gene (*CPS1*) the product of which appears to be a general pathogenicity determinant (Lu 1998). This gene has the following features: 1) it is found in both races of *C. heterostrophus*. Disruption causes reduced virulence (lesion size half of wild type) in race O <u>and</u> race T even though the latter produce wild type levels of T-toxin. 2) DNA gel blot analysis shows that *CPS1* has homologues in many fungal species including all *Cochliobolus* spp., species in all closely-related genera such as *Pleospora, Pyrenophora, Setosphaeria, Bipolaris* and *Alternaria,* and in a number of pyrenomycetes species. 3) Disruption of a *CPS1* homologue in *C. victoriae* also causes drastically reduced virulence, even though transformants produce wild type levels of victorin (Section 3.2.6.3). 4) *CPS1* is part of a large gene cluster. At least 15 ORFs have been identified in the *CPS1* region (~35 kb), six of which encode proteins with known functions (e.g., thioesterase, coenzyme A transferase, decarboxylase, laccase, DNA-binding protein and transport protein). Some of these ORFs (e.g., the ORF encoding a putative DNA-binding protein) have been disrupted, yielding transformants with the same reduced pathogenicity phenotype as the *CPS1-* mutant.

Thus, it is clear that pathogenesis by *C. heterostrophus* to corn involves at least two secondary metabolites: T-toxin, a host specific factor that determines high virulence on a particular host, T-corn and the hypothetical CPS1 peptide, a general virulence factor that contributes to basic mechanisms underlying disease establishment by the fungus on appropriate host plants.

3.3. THE EVOLUTION OF POLYKETIDE-MEDIATED FUNGAL SPECIFICITY FOR T-CYTOPLASM CORN

Initial studies on the epidemiology of Southern Corn Leaf Blight disease led to the hypothesis that race T arose from race O (Alcorn 1975; Leonard 1973; 1977), and that the origin was recent. Evidence for the recent origin of race T includes the following: 1) When race T was first isolated, all isolates were *MAT-1* while *MAT* distribution in race O was 50% *MAT-1*/50% *MAT-2*. A few years later, presumably after mating with race O, race T was also 50% *MAT-1*/50% *MAT-2* (Leonard 1973). 2) A spore killer gene was identified in race O (~50% of field isolates) and not in race T, suggesting long term interbreeding among race O, but not race T, isolates. 3) The extant collection of race T field isolates shows fewer RFLPs than a corresponding collection of race O. 4) All field isolates of race T show the same reciprocal translocation and 'extra' DNA linked to the *Tox1* locus (Chang and Bronson 1996).

Did race T originate in the Philippines? Increased susceptibility of corn to *C. heterostrophus* was reported there prior to the US epidemic (Mercado and Lantican 1961), however this report did not infer a new race. It has been suggested that race T may have been in the field in the Philippines prior to the introduction of T cytoplasm corn in 1957, since race T symptoms appeared so soon after the host arrived (Mercado and Lantican 1961; Wise et al. 1999). In contrast, it is considered unlikely that race T was in the US field population prior to the widespread planting of T cytoplasm corn, unless it occupied a specific, as yet unknown, ecological niche, since it is less fit than race O as evidenced by its rapid decline after the withdrawal of its highly susceptible host (Wise et al. 1999). Did race T originate just once, in the Philippines, and was it introduced from there to the US? Alternatively, were there multiple origins of the race T phenotype? The ideal approach to answering this question would come from direct examination of isolates of race T collected in the Philippines prior to the US epidemic, and comparison of these isolates with those from the US. Unfortunately all Philippine isolates of race T apparently have been lost. An effort is currently underway to 'trap' race T in the Philippines, by planting T-cms corn. If successful, the isolates will be examined for similarities to the extant collection of race T. Will they all have the 1.2 Mb insertion and the reciprocal translocation? If they do not, yet are still race T (producing T-toxin and highly virulent on T-cms corn), this will be evidence that race T arose more than once and that Philippines isolates may not have resulted from the same molecular events as the isolate(s) that caused the US epidemic. If they do, we will need to prove that these isolates are endemic to the Philippines and not imports from the US.

    Cloning of genes at *Tox1* has contributed critical information to our understanding of the evolution of race T in the US. The data are consistent with the hypothesis that race T evolved from race O by acquiring *Tox1* DNA *via* horizontal gene transfer from an unknown source rather than by inheriting it from an ancestor. How solid is the evidence for this? 1) The most compelling evidence is that *Tox1* genes hybridize exclusively with DNA from race T isolates; race O lacks alleles or close homologues of these genes, which are also not detected in the genomes of other *Cochliobolus* spp. or other related Loculoascomycete genera (*Pyrenophora, Setosphaeria, Alternaria*) examined. The simplest hypothesis would be that race T alone acquired *Tox* DNA, rather than the more elaborate alternative that *Tox* DNA was present an ancestral strain and that all genera derived from it lost the DNA (including race O), except for race T. This latter scenario seems unlikely, as it would require multiple losses. 2) The G+C content of *PKS1* is only 50%, *DEC1* is 46% and *RED1* is 44%, all lower than that of most *Cochliobolus* genes (54 to 62%). 3) The codon bias of the *Tox1* genes is atypical: unlike most *Cochliobolus* genes and genes from other fungi, *PKS1, DEC1* and *RED1* have no preference for a C residue in the third position for all eight amino acids that have the option of all four bases at this (Yang et al. 1996). The highest nucleotide, amino acid, and codon bias similarity is between *PKS1* from *C. heterostrophus* and *PKS1* from *M. zeae maydis* (Yun 1998), while the highest sequence similarity from public database searches are to two fungal polyketide synthases, *Aspergillus terreus LovF*, encoding an enzyme for lovastatin biosynthesis and *Gibberella fujikuroi FUM5*, encoding an enzyme for fumonisin biosynthesis. The G+C content and codon bias of these two genes is different from *ChPKS1*. Codon bias of *ChPKS1* is not unlike that of the *pksE* gene from the filamentous bacterium *Mycobacterium leprae*, however G+C content of the *M. leprae* gene is much higher. 4) DNA on both flanks of *PKS1* and on one flank of *DEC1* is non-coding, A+T rich (>70%) and highly (Rose 1996; Yang et al. 1996).

    The source of the 'alien' DNA is unknown. The presence of 'fungal type' introns in *PKS1* suggests a fungal origin however, as noted above, codon usage does not fit. Until we know more about the overall genomic organization, this latter point is weak; perhaps genes for secondary metabolism in fungi have a different codon bias and/or G+C content than genes for primary metabolism.

    A model for race T evolution has been proposed based on *MAT* gene distribution over time, the structural features of the *Tox1* genes, the unusual chromosome rearrangement at *Tox1*, and the lack of these genes in any other species (Fig. 9, (Turgeon and Berbee 1998; Yang et al. 1996). Perhaps *Tox1* DNA inserted, *via* horizontal gene transfer from an unknown source, into chromosome 6 or 12 of a *C. heterostrophus MAT-1* race O isolate. The large insertion (about two thirds the size of the race O chromosome) then caused instability, promoting a reciprocal translocation between chromosomes 6 and 12, which placed *Tox1A* on one chromosome (12;6) and *Tox1B* on another (6;12), and created the extant form we call race T. (Note that none of the *Tox1* genes found thus far is physically at the translocation breakpoint).

    Since *M. zeae-maydis* and *C. heterostrophus* race T appeared simultaneously, it was not unreasonable, before the genes responsible for polyketide production were available, to expect that these two fungi shared genes for toxin biosynthesis. The evidence, however, does not support this idea. First, although the general structures of *ChPKS1* and *MzPKS1* are similar, the corresponding proteins have only 60% amino acid identity. Second, the neighborhoods of these genes are different. *MzPKS1* is part of a cluster that includes at least two genes (*MzRED1* and *MzRED2*) encoding two different keto-reductases, both of which are required for PM-toxin production; *ChRED1* has no nucleotide or amino acid similarity with *MzRED1* or *MzRED2* and, in contrast to *M. zeae-maydis*, the phenotype of a *C. heterostrophus* mutant with a *ChRED1* disruption appears to be Tox[+]. When *M. zeae-maydis* genomic DNA is probed with *ChDEC1*, no signal can be detected even at low stringency. Finally, when *ChPKS1* (Zhu 1999) was transformed into a *pks1[-]* mutant of *M. zeae maydis*, none of the transformants produced detectable PM-toxin, suggesting that these two PKSs are not cross functional.

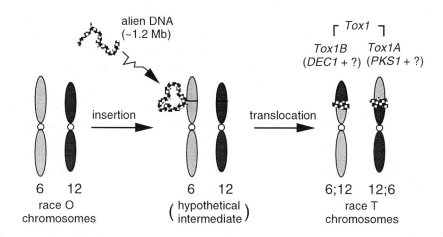

Figure 9. Diagrammatic representation of a model proposed for the evolution of race T. Race T could have evolved from race O *via* two associated genetic rearrangements: i) about 1.2 Mb of alien DNA inserted into chromsome 6 (or 12) in race O *via* horizontal gene transfer. ii) after insertion, a reciprocal translocation between chromosomes 6 and 12 (the breakpoints are hypothesized to be in the inserted alien DNA) occurred, generating two unlinked loci each on a different chromosome (6;12 and 12;6) in race T. The two *Tox1* loci are designated *Tox1A* and *Tox1B* (Yoder et al. 1997); both loci are thought to be gene clusters required for T-toxin production (see text). The chromosomes are not drawn to scale and the location of the insertion relative to the centromere is hypothetical.

Are the *Tox* genes of ancient or recent origin? If we assume that both fungi acquired the genes for polyketide toxin production by horizontal transfer, as the evidence suggests for race T (Yang et al. 1996), the discovery that ChPKS1 and MzmPKS1 are only 60% similar allows us to rule out the possibility that the two fungi acquired genes for toxin production from the same organism recently (e.g., at the time of the epidemic in 1970) and also to rule out the possibility that one of these fungi transferred genes to the other recently. If the transfer was recent, there were two different sources of genes. If the transfer was ancient, or if the genes were present in the ancestors of both *C. heterostrophus* and *M. zeae-maydis*, the genes have diverged substantially, subsequently.

3.4. UNDERSTANDING THE EVOLUTION OF PATHOGENESIS

Within the ascomycete genus *Cochliobolus* at least thirty species are recognized (Sivanesan 1984). Of these, three (*C. heterostrophus, C. carbonum, C. victoriae*) have been studied extensively because they cause plant diseases that have had (*C. heterostrophus* and *C. victoriae*), or could have had (*C. carbonum*) devastating agronomic consequences. Each species is capable of high virulence on a different host, caused by production of a chemically distinct, host-specific toxin (Yoder 1980; Yoder et al. 1997).

    The study of secondary metabolism provides insight into molecular mechanisms of fungal pathogenesis. Our focus in the past has been host-specific toxins, and we, as well as others, have provided convincing evidence that these toxins are not only necessary for pathogenicity itself or for some level of virulence, but also determine fungal host range (Yoder et al. 1997). The time is ripe, therefore, to consider the relationship of so-called 'toxins' to other factors known to be important in pathogenesis (Yoder and Turgeon 1996). For example, diseases mediated by host-specific toxins and diseases involving 'gene-for-gene' relationships are sometimes regarded as two different classes of fungal/plant interaction. This is largely because the key molecular recognition event in so-called 'toxin' diseases leads to compatibility, whereas the corresponding event in 'gene-for-gene' diseases leads to incompatibility. Yet the race specific elicitors produced by the 'gene-for-gene' fungi *Cladosporium fulvum* (DeWit et al. 1995) and *Rhynchosporium secalis* (Rohe et al. 1995) are host-specific toxins by the same definition that is used to describe the pathologically significant secondary metabolites produced by species of *Cochliobolus* (Kohmoto and Otani 1991; Scheffer and Yoder 1972; Yoder 1980) and certain other fungi: in all cases the relevant molecules are toxic and the activity of each is limited to just one genotype of plant. This suggests that rather than a fundamental difference between these types of disease, there is a fundamental similarity, i.e., genotype-specific fungal molecules determine host range in both cases, and the outcome of molecular recognition can lead either to disease or to disease resistance.

    Toxin-mediated disease systems lend themselves to studies of the evolution of pathogenic capabilities. Fig. 10 is a simple model to illustrate how new pathogenic forms might arise within the genus *Cochliobolus*. Most *Cochliobolus* spp. are weak pathogens or saprophytes (Sivanesan 1984), and represent the 'low general virulence' pool of germplasm. One of our objectives is to determine how widespread the putative general virulence determinant *CPS1* (Section 3.2.7) is in this pool. The model predicts that a benign member of the pool can suddenly turn vicious upon importation, by horizontal transfer, of a gene cluster conferring ability to produce an exotic secondary metabolite, e.g., a toxin, that adversely affects one or more genotypes of plant. The *Tox1* and *Tox2* gene 'clusters' serve to make the point; the hypothesis is that acquisition of *Tox1* by mildly virulent *C. heterostrophus* race O gave rise to race T, and that acquisition of *Tox2* by *C. carbonum* race 2 (which is virtually nonpathogenic) gave rise to race 1. By extension, *Tox3*, which has not yet been cloned, would give rise to *C. victoriae* after being imported into a strain of *C. carbonum* race 2, its likely progenitor (Christiansen et al. 1998).

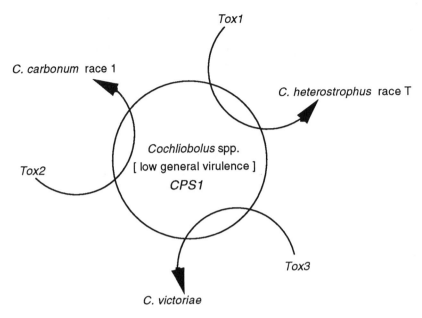

Figure 10. Model for evolution of fungal pathogenesis. Individuals from a pool of fungal germplasm (normally weakly virulent) can acquire fragments of DNA by horizontal transfer from unrelated donors. If a DNA fragment carries a set of genes controlling biosynthesis of a secondary metabolite effective against a particular plant genotype, the strain possessing that fragment abruptly becomes highly virulent, specifically on the plant genotype sensitive to the newly produced metabolite. Examples shown here are *Tox1* (1.2 Mb) causing high virulence to T-cytoplasm corn and *Tox2* (>540 kb) causing high virulence to *hmhm* corn; the role of *Tox3* in high virulence of *C. victoriae* to *Vb* oats is hypothetical, since *Tox3* has not been cloned. If correct, this model may have general relevance to the molecular explanation of the sudden appearance in field populations of new, unusually virulent, races.

4. The Future: The Power of Genomics

It took approximately thirty years from the identification of *C. heterostrophus* race T to the determination of the molecular nature of the genes at *Tox1* and the discovery that *Tox1* is on two different chromosomes. Furthermore, analysis of *Tox1* is not finished, as there is good evidence that *Tox1A* and *Tox1B* are each gene clusters; to date, we have identified only one gene at each of these loci that is required for T-toxin biosynthesis. We suspect that the additional genes are scattered throughout the ~1.2 Mb of 'extra, foreign' DNA at the *Tox1* locus in race T. Clearly, identifying these genes is tough problem.

In addition, we know that the ability to produce toxin renders the fungus highly virulent on a particular host, however, we know very little about the other molecules required to make it a pathogen. How many genes are involved in development of pathogenicity? What is the difference between a pathogen and a saprophyte? Fortunately, we are in a period of high throughput inquiry for biological investigation and can apply the powerful techniques of genomics to a genome wide determination of the complete set of genes whose products are directly and causally involved in pathogenicity. Thus we no longer need to grind our way, gene by gene, towards an answer to these questions. Sequencing of the *C. heterostrophus* genome will provide the database for functional analyses, which could take many forms. The simplest might be genome-wide targeted deletion of all predicted ORFs, followed by plant tests with each mutant, to identify genes required for pathogenicity. Expression profiling, to identify the set of genes turned on in the disease interaction, could be achieved by hybridizing the set of DNAs representing all *C. heterostrophus* ORFs with RNA from infected plant tissue *versus* RNA from the fungus in culture. Similarly, the set of genes whose products affect production and secretion of T-toxin might be identified by probing with RNA from race O *versus* race T. Whatever our strategy, genomics technologies permit quick identification of gene products that are required, either directly or indirectly, for pathogenesis. Thus ends three decades of biochemical, classical genetic, and molecular genetic investigation of one virulence factor- T-toxin and thus begins our global investigation of the state of pathogenicity itself.

References

Ahn, J.H. and Walton, JD. (1998) Regulation of cyclic peptide biosynthesis and pathogenicity in *Cochliobolus carbonum* by TOXEp, a novel protein with a bZIP basic DNA-binding motif and four ankyrin repeats, *Molecular and General Genetics* **260**, 462-469.

Ahn, J.H. and Walton, JD. (1996) Chromosomal organization of *TOX2*, a complex locus controlling host-selective toxin biosynthesis in *Cochliobolus carbonum*, *Plant Cell* **8**, 887-897.

Ahn, J.H. and Walton, J.D. (1997) A fatty acid synthase gene in *Cochliobolus carbonum* required for production of HC-toxin, cyclo(D-prolyl-L-alanyl-D-alanyl-L-2-amino-9,10-epoxi-8-oxodecanoyl), *Mol. Plant Microbe Interact.* **10**, 207-214.

Akamatsu, H., Itoh, Y., Kodama, M., Otani, H., and Kohmoto, K. (1997) AAL-toxin-deficient mutants of *Alternaria alternata* tomato pathotype by restriction enzyme-mediated integration, *Phytopathology* **87**, 967-972.

Alcorn, J.L. (1975) Race-mating type associations in Australian populations of *Cochliobolus heterostrophus*, *Plant Dis. Reptr.* **59**, 708-711.

Arny, D.C. and Nelson, R.R. (1971) *Phyllosticta maydis* species nova, the incitant of Yellow Leaf Blight of maize, *Phytopathology* **61**, 1170-1172.

Arny, D.C., Worf, G.L., Ahrens, R.W., and Lindsey, M.F. (1970) Yellow Leaf Blight of maize in Wisconsin: Its history and the reactions of inbreds and crosses to the inciting fungus (*Phyllosticta* sp.), *Plant Dis. Reptr.* **54**, 281-285.

Bhullar, B., Daly, J., and Rehfeld, D. (1975) Inhibition of dark carbon dioxide fixation and photosynthesis in leaf discs of corn susceptible to the host-specific toxin produced by *Helminthosporium maydis* Race T, *Plant Physiol.* **56**, 1-7.

Bisang, C., Long, P.F., Cortes, J., Westcott, J., Crosby, J., Matharu, A.L., Cox, R.J., Simpson, T.J., Staunton, J., and Leadlay, P.F. (1999) A chain initiation factor common to both modular and aromatic polyketide synthases, *Nature* **401**, 502-505.

Boelker, M., Boehnert, H.U., Braun, K.H., Goerl, J., and Kahmann, R. (1995) Tagging pathogenicity genes in *Ustilago maydis* by restriction enzyme-mediated integration (REMI), *Mol. Gen. Genet.* **248**, 547-552.

Briggs, S.P. and Johal, G.S. (1994) Host-selective toxins and gene-for-gene interactions: Mutually exclusive or faces of a coin?, in K. Kohmoto and O. C. Yoder (eds.), *Host-Specific Toxin: Biosynthesis, Receptor and Molecular Biology*, Faculty of Agriculture, Tottori University, Tottori. pp. 219-226.

Bronson, C.R. (1988) Ascospore abortion in crosses of *Cochliobolus heterostrophus* heterozygous for the virulence locus *Tox1*, *Genome* **30**, 12-18.

Bronson, C.R., Taga, M., and Yoder, O.C. (1990) Genetic control and distorted segregation of T-toxin production in field isolates of *Cochliobolus heterostrophus*, *Phytopathology* **80**, 819-823.

Brosch, G., Ransom, R., Lechner, T., Walton, J.D., and Loidl, P. (1995) Inhibition of maize histone deacetylases by HC toxin, the host-selective toxin of *Cochliobolus carbonum*, *Plant Cell* **7**, 1941-1950.

Chang, H.R. and Bronson, C.R. (1996) A reciprocal translocation and possible insertion(s) tightly associated with host-specific virulence in *Cochliobolus heterostrophus*, *Genome* **39**, 549-557.

Christiansen, S.K., Wirsel, S., Yoder, O.C., and Turgeon, B.G. (1998) The two *Cochliobolus* mating type genes are conserved among species but one of them is missing in *C. victoriae*, *Mycol. Res.* **102**, 919-929.

Churchill, A.C.L., Lu, S.W., Turgeon, B.G., Yoder, O.C., and Macko, V. (1995) Victorin-deficient REMI mutants of *Cochliobolus victoriae* demonstrate a requirement for victorin in pathogenesis, *Fungal Genet. Newsl.* **42A**, 41.

Ciuffetti, L.M., Yoder, O.C., and Turgeon, B.G. (1992) A microbiological assay for host-specific fungal polyketide toxins, *Fungal Genet. Newsl.* **39**, 18-19.

Comstock, J., Martinson, C., and Gengenbach, B. (1972) Characteristics of a host-specific toxin produced by *Phyllosticta maydis*, **62**, 1107.

Comstock, J.C., Martinson, C.A., and Gengenbach, B.G. (1973) Host specificity of a toxin from *Phyllosticta maydis* for Texas cytoplasmically male sterile maize, *Phytopathology* **63**, 1357-1360.

Danko, S.J., Kono, Y., Daly, J.M., Suzuki, Y., Takeuchi, S., and McCrery, D.A. (1984) Structural and biological activity of a host-specific toxin produced by the fungal corn pathogen *Phyllosticta maydis*, *Biochemistry* **23**, 759-766.

Dewey, R.E., Siedow, J.N., Timothy, D.H., and Levings, C.S.I. (1988) A 13-kilodalton maize mitochondrial protein in *E. coli* confers sensitivity to *Bipolaris maydis* toxin, *Science* **239**, 293-295.

DeWit, P.J.G.M., Joosten, M.H.A.J., Honee, G., Vossen, P.J.M.J., Cozijnsen, T.J., Kooman Gersmann, M., and Vogelsang, R. (1995) Molecular aspects of avirulence genes of the tomato pathogen *Cladosporium fulvum*, *Can. J. Bot.* **73**, S490-S494.

Drechsler, C. (1925) Leafspot of maize caused by *Ophiobolus heterostrophus* n. sp., the ascigerous stage of a *Helminthosporium* exhibiting bipolar germination, *J. Agr. Res.* **31**, 701-726.

Drechsler, C. (1927.) An emendation of the description of *Ophiobolus heterostrophus*, *Phytopathol.* **17**, 414.

Drechsler, C. (1934) Phytopathological and taxonomic aspects of *Ophiobolus, Pyrenophora, Helminthosporium*, and a new genus, *Cochliobolus, Phytopathol.* **24**, 953-983.

Forde, B.G., Oliver, R.J.C., and Leaver, C.J. (1978) Variation in mitochondrial translation products associated with male-sterile cytoplasms in maize, *Proc. Natl. Acad. Sci.* **75**, 3841-3845.

Gengenbach, B., Koeppe, D., and Miller, R. (1972) Mitochondrial reactions indicating the influence of kaempferol on corn blight toxin effects, *Plant Physiol.* **49**, 10.

Gracen, V., Forster , M., and Grogan, C. (1971) Reactions of corn (zea mays) genotypes and cytoplasms to *Helminthosporum maydis* toxin, **55**, 938-941.

Gross, M.L., McCrery, D.A., Crow, F., Tomer, D.B., Pope, M.R., Ciuffetti, L.M., Knoche, H.W., Daly, J.M., and Dunkle, L.D. (1982) The structure of the toxin from *Helminthosporium carbonum*, *Tet. Lett.* **23**, 5381-5384.

Hooker, A., Smith, D., Lim, S., and Beckett, J. (1970) Reaction of corn seedlings with male-sterile cytoplasm to *Helminthosporium maydis*, **54**, 708-712.

Hooker, A.L. (1974) Cytoplasmic susceptibility in plant disease, *Ann. Rev. Phytopathol.* **12**, 167-179.

Hopwood, D.A. and Sherman, D.H. (1990) Molecular genetics of polyketides and its comparison to fatty acid biosynthesis, *Annu. Rev. Genet.* **24**, 37-66.

Horwitz, B.A. Sharon, A. Lu, S.W. Ritter, V. Sandrock, T.M. Yoder, O.C. and Turgeon, B.G. (1999.) A G protein alpha subunit from *Cochliobolus heterostrophus* involved in mating and appressorium formation., *Fungal Genet. Biol.* **26**, 19-32.

Huang, J., Lee, S.H., Lin, C., Medici, R., Hack, E., and Myers, A.M. (1990) Expression in yeast of the T-URF13 protein from Texas male-sterile maize mitochondria confers sensitivity to methomyl and to Texas-cytoplasm-specific fungal toxins, *EMBO. J.* **9**, 339-347.

Johal, G.S. and Briggs, S.P. (1992) Reductase activity encoded by the *HM1* disease resistance gene in maize, *Science* **258**, 985-987.

Karr, D., Karr, A., and Strobel, G. (1975) The toxins of *Helminthosprium maydis* race T a calorimetric determinaion of the toxins, their appearance in culture and in infected plants, **55**, 727-730.

Keller, N.P., Bergstrom, G.C., and Yoder, O.C. (1990) Effects of genetic transformation on fitness of *Cochliobolus heterostrophus*, *Phytopathology* **80**, 1166-1173.

Kim, Y.J. and Baker, B.S. (1993) Isolation of RRM-type RNA-binding protein genes and the analysis of their relatedness by using a numerical approach., *Molecular and Cellular Biology* **13**, 174-183.

Kimura, M., Kamakura, T., Tao, Q.Z., Kaneko, I., and Yamaguchi, I. (1994) Cloning of the blasticidin S deaminase gene (BSD) from *Aspergillus terreus* and its use as a selectable marker for *Schizosaccharomyces pombe* and *Pyricularia oryzae*, *Mol. Gen. Genet.* **242**, 121-129.

Klittich, C.R.J., and Bronson, C.R. (1986) Reduced fitness associated with *Tox1* of *Cochliobolus heterostrophus*, *Phytopathology* **76**, 1294-1298.

Kodama, M., Yoder, O.C., and Turgeon, B.G. (1999) The translocation-associated "*Tox1* locus" of *Cochliobolus heterostrophus* is two genetic elements on two different chromosomes, *Genetics* **151**, 585-596.

Kohmoto, K. and Otani, H. (1991) Host recognition by toxigenic plant pathogens, *Experientia* **47**, 755-764.

Kohmoto, K. and Yoder, O.C. (1997) Molecular Genetics of Host-Specific Toxins in Plant Disease, in K. Kohmoto and O. C. Yoder (eds.), *Molecular Genetics of Host-Specific Toxins in Plant Disease*, Kluwer, Dordrecht, **13**, pp 1-413

Kono, Y. and Daly, J.M. (1979) Characterization of the host-specific pathotoxin produced by *Helminthosporium maydis* race T affecting corn with Texas male sterile cytoplasm, *Bioorg. Chem.* **8**, 391-397.

Kono, Y., Danko, S.J., Suzuki, Y., Takeuchi, S., and Daly, J.M. (1983) Structure of the host-specific pathotoxins produced by *Phyllosticta maydis*, *Tetrahedron Lett.* **24**, 3803-3806.

Kono, Y., Takeuchi, S., Kawarada, A., Daly, J.M., and Knoche, H.W. (1981) Studies on the host-specific pathotoxins produced in minor amounts by *Helminthosporium maydis* race T, *Bioorg. Chem.* **10**, 206-218.

Korth, K.L. and Levings, C.S. (1993) Baculovirus expression of the maize mitochondrial protein URF13 confers insecticidal activity in cell cultures and larvae, *Proc Natl Acad Sci USA* **90**, 3388-3392.

Kuspa, A. and Loomis, W.F. (1992) Tagging developmental genes in *Dictyostelium* by restriction enzyme-mediated integration of plasmid DNA, *Proc. Natl. Acad. Sci. USA* **89**, 8803-8807.

Leach, J., Lang, B.R., and Yoder, O.C. (1982) Methods for selection of mutants and *in vitro* culture of *Cochliobolus heterostrophus*, *J. Gen. Microbiol.* **128**, 1719-1729.

Leonard, K.J. (1973) Association of mating type and virulence in *Helminthosporium maydis*, and observations on the origin of the race T population in the United States, *Phytopathology* **63**, 112-115.

Leonard, K.J. (1977) Races of *Bipolaris maydis* in the Southeastern U. S. from 1974-1976, *Plant Dis. Reptr.* **61**, 914-915.

Levings, C.S. (1990) The Texas cytoplasm of maize: Cytoplasmic male sterility and disease susceptibility, *Science* **250**, 942-947.

Levings, C.S. and Siedow, J.N. (1992) Molecular basis of disease susceptibility in the Texas cytoplasm of maize, *Plant Mol. Biol.* **19**, 135-147.

Levings, C.S.I., Rhoads, D.M., and Siedow, J.N. (1995) Molecular interactions of *Bipolaris maydis* T-toxin and maize, *Can. J. Bot.* **73**, S483-S489.

Lim, S. and Hooker, A. (1971) Southern corn leaf blight: genetic control of pathogenicity and toxin production in race T and race O of *Cochliobolus heterostrophus*, **69**, 115-117.

Lim, S. and Hooker, A. (1972) A preliminary characterization of *Helminthosporium maydis* toxins, **56**, 805-807.

Litzenberger, S.C. (1949) Nature of susceptibility to *Helminthosporium victoriae* and resistance to *Puccinia coronata* in Victoria oats, *Phytopathology* **39**, 300-318.

Liu, Y.G. and Whittier, F. (1995) Thermal Asymmetric Interlaced PCR: Automatable amplification and sequencing of insert end fragment from P1 and YAC clones for chromosome walking, *Genomics* **25**, 674-681.

Lu, S.W. 1998. Molecular-genetic analysis of general and specific pathogenesis factors in *Cochliobolus heterostrophus*, PhD Thesis, Cornell Univ., Ithaca, 344 pages.

Lu, S.W., Lyngholm, L., Yang, G., Bronson, C., Yoder, O.C., and Turgeon, B.G. (1994) Tagged mutations at the *Tox1* locus of *Cochliobolus heterostrophus* using restriction enzyme-mediated integration, *Proc. Natl. Acad. Sci. USA* **91**, 12649-12653.

Macko, V., Wolpert, T.J., Acklin, W., Jaun, B., Seibl, J., Meili, J., and Arigoni, D. (1985) Characterization of victorin C, the major host-selective toxin from *Cochliobolus victoriae*: structure of degradation products, *Experientia* **41**, 1366-1370.

Mallet, L., Bussereau, F., and Jacquet, M. (1995) A43 cntdot 5 kb segment of yeast chromosome XIV, which contains MFA2, MEP2, CAP/SRV2, NAM9, FKB1/FPR1/RBP1, MOM22 and CPT1, predicts an adenosine deaminase gene and 14 new open reading frames, *Yeast* **11**, 1195-1209.

Meehan, F.L. and Murphy, H.C. (1946) A new *Helminthosporium* blight of oats, *Science* **104**, 413-414.

Meeley, R.B., Johal, G.S., Briggs, S.P., and Walton, J.D. (1992) A biochemical phenotype for a disease resistance gene of maize, *Plant Cell* **4**, 71-77.

Meeley, R.B. and Walton, J.D. (1991) Enzymatic detoxification of HC-toxin, the host-selective cyclic peptide from *Cochliobolus carbonum*, *Plant Physiol.* **97**, 1080-1086.

Mercado, A.C. and Lantican, R.M. (1961) The susceptibility of cytoplasmic male-sterile lines of corn to *Helminthosporium maydis*, *Philippine Agriculturist* **45**, 235-243.

Miller, P., Wallin, J., and Hyre, R. (1970) Plans for forecasting corn blight epidemics, **54**, 1134-1136.

Miller, R.J. and Koeppe, D.E. (1971) Southern Corn Leaf Blight: Susceptible and resistant mitochondria, *Science* **173**, 67-69.

Moore, W. (1970) Origin and spread of southern corn leaf blight in 1970, **54**, 1104-1108.

Mukunya, D.M. and Boothroyd, C.W. (1973) *Mycosphaerella zeae-maydis* sp. n., the sexual stage of *Phyllosticta maydis*, *Phytopathology* **63**, 529-532.

Mullin, P.G., Turgeon, B.G., and Yoder, O.C. (1993) Complementation of *Cochliobolus heterostrophus trp-* mutants produced by gene replacement, *Fungal Genet. Newsl.* **40**, 51-53.

Murphy, H. and Meehan, F. (1946) Reaction of oat varieties to a new species of *Helminthosporium*, *Phytopathology* **36**, 407.

Navarre, D.A. and Wolpert, T.J. (1995) Inhibition of the glycine decarboxylase multienzyme complex by the host-selective toxin victorin, *Plant Cell* **7**, 463-471.

Navarre, D.A. and Wolpert, T.J. (1999a) Effects of light and CO2 on victorin-induced symptom development in oats., *Physiological and Molecular Plant Pathology* **55**, 237-242.

Navarre, D.A. and Wolpert, T.J. (1999b) Victorin induction of an apoptotic/senescence-like response in oats., *Plant Cell* **11**, 237-249.

Nelson, R. (1959) Genetics of *Cochliobolus heterostrophus* i. variability in degree of compatibility, *Mycologia* **51**, 18-23.

Nelson, R. and Hebert, T. (1960) The inheritance of pathogenicity and mating type crosses of *Helminthosporium carbonum* and *Helminthosporium victoriae*, **50**, 649.

Nelson, R.R. (1957) Heterothallism in *Helminthosporium maydis*, *Phytopathology* **47**, 191-192.

Orillo, F. (1952) Leafspot of maize caused by *Helminthosporium maydis*, **36**, 327-392.

Panaccione, D.G., Scott-Craig, J.S., Pocard, J.A., and Walton, J.D. (1992) A cyclic peptide synthetase gene required for pathogenicity of the fungus *Cochliobolus carbonum* on maize, *Proc. Natl. Acad. Sci. USA* **89**, 6590-6594.

Pope, M.R., Ciuffetti, L.M., Knoche, H.W., McCrery, D., Daly, J.M., and Dunkle, L.D. (1983) Structure of the host-specific toxin produced by *Helminthosporium carbonum*, *Biochemistry* **22**, 3502-3506.

Proctor, R.H., Desjardins, A.E., Plattner, R.D., and Hohn, T.M. (1999) A polyketide synthase gene required for biosynthesis of fumonisin mycotoxins in *Gibberella fujikuroi* mating population A, **27**, 100-112.

Punithalingam, E. (1990) CMI Descriptions of Fungi and Bacteria No. 1015: *Mycosphaerella zeae-maydis*, *Mycopathologia* **112**, 49-50.

Ransom, R.F. and Walton, J.D. (1997) Purification and characterization of extracellular beta-xylosidase and alpha-arabinosidase from the plant pathogenic fungus *Cochliobolus carbonum*, *Carbohydrate Research*. **297**, 357-364.

Rohe, M., Gierlich, A., Hermann, H., Hahn, M., Schmidt, B., Rosahl, S., and Knogge, W. (1995) The race-specific elicitor, NIP1, from the barley pathogen, *Rhynchosporium secalis*, determines avirulence on host plants of the *Rrs1* resistance genotype, *EMBO J.* **14**, 4168-4177.

Rose, M.S. 1996. Molecular genetics of polyketide toxin production in *Cochliobolus heterostrophus*. PhD Thesis, Cornell Univ., Ithaca, 217 pages.

Rose, M.S., Yoder, O.C., and Turgeon, B.G. (1996) A decarboxylase required for polyketide toxin production and high virulence by *Cochliobolus heterostrophus*, *8th Int. Symp. Mol. Plant-Microbe Int., Knoxville, p. J-49*

Scheffer, R.P., Nelson, R.R., and Ullstrup, A.J. (1967) Inheritance of toxin production and pathogenicity in *Cochliobolus carbonum* and *Cochliobolus victoriae*, *Phytopathology* **57**, 1288-1291.

Scheffer, R.P. and Ullstrup, A.J. (1965) A host-specific toxic metabolite from *Helminthosporium carbonum*, *Phytopathology* **55**, 1037-1038.

Scheffer, R.P. and Yoder, O.C. (1972) Host-specific toxins and selective toxicity, in R. K. S. Wood, A. Ballio and A. Graniti (eds.), *Phytotoxins in Plant Diseases*, Academic Press, London. pp. 251-272.

Scheifele, G., Whitehead, W., and Rowe, C. (1970) Increased susceptibility to southern leaf spot (*Helminthosporium maydis*) in inbred lines and hybrids of maize with Texas male-sterile cytoplasm, **54**, 501-503.

Scheifele, G.L. and Nelson, R.R. (1969) The occurrence of *Phyllosticta* leaf spot of corn in Pennsylvania, *Plant Dis. Reptr.* **53**, 186-189.

Schiestl, R.H. and Petes, T.D. (1991) Integration of DNA fragments by illegitimate recombination in *Saccharomyces cerevisiae*, *Proc. Natl. Acad. Sci. USA* **88**, 7585-7589.

Scott-Craig, J.S., Panaccione, D.G., Pocard, J.A., and Walton, J.D. (1992) The cyclic peptide synthetase catalyzing HC-toxin production in the filamentous fungus *Cochliobolus carbonum* is encoded by a 15.7-kilobase open reading frame, *J. Biol. Chem.* **267**, 26044-26049.

Scott-Craig, J.S., Pitkin, J.W., Ahn, J.H., and Walton, J.D. 1995. Molecular genetic analysis of the *TOX2* locus in *Cochliobolus carbonum*. Tottori Japan, abstract:

Sivanesan, A. (1984) The Bitunicate Ascomycetes and Their Anamorphs, Strauss & Cramer, Hirschberg.

Smedegard-Peterson, V. and Nelson, R. (1969) The production of a host-specific pathotoxin by *Cochliobolus heterostrophus*, **47**, 951-958.

Smith, D., Hooker, A., and Lim, S. (1970) Physiologic races of *Helminthosporium maydis*, **54**, 819-822.

Straubinger, B., Straubinger, E., Wirsel, S., Turgeon, G., and Yoder, O. (1992) Versatile fungal transformation vectors carrying the selectable *bar* gene of *Streptomyces hygroscopicus*, *Fungal Genet. Newsl.* **39**, 82-83.

Sugimoto, K., Matsumoto, K., Kornberg, R.D., Reed, S.I., and Wittenberg, C. (1995) Dosage suppressors of the dominant G1 cyclin mutant CLN3-2: Identification of a yeast gene encoding a putative RNA/ssDNA binding protein, *Molecular & General Genetics* **248**, 712-718.

Suzuki, Y., Danko, S.J., Daly, J.M., Kono, Y., Knoche, H.W., and Takeuchi, S. (1983) Comparison of activities of the host-specific toxin of *Helminthosporium maydis* race T and a synthetic C-41 analog, *Plant Physiol.* **73**, 440-444.

Suzuki, Y., Tegtmeier, K.J., Daly, J.M., and Knoche, H.W. (1982) Analogs of host-specific phytotoxin produced by *Helminthosporium maydis* race T. II. Biological activities, *Bioorg. Chem.* **11**, 313-321.

Sweigard, J. (1996) A REMI primer for filamentous fungi, *IS-MPMI Reporter* **Spring**, 3-5.

Taga, M., Bronson, C.R., and Yoder, O.C. (1985) Nonrandom abortion of ascospores containing alternate alleles at the *Tox1* locus of the fungal plant pathogen *Cochliobolus heterostrophus*, *Can. J. Genet. Cytol.* **27**, 450-456.

Tasma, I.M. and Bronson, C.R. (1998) Genetic mapping of telomeric DNA sequences in the maize pathogen *Cochliobolus heterostrophus*, *Current Genetics* **34**, 227-233.

Tegtmeier, K.J., Daly, J.M., and Yoder, O.C. (1982) T-toxin production by near-isogenic isolates of *Cochliobolus heterostrophus* races T and O, *Phytopathology* **72**, 1492-1495.

Toda, T., Iida, A., Miwa, T., Nakamura, Y., and Imai, T. (1994) Isolation and characterization of a novel gene encoding nuclear protein at a locus (D11S636) tightly linked to multiple endocrine neoplasia type 1 (MEN1), *Human Molecular Genetics* **3**, 465-470.

Turgeon, B.G. and Berbee, M.L. (1998) Evolution of pathogenic and reproductive strategies in *Cochliobolus* and related genera, in K. Kohmoto and O. C. Yoder (eds.), *Molecular Genetics of Host-Specific Toxins in Plant Disease*, Kluwer, Dordrecht. pp. 153-163.

Turgeon, B.G., Bohlmann, H., Ciuffetti, L.M., Christiansen, S.K., Yang, G., Schafer, W., and Yoder, O.C. (1993) Cloning and analysis of the mating type genes from *Cochliobolus heterostrophus*, *Mol. Gen. Genet.* **238**, 270-284.

Turgeon, B.G., Garber, R.C., and Yoder, O.C. (1985) Transformation of the fungal maize pathogen *Cochliobolus heterostrophus* using the *Aspergillus nidulans amdS* gene, *Mol. Gen. Genet.* **201**, 450-453.

Turgeon, B.G., Garber, R.C., and Yoder, O.C. (1987) Development of a fungal transformation system based on selection of sequences with promoter activity, *Mol. Cell. Biol.* **7**, 3297-3305.

Turgeon, B.G., Kodama, M., Yang, G., Rose, M.S., Lu, S.W., and Yoder, O.C. (1995) Function and chromosomal location of the *Cochliobolus heterostrophus Tox1* locus, *Can. J. Bot.* **73**, S1071-S1076.

Turgeon, B.G., MacRae, W.D., Garber, R.C., Fink, G.R., and Yoder, O.C. (1986) A cloned tryptophan synthesis gene from the Ascomycete *Cochliobolus heterostrophus* functions in *Escherichia coli,* yeast and *Aspergillus nidulans, Gene* **42**, 79-88.

Tzeng, T.H., Lyngholm, L.K., Ford, C.F., and Bronson, C.R. (1992) A restriction fragment length polymorphism map and electrophoretic karyotype of the fungal maize pathogen *Cochliobolus heterostrophus*, *Genetics* **130**, 81-96.

Ullstrup, A.J. (1944) Further studies on a species of *Helminthosporium* parasitizing corn, *Phytopathology* **34**, 214-222.

Ullstrup, A.J. (1970) History of Southern Corn Leaf Blight, *Plant Dis. Reptr.* **54**, 1100-1102.

Ullstrup, A.J. (1972) The impacts of the Southern Corn Leaf Blight epidemics of 1970-1971, *Ann. Rev. Phytopathol.* **10**, 37-50.

Vonallmen, J.M., Rottmann, W.H., Gengenbach, B.G., Harvey, A.J., and Lonsdale, D.M. (1991) Transfer of methomyl and HmT-toxin sensitivity from T-cytoplasm maize to tobacco, *Mol. Gen. Genet.* **229**, 405-412.

Walton, J. and Earle, E. (1984) Characterization of the host-specific phytotoxin victorin by high-pressure liquid chromatography, *Plant Science Letters* **34**, 231-238.

Walton, J.D. (1996) Host-selective toxins: agents of compatibility, *Plant Cell* **8**, 1723-1733.

Walton, J.D., Akimitsu, K., Ahn, J.H., and Pitkin, J.W. (1994) Towards an understanding of the *Tox2* gene of *Cochliobolus carbonum*, in K. Kohmoto and O. C. Yoder (eds.), *Host-Specific Toxin: Biosynthesis, Receptor and Molecular Biology,* Faculty of Agriculture, Tottori University, Tottori. pp. 227-237.

Walton, J.D., Earle, E.D., and Gibson, B.W. (1982) Purification and structure of the host-specific toxin from *Helminthosporium carbonum* race 1, *Biochem. Biophys. Res. Comm.* **107**, 785-794.

Wheeler, H. (1977) Ultrastructure of penetration by *Helminthosporium maydis, Physiol. Plant. Pathol,* **11**, 171-178.

Wilson, SM., Datar, KV., Paddy, M.R., Swedlow, J.R., and Swanson, M.S. (1994) Characterization of nuclear polyadenylated RNA-binding proteins in *Saccharomyces cerevisiae, Journal of Cell Biology* **127**, 1173-1184.

Wirsel, S., Turgeon, B.G., and Yoder, O.C. (1996) Deletion of the *Cochliobolus heterostrophus* mating type (*MAT*) locus promotes function of *MAT* transgenes, *Curr. Genet.* **29**, 241-249.

Wise, R. P., Bronson, C.R., Schnable, P.S., and Horner, H.T. (1999) The genetics, pathology and molecular biology of T-cytoplasm male sterility in maize, *Advances in Agronomy* **65**, 79-130.

Wolpert, T. J., Macko, V., Acklin, W., Jaun, B., Seibl, J., Meili, J., and Arigoni, D. (1985) Structure of victorin C, the major host-selective toxin from *Cochliobolus victoriae, Experientia* **41**, 1524-1529.

Wolpert, T.J., Navarre, D.A., and Lorang, J.M. (1998) Victorin-induced oat cell death, in K. Kohmoto and O. C. Yoder (eds.), *Molecular Genetics of Host-Specific Toxins in Plant Disease,* Kluwer, Dordrecht. pp. 105-114.

Wolpert, T.J., Navarre, D.A., Moore, D.L., and Macko, V. (1994) Identification of the 100-kD victorin binding protein from oats, *Plant Cell* 6, 1145-1155.

Wrehlke, C., Schmitt-Wrede, H.P., Qiao, Z., and Wunderlich, F. (1997) Enhanced expression in spleen macrophages of the mouse homolog to the human putative tumor suppressor gene ZFM1, *DNA and Cell Biology* 16, 761-767.

Yang, G., Rose, M.S., Turgeon, B.G., and Yoder, O.C. (1996) A polyketide synthase is required for fungal virulence and production of the polyketide T-toxin, *Plant Cell* 8, 2139-2150.

Yang, G., Turgeon, B.G., and Yoder, O.C. (1994) Toxin-deficient mutants from a toxin-sensitive transformant of *Cochliobolus heterostrophus*, *Genetics* 137, 751-757.

Yoder, O. and Mukunya, D. (1972) A host-specific toxic metabolite produced by *Phyllosticta maydis*, *Phytopathology* 62, 799.

Yoder, O., Payne, G., Gregory, P., and Earle, E. (1976) Relative sensitivities of bioassays for *Helminthosporium maydis* race T toxin, *Proc Amer Phytopathol Soc* 3, 281.

Yoder, O.C. (1973) A selective toxin produced by *Phyllosticta maydis*, *Phytopathology* 63, 1361-1365.

Yoder, O.C. (1976) Evaluation of the role of *Helminthosporium maydis* race T toxin in Southern Corn Leaf Blight, in K. Tomiyama, J. M. Daly, I. Uritani, H. Oku and S. Ouchi (eds.), *Biochemistry and Cytology of Plant Parasite Interaction*, Elsevier, New York. pp. 16-24.

Yoder, O.C. (1980) Toxins in pathogenesis, *Ann. Rev. Phytopathol.* 18, 103-129.

Yoder, O.C. (1988) *Cochliobolus heterostrophus*, cause of Southern Corn Leaf Blight, in G. S. Sidhu (eds.), *Genetics of Plant Pathogenic Fungi*, Academic Press, San Diego. pp. 93-112.

Yoder, O.C. (1997) A mechanistic view of the fungal/plant interaction based on host-specific toxin studies, in K. Kohmoto and O. C. Yoder (eds.), *Molecular Genetics of Host-Specific Toxins in Plant Disease*, Kluwer, Dordrecht (in press). pp.

Yoder, O.C. and Gracen, V.E. (1975) Segregation of pathogenicity types and host-specific toxin production in progenies of crosses between races T and O of *Helminthosporium maydis* (*Cochliobolus heterostrophus*), *Phytopathology* 65, 273-276.

Yoder, O.C. and Gracen, V.E. (1977) Evaluation of a chemical method for assay of *Helminthosporium maydis* race T toxin, *Plant Physiol.* 59, 792-794.

Yoder, O.C., Macko, V., Wolpert, T.J., and Turgeon, B.G. (1997) *Cochliobolus* spp. and their host-specific toxins, in G. Carroll and P. Tudzynski (eds.), *The Mycota Vol. 5: Plant Relationships, Part A*, Springer-Verlag, Berlin. pp. 145-166.

Yoder, O.C., Payne, G.A., Gregory, P., and Gracen, V.E. (1977) Bioassays for detection and quantification of Helminthosporium maydis race T toxin: A comparison, *Physiol Plant Pathol* 10, 237-245.

Yoder, O.C. and Turgeon, B.G. (1996) Molecular-genetic evaluation of fungal molecules for roles in pathogenesis in plants, *J. Genet.* 75, 425-440.

Yoder, O.C., Turgeon, B.G., Ciuffetti, L.M., and Schäfer, W. (1989) Genetic analysis of toxin production by fungi, in A. Graniti, R. Durbin and A. Ballio (eds.), *Phytotoxins and Plant Pathogenesis*, Springer-Verlag, Berlin. pp. 43-60.

Yoder, O. C., Valent, B., and Chumley, F. (1986) Genetic nomenclature and practice for plant pathogenic fungi, *Phytopathology* 76, 383-385.

Yoder, O.C., Yang, G., Adam, G., DiazMinguez, J.M., Rose, M., and Turgeon, B.G. (1993) Genetics of polyketide toxin biosynthesis by plant pathogenic fungi, in R. H. Baltz, G. D. Hegeman and P. L. Skatrud (eds.), *Industrial Microorganisms: Basic and Applied Molecular Genetics*, Amer. Soc. Microbiol., Washington DC. pp. 217-225.

Yoder, O.C., Yang, G., Rose, M.S., Lu, S.W., and Turgeon, B.G. (1994) Complex genetic control of polyketide toxin production by *Cochliobolus heterostrophus*, in M. J. Daniels, J. A. Downie and A. E. Osbourn (eds.), *Adv. Mol. Genet. Plant-Microbe Int.*, Kluwer, Dordrecht. pp. 223-230.

Yu, T. (1933) Studies on *Helminthosporium* leaf spot of maize, 3, 273-318.

Yun, S.H. 1998. Molecular genetics and manipulation of pathogenicity and nating determinants in *Mycosphaerella zeae-maydis* and *Cochliobolus heterostrophus*. PhD Thesis, Cornell Univ., Ithaca, 285 pages.

Yun, S.H., Turgeon, B.G., and Yoder, O.C. (1998) REMI-induced mutants of *Mycosphaerella zeae-maydis* lacking the polyketide PM-toxin are deficient in pathogenesis to corn, *Physiol. Mol. Plant Pathol.* 52, 53-66.

Zhu, X. 1999. Molecular analysis of loci controlling T-toxin biosynthesis in *Cochliobolus heterostrophus*. PhD Thesis, Cornell Univ., Ithaca, 174 pages.

COLLETOTRICHUM

Martin B. Dickman
University of Nebraska
Department of Plant Pathology
Lincoln, Nebraska 68583
USA

1. INTRODUCTION

The genus *Colletotrichum* represents a large number of economically important Ascomycete fungi which collectively cause anthracnose disease or leaf blights on all significant agricultural crops and ornamental plants around the world. *Colletotrichum* spp. often serve as models in studies ranging from pathogenic development and differentiation to plant-microbe interactions. During establishment and colonization of host plants, members of this genus uniquely acquire nutrients via biotrophy and necrotrophy. Thus in a single interaction, these fungal pathogens exhibit two modes of nutrient acquisition; initially nutrients are acquired from living cells after which a necrotrophic phase is initiated and the same fungus then obtains nutrients from dead cells which have been killed by the fungus. A given species of this genus can exhibit both of these strategies separately or concurrently. In addition, these fungi develop a series of specialized infection related structures including germ tubes, appressoria, primary intracellular hyphae and secondary necrotrophic hyphae. *Colletotrichum* species thus are experimentally attractive organisms to study the molecular, biochemical and cellular basis of fungal pathogenicity, development and signal transduction (see Bailey and Jeger, 1992; Prusky et al., 1999).

Despite significant developments over the last ten years, the taxonomy of *Colletotrichum* is in a state of flux. Many uncertainties exist in the systematics of fungal pathogens from this genus, depending on the taxonomic "guides" the number of species can range from 29 to over 700 (von Arx, 1957; Sutton, 1992). One of the most confusing species is *C. gloeosporioides*. For example, 594 species of *Colletotrichum* were reclassified by Von Arx as synonyms of *C. gloeosporioides*. Identification has relied primarily on morphological characteristics of conidia, vegetative and sexual structures, and on host specificity and cultural characteristics (von Arx, 1957). Unfortunately conidial morphology and colony characteristics vary even within isolates. Consequently, classical methods have been supplemented with biochemical and molecular technologies coupled with classical methods

J. W. Kronstad (ed.), Fungal Pathology, 127–148.

to differentiate *Colletotrichum* species. For a more complete discussion of this topic, see Sherriff et al. (1994).

This review focuses on current research into the infection process of *Colletotrichum* species with overall emphasis on how signal transduction influences prepenetration morphogenesis including conidiation, conidial attachment, conidial germination, germ tube formation, and appressorial differentiation. Post-penetration biotrophic and necrotrophic growth patterns will also be described. In addition, intriguing studies involving host specificity and chromosome transfer will be summarized.

2. APPROACHES TO IDENTIFYING *COLLETOTRICHUM* DIFFERENTIATION AND PATHOGENICITY GENES AND PROTEINS

Differentiation of *Colletotrichum* infection structures and *Colletotrichum*-plant interactions have been studied by several approaches. Examples of these approaches will be illustrated throughout this chapter. Genes which are differentially expressed during infection have been identified by subtractive hybridization, differential screening and differential display reverse transcriptase-PCR. For example, several genes expressed during appressorium formation by *C. gloeosporioides* have been cloned by these techniques (Hwang et al., 1995). Differential screening of a cDNA library prepared from nitrogen-starved axenic cultures of *C. gloeosporioides* led to the isolation of a glutamine synthetase gene that is upregulated during infection of *Stylosanthes guianensis* (Stephenson et al., 1997).

Complementation of genes in model organisms is a powerful technique for characterizing signaling protein function. Two notable signaling proteins from fungal plant pathogens which complement defects in "tester" organisms are the *tb3* protein kinase gene from *C. trifolii* (Buhr et al., 1996), and the *PMK1* MAP kinase gene from *M. grisea* (Xu and Hamer, 1996). The TB3 kinase complemented the colonial phenotype in *N. crassa* (due to disruption of the *cot-1* protein kinase gene) and PMK1 complemented a mating defect in *S. cerevisiae* (due to deletion of the *FUS3/KSS1* MAP kinase genes).

A serine/threonine kinase gene *(clk1)* expressed by *C. lindemuthianum* and required for pathogenicity on *P. vulgaris* has been identified by insertional mutagenesis (Dufresne et al., 1998). The gene is constitutively expressed, but the mutant cannot penetrate. Mutants in *C. graminicola* with unpigmented spores or weakened cell walls have recently been identified by restriction enzyme mediated integration (REMI) (Epstein et al., 1998).

Developmentally regulated proteins in *C. lindemuthianum* have been identified and characterized with monoclonal antibodies (Mabs) (Green et al., 1995). Mabs were raised either to homogenates of germlings grown in liquid culture (Pain et al., 1992), or to infection structures isolated from bean leaves (Pain et al., 1994a, b). Recently, one of these Mabs (UB25), was used as a probe for expression cloning and a gene was isolated encoding a fungal glycoprotein present at the biotrophic interface in host cells (Perfect et al., 1998, see later).

3. HOST INFECTION STRATEGIES

Colletotrichum spp. can establish a compatible relationship with its host either by subcuticular growth or intracellularly. The initial stages of *Colletotrichum* infection generally include the following: conidia adhere to, and germinate on plant surfaces, produce germ-tubes which differentiate to form melanized appressoria which are required for breaching the cuticular barrier. Following penetration, a hyphal network forms and both inter- and intracellular hyphae spread rapidly throughout the tissue, killing in advance of mycelial spread (Bailey et al., 1992). Examples of subcuticular, necrotrophic pathogens include *C. capsici* on cowpea (*Vigna unguiculata*: Bailey et al., 1992; Pring et al., 1995) and cotton (*Gossypium hirsutum L.*; Roberts and Snow, 1984), and *C. circinans* on onion (Walker, 1921).

Commonly, *Colletotrichum* species become established by intracellular colonization. This is illustrated by the infection process of *C. lindemuthianum* on bean. The mycelia grow within the cell lumen without perturbing the host membrane; in other words, growth is between plant plasma membranes and plant cell walls. These intracellular biotrophic hyphae then produce secondary necrotrophic hyphae (O'Connell et al., 1985; Bailey et al., 1992; Latunde-Dada et al., 1996). This initial feeding on living host cells prior to subsequent switching to necrotrophy is the reason that *Colletotrichum* species are considered hemibiotrophic or facultative biotrophs. During initial colonization by the fungus, whether biotrophic or not, the host plant appears not to recognize the pathogen , or the pathogen avoids host recognition; in either case, there is no specific resistance response. Biotrophic stages are well-characterized in the following interactions: *C. lindemuthianum* and bean (O'Connell et al., 1985; O'Connell, 1987), *C. truncatum* and pea (O'Connell et al., 1993), *C. destructivum* and cowpea (Latunde-Dada et al., 1996) and *C. sublineolum* and sorghum (Wharton and Julian, 1996).

Host colonization and pathogenesis are well studied for several species for *Colletotrichum* and appears to be similar in the genera (O'Connell et al., 1985; Green et al., 1995; Kolattukudy et al., 1995). Initiation of disease in fungal-plant interactions requires adhesion of spores to host plants. The spores of *Colletotrichum* species can be dispersed by rain and when a suitable host substrate is encountered, rapidly adhere to aerial parts of plants (Nicholson, 1992; Nicholson, 1996; Mercure et al., 1994a). Conidia are produced in acervuli. These acervuli are embedded in a mucilaginous matrix composed of high molecular weight glycoproteins (Nicholson, 1992). This mucilage also contains germination inhibitors (common in *Colletotrichum spp.*) as well as enzymes including cutinases, lipases and DNases (Nicholson, 1992).

The initial attachment of *Colletotrichum* spores to the plant cuticle appears to be a requisite for colonization and involves hydrophobic interactions, as also occurs for many other plant pathogenic fungi (Nicholson and Epstein, 1991; Mercure et al., 1994a). Proteolytic enzymes inhibited conidial adhesion in *C. musae* and *C. graminicola* suggesting the involvement of surface proteins in spore attachment. In addition, it appears that spore adhesion in *C.*

lindemuthianum, C. musae and *C. graminicola* also requires active metabolism, including protein synthesis (Young and Kauss, 1984; Sela-Buurlage et al., 1991; Mercure et al., 1994b).

4. SIGNAL TRANSDUCTION

In *Colletotrichum* spp., as well as in other fungi, activation of a series of intracellular signal transduction pathways in response to external stimuli stimulates cell growth, differentiation, and infection related development. A growing body of evidence shows that both the plant and pathogen produced signals determine the eventual outcome of a parasitic relationship (e.g. Dickman, 1999). Each participant recognizes and utilizes these signal molecules leading to the compatible/incompatible relationship. Moreover, such signal elements are common in species ranging from mammals to yeast and bacteria, as well as phytopathogenic fungi. The successful, functional interchange of kinase genes among organisms (e.g. Kincaid, 1991; King, et al., 1990; Neimann, 1993; Buhr, et al., 1996, Yang and Dickman, 1999b) illustrates the conservation of these genes. In particular, reports of the involvement of ser/thr protein kinases and phosphatases as well as G-proteins in fungal morphogenesis and pathogenicity are accumulating (e.g. Dickman and Yarden, 1999). It is evident that inappropriate regulation of signal transduction pathways can interfere with infection related morphogenesis and pathogenicity (see also *Magnaporthe, Cryphonectria* chapters).

Molecular communication begins as soon as a fungal conidium lands on a plant surface. Since this contact is often on the plant cuticle, cuticular components are ideally located to influence plant-fungus interactions. The consequences of this early interaction can be crucial for the survival of both pathogen and host.

4.1. Germination and Appressorium Formation

Colletotrichum spores sense both physical and chemical signals from the plant surface triggering germination and differentiation into appressoria. Chemical inducers have been identified for *C. gloeosporioides* infecting avocado and *C. musae* infecting banana (Podila et al., 1993; Kolattukudy et al., 1995). The surface wax of the host avocado selectively triggers germination and appressorium formation by *C. gloeosporioides* spores. The signal is specific since the waxes of other plants cannot substitute for avocado wax. Ethylene, the fruit ripening hormone induces germination and appressorial formation in both *C. gloeosporioides* and *C. musae*, pathogens that attack climacteric fruits, but not in other *Colletotrichum* species that normally infect non-climacteric fruits. Thus, spores landing on developing fruit germinate and penetrate after triggering by the wax signal, but remain latent until the fruit ripens (Kolattukudy et al., 1995).

Conidia of *Colletotrichum* species require contact with a hard surface before they can recognize host signals. For example, *C. gloeosporioides* and *C. trifolii* germinate on soft agar but do not differentiate to form appressoria, suggesting physical contact is required for appropriate morphogenesis. Hard surface contact may be analogous to the touch response in higher plants where touch stimulates transcriptional activation of calmodulin *(cam)*-like genes. The highly conserved *cam* gene has been cloned from *C. gloeosporioides* and *C. trifolii* by PCR (Kim et al., 1998; Warwar and Dickman, submitted). At the nucleotide level the *cam*-encoded polypeptide of *C. gloeosporioides* was 87% and 88% identical with *N. crassa* and *C. trifolii* proteins respectively. The *C. trifolii* sequence showed 97% identity with the *cam*-encoded protein from *N. crassa*. Southern analysis indicated a single *cam* gene in both the *C. gloeosporioides* and *C. trifolii* genomes. In *C. gloeosporioides,* the *cam* transcript levels increased 11-fold reaching a maximum at 2 hr after surface contact and subsequently decreased. A CaM antagonist, compound 48/80, severely inhibited both germination and appressorium formation in both *C. gloeosporioides* and *C. trifolii* consistent with the importance of calmodulin (Kim et al., 1998, Warwar and Dickman, 1996). In addition, antisense expression of the *C. trifolii cam* transcript impaired appressorium development (Warwar and Dickman, submitted). Cell signaling pathways operating in spore germination and appressorium formation in *Colletotrichum* have been studied with inhibitors or by monitoring expression of genes encoding putative signaling components, e.g. calmodulin (CaM) and protein kinases. In the *C. gloeosporioides* system the protein kinase inhibitors H7 and genistein inhibit ethylene-indued appressorium formation and phosphorylation of proteins, suggesting a possible role for the latter in differentiation (Flaishman et al., 1995). Ethylene and avocado wax induced phosphorylation of proteins at 29 kDa and 43 kDa. A putative CaM kinase (CaMK) cDNA has been cloned from *C. gloeosporioides* and inhibition of this enzyme by KN93 reduced germination, appressorium formation and melanization (Kim et al., 1998).

If CaM and CaMK signaling triggers germination and appressorium formation in *C. gloeosporioides*, Ca^{+2} involvement is to be expected. Chelation of exogenous Ca^{+2} by EGTA in the medium severely inhibited germination and appressorium formation. Release of internal Ca^{+2} sources by inositol triphosphate (IP_3) generated by phospholipase C may also trigger germination and appressorium formation. In fact, an inhibitor of phospholipase C, U73122, at nanomolar concentrations severely inhibited both. Both EGTA and U73122 inhibited germination and appressorium formation but only if added during the early (a few hrs) phase of hard surface contact. Ca^{+2}, CaM, and CaMK signaling probably plays a highly significant role in the early phase of interaction between *C. gloeosporioides* and its host, and a similar scenario may occur in *C. trifolii*.

One way to explore the molecular events of the early phase of contact of the conidium with a hard surface is to examine gene expression triggered by hard surface contact. Unique transcripts appearing in the conidia of *C. gloeosporioides* during a 2 hr contact with hard surface were detected by differential display (Kolattukudy et al., 1999). DNA from the differential display representing the up-regulated genes were sequenced and probed to northern blots to confirm induction by hard surface contact. The probes also identified clones in a cDNA library made with RNA from hard surface-treated conidia. The isolated hard surface specific cDNA clones were sequenced. This approach yielded seven unique clones designated as *chip* genes. *Chip1* had a 147-amino acid protein of 16.2 kDa. This protein closely resembles ubiquitin-conjugating enzymes. Northern analysis showed the transcript for *Chip1* of 1 kb was detectable within 2 hr of surface contact of the conidia; the level increased up to 6 hr and then decreased. *Chip 1* effectively complemented a conjugating enzyme deficient mutant of *S. cerevisae*. When the yeast mutant cells were transformed with plasmids containing *Chip1*, growth deficiency and heat sensitivity of the mutant were overcome, demonstrating that the *Chip1* encodes a ubiquitinylating enzyme. Selective protein degradation by the ubiquitin-proteosome system plays a critical role in many situations where protein synthesis is reprogrammed. *Chip* therefore could mediate ubiquitin-dependent protein degradation associated with germination and differentiation into appressoria. Other *chip* gene products remain to be identified as they do not resemble proteins currently in the data base.

4.2. Appressorium Differentiation and Development

Following conidial germination, *Colletotrichum* speies differentiate into appressoria when presented with an inductive surface. Germ tube elongation ceases, tip swelling occurs and becomes delimited by a septum. Maturation of the appressorium involves formation of a penetration pore in the base of the cell, deposition of new wall layers and secretion of extracellular matrix materials. Melanin is subsequently deposited in a layer of the cell wall close to the plasma membrane (Bailey et al., 1992). In some species, e.g. *C. lindemuthianum*, the penetration pore is surrounded by an appressorial cone. This structure contains neither chitin or melanin and is continuous with the wall of the penetration peg. Other species lack appressorial cones but the cell wall forms a thickened ring around the penetration pore. Apical growth resumes with the emergence of the penetration peg through the pore and penetration of the plant cuticle and cell wall may involve a combination of mechanical force, produced by high turgor pressure and enzymatic degradation.

To identify genes switched on during appressorium formation in *C. gloeosporioides*, a subtracted cDNA library was constructed using mRNA from non-germinating conidia and appressorium-forming conidia. Differential screening yielded four cDNA clones specific to appressorium-forming spores (Hwang et al., 1995; Hwang and Kolattukudy, 1995). Two of these clones *cap3* and *cap5*, encode 26 and 27-amino acid cysteine-rich polypeptides, respectively, which resemble metallothionins. Two other genes are uniquely expressed during appressorium formation after 4 h exposure of spores to wax. *Cap22* encodes a 22 kDa protein which is probably glycosylated, since antibody to the *E. coli* expressed protein recognizes a 43 kDa protein. *Cap20* encodes a 20 kDa protein. Spores with a disrupted *cap20* gene germinated and produced appressoria of normal morphology, but failed to produce lesions on avocado and tomato. Immunogold labelling with antibodies against cap20 and cap22 proteins showed that both of these gene products were localized in the appressorial wall (Hwang et al., 1995; Hwang and Kolattukudy, 1995).

Melanization is apparently essential to generate appressorial turgor and is required for mechanical penetration by *C. lagenarium* and *C. lindemuthianum* (Kubo and Furusawa, 1991). Melanin biosynthesis begins with synthesis of pentaketide and formation of scytalone. Polyketide synthase, encoded by *PKS1*, is involved in this early step, which is followed by two dehydrations and a reduction step: the dehydration of scytalone to 1,3,8-trihydroxynaphthalene (1,3,8-THN) and vermalone to 1,8-dihydroxynaphthalene are performed by scytalone dehydratase (encoded by *SCD1*). Reduction of 1,3,8-THN to vermalone is performed by 1,3,8-THN reductase (encoded by *THR1*). 1,8-dihydroxynaphthalene is then polymerized and oxidized to melanin. Generating melanin-deficient mutants by treatment of conidia with N-methyl-N'-nitro-N-nitrosoguanidine and screening homologues of known genes, four melanin biosynthesic genes (*PKS1, SCD1, THR1, and CMR1*) have been cloned and characterized from *C. lagenarium* (Kubo et al., 1991; Takano et al., 1995; Kubo et al., 1996; Perpetua et al., 1996). The first three genes are required for melanin synthesis and defective mutants generated by gene disruption cannot penetrate or infect plants. During appressorium differentiation, de novo transcripts of these three genes accumulated 1-2 h after the start of conidial incubation but then decreased after 6 h (Takano et al., 1997). Appressorium formation is repressed by complex nutrients (e.g. yeast extract and tryptone) and expression of these three genes could not be induced in these media. Similar expression studies have been repeated with *C. trifolii*. *THR1* and *SCD1* were also expressed prior to, and during, appressorium formation using inductive conditions (2-6 h; Buhr and Dickman, 1997). *CMR1* is a regulatory gene encoding for Zn (II) 2Cys6 transcriptional activator with DNA binding and activation domains which is involved in the expression of *SCD1* and *THR1* during melanin biosynthesis. In *CMR1* disruptants, only the *PKS1* transcript was observed; *SCD1* and *THR1* were barely detectable.

In *C. trifolii* inhibitor studies (e.g. using KT5720 (a specific inhibitor of cAMP dependent protein kinase; PKA) have suggested that cAMP generation and PKA are involved in germination and appressorium formation (Yang and Dickman, 1997) (see below). In addition, cAMP induced appressorium formation on non-inductive surfaces in conditions of nutrient starvation (Yang and Dickman, 1997). The cloning and characterization of the genes for the PKA catalytic and regulatory subunits has been accomplished to more precisely evaluate the role of PKA in *C. trifolii* morphogenesis (Yang and Dickman 1999 a, b). The gene encoding the catalytic (C) subunit of PKA (Ct-PKAC) was sequenced and insertionally inactivated by gene replacement (Yang and Dickman, 1999a). Southern blot analysis with *C. trifolii* genomic DNA suggested that Ct-PKAC is a single copy gene. Northern analysis with total RNA from different fungal growth stages indicated regulated expression of this gene. Ct-PKAC is a functional kinase as it complemented a *Schizosaccharomyces pombe* PKA mutant. When Ct-PKAC was insertionally inactivated by gene replacement, the transformants grew slower than the wild type, and conidiation patterns were altered. Importantly, PKA deficient strains were unable to infect the host, alfalfa, though only minor differences were observed between the developmental profiles of the wild type and Ct-PKAC disruption mutants. Moreover, these mutants could colonize host tissue following artificial wounding, resulting in typical anthracnose disease lesions. Coupled with microscopy, these data suggest that the defect in pathogenicity is likely due to a failure in penetration. These results suggest similarities with appressorium development in *Magnaporthe grisea*, where it has been established that appressorium development requires cAMP and PKA (Dean, 1997). PKA appears to be essential for pathogenic development in *C. trifolii* and is required for regulating the transition between vegetative and reproductive growth. Evidence suggests that signaling pathways involving cAMP, PKA and CaM operate during differentiation of *Colletotrichum* infection structures. However, the details of these signaling events remain to be elucidated.

4.2.A. *C. trifolii Lipid Activated Protein Kinase*

Members of the protein kinase C (PKC) family of phospholipid dependent, serine/thronine-specific kinases respond to extracellular signals which stimulate receptor mediated hydrolysis of membrane lipids (phosphatidylinositol biphosphate) generating diacylglycerol (DAG) and inositol triphosphate. DAG is a second messenger to activate PKC. These kinases are key components in the phosphoinositol cascade, which in various animal cell types evokes a variety of responses such as cell proliferation, gene expression, membrane transport and organization of the cytoskeleton (Nishizuka 1988; Azzi et al., 1992). In addition PKC is involved with mammalian disease, (Jarvis et al., 1996) and can be oncogenic.

The association of PKC with numerous mammalian developmental and disease related phenomena promoted a search for functional homolog in *C. trifolii*. PCR primers were designed to amplify the highly conserved C1 domain (Nishizuka, 1988), which is a hallmark of all PKC's, and on domain VIII in the catalytic domain which is a more general motif for both PKA or PKC (Hanks and Quinn, 1991). By manipulating primer length and levels of degeneracy, we obtained unique fragments which were used to screen a *C. trifolii* cDNA library. A full length clone was isolated and designated *lapk*. Sequence analysis of *lapk* identified conserved sequences characteristic of the PKC family. Using model substrates and fungal proteins we have been unable to stimulate phosphorylation by diacylglycerol or phorbol esters. "Classic" PKC induction also does not occur in *S. cerevisiae* and *Metarhizium anisopliae* (Levin et al., 1990; St. Leger et al., 1990).

In order to establish that LAPK is a functional protein kinase, LAPK was tested for autophosporylation, a characteristic of protein kinases. LAPK was expressed in *E. coli* and incubated in a kinase assay mixture. LAPK protein was immunoprecipitated with the PKC ε antibody. Following autoradiography, results clearly show that LAPK is labeled and therefore capable of autophosphorylation.

The kinase inhibitor staurosporine affected both conidial germination and appressorial differentiation (Yang, et al., in preparation). Staurosporine completely inhibited appressorial formation at 100nM, while at 1uM staurosporine completely prevented conidial germination. GO976, a more specific PKC inhibitor, affected these processes only at 10 fold higher concentrations. This is not altogether surprising as this inhibitor reacts primarily with the classical PKC regulatory domain which is absent in LAPK. It should be noted that while often considered a PKC specific inhibitor, staurosporine also inhibits other protein kinases. Thus, these studies do not confirm LAPK is in the PKC family, but do indicate that this protein is involved in morphogenesis. Following staurosporine treatment, Western blots using the PKC ε antibody show greatly reduced levels of LAPK (data not shown). These data illustrate the problem of classifying genes based on primary sequence data. In addition, even though a mammalian PKC ε antibody immunoprecipitated LAPK, based on the location of the relevant LAPK sequence (3' end), it is unlikely that this is biologically significant (i.e., it is a false positive). Despite the inability to accurately classify this gene, the inhibitor data suggest involvement of LAPK in appressorium formation; thus we were interested in characterizing this gene in more detail.

To establish conditions for synchronizing large scale *C. trifolii* populations for developmental studies (Buhr and Dickman, 1997), we found that cutin influenced spore germination. Interestingly, addition of cutin to an aqueous spore suspension, prior to placing on a contact surface, (which is required for appressorial differentiation) significantly enhanced LAPK expression. Northern analysis of spores in various nutritional regimes confirmed that LAPK is rapidly induced (30 minutes) by plant cutin. Thus, plant cutin specifically activates *lapk* gene expression.

Overexpression of LAPK in *C. trifolii* results in multiple appressoria emerging from a single conidium (>75% of conidia forming appressoria had at least two appressoria) (Yang et al., in preparation). We are currently seeking null allelles of LAPK by gene replacement. Based on available data, plant cutin fatty acids possibly released by a cutinase, activated LAPK which is important for infection structure development. Thus *C. trifolii* uses host surface components as signals to activate gene expression required for pathogenic development.

4.2.B. *G-Proteins*
Heterotrimeric G proteins are evolutionarily conserved GTP-binding proteins which regulate diverse signaling pathways, including pathways for growth and development, in eukaryotes (Gilman, 1987; Neer, 1995). The importance of G protein signaling in filamentous fungi has recently been demonstrated, following the cloning and disruption of a number of fungal Gα subunit genes (Turner and Borkovich, 1993; Choi et al., 1995; Gao and Nuss, 1996; Regenfelder et al., 1997; Liu and Dean, 1997; Alspaugh et al., 1997). To study how heterotrimeric G proteins mediate signals for growth and development in the filamentous fungal pathogen *Colletotrichum trifolii*, causal agent of alfalfa anthracnose, a gene encoding a Gα subunit homolog, designated *ctg-1*, was isolated and characterized. The deduced product encoded by *ctg-1* is nearly identical to Gα-encoding genes from other filamentous fungi, and is essentially identical to *cpg-1* from *C. parasitica* (Choi et al., 1995), *MagB* from *M. grisea* (Liu and Dean, 1997), and *gna-1* from *N. crassa* (Turner and Burkovich, 1993).

Southern blot analysis showed that *C. trifolii* contains a single copy of *ctg-1*. On northern blots, *ctg-1* transcripts accumulate in germinating conidia, suggesting its involvement during this developmental stage. Fungi harboring a gene replacement of *ctg-1* with a null allele, resulted in transformants whose conidia fail to germinate, demonstrating that *ctg-1* is essential for an early stage in the pathogenic cycle of this fungus (Truesdell et al., submitted).

When *ctg-1* disruption mutants were evaluated further, a number of phenotypic changes were observed, including decreased growth rate in both liquid and solid medium, a significant decrease in percentage of conidial germination and appressorial formation on both glass slides and alfalfa leaves, as well as reduced pathogenicity on the host plant alfalfa.

In other pathogenic fungi, inactivation of a specific Gα-encoding gene also reduces virulence. For example, Gα-deficient *gpa3* mutants of *U. maydis* are nonpathogenic on corn plants (Regenfelder et al., 1997). The nonpathogenic phenotype is specific to *gpa3*, i.e., mutations in three other Gα-encoding genes (*gpa1*, *gpa2*, and *gpa4*) had no obvious phenotype, a result consistent with the possibility of a pathway dedicated in part to pathogenicity. Similar conclusions can be drawn from experiments with *C. parasitica*. Disruption of the Gα-encoding gene *cpg1* eliminated conidiation and pathogenicity of the fungus to chestnut trees, whereas disruption of the homolog *cpg2* did not (Gao and Nuss 1996), a result reminiscent of the disruption of *magB* vs. *magA* and *magC* in *M. grisea* (Liu and Dean, 1997). Preliminary data indicates that *C. trifolii* has additional G-protein sequences (Truesdell and Dickman, unpublished), but we have not yet determined the functional significance of these observations.

4.2.C. *Ras*

Ras proteins are small (21-24 kDa), monomeric GTP-binding proteins that transduce signals for growth and differentiation in eukaryotic organisms. *Ras* genes were first identified as oncogenes of the Harvey and Kirsten strains of rat sarcoma viruses (Ellis et al., 1981). Cellular *ras* genes were isolated later, after the discovery in 1982 that certain human tumors harbor mutated *ras* alleles capable of transforming mouse NIH 3T3 cells in gene transfer assays (see below). Ras has since been intensely studied, and genes encoding Ras homologs have been identified in evolutionarily diverse organisms. In mammals, the importance of Ras in regulating growth is underscored by the observation that activating mutations in *ras*, together with inactivation of the tumor suppressors p53 and p16, are the most prevalent mutations leading to human tumors (Hiram et al., 1995, Hollstein et al., 1996).

Ras proteins are synthesized as cytosolic precursors which undergo post-translational modification, including farnesylation and palmitoylation, for proper membrane localization and function (Clark, 1992; Casey, 1994). Once localized, Ras serves as a molecular switch, coupling activated membrane receptors to downstream signaling molecules by alternating between GTP-bound (active) and GDP-bound (inactive) conformations (Bourne et al., 1990; Lowy and Willumsen, 1993). Ras-GTP associates with and stimulates target effectors, whereas Ras-GDP cannot stimulate effectors but specifically interacts with upstream regulators. The rate of conversion between the active and inactive conformations is modulated by the weak intrinsic GTPase activity of Ras and two kinds of regulatory proteins, guanine nucleotide exchange factors (GEFs) and GTPase-activating proteins (GAPs) (Boguski and McCormick, 1993). GEFs activate Ras by increasing the dissociation rate of guanine nucleotides, allowing GTP, which is more prevalent in the cell, to bind Ras. GAPs inactivate Ras by dramatically increasing the slow intrinsic rate of GTP hydrolysis. Dominant activating (oncogenic) mutations occur at amino acids critical for guanine nucleotide coordination (Sheffzek et al., 1997). Oncogenic Ras mutants have impaired intrinsic GTPase and are insensitive to GAPs, and are thus unable to switch off transmitted signals (Lowy and Willumsen, 1993; Sheffzek et al., 1997).

On the basis of a "comparative pathobiology" hypothesis that homologs of animal signaling proteins likewise regulate growth and development in the phytopathogenic fungus *C. trifolii*, a *ras* gene fragment was amplified from *C. trifolii* genomic DNA by the polymerase chain reaction (PCR) with degenerate primers corresponding to conserved regions in animal and fungal *ras* genes. A vegetative mycelia-specific cDNA library was screened with a PCR product of about 100 base pairs (bp), whose sequence resembled *ras* genes in GenBank. One full-length clone, designated *Ct-ras* was analyzed in detail (Truesdell et al., 1999).

DNA blot analysis revealed a single copy of *CT-ras* in the *C. trifolii* genome. On RNA blots, *CT-ras* hybridized to a 2.2 kilobase (kb) transcript in tissues from all developmental stages of *C. trifolii* (data not shown). Transcript accumulation was high in germinating conidia and vegetatively growing and conidiating hyphae, and relatively low in conidia and appressoria, suggesting CT-Ras influence morphogenesis and proliferation of the fungus.

In mammals, oncogenic *ras* alleles commonly have a single activating mutation in codon 12, 13, 59 or 61(Barbacid, 1987; Lowry and Willumsen, 1993). Ras oncoproteins have an impaired intrinsic GTPase and are insensitive to regulation by GAP (Sheffzek et al., 1997). The oncoproteins are thus constitutively active and promote unregulated growth in some cell types. Two common transforming mutations in human *H-ras* are replacements of glycine-12 with valine (G12V) and glutamine-61 with leucine ((Q61L) (Seeburg et al., 1984; Der et al., 1986). Residues corresponding to G12 and Q61 in H-Ras are conserved in CT-Ras and occur at positions G17 and Q66 respectively. Appropriate site directed mutations were made (CT-*ras*^G17V or CT-*ras*^Q66L) and transformed into NIH 3T3 cells. Cells transfected with either of these two constructs displayed the following characteristics *in vitro*: altered cell morphology, increased growth rate, reduced serum dependence, loss of density-dependent growth inhibition, and anchorage independent growth. Importantly, the transfected cells also formed tumors in animals, directly demonstrating that mutationally activated CT-*ras* functions as an oncogene in mammalian cells. CT-*ras* is therefore functionally homologous to mammalian ras. Thus genetically well-characterized mammalian cell lines are useful test systems for genes and proteins of plant pathogenic fungi.

To determine the effects of oncogenic Ras in *C. trifolii*, we inserted the *Ct-ras*, *Ct-ras* (G17V), and *Ct-ras* (Q66L) coding sequences downstream of the constitutive *A. nidulans* glyceraldehyde-3-phosphate dehydrogenase (gpd) promoter (Roberts et al., 1989) and introduced the genes into *C. trifolii*. In *C. trifolii* mutationally active CT-Ras induced abnormal hyphal proliferation, defects in polarized growth and significantly reduced differentiation in a nutrient-dependent manner (conidiation and appressorium formation). That is, in rich nutritional conditions (e.g. potato dextrose agar), activated *ras* transformants were phenotypically identical to wild type; however, under conditions of nutrient deprivation (minimal medium) hyphal branches became curled and distorted indicating a defect in polarized growth, although more hyphae (dry weight) were produced compared to wild type controls. While addition of yeast extract or peptone reverses the phenotype, addition of various carbon sources, nitrogen sources, carbon plus nitrogen, heat treatment, or addition of osmotic stabilizers were unable to reverse the phenotype. Staining of these transformants with 4', 6-diamidino-2-phenylindole (DAPI) and the cell wall-specific dye calcofluor, which labels chitin and other linked polysaccharides in fungal cell walls (Pringle et al., 1989), revealed abnormally high numbers of nuclei and septa in the tip regions. The periphery of the distended tips also stained more heavily than normal with calcofluor.

Our results are consistent with a model in which CT-Ras regulates a nutrient responsive signal transduction pathway. CT-Ras regulates fungal cellular responses to the nutrient environment is reminiscent of the situation in mammalian cells (e.g. NIH 3T3 cells) where ras transmits growth factor signals. When growth factors are removed (a nutrient limiting situation), wild-type cells arrest growth whereas cells expressing constitutively active Ras continue cell division. Under nutrient limiting conditions in *C. trifolii*, wild-type cells arrest vegetative growth (hyphal elongation) and differentiate (conidiate), whereas constitutively active CT-Ras prevents differentiation, and vegetative growth continues but cells are impaired in polarized, asymmetric growth.

5. VIRULENCE AND HOST SPECIFICITY

In an attempt to better understand the functions that fungi from the genus *Colletotrichum* need to elaborate disease, Rodriguez and colleagues employed a mutagenesis approach using the cucurbit pathogen *C. magna*. One mutant, path-1, was isolated by UV and chemical mutagenesis. Path-1 caused no disease symptoms or mortality in watermelon seedlings despite wild type (wt) levels of sporulation, adhesion, appressorial formation, and importantly it still colonized its host (Freeman and Rodriguez, 1993). Furthermore, path-1 maintained the wild type *C. magna* host range, suggesting that pathogenicity is distinct and thus separable from host specificity. Genetic analysis of a sexual cross between path-1 and a compatible wt strain indicated mutation of a single genetic locus. Moreover, susceptible watermelon seedlings colonized by path-1 also were protected against the pathogenic wild type isolate of *C. magna* as well as an isolate of the wilt pathogen, *Fusarium niveum*, which differs in infection mechanism. It is plausible to propose therefore, that path-1 was continuously stimulating or priming low levels of plant defense responses since no visible hypersensitive response was evident during colonization (Freeman and Rodriguez, 1993).

The ability of path-1 to grow through host tissue without eliciting disease may be due to suppression or avoidance of host defense systems. The fact that path-1 colonized plants are resistant to virulent wild type fungi (*C. magna* and *F. niveum*) suggests that path-1 is not suppressing the host defense systems. Path-1 caused a slight increase in phenylalanine ammonia lyase (PAL) activity but had no effect on peroxidase expression. In addition, peroxidase and PAL activities were significantly higher in wild type inoculated versus path-1 colonized or non-inoculated plants. In plants inoculated with path-1 and challenged with wild type strains, peroxidase and PAL activities reached the levels of plants inoculated with the wild type alone. The deposition of lignin, which is also correlated to disease resistance, corroborated the peroxidase and PAL results with significantly more accumulation observed in wild type and path-1 inoculated/wild type challenged plants than in either path-1 or non-inoculated control plants (Redman et al., 1999). These experiments suggest that path-1 is activating an early step in the defense response but without detectable negative effects on path-1-colonized plants. This activation may allow path-1 colonized plants to respond more rapidly to virulent fungi and terminate what would normally be a compatible pathogen interaction.

Rodriguez and colleagues hypothesized that plants colonized by path-1 express defense-related compounds at enhanced levels thus "priming" the host to resist pathogen invasion, although it must be pointed out that the mechanisms behind path-1 protection are far from conclusive.

To further understand the genetic basis of host specificity in *C. magna,* and to determine if pathogenicity is linked to host colonization, additional mutants were generated (R. Rodriguez, personal communication). By generating and screening 15,000 REMI (restriction enzyme mediated insertion) transformants for loss of virulence and the ability to protect plants against fungal disease, 150 non-pathogenic REMI mutants (path mutants) were isolated that expressed three phenotypes:

> A- colonization and full protection of watermelon plants against fungal disease
> B- colonization and partial disease protection, and
> C- colonization but no disease protection.

Molecular and genetic segregation analysis of two pathogenicity mutants indicated that their phenotypes (A type) arose from single site integrations. The integrated vector and 4.5 kb of flanking DNA from one such mutant was cloned and designated pGMR1. To verify that pGMR1 contained a pathogenicity gene, a wild type isolate was transformed with pGMR1 to induce gene disruptions. About one half of the pGMR1 transformants were nonpathogenic and expressed the type A mutant phenotype. Host range of the path mutants was the same as the wild type on cucurbit species. However, cultivar specificity differed with resistant cucurbit varieties thwarting colonization by the wild type but not the path or gene knock-out mutants. Both the path mutants and the wild type colonized some "non-host" plants (e.g. tomato) without causing disease symptoms. In addition, the path mutants and wild type protected "non host" plants against lethal challenge by virulent *Colletotrichum* and *Fusarium* species. The data indicate that 1) pathogenicity and species specificity are genetically distinct, 2) cultivar specificity may be linked to pathogenicity, and 3) host specificity of *Colletotrichum* species is poorly defined and may require a more thorough examination. The functional characterization of this interesting gene is anticipated with great interest.

6. CHROMOSOME EXCHANGE

Anthracnose is the most important disease of the tropic pasture legumes in the genus *Stylosanthes*. These legumes originate from South America and are grown extensively for cattle production in Northern Australia and South-East Asia. Anthracnose diseases of *Stylosanthes* spp. are currently controlled by resistant varieties (Manners et al., 1992). However, the pathogen is genetically variable and shows considerable pathogenic specialization.

Two biotypes (A and B) of *C. gloeosporioides* infect the tropical legumes *Stylosanthes* spp. in Australia. Previous research in the laboratory of John Manners indicates that the two biotypes are asexual, vegetatively incompatible and genetically distinct. However, isolates of biotype B carrying a supernumerary 2 Mb chromosome (thought to originate from biotype A) have been observed in the field (Masel et al., 1996). Transfer of the 2 Mb chromosome from biotype A to biotype B was investigated under controlled laboratory conditions. Initial studies using *nit* mutants and standard vegetative compatibility tests did not work; chromosomes were not transferred. Using an alternative strategy, selectable marker genes conferring resistance to hygromycin and phleomycin were introduced into representative isolates of A and B biotypes. A transformant of biotype A with hygromycin resistance integrated only on the 2 Mb chromosome was identified and grown in a mixed culture with a phleomycin resistant transformant of biotype B. Conidia from this mixed culture were collected, germinated and double antibiotic resistant colonies were isolated. Molecular analysis using RFLPs, RAPDs and karyotype patterns showed that these colonies contained the 2 Mb chromosome in a biotype B genetic background. These results demonstrate that the 2 Mb chromosome can introgress vegetatively from biotype A to biotype B, showing that supernumerary chromosomes in fungi may arise by non-sexual transfer from otherwise genetically incompatible genotypes (He et al., 1998).

These results are important for two reasons. First, it suggests a possible origin for supernumerary chromosomes in filamentous fungi and secondly it suggests that the standard techniques for investigating vegetative compatibility groups (e.g., *nit* mutant complementation) may not fully measure the potential for asexual recombination. It was suggested that the 2 Mb chromosome may lack incompatibility loci and therefore may escape processes that constrain the recombination of biotype A and B genomes. Two mechanisms for selective transfer of the 2 Mb chromosome have been proposed. In both models a transient or slow growing heterokaryon is formed between the biotypes. In one mechanism, the nuclei fuse and all biotype A chromosomes are lost with the exception of the 2 Mb chromosome. Alternatively, the 2 Mb chromosome may be transferred between nuclei in a transient heterokaryon.

It is evident that the 2 Mb supernumerary chromosome of *C. gloeosporioides* infecting *Stylosanthes* may be mobile and compatible with different genetic backgrounds in *C. gloeosporioides*. The function of this chromosome is now under study by the Manners group. So far there is little evidence to suggest that this chromosome affects either pathogenicity or virulence on *Stylosanthes* (Masel et al., 1996, He et al., 1998). The observation of selective chromosome transfer and survival of recipients in the field (Masel et al., 1996) suggests that the 2 Mb chromosome may provide a selective advantage in the natural environment.

7. BIOTROPHIC DEVELOPMENT

Biotrophic plant pathogens are important economically, in terms of disease, and scientifically due to the unique relationship between host and fungus. There are a limited number of examples of specific molecules known to be associated with biotrophy.

A putative amino acid permease has been cloned from *Uromyces fabae*, a rust fungus

(Hahn et al., 1997; see chapter by Hahn). This gene was identified using differential hybridization and is specifically located in the fungal haustorial plasmic membrane, although its precise function particularly with respect to biotrophy is as yet unresolved. As mentioned previously, a number of *Colletotrichum* spp. have a hemibiotrophic phase in their life cycle (e.g. *C. lindemuthianum*), and unlike true biotrophs, offer the experimental advantages of culturability on synthetic media, as well as the ability to be genetically transformed. Although there is considerable information on pre-penetration and morphogenesis in *Colletotrichum,* there is a paucity of information about the biotrophic phases of *Colletotrichum* growth and development.

While Colletotrichum biotrophy is not identical to haustorial pathogens (see O'Connell, 1987), there are some common features, particularly with respect to the interface area between plant and fungus (Green et al., 1995). This region is likely to be a key area, both for establishment of biotrophy and for avoidance or suppression of host defense response.

The most significant research efforts in this area come from the laboratory of Richard O'Connell, Jon Green and colleagues who have been utilizing an antibody/cell biology approach. Working with *C. lindemuthianum*, monoclonal antibodies have been generated against fungal infection structures isolated from bean leaves. In this way proteins were identified that were specifically associated with intracellular hyphae during fungal interaction with its plant host (Paine et al., 1994, a,b; Green et al., 1995; O'Connell et al., 1996). One antibody (UB25) recognized a group of N-linked glycoproteins that are only expressed during the hemibiotrophic growth stage, and that are absent in all other developmental stages.

Recently, a cDNA library prepared from infected bean tissue has been screened with the UB25 monoclonal antibody and the gene encoding this glycoprotein was isolated and sequenced. The cloned gene, *CIH1* (Colletotrichum *I*ntracellular *H*ypha 1), is present as a single copy of *C. lindemuthianum*. The presence of a putative signal peptide at the N terminus is in agreement with electron microscope studies which showed that this glycoprotein is secreted through the wall of the intracellular hyphae, thus putting it at plant-hemibiotrophic fungal interface.

An important aspect of biotrophic interactions is that the fungus avoids triggering, or in some way suppresses, host defense responses (e.g. hypersensitive cell death) (Heath and Skalamera, 1997). The CIH1 glycoprotein seems to be a major structural component of the interface of *C. lindemuthianum* with bean and thus may provide a buffer zone in order to physically separate the plant plasma membrane from the fungal cell wall. This protein may facilitate avoidance of recognition of the fungus by the plant. The CIH1 glycoprotein could also function as a physical barrier for the fungus against host defense response. The sequence of CIH1shows similarity to plant cell-wall proteins, thus the fungus may mimic plant cell wall to the host, and delay or prevent recognition. The CIH1 gene product may be the first fungal component described which is involved in the biotrophic stage in a *Colletotrichum* host interaction.

8. CONCLUSIONS

Colletotrichum continues to be a significant and important genus of plant pathogenic fungi. The work described in this review has demonstrated the considerable progress that has been made not only in the identification of genes and gene products relevant to fungal pathogenic development and host interaction, but importantly, functional examination of these molecules has been initiated. While it is clear that pathways involved with these processes will be delineated in the future; on the other hand, it must be stressed that still relatively few genes have been identified as bona fide pathogenicity determinants. It is evident that there is a need to continue to search for additional genes involved in *Colletotrichum* pathogenesis. Fortunately there has been an increase in the number of species which are amenable to powerful molecular genetic technologies, such as targeted gene inactivation.

Although pre-penetration morphogenesis of *Colletotrichum* species appear similar, there are distinct differences between species for example, in the mechanism of conidial adhesion and the relative importance of cutinases and melanization with respect to appressorial penetration. Thus caution should be exercised when making comparisons. Interesting parallels between *Colletotrichum* and *Magnaporthe* have also surfaced, suggestive of the possibility of similar signaling pathways with respect to pre-penetration morphogenesis. Thus it is important to continue these studies which will hopefully be interfaced with genomic projects (yet to be initiated) both to determine common threads in the infection process as well as to identify clear differences. Both avenues hold considerable promise in future studies.

Acknowledgements

I thank the members of the Dickman lab for discussions. Work in the Dickman lab is supported by BARD, USDA and the DOE/NSF/USDA Collaborative Research Program in Plant Biology.

References

Alspaugh, J. A., Perfect, J. R. and Heitman, J. (1997). *Cryptococcus neoformans* mating and virulence are regulated in the G-protein α subunit GPA1 and cAMP. Genes and Dev. **11**, 3206-3217.

Azzi, A., Boscoboinik, D. and Hensey, D. (1992). The protein kinase C family. Eur. J. Biochem. **208**, 547-557.

Bailey, J. A. and Jeger, M. J. (1992). *Colletotrichum*: Biology, Pathology and Control. Wallingford: CAB International.

Bailey, J. A., O'Connell, R. J., Pring, R. J. and Nash, C. (1992). Infection strategies of *Colletotrichum* species. In *Colletotrichum*: Biology, Pathology and Control (J. A. Bailey and M. J. Jeger, Eds.). Pp. 88-120. Wallingford: CAB International.

Birkeland, N. K. (1994). Cloning, molecular characterization and expression of the genes encoding lytic functions of lactococcal bacteriophase LC3: a dual lysis system of modular design. Can. J. Microbiol. **40**, 658-665.

Boguski, M. S. and McCormick, F. (1993). Proteins regulating Ras and its relatives. Nature **366**, 643-663.

Bourne, H. R., Sanders, D. A. and McCormick, F. (1990). The GTPase superfamily: a conserved switch for diverse cell functions. Nature **348**, 125-132.

Buhr, T. L. and Dickman, M. B. (1997). Gene expression analysis during conidial germ-tube and appressorium development in *Colletotrichum trifolii*. Appl. Env. Microbiol. **63**, 2378-2383.

Buhr, T. L., Oved, S., Truesdell, G. M., Huang, C., Yarden, O. and Dickman, M. B. (1996). A kinase-encoding gene from *C. trifolii* complements a colonial growth mutant of Neurospora crassa. Mol. Gen. Genet. **251**, 565-572.

Casey, P. J. (1994). Lipid modifications of G proteins. Curr. Opin. Cell Biol. **6**, 219-225.

Choi, G. H., Chen, B. and Nuss, D. L. (195). Virus-mediated or transgenic suppression of a G-protein α subunit and attenuation of fungal virulence. Proc. Natl. Acad. Sci. USA **92**, 305-309.

Clark, S. (1992). Protein isoprenylation and methylation at carboxyl-terminal cysteine residues. Annu. Rev. Biochem. **61**, 335-386.

Dean, R. A. (1997). Signal pathways and appressorium morphogenesis. Annu. Rev. Phytopathol. **35**, 211-234.

Dickman, M. B. and Yarden, O. (1999). Serine-theonine kinases and phosphatases in filamentous fungi. Fungal Genetics and Biology (in press).

Dickman, M. B. (1999). Signal exchange during *Colletotrichum trifolii*-alfalfa interactions. In: *Colletotrichum*: Host specificity, pathology and host-pathogen interaction (M. Dickman, S. Freeman, D. Prusky, eds.). APS Press, St. Paul, Minnesota (in press).

Dufresne, M., Bailey, J. A., Dron, M. and Langin, T. (1998). *Clkl*, a serine/threonine protein kinase-encoding gene, is involved in pathogenicity of *C. lindemuthianum* on common bean. Mol. Plant Microbe Int. **11**, 99-108.

Ellis, R. W., DeFoe, D., Shih, T. Y., Gonda, M. A., Young, H. A., Tsuchida, N., Lowy, D. R. and Scolnick, E. M. (1981). P21 *src* genes of Harvey and Kirsten sarcoma viruses originate from divergent members of a family of normal vertebrate genes. Nature **292**, 506-511.

Epstein, L., Lusnak, K. and Kaur, S. (1998). Transformation-mediated developmental mutants of *Glomerella graminicola* (*Colletotrichum graminicola*). Fungal Genet. Biol. **23**, 189-103.

Freeman, S. and Rodriguez, R. J. (1993). Genetic conversion of a fungal plant pathogen to a nonpathogenic, endophytic mutualist. Science **260**, 75-78.

Gao, S. and Nuss, D. L. (196). Distinct roles for two G protein α subunits in fungal virulence, morphology and reproduction revealed by targeted gene disruption. Proc. Natl. Acad. Sci. USA **93**, 14122-14127.

Gilman, A. G. (1987). G proteins: Transducers of receptor-generated signals. Annu. Rev. Biochem. **56**, 615-649.

Glaishman, M. A., Hwang, C. S. and Kolattukudy, P. E. (1995). Involvement of protein phosphorylation in the induction of appressorium formation in *Colletotrichum gloeosporioides* by its host surface wax and ethylene. Physiol. Mol. Plant Pathol. **47**, 103-117.

Green, J. R., Pain, N. A., Cannell, M. E., Jones, G. L., Leckie, C. L., McCready, S., Mendgen, K., Mitchell, A. J., Callow, J. A. and O'Connell, R. J. (1995). Analysis of differentiation and development of the specialized infection structures formed by biotrophic fungal plant pathogens using monoclonal antibodies. Can. J. Bot. **73 (suppl)**, S408-S417.

M. B. Dickman

Hahn, M., Neef, U., Struck, C., Gottfert, M. and Mendgen, K. (1997). A putative amino acid transporter is specifically expressed in haustoria of the rust fungus *Uromyces fabae*. Mol. Plant Microbe Inter. **10**, 438-445.

Hanks, S. K. and Quinn, A. M. (1991). Protein kinase catalytic domain sequence database: identification of conserved features of primary structure and classification of family members. Methods Enzymol. **200**, 38-62.

He, C., Rusu, A. G., Poplawski, A. M., Irwin, J. A. G. and Manners, J. M. (1998). Transfer of a supernumerary chromosome between vegetatively incompatible biotypes of the fungus *Colletotrichum gloeosporioides*. Genetics **150**, 1459-1466.

Heath, M. C. and Skalamera, D. (1997). Cellular interactions between plants and biotrophic fungal parasites. Adv. Bot. Res. **24**, 196-225.

Hirama, T. and Koeffler, H. P. (1995). Role of the cyclin-dependent kinase inhibitors in the development of cancer. Blood **86**, 841-854.

Hollstein, M., Shomer, B., Greenblatt, M., Soussi, T., Hovig, E., Motesano, R. and Harris, C. C. (1996). Somatic point mutations in the p53 gene of human tumors and cell lines: update compilation. Nucleic Acid Res. **24**, 141-146.

Hwang, C. H., Flaishman, M. A. and Kolattukudy, P. E. (1995). Cloning of a gene expressed during appressorium formation by *Colletotrichum gloeosporioides* and a marked decrease in virulence by disruption of this gene. Plant Cell **7**, 183-193.

Hwang, C. H. and Kolattukudy, P. E. (1995). Isolation and characterization of genes expressed uniquely during appressorium formation by *Colletotrichum gloeosporioides* conidia induced by the host surface wax. Mol. Gen. Genet. **247**, 282-294.

Jarvis, W. D., Turner, A. J., Povirk, L. F., Traylor, R. S. and Grant, S. (1994). Induction of apoptotic DNA fragmentation and cell death in HL-60 human promyelocytic leukemia cells by pharmacological inhibitor of protein kinase C. Cancer Res. **54**, 1707-1714.

Kim, Y. K., Li, D. and Kolattukudy, P. E. (1998). Induction of Ca $^{2+}$ -calmodulin signaling by hard surface contact primes *Colletotrichum gloeosporioides* conidium to germinate and form appressorium. J. Bacteriol. **180**, 5144-5150.

Kincaid, R. L. (1991). Signaling mechanisms in microorganisms: common themes in the evolution of signal transduction pathways. Adv. Second Messenger Phosphoprotein Res. **23**, 165-183.

King, K., Dohlmen, H. G., Thorner, J., Caron, M. G. and Lefkowitz, R. J. (1990). Control of yeast mating signal transduction by a mammalian B$_2$ -andrenergic receptor and G$_s\alpha$ subunit. Science **250**, 121-123.

Kolattukudy, P. E., Rogers, L. M., Li, D., Hwang, C. S. and Flaishman, M. A. (1995). Surface signaling in pathogenesis. Proc. Natl. Acad. Sci. USA **92**, 4080-4087.

Kolattukudy, P. E., Kim, Y-K, Li, D., Liu, Z-M, and Rogers, L. (1999). Early molecular communication between *Colletotrichum gloeosporioides* and its host. In D. Drusky, S. Freeman and M. Dickman (eds.), colletotrichum: Host Specificity, Pathology and host Pathogen Interactions. APS Press, St. Paul, Minnesota (in press).

Kubo, Y., Nakamura, H., Kobayashi, K., Okuno, T. and Furusawa, I. (1991). Cloning of a melanin biosynthetic gene essential for appressorial penetration of *Colletotrichum lagenarium*. Mol. Plant Microbe Inter. **5**, 440-445.

Latunde-Dada, A. O., O'Connell, R. J., Nash, C., Pring, R. J., Lucas, J. A. and Bailey, J. A. (1996). Infection process and identity of the hemibiotrophic anthracnose fungus (*Colletotrichum destructivum* O'Gara) from cowpea (*Vigna unguiculata* (L.) Walp.) Mycol. Res. **100**, 1133-1141.

Levin, D. E., Fields, F. O., Kunisawa, R., Bishop, J. M. and Turner, J. (1990). A candidate protein kinase C gene, PKC1, is required for the *S. cerevisae* cell cycle. Cell **62**, 212-224.

Liu, S. and Dean, R. A. (1997). G protein α subunit genes control growth, development and pathogenicity of *Magnaporthe grisea* Mol. Plant-Microbe Interact. 10, 1075-1086.

Lowy, D. R. and Willumsen, B. M. (1993). Function and regulation of RAS. Annu. Rev. Biochem. **62**, 851-891.

Manners, J. M., Masel, A. M., Braithwaite, K. S. and Irwin, J. A. G. (1992). Molecular analysis of *Colletotrichum gloeosporioides* pathogenic on the tropical pasture legume *Stylosanthes spp.* Pp. 250-268 in: *Colletotrichum*, Biology, Pathology and Control (J. A. Bailey and M. Jeger Eds.). CAB International, Wallingford, UK.

Masel, A. M., He, C., Poplawski, A. M., Irwin, J. A. G. and Manners, J. M. (1996). Molecular evidence for chromosome transfer between biotypes of *Colletotrichum gloeosporioides*. Mol. Plant Microbe Inter. 9:339-348. morphology and reproduction revealed by targeted gene disruption. Proc. Natl. Acad. Sci. USA **93**, 14122-14127.

Mercure, E. W., Kunoh, H. and Nicholson, R. L. (1994b). Adhesion of *Colletotrichum graminicola* conidia to corn leaves: a requirement for disease development. Physiol. Mol. Plant Pathol. **45**, 407-420.

Mercure, E. W., Leite, B. and Nicholson, R. L. (1994a). Adhesion of ungerminated conidia of *Colletotrichum graminicola* to artificial hydrophobic surfaces. Physiol. Mol. Plant Pathol. **45**, 421-440.

Neer, E. J. (1995). Heterotrimeric G proteins: organizers of transmembrane signals. Cell **80**, 249-257.

Nicholson, R. L. (1992). *Colletotrichum graminicola* and the anthracnose disease of corn and sorghum. In *Colletotrichum*: Biology, Pathology and Control (J. A. Bailey and M. J. Jeger, Eds.). Pp. 186-202. Wallingford: CAB International.

Nicholson, R. L. (1996). Adhesion of fungal propagules. In Histology, Ultrastructure and Molecular Cytology of Plant-Microorganism Interactions (Nicole, M. and Gianinazzi-Pearson, V., Eds.). Pp. 117-134. Kluwer academic Publishers.

Nicholson R. L., Epstein, L. (1991). Adhesion of fungi to the plant surface: prerequisite for pathogenesis. In G. T. Cole and H. C. Hoch, Eds. The Fungal Spore and Disease Initiation in Plants and Animals. Plenum Press, New York, NY, pp. 3-23.

Nieman, A. M. (1993). Conservation and reiteration of a kinase cascade. Trends in Genetics **9**, 390-394.

(1988). The molecular heterogeneity of protein kinase C and implications of cellular regulation. Nature **334**, 661-665.

Nishizuka, Y. O'Connell, R. J. (1987). Absence of a specialized interface between infection hyphae of *Colletotrichum lindemuthianum* and *Phaseolus vulgaris*. New Phytol. **107**, 725-734.

O'Connell, R. J., Bailey, J. A. and Richmond, D. V. (1985). Cytology and physiology of infection of *Phaseolus vulgaris* by *Colletotrichum lindemuthianum*. Physiol. Plant Pathol. **27**, 75-98.

O'Connell, R. J., Uronu, A. B., Waksman, G., Nash, C., Keon, J. P. R. and Bailey, J. A. (1993). Hemibiotrophic infection of *Pisum sativum* by *Colletotrichum truncatum*. Plant Pathol. **42**, 774-783.

O'Connell, R. J., Pain, N. A., Hutchison, K. A., Jones, G. L. and Green, J. R. (1996). Ultrastructure and composition of the cell surfaces of infection structures formed by the fungal plant pathogen *Colletotrichum lindemuthianum*. J. Micros. **181**, 204-212.

Pain, N. A., O'Connell, R. J., Bailey, J. A., Green, J. R. (1992). Monoclonal antibodies which show restricted binding to four *Colletotrichum* species: *C. lindemuthianum*, *C. malvarum*, *C. orbiculare* and *C. trifolii*. Physiol. Mol. Plant Pathol. **40**, 111-126.

Pain, N. A., Green, J. R., Gammie, F. and O'Connell, R. J. (1994a). Immunomagnetic isolation of viable intracellular hyphae of *Colletotrichum lindemuthianum* from infected bean leaves using a monoclonal antibody. New Phytol. **127**, 223-232.

Pain, N. A., O'Connell, R. J., Mendgen, K. and Green, J. R. (1994b). Identification of glycoproteins specific to biotrophic intracellular hyphae formed in the *Colletotrichum*-bean interaction. New Phytol. **127**, 233-242.

Perfect, S. E., O'Connell, R. J., Green, E. F., Doering-Saad, C. and Green J. R. 1998. Expression cloning of a fungal proline-rich glycoprotein specific to the biotrophic interface formed in the *Colletotrichum*-bean interaction. Plant J. **15**, 273-279.

Perpetua, N. S., Kubo, Y., Yasuda, N., Takano, Y. and Furusawa, I. (1996). Cloning and characterization of a melanin biosynthetic *THR1* reductase gene essential for appressorial penetration of *Colletotrichum lagenarium*. Mol. Plant Microbe Inter. **9**, 323-329.

Podila, G. K., Rogers, L. M. and Kolattukudy, P. E. (1993). Chemical signals from avocado surface wax trigger germination and appressorium formation in *Colletotrichum gloeosporioides*. Plant Physiol. **103**, 267-272.

Pring, R. J., Nash, c. Zakaria, M. and Bailey, J. A. (1995). Infection processes and host-range of *Colletotrichum capsici*. Physiol. Mol. Plant Pathol. **46**, 137-152.

Pringle, J., Preston, R. A., Adams, A., Stearns, T., Drubin, D. G., Haarer, B. K. and Jones, E. W. (1989). Fluorescencemicroscopy methods for yeast. Meth. Cell. Biol. **31**, 357-435.

Prusky, D., Freeman, S., Dickman, M. B. (1999). *Colletotrichum*: Host Specificity, Pathology and Host Pathogen Interactions. APS Press, St. Paul, Minnesota (in press).

Redman, R. S., Freeman, S., Clifton, D. R., Mueller, G., Morrel, J., Brown, G. and Rodriguez, R. J. (1999). Defining the basis of plant protection afforded by a non-pathogenic endophytic mutant of *Colletotrichum magna*. Plant Physiology (in press).

Roberts, R. G. and Snow, J. P. (1984). Histopathology of cotton boll rot caused by *Colletotrichum capsici*. Phytopathol. **74**, 390-397.

Rodriguez, R. J. and Redman, R. S. (1992). Molecular transformation and genome analysis of *Colletotrichum* species. In *Colletotrichum*: Biology, Pathology and Control (J. A. Bailey and M. J. Jeger, Eds.). Pp. 47-76. Oxford: CAB.

Rodriguez, R. J. and Redman, R. S. (1999). *Colletotrichum* as a model system for defining the genetic basis of fungal symbiotic lifestyles. In *Colletotrichum*: Host Specificity, Pathology and Host Pathogen Interactions (D. Prusky, B. Freeman, M. B. Dickman, Eds.) APS Press, St. Paul, Minnesota (in press).

Scheffzek, K., Ahmadian, M. R., Kabsch, W., Wiesmuller, L., Lautwein, A., Schmitz, F. and Wittinghofer, A. (1997). The Ras-Ras GAP Complex: Structural basis for GTPase activation and its loss in oncogenic Ras mutants. Science **277**, 333-338.

Sela-Buurlage, M. B., Epstein, L. and Rodriguez, R. J. (1991). Adhesion of ungerminated *Colletotrichum musae* conidia. Physiol. Mol. Plant Pathol. **39**, 345-352.

Sherriff, C., Whelan, M. J., Arnold, G. M., Lafay, J. F., Brygoo, Y. and Bailey, J. A. (1994). Ribosomal DNA sequence analysis reveals new species groupings in the genus *Colletotrichum*. Exp. Mycol. 18, 121-138.

St. Leger, R. J., Lacetti, L. B., Staples, R. C. and Roberts, D. W. (1990). Protein kinases in the entomopathogenic fungus *Metarhizium anisopliae*. J. Gen. Microbiol. **136**, 1401-1411.

Stephenson, S. A., Green, J. R., Manners, J. M. and Maclean, D. J. (1997). Cloning and characterization of glutamine synthetase from *Colletotrichum gloeosporioides* and demonstration of elevated expression during pathogenesis on *Stylosanthes guianensis*. Curr. Genet. **31**, 447-454.

Sutton, B. C. (1992). The genus *Glomerella* and its anamorph *Colletotrichum*. Pages 1-26 in: *Colletotrichum*: Biology, Pathology and Control. J. A. Bailey and M. J. Jeger, eds. CAB Int., Wallingford, UK.

Takano, Y., Kubo, Y., Kuroda, I. and Furusawa, I. (1997). Temporal transcriptional pattern of three melanin biosynthesis genes, *PKS1, SCD1* and *THR1* in appressorium-differentiating and nondifferentiating conidia of *Colletotrichum lagenarium*. Appl. Env. Microbiol. **63**, 351-354.

Takano, Y., Kubo, Y., Shimizu, K., Mise, T., Okuno, T. and Furusawa, I. (1995). Structural analysis of *PKS1*, a polyketide synthase gene involved in melanin biosynthesis of *Colletotrichum* lagenarium. Mol. Gen. Genet. **249**, 162-167.

Truesdell, G. M., Jones, C., Holt, T., Henderson, G. and Dickman, M. B. (1999). Defects in hyphal growth polarity and mammalian tumors induced by RAS from a phytopathogenic fungus. Mol. Gen. Genet. **262**, 46-54..

Turner, G. E. and Borkovich, K. A. (1993). Identification of a G protein α subunit from *Neurospora crassa* that is a member of the Gi family. J. Biol. Chem. **268**, 14805-14811.

Von Arx, J. A. (1957). Die Arten der Gattung *Colletotrichum* Cda. Phytopathol. Z. **29**, 413-468.

Walker, J. C. (1921). Onion smudge. J. of Agric. Res. **20**, 685-721.

Warwar, V. and Dickman, M. B. (1996). Calcium-calmodulin mediated appressorium development in *Colletotrichum trifolii*. Appl. Env. Microbiol. **62**, 74-79.

Wharton, P. S. and Julian A. M. (1996). A cytological study of compatible and incompatible interactions between *Sorghum bicolor* and *Colletotrichum sublineolum*. New Phytol. **14**, 25-34.

Xu, J. R. and Hamer, J. E. (1996). MAP kinase and cAMP signaling regulate infection structure formation and pathogenic growth in the rice blast fungus *Magnaporthe grisea*. Genes and Dev. **10**, 2696-2706.

Yang, Z. and Dickman, M. B. (1997). Regulation of cAMP and cAMP dependent protein kinase during conidial germination and appressorium formation in *Colletotrichum trifolii*. Physiol. Mol. Plant Pathol. 50, 117-127.

Yang, Z. and Dickman, M. B. (1999a). *Colletotrichum trifolii* mutants disrupted in the catalytic subunit of cAMP-dependent protein kinase are non-pathogenic. Mol. Plant Microbe Inter. **12**, 430-439.

Yang, Z. and Dickman, M. B. (1999b). Molecular cloning and characterization of Ct-PKAR, a gene encoding the regulatory subunit of cAMP-dependent protein kinase in *Colletotrichum trifolii*. Archives of Microbiology **171**, 249-256.

Young, D. H. and Kauss, H. (1984). Adhesion of *Colletotrichum lindemuthianum* spores to *Phaseolus vulgaris* hypocotyls and to polystyrene. Appl. Env. Microbiol. **47**, 616-619.

HYPOVIRULENCE AND CHESTNUT BLIGHT: FROM THE FIELD TO THE LABORATORY AND BACK

DONALD L. NUSS
Center for Agricultural Biotechnology
University of Maryland Biotechnology Institute
5115 Plant Sciences Building
College Park, Maryland 20742-4450

1. Introduction

Most of the pathogenic fungi discussed in this book have been reported to harbor viral-like double-stranded (ds) RNAs (Buck, 1986; Nuss and Koltin, 1990; Ghabrial, 1998). Correlation between the presence of extrachromosomal dsRNAs and attenuation of virulence (hypovirulence) for some pathogenic fungi have stimulated interest in these elements as potential biological control agents. Among these, the dsRNAs associated with hypovirulence of the chestnut blight fungus, *Cryphonectria parasitica* (Murr.) Barr, are the best characterized (Nuss, 1992; Hillman et al., 1995; Nuss, 1996). However, interest in hypovirulence-associated dsRNAs extends past biological control potential to their utility as unique experimental tools for uncovering fundamental processes underlying fungal pathogenesis. The primary goals of this chapter are to provide the reader with a perspective of the journey that has seen *C. parasitica* hypovirulence progress from the field to the laboratory and to indicate how the unexpected fundamental insights derived from the exploitation of hypovirulence as an experimental system are being applied to facilitate a return to the field in the form of enhanced biological control potential.

2. From the Field to the Laboratory

Relevant details of the devastation caused by the North American chestnut blight epidemic, the discovery of hypovirulent *C. parasitica* field isolates in southern Italy in the 1950s, accounts of efficacious biological control of chestnut blight by

J. W. Kronstad (ed.), Fungal Pathology, 149–170.
© 2000 *Kluwer Academic Publishers. Printed in the Netherlands.*

hypovirulence in European forests and orchards, and factors contributing to the less successful record of biological control in North American forest ecosystems have been extensively covered in several excellent reviews (Anagnostakis, 1982; Van Alfen, 1986; MacDonald and Fulbright, 1991; Nuss, 1992; Heiniger and Rigling, 1994). Reports by Day and co-workers (Day et al., 1977; Day and Dodds, 1979) correlating the presence of dsRNA genetic elements with hypovirulence are often credited with providing the conceptual framework for the subsequent development of transmissible hypovirulence as an experimental system. Although a number of different types of dsRNAs have been recovered from hypovirulent *C. parasitica* strains since those reports (Hillman et al., 1992,1994; Polashoch and Hillman, 1994; Enebak et al., 1994), it is the commonly found elements assigned to the genus *hypovirus* within the newly described virus family Hypoviridae (Hillman et al., 1995) that have been studied in greatest detail. As indicated in Figure 1, members of this genus are characterized by a genome of 12-12.7 kbp in length containing two contiguous open reading frames (ORFs A and B) within a polyadenylated coding strand, each specifying polyproteins that undergo proteolytic processing (Shapira et al., 1991).

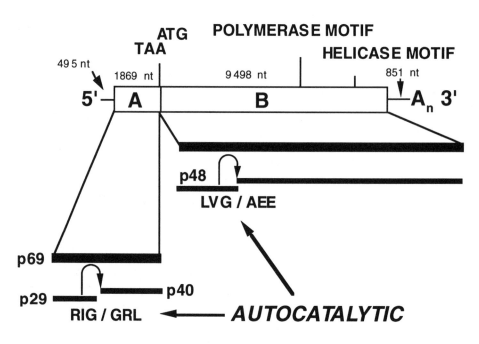

Figure 1. Genetic organization and basic expression strategy for prototypic hypovirus CHV1-EP713. The coding strand RNA of hypovirus CHV1-EP713 consists of 12,712 nucleotides, excluding the poly(A) tail (An) (Shapira et al., 1991). The 5' proximal coding domain, ORF A (622 codons) encodes two polypeptides, p29 and p40, that are released from a polyprotein, p69, by an autoproteolytic event mediated

by p29. Cleavage occurs between Gly-248 and Gly-249 during translation and is dependent on Cys-162 and His-215 within the putative p29 catalytic site (Choi et al.,1991a; 1991b). Expression of ORF B (3165 codons) also involves an autoproteolytic event in which a 48-kD polypeptide, p48, is released from the N-terminal portion of the encoded polyprotein. Cleavage, in this case, occurs between Gly-418 and Ala-419 (Shapira and Nuss, 1991). The junction between ORF A and ORF B consists of the sequence 5'UAAUG-3'. Translational mapping studies indicated that the UAA portion of the pentanucleotide serves as a termination codon of ORF A, whereas the AUG portion is the 5 proximal initiation codon for ORF B, as indicated by the overlapping TAA/ATG codons. Computer-assisted analysis revealed five distinct domains within the CHV1-EP713 coding regions that showed significant sequence similarity to previously described domains, including polymerase and helicase domains, within plant potyvirus-encoded polyproteins (Koonin et al., 1991). This figure has been adapted from Shapira et al., 1991.

Completion of the nucleotide sequence for the prototypic hypovirus CHV1-EP713 in 1991 (Shapira et al., 1991) paved the way for the development of an infectious hypovirus cDNA clone. This key milestone was made possible by the development of an efficient DNA-mediated transformation protocol for *C. parasitica* (Churchill et al., 1990) that allowed Choi and Nuss (1992) to install a full-length cDNA copy of CHV1-EP713 RNA into the chromosome of virus-free *C. parasitica* strains. The resulting transformants were indistinguishable from CHV1-EP713-infected strains, including the presence of cytoplasmically replicating viral dsRNA. Chen et al (1994b) showed that the viral dsRNA present in these "transgenic" hypovirulent strains originated as large cDNA-derived nuclear transcripts that were subsequently trimmed of nonviral vector nucleotides in a very precise manner. This observation suggested that it should be possible to bypass chromosomal installation of the hypovirus cDNA and initiate infection by introduction of a synthetic copy of the hypovirus coding strand RNA directly into fungal spheroplasts. Chen et al (1994a) confirmed this prediction by developing an efficient electroporation-based transfection protocol utilizing *in vitro* synthesized coding strand transcripts derived from the full-length CHV1-EP713 cDNA clone. Combined, the hypovirus transformation and transfection protocols have been used to extend hypovirus host range to fungal species taxonomically related to *C. parasitica* (Chen et al., 1994a;1996a) and to identify the hypovirus-encoded protease p29 as a symptom determinant (Craven et al., 1993). The utility of both protocols for mapping viral symptom and hypovirulence determinants and for the engineering of an enhanced form of biological control will be discussed in a subsequent section.

Progress in developing protocols for the efficient manipulation of the hypovirus genome has been complemented by additional refinement of the *C. parasitica* DNA-mediated transformation system. A variety of plasmid transformation vectors have been designed to provide transformation frequencies in excess of 1,000 transformants/ μg DNA and allow expression of endogenous or foreign genes (Choi et al., 1993; Zhang et al., 1993; Wang et al., 1998). The haploid nature of the *C. parasitica* genome makes targeted gene disruption an effective and routine procedure for functional analysis (Zhang et al., 1993; Kim et al., 1995; Gao et al., 1996; Gao and Nuss, 1996; Kasahara and Nuss, 1997). Efficient genetic complementation vectors have been designed for purposes of gene mapping (Kasahara, Wang and Nuss, unpublished results) and *in vivo* structure/function analyses (Gao and Nuss, 1998). Transformation studies also benefit from the ability to preserve frozen viable, competent *C. parasitica* spheroplasts for long time periods and the ability to obtain nuclear homogeneity by simply selecting single conidial isolates from transformed colonies. These combined technical advances have provided the hypovirus/*C. parasitica* system with the rare capability to efficiently manipulate the genomes of both a eukaryotic virus and its host.

3. Exploiting Hypovirulence as an Experimental System

It became apparent early on in the characterization of hypovirulent *C. parasitica* field strains that phenotypic changes associated with hypovirus infection were not confined to virulence attenuation. Hypovirulence-associated traits included reduced asexual sporulation, female infertility, reduced pigmentation, diminished accumulation of certain metabolites and altered colony morphology (Elliston, 1978,1985; Anagnostakis and Day, 1979; Anagnostakis, 1982,1984). Consistent with these observed changes in multiple fungal biological processes, the accumulation of several fungal gene transcripts and protein products was also reported to be altered following hypovirus infection (Powell and Van Alfen, 1987a, 1987b; Rigling and Van Alfen, 1991; Varley et al., 1992; Zhang et al., 1993,1994; Gao and Shain, 1995; Wang and Nuss, 1995; Gao et. al., 1996; Kazmierczak et al., 1996). The pleiotrophic nature of these stable hypovirus-mediated changes in fungal phenotype and gene expression profiles suggested the possibility that hypovirus infection resulted in the perturbation of one or more key regulatory pathways. These considerations stimulated experimentation that eventually revealed a crucial role for G-protein signal transduction in a wide range of vital fungal physiological processes, including fungal pathogenesis. The line of investigation leading to these unexpected revelations and recent extensions of those studies provide an excellent illustration of the successful exploitation of transmissible hypovirulence as an experimental system.

3.1 EARLY EVIDENCE LINKING G-PROTEIN REGULATED cAMP-MEDIATED SIGNAL TRANSDUCTION, FUNGAL VIRULENCE AND HYPOVIRUS-MEDIATED HYPOVIRULENCE

In retrospect, the notion that G-protein signal transduction would play a role in fungal pathogenesis seems obvious. Heterotrimeric GTP-binding proteins serve as a critical link in signal transduction pathways that enable eukaryotic organisms to respond to a variety of extracellular events (Gilman, 1987). One could easily envision the use of this highly conserved signaling pathway by pathogenic fungi to monitor and respond to extracellular cues derived from the dynamic interaction between a fungal pathogen and its host. However, it was not until 1995 that Choi et al reported a correlation between reduced accumulation of a specific Gα subunit, *C. parasitica* CPG-1, and virulence attenuation, thus providing some of the first evidence linking this regulatory pathway with fungal pathogenesis. This publication reported that CPG-1 levels were depressed following hypovirus infection and that transgenic co-suppression of CPG-1 accumulation in the absence of virus infection also resulted in attenuation of fungal virulence. Chen et al. (1996a) subsequently reported that both hypovirus infection and CPG-1 transgenic co-suppression resulted in significantly elevated cAMP levels, consistent with the prediction that CPG-1, like related mammalian Giα subunits, may function to negatively regulate adenylyl cyclase. Using mRNA differential display, these authors also provided evidence for an extensive and stable alteration in the pattern of fungal gene expression following hypovirus infection and showed that the majority of these changes could be attributed to disruption of the CPG-1/cAMP signaling pathway. Finally, targeted disruption of the gene encoding CPG-1, *cpg-1*, was shown to result in elevated cAMP levels, a complete loss of virulence and a set of phenotypic changes similar to, but more severe than, those caused by hypovirus infection (Gao and Nuss, 1996). Disruption of a second *C. parasitica* Gα gene, *cpg-2*, in contrast, resulted only in a slight reduction in growth rate and asexual sporulation and no significant reduction in virulence or other hypovirulence-associated traits. These combined results established the requirement of an intact CPG-1 signaling pathway for optimal execution of a number of important fungal physiological processes, including virulence, and provided strong support for the view that a primary mechanism by which hypoviruses alter fungal virulence and phenotype involves disruption of CPG-1 signaling.

The subsequent cloning and targeted disruption of Gα genes for a number of pathogenic fungi, including several described in this book, have confirmed a broad role for G-protein signaling in fungal pathogenesis. Liu and Dean (1997) reported the cloning of three independent Gα genes for the rice pathogen *Magnaporthe grisea* . Two of the genes showed a high level of sequence conservation with the *C. parasitica* Gα genes; *mag*A shares 91% identity with *cpg-2* while *mag*B is virtually identical to *cpg-1* (99% identity). The third Gα gene, *mag*C is most closely related to *gna-2* from *Neurospora crassa*, one of the first Gα genes to be cloned from a filamentous fungus (Turner and Borkovich, 1993). Targeted disruption of these genes gave results similar to those described by Gao and Nuss (1996) for *C. parasitica*. Disruption of the Giα family member, *mag*B, significantly reduced vegetative growth, conidiation, appressorium formation and the ability to infect and colonize susceptible rice leaves, while disruption of *mag*A or *mag*C, had little to no effect on these processes. Interestingly, disruption of the Giα homologue in the corn pathogen *Cochliobolus heterostrophus* resulted in the production of infertile pseudothecia and a loss in the ability to form appressoria, but had no effect on virulence (Horowitz et al., 1999).

Roles in fungal virulence are not restricted to homologues of the Giα family. While disruption of three Gα subunits, including the Giα homolog *gpa1*, in the basidiomycetous corn pathogen *Ustilago maydis* caused no discernible change in phenotype, the null mutant of the fourth Gα gene, *gpa3*, was found to be defective in both mating and pathogenesis (Regenfelder, et al., 1997). Interestingly, the Gα gene most closely related to *U. maydis gpa3* is *GPA1* of the human fungal pathogen *Cryptococcus neoformans*, also a basidiomycete. Alspaugh et al (1997) recently confirmed the requirement of *GPA1* for mating and virulence.

While a conserved role for G-protein signaling in fungal pthogenesis has become firmly established, details remain sketchy concerning the nature of the ligands, receptors, effectors or second messenger systems that filamentous fungi utilize to mediate pathogenic responses. A report that exogenous cAMP can suppress the *C. neoformans GPA1* null mutant phenotype (Alspaugh et al., 1997), is consistent with the suggestion that this Gα subunit is a positive regulator of adenylyl cyclase. The picture is much less clear for the highly conserved members of the fungal Giα family, *C. parasitica cpg-1* and *M. grisea magB*. Disruption of *cpg-1* results in a greater than 2.5 fold constitutive increase in intracellular cAMP levels (Gao and Nuss, 1996), consistent with reports that mammalian Giα family members function as negative regulators of adenylyl cyclase. Although cAMP levels were not directly measured for the *magB* null mutants, appressorium formation was reported to be restored by addition of exogenous cAMP (Liu and Dean, 1997) consistent with a role for *magB* as a positive regulator of adenylyl cyclase. A decrease in intracellular cAMP levels has been reported following disruption of the Giα subunit *gna-1* from the nonpathogenic fungus *N. crassa* (Yang and Borkovich, 1999). Thus, conservation of primary amino acid sequence among Gα genes in filamentous fungi may not necessarily correlate with conservation of function. This complexity clearly emphasizes the need for additional detailed mechanistic studies of heterotrimeric G-protein signaling in filamentous fungi.

3.2 CLONING AND CHARACTERIZATION OF A *C. PARASITICA* Gβ SUBUNIT GENE

Efforts to assign an individual Gα subunit to specific functional roles is further complicated by the fact that the Gβγ subunit partners of a Gα subunit can also influence effector molecules leading to the activation of different signaling cascades. The *C. parasitica* system is also beginning to provide some insight into this complexity with the cloning and characterization of the first filamentous fungal Gβ subunit gene, *cpgb-1* (Kasahara and Nuss, 1997). Low stringency Southern hybridization analysis with *cpgb-1* as probe failed to detect any related genes suggesting that the *C. parasitica* genome, as reported for several other lower eukaryotes (Kim et al., 1996; Lilly et al., 1993), may contain a single Gβ-subunit gene. The protein product of this gene, CPGB-1, was found to be approximately 66% identical at the amino acid level with Gβ subunits from human, *Drosophila* and *Dictyostelium* origins and to share only 39.7% identity with the *Saccharomyces cerevisiae* homologue Ste4p. This observation and the apparent absence of a Gαi subunit homologue gene in the *S. cerevisae* or *Schizosaccharomyces pombe* genomes (Ivey et al., 1996) suggests that filamentous fungi may provide more appropriate experimental systems than other commonly studied lower eukaryotes for examining general mechanisms underlying G-protein signaling.

Disruption of *cpgb-1* also resulted in some of the same phenotypic changes that were observed after hypovirus infection or *cpg-1* disruption, i.e., reduced pigmentation, reduced conidiation and reduced virulence. These results confirmed the importance of G-protein signaling in multiple vital biological processes in *C. parasitica*. A noteworthy difference between the *cpg-1* and *cpgb-1* disruptants was the increased growth phenotype exhibited by the latter. This result is not inconsistent with the expected differences in the influence that Gα and Gβγ subunits can have on different effector molecules and signaling pathways. Additionally, it clearly indicates that reduced virulence resulting from alteration of G-protein signaling in *C. parasitica* is not simply due to a general decrease in vegetative growth or viability. The cloning of what appears to be a unique *C. parasitica* Gβ subunit gene and the availability of *cpgb-1* disruptant strains now allows detailed studies of the dynamics and possible coordination of G-protein subunit synthesis and accumulation as well as their relative contribution to complex fungal physiological processes.

3.3 DEVELOPMENT OF A FUNGAL Gα NULL MUTANT COMPLEMENTATION SYSTEM FOR STRUCTURE/FUNCTION STUDIES

Advantage has already been taken of the *cpg-1* null mutant to examine the *in vivo* consequences of mutations of conserved CPG-1 lipid modification sites. Gao and Nuss (1998) reported the construction of a general *cpg-1* complementation vector pGPEX1 that allows complementation of a characterized *cpg-1* null mutant with wild type and mutant cDNA copies of the *cpg-1* coding domain under control of the *cpg-1* genomic promoter and terminator elements. Stable transformation with a wild-type *cpg-1* cDNA fully rescued deficiencies in mycelial growth, orange pigmentation, asexual sporulation and virulence caused by *cpg-1* disruption (Table 1). Moreover, *cpg-1* transcript levels were shown to accumulate to comparable levels in control untransformed *C. parasitica* strains and in the transformants of *cpg-1* null mutant strains complemented with wild-type or a variety of mutant *cpg-1* cDNAs. Thus, it was concluded that the wild-type *cpg-1* cDNA copy used in that study encoded a completely functional CPG-1 and that the pGPEX1 vector contained all the information required for appropriately regulated synthesis of *cpg-1* transcripts from an ectopic integration site.

CPG-1 and related members of the proposed fungal Giα subfamily contain conserved putative myristoylation (G2) and palmitoylation (C3) sites. Previous studies have revealed that mutations of these acylation sites for mammalian $Gi\alpha$, $G\alpha0$ and $G\alpha z$ subunits altered their ability to associate with the cell membrane and greatly reduced their affinity for $\beta\gamma$ dimers and effector molecules (eg., Mumby, et al., 1990; Linder et al., 1991; Gallego, et al., 1992; Galbiati, et al., 1994). The *cpg-1* complementation system provided an opportunity to evaluate the importance of these two putative acylation sites for *C. parasitica* virulence and the additional complex fungal biological processes regulated through the CPG-1 signaling pathway. Transformation of the *cpg-1* null mutant with a *cpg-1* cDNA mutated at the putative N-terminal palmitoylation site (C3A) restored mycelial growth and virulence to near wild-type levels, but caused only a slight increase in conidiation and little increase in orange pigmentation (Table 1; see Fig. 2 in Gao and Nuss, 1998 for color photographs). In contrast, transformation with the *cpg-1* cDNA mutated at the putative myristoylation site (G2A) resulted in no significant increase in mycelial growth or conidiation, but did cause a slight increase in the ability to expand on chestnut tissue and completely restored orange pigmentation. Transformants containing *cpg-1* cDNA mutated at both putative acylation sites (GCAA) resembled the G2A transformant in colony morphology, sporulation levels and orange pigmentation, but grew slightly more slowly.

Table 1.   Traits exhibited by virulent *C. parasitica* strain EP155 and *cpg-1* disruptant (Δcpg-1) complemented with wild-type *cpg-1* cDNA (cpg-1c) or with *cpg-1* cDNA mutated at the putative myristoylation (G2A), the putative palmitoylation site (C3A), or both sites (CGAA). ND = not determined (See Gao and Nuss, 1998 for data).

TRAIT	EP155	Δcpg-1	cpg-1c	C3A	G2A	CGAA
Growth	100%	20%	100%	90%	30%	25%
Pigmentation	(+)	(-)	(+)	(-)	(+)	(+)
Conidiation	4×10^9	0	3×10^9	4×10^5	0	0
Mating	(+)	(-)	(+)	ND	ND	ND
Virulence	100%	0	100%	100%	20%	10%

Consistent with reports for mammalian $G\alpha$ subunits, mutations of the putative acylation sites also altered CPG-1 subcellular location. Whereas CPG-1 was found almost exclusively in the membrane fraction in the untransformed control and in wild-type *cpg-1* complemented transformants, mutation of the putative palmitoylation site resulted in the accumulation of approximately 60% of the CPG-1 in the soluble cytosolic fraction. Mutation of the putative myristoylation site and the double-site mutation also caused a significant reduction in the proportion of CPG-1 that was membrane associated. Surprisingly, this mutation also caused a greater than five-fold increase in the accumulation of CPG-1. Since *cpg-1* transcripts accumulated to equivalent levels in untransformed wild-type control strains and in all complemented transformants characterized, the unexpected increase in CPG-1 observed for the myristoylation and double mutant suggested a possible role for the putative myristoylation site in post-transcriptional regulation of CPG-1 accumulation. Moreover, this phenomenon appeared not to be a simple function of altered membrane association since the decreased accumulation of CPG-1 in the membrane fraction observed for the palmitoylation mutant was not accompanied by a change in the overall level of CPG-1 accumulation.

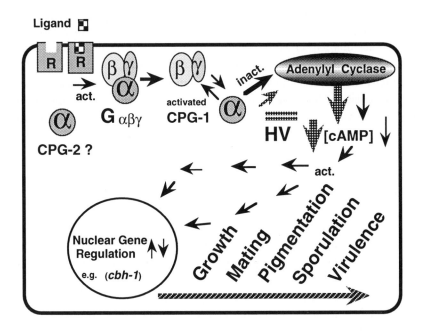

Figure 2. Cartoon depicting model of CPG-1-mediated signal transduction in virulent and hypovirulent strains of *C. parasitica*. CPG-1 is viewed as residing at the apex of branched interconnected signaling pathways that regulate a broad range of important physiological processes including mycelial growth, mating, pigmentation, conidiation and virulence. According to the model, the level of activated CPG-1 in virus-free strains is determined throughout the infection process by events at the cell surface involving as yet undefined G-protein-coupled receptors and unidentified ligands. By negatively regulating adenylyl cyclase, activated CPG-1 modulates cAMP levels (solid arrows) which, in turn, modulates gene expression to elicit appropriate adaptive responses, e.g., cellulase induction (Wang and Nuss, 1995). By directly or indirectly suppressing CPG-1 accumulation, hypovirus infection constitutively elevates cAMP levels by relieving the negative regulation of adenylyl cyclase (stippled arrows). This effectively compromises the ability of the invading fungus to respond appropriately to events at the fungus-plant interface, thereby impeding penetration, canker formation and fungal reproduction. Adapted in part from Chen et al., 1996c.

The observation that independent mutations of two putative N-terminal CPG-1 acylation sites differentially altered complex fungal phenotypic traits has led to modifications in the minimal working model of CPG-1-mediated signal transduction as shown in Figure 2. CPG-1 is viewed as being positioned at the apex of a branched, interconnected signaling pathway that regulates a broad range of important physiological processes. In this position, it is likely to be involved in the transduction of a variety of different stimuli and to serve as a common component of multiple regulatory pathways that may also be subject to cross talk. Accordingly, it is not unreasonable to expect that different CPG-1 regulated processes, ranging from virulence to conidiation, would exhibit different sensitivities to modulations in the output level of the CPG-1 signaling pathway. Future studies are likely to focus on the identification of environmental and chemical cues that are sensed by, and response factors that are activated through, the CPG-1 signaling pathway. These studies will be facilitated by the recent development of reporter/promoter expression plasmids composed of the green fluorescent protein (Parsley, Geletka, Chen and Nuss, unpublished observations) and response elements derived from CPG-1-regulated genes initially identified by mRNA differential display (Chen et al., 1996b).

4. Engineering Hypoviruses to Fine Tune the Interaction of a Fungal Pathogen with its Host

If, a) hypovirus infection alters growth, pigmentation, sporulation and virulence primarily by modulating the CPG-1 signaling pathway and b) subtle perturbations of the CPG-1 signaling pathway (e.g., point mutations at putative CPG-1 acylation sites) differentially affect these same processes, could one genetically modify hypoviruses to fine tune the effect of hypovirus infection on virulence and associated processes ?

A tentative answer to this question was supplied by Craven et al. (1993) with the reported deletion of the ORF A-encoded papain-like protease p29 from the CHV1-EP713 infectious cDNA clone. The resulting viral RNA (recombinant virus Δp29) was found to be replication competent upon transfection into virus-free *C. parasitica* strains but to cause a less severe reduction than wild-type CHV1-EP713 viral RNA in fungal pigment production, asexual sporulation and laccase production. Importantly, deletion of p29 had no effect on the level of virus-mediated hypovirulence. These results clearly demonstrated the feasibility of manipulating an infectious hypovirus cDNA clone to modulate the phenotypic consequences of hypovirus infection. However, the observation that p29 deletion partially relieved only a subset of the symptoms associated with CHV1-EP713 infection clearly indicated the involvement of multiple viral-encoded factors in symptom expression. In an effort to define these additional symptom determinants, Chen and Nuss (1999) exploited the considerable diversity among natural hypovirus isolates by developing a comparative virology system based on the construction of an infectious cDNA clone of a second hypovirus, CHV1-Euro7.

In analogy with plant viruses, CHV1-EP713 and CHV1-Euro7 can be viewed as severe and mild hypovirus isolates, respectively. *Cryphonectria parasitica* strain Euro7, the source of hypovirus isolate CHV1-Euro7, actually grows faster than the corresponding virus-free strain on synthetic media, while strain EP713, the source of prototypic hypovirus CHV1-EP713, grows more slowly. Strain EP713 is considered highly hypovirulent, forming small, superficial cankers on chestnut tissue and producing few to no asexual spores either on synthetic medium or on chestnut tissue. Canker formation and morphology are quite different for Euro7. An initial aggressive burst of colonization and expansion slows or ceases, concomitant with heavy callus formation at the canker margins. Moreover, Euro7 produces many more asexual spores than strain EP713 on synthetic medium and especially on canker surfaces. As will be discussed in a later section, these properties are predicted to positively contribute to biological control potential by enhancing persistence and dissemination.

The development of an infectious cDNA clone of CHV1-Euro7 allowed an examination of whether the differences in phenotypic traits exhibited by strains Euro7 and EP713 were due to the relative contributions of the two viral genomes or to additional contributions by the genome of the two fungal hosts. This was accomplished by independently transfecting virus-free strains corresponding to strains Euro7 and EP713, Euro7(-v) and EP155, respectively, with synthetic transcripts of the two viruses. Colonies transfected with CHV1-EP713 transcripts closely resembled strain EP713 in colony morphology, growth rate and sporulation level irrespective of the fungal genetic background, while colonies transfected with CHV1-Euro7 transcripts resembled Euro7 in these parameters (See Chen and Nuss, 1999 for color photographs). The relative contributions of the two viral genomes was even more dramatically observed on dormant chestnut tissue. As shown in Figure 3, transfection with CHV1-EP713 transcripts severely reduced the ability of both virus-free strains to expand on chestnut tissue, resulting in the production of small, superficial cankers that were generally devoid of spore-bearing stromal pustules as previously described for strain EP713 (Choi and Nuss; 1992). As described for strain Euro7, CHV1-Euro7 transfectants were more aggressive in canker formation, producing cankers with distinctive, ridged margins suggestive of callus formation and covered with stromal pustules containing viable asexual spores (Table 2). Thus, canker morphology, canker expansion and asexual sporulation levels on chestnut stem tissue appear to be controlled to a much greater extent by the hypovirus genome than by the genome of the fungal host.

The real utility of this comparative system is derived from the ability to construct viable recombinant chimeric viruses provided by the high level of sequence identity shared by the two viruses (approximately 90% at the nucleotide level). This was initially demonstrated by interchanging the two viral polyprotein coding domains, ORF A and ORF B, to form chimeras designated AE7B713 and A713BE7 (Figure 4). Transfection of either EP155 or Euro7(-v) with the AE7B713 chimera resulted in a colony morphology similar to that of CHV1-EP713 transfectants, while transfection with the A713BE7 chimera resulted in a colony morphology similar to that of CHV1-Euro7 transfectants (See Chen and Nuss, 1999, Figure 4 for color photographs). Therefore, the ORF B portion of the chimera generally determines colony morphology.

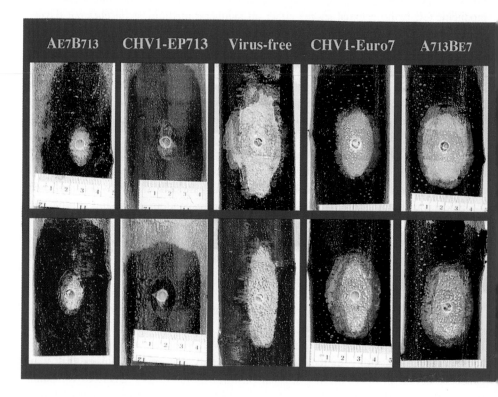

Figure 3. Gallery of representative cankers from virus-free and transfected *C. parasitica* strains (Top row) Typical cankers formed by virus-free strain EP155 (center) and strain EP155 transfected with CHV1-EP713, CHV1-Euro7, chimeric virus AE7B713, and chimeric virus A713BE7. (Bottom row) Cankers formed by virus-free strain Euro7(-v) and the corresponding set of Euro7(-v) transfectants. Stromal pustules (stromata that contain asexual spore-forming bodies termed pycnidia) are prominent features on the surface of cankers caused by virus-free strains EP155 and Euro7(-v) as well as the CHV1-Euro7 and A713BE7 transfectants. Spiral structures, termed ceri, composed of conidia are seen exuding from some stromata. These structures are rarely observed on the surface of cankers formed by CHV1-EP713 or AE7B713 transfectants. Adapted from Chen and Nuss, 1999.

Similarly, differences in growth rates and sporulation levels on synthetic medium mapped to ORF B. These trends also extended to canker morphology but in a more pronounced fashion. Transfectants containing the chimeric viruses with a CHV1-Euro7 ORF B produced cankers strikingly similar in morphology to those caused by wild-type CHV1-Euro7, including the formation of callused margins and extensive production of stromal pustules. Consistent with these results, chimeras containing an ORF B from CHV1-EP713 produced cankers with morphologies similar to those caused by wild-type CHV1-EP713 transfectants. Therefore, the contribution of ORF B to differences in host phenotypic changes caused by CHV1-EP713 and CHV1-Euro7 extended to canker morphology, size and spore production (Table 2).

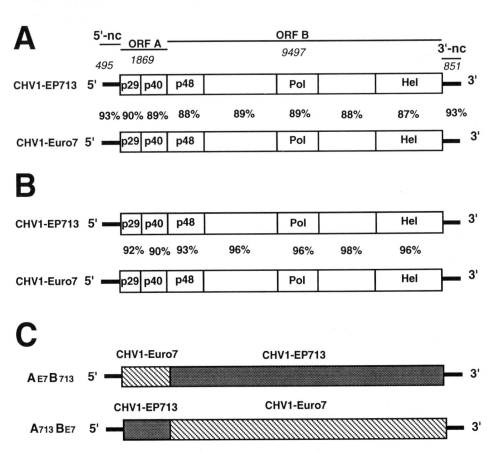

Figure 4. Sequence similarity between CHV1-EP713 and CHV1-Euro7. (A) Similarities at the nucleotide levels. The percent nucleotide identity for different coding and noncoding (nc) regions is indicated between representations of the two viral genomes. (B) Similar information at the deduced amino acid levels. (C) Cartoon of chimeric viruses AE7B713 and A713BE7, respectively. Adapted in part from Chen and Nuss, 1999)

D.L. Nuss

TABLE 2. Effect of transfection with wild-type and chimeric hypovirus transcripts on canker expansion and production of asexual spores on cankered tissue.

STRAIN	CANKER SIZE (cm^2)		SPORULATION[1]	
	DAY 21	DAY 31	STROMATA PER CANKER	CONIDIA PER CANKER
EP155	17.0±6.7	40±15.7	246.7±115.3	7.4 x 10^8±8.7 x 10^8
Euro7(-v)	8.3±3.9	21.0±9.3	212.0±71.7	1.1 x 10^9±8.8 x 10^8
EP155/CHV1-EP713	2.6±1.2	3.2±1.2	0.8±1.1	0
Euro7(-v)/CHV1-EP713	2.3±0.5	2.4±0.6	0.2±0.3	0
EP155/AE7B713	2.9±0.8	3.3±0.9	26.5±12.8	1.4 x 10^5±6.0 x 10^4
Euro7(-v)/AE7B713	3.5±0.5	3.7±0.5	33.3±10.0	3.4 x 10^5±3.5 x 10^5
EP155/CHV1-Euro7	6.4±2.8	12.1±6.4	101.2±23.2	3.1 x 10^6±2.3 x 10^6
Euro7(-v)/CHV1-Euro7	6.5±1.7	18.2±5.1	172.2±44.5	2.6 x 10^9±1.1 x 10^9
EP155/A713BE7	8.7±2.6	25.4±8.4	106.0±38.7	7.3 x 10^6±3.8 x 10^6
Euro7(-v)/A713BE7	6.6±2.2	17.1±6.3	135.8±58.8	1.7 x 10^9±1.2 x 10^9

1. The number of stromata (stromal pustules) per canker was determined at day 30. A subset of 10 stromata was identified and removed from the canker. The stromata were placed between two cleaned microscope slides in 100 µl of 0.15% Tween 80 and genetly crushed. The crushed stromata material was then diluted into a total of 4 ml of 0.15% Tween 80 and the number of conidia determined with the aid of a hemacytometer (See Chen and Nuss, 1999 for details and color photographs).

An extension of the chimeric mapping approach is anticipated to provide an efficient route to the identification of hypovirus determinants for reduced fungal virulence and altered phenotype. Indeed, chimeric hypoviruses that cause infected *C. parasitica* strains to produce very small cankers densely covered with fruiting spore bodies have now been constructed (Chen and Nuss, unpublished results). The information gained from these mapping studies will, in turn, allow one to more intelligently engineer hypoviruses with the view of modifying specific signal transduction pathways, e.g., the CPG-1 pathway and its branches, in order to rationally fine tune the interactions between a fungal pathogen and its host. This is just one example of how knowledge of signal transduction pathways that underlie fungal pathogenesis might be used in the future to manipulate fungal pathogens.

5. Back to the Field

The construction of a biologically active infectious hypovirus cDNA also has practical implications for biological control. Although natural hypovirulence-mediated control of chestnut blight has been reported for European forests and orchards (Heiniger and Rigling, 1994; Bissegger et al., 1997), hypovirulence has been generally ineffective in North American forest ecosystems. Factors contributing to this lower efficacy include barriers to cytoplasmic spread of hypoviruses due to the abundance of different vegetative compatibility (VC) types present in North American *C. parasitica* populations relative to that found in Europe (Anagnostakis and Kranz, 1987; Liu and Milgroom, 1996; Bissegger et al., 1997; Anagnostakis et al., 1998). Fungal strains of the same or closely related VC types readily anastomose (fuse) allowing transmission of cytoplasmically replicating hypoviruses, while strains of distantly related VC types fail to fuse, preventing hypovirus transmission. Because transgenic hypovirulent strains contain a chromosomally integrated viral cDNA, it was predicted that, unlike the situation for natural hypovirulent strains, viral genetic information would be inherited in a Mendelian fashion by ascospore progeny of a sexual cross. Additionally, these hypovirus-containing progeny should represent a spectrum of different VC types due to allelic rearrangement at the genetic loci controlling vegetative compatibility. Furthermore, essentially all asexual spores generated from a transgenic hypovirulent *C. parasitica* strain should carry the integrated viral cDNA and the derived, cytoplasmically replicating viral RNA. These combined transmission properties were predicted to circumvent barriers posed by the VC system leading to enhanced dissemination of the hypovirulence phenotype and possibly enhanced biological control of *C. parasitica* populations having high VC diversity.

Following laboratory confirmation that transgenic hypovirulent strains did indeed exhibit a novel mode of hypovirus transmission (Chen et al., 1993), permission was requested to test the performance of these strains under actual field conditions. A transgenic hypovirulent *C. parasitica* strain containing the CHV1-EP713 cDNA was introduced into a Connecticut forest site in July of 1994 as a single season, limited environmental release (Anagnostakis et al., 1998). Subsequent analysis over the following three years confirmed hypovirus transmission from transgenic hypovirulent strains to ascospore progeny under actual field conditions. Persistence of the released transgenic hypovirulent strains was demonstrated for as long as two years after introduction. Additionally, evidence was obtained for dissemination of transgenic cDNA-derived hypovirus RNA to nontransgenic endogenous *C. parasitica* strains.

While the transgenic strains performed as predicted in terms of hypovirus transmission, the limited persistence of these strains in the ecosystem raises important issues.   MacDonald and Fulbright (1991) have noted that most North American biological control efforts have employed highly hypovirulent strains that harbored viruses with properties similar to CHV1-EP713, i.e., strains that , while highly curative, were quite debilitated in their ability to colonize and produce spores on chestnut tissue.   These authors further emphasized the view that successful hypovirulence-mediated biological control is likely to require a continual reservoir of hypovirulent inoculum. This prediction is currently being examined by testing an intense, multiyear deployment protocol employing CHV1-EP713 transgenic strains (Chen, Balbalian, Geletka, Anagnostakis and Nuss, unpublished results). An alternative strategy to the problem of persistence and sustainability of hypovirulence is suggested by results of the comparative hypovirus studies described above.

As noted in a previous section, CHV1-Euro7-infected *C. parasitica* strains differ from strains infected with CHV1-EP713 in precisely those properties that are expected to have a direct impact on persistence: colonization of and spore production on bark tissue. MacDonald and coworkers have observed high rates of dissemination after introduction of strain Euro7 into several forest ecosystems (personal communication). Additionally, a recent survey of European hypoviruses (Allemann et al., 1999) indicated a wide distribution and dominance of CHV1-Euro7-related hypovirus isolates of the Italian subtype and a low incidence of CHV1-EP713-related isolates of the French subtype. This result is consistent with a greater ecological fitness of the former group of hypoviruses. Thus, the full-length infectious CHV1-Euro7 cDNA clone provides the means of constructing second generation transgenic hypovirulent strains that combine properties of enhanced colonization and spore production with a novel mode of virus transmission to ascospore progeny.   One can easily imagine extending this theme to the construction of transgenic hypovirulent strains with chimeras of CHV1-EP713 and CHV1-Euro7 or additional infectious hypovirus cDNAs as they become available. These efforts will be complemented by additional insights into the precise mechanisms by which hypovirus-encoded determinants impact fungal signaling pathways and their impact on fungal-plant host interactions. In many ways, the engineering of hypoviruses for enhanced biological control can be compared to the development of viral vectors in contemporary medical gene therapy.

6. Future Perspectives

The journey described in this chapter continues.   Many of the unique capabilities provided by the hypovirus/*C. parasitica* experimental system were unanticipated and evolved over time. The opportunity now exists to take full advantage of these capabilities to examine fundamental issues ranging from virus-host interactions to signal transduction underlying fungal pathogenesis. The system will be further strengthened by the recent development of promoter/reporter constructs that allow real time nondestructive monitoring of specific signal transduction pathways, the application of expressed sequence tagging approaches for gene discovery and the continued exploitation of hypovirus natural diversity. The most challenging portion of the journey, the return to the field of hypovirulent strains with enhanced biological control properties, will necessitate a multidisciplinary approach involving cooperation from forest pathologists, ecologists, molecular biologists and regulatory agencies. Recent progress in the molecular characterization of virulence modulating viruses from other pathogenic fungi (Hong et al., 1998a, 1998b; Jian et al., 1997, 1998; Lakshman et al., 1998) and the potential for expanding host range for hypoviruses (Chen et al., 1994a, 1996a) suggest that the concept of engineering mycoviruses for purposes of understanding and controlling fungal pathogenesis may find broad application.

7. Acknowledgments

Portions of this review were adapted in part from Kasahara and Nuss (1997), Gao and Nuss (1998) and Chen and Nuss (1999). I thank the following members of our laboratory for their critical reading of this review: Clarissa Balbalian, Baoshan Chen, Lynn Geletka and Shin Kasahara. Support from the NIH (GM55981) and the USDA (NRICGP 95-37312-1638) is also gratefully acknowledged.

8. References

Allemann, C., Hoegger, P., Heiniger, U., and Rigling, D. (1999) Genetic variation of Cryphonectria hypoviruses (CHV1) in Europe, assessed using RFLP markers, *Journal of Molecular Ecology* in press.

Alspaugh, J. A., Perfect, J. R., and Heitman, J. (1997) *Cryptococcus neoformans* mating and virulence are regulated by the G-protein α subunit GPA1 and cAMP, *Genes and Development* **11**, 3206-3216.

Anagnostakis, S. L. (1982) Biological control of chestnut blight, *Science* **215,** 466-471.

Anagnostakis, S.L. (1984) The mycelial biology of *Endothia parasitica* I: Nuclear and cytoplasmic genes that determine morphology and virulence, in D.H. Jennings and A.D.M. Rayner (eds), *The Ecology and Physiology of the Fungal mycelium* , Cambridge University Press, Cambridge, pp. 353-366.

Anagnostakis, S. L., Chen, B., Geletka, L. M., and Nuss, D. L. (1998) Hypovirus transmission to ascospore progeny by field-released transgenic hypovirulent strains of *Cryphonectria parasitica,* *Phytopathology* **88**, 598-604.

Anagnostakis, S. L. and Day, P. R. (1979) Hypovirulence conversion in *Endothia parasitica,* *Phytopathology* **69**, 1226-1229.

Anagnostakis, S. L. and Kranz, J. (1987) Population dynamics of *Cryphonectria parasitica* in a mixed-hardwood forest in Connecticut, *Phytopathology* **77**, 751-754.

Bissegger, M., Rigling, D., and Heiniger, U. (1997) Population structure and disease development of *Cryphonectria parasitica* in European chestnut forests in the presence of natural hypovirulence, *Phytopathology* **87**, 50-59.

Buck, K. W. (1986) Fungal virology - an overview, in K. W. Buck (ed.), *Fungal Virology*, CRC Press, Boca Raton, Fla., pp.2-84.

Chen, B., Chen, C-H, Bowman, B. H. and Nuss, D. L. (1996a) Phenotypic changes associated with wild-type and mutant hypovirus RNA transfection of plant pathogenic fungi phylogenetically related to *Cryphonectria parasitica*, *Phytopathology* **86**, 301-310.

Chen, B., Choi, G.H., and Nuss, D.L. (1993) Mitotic stability and nuclear inheritance of integrated viral cDNA in engineered hypovirulent strains of the chestnut blight fungus, *EMBO J.* **12**, 2991-2998.

Chen, B., Choi, G.H., and Nuss, D.L. (1994a) Attenuation of fungal virulence by synthetic infectious hypovirus transcripts, *Science* **264**, 1762-1764.

Chen, B., Craven, M.G., Choi, G.H., and Nuss, D.L. (1994b) cDNA-derived hypovirus RNA in transformed chestnut blight fungus is spliced and trimmed of vector nucleotides, *Virology* **202**, 441-448.

Chen, B., Gao, S., Choi, G.H., and Nuss, D.L. (1996b) Extensive alteration of fungal gene transcript accumulation and alteration of G-protein regulated cAMP levels by virulence-attenuating hypovirus, *Proc. Natl. Acad. Sci. USA*, **93**, 7996-8000.

Chen, B., Gao, S., Geletka, L. M., Kasahara, S., Wang, P., and Nuss, D. L. (1996c) Review of evidence linking hypovirulence-mediated disruption of cellular G-protein signal transduction and attenuation of fungal virulence, in G. Stacey, B. Mullin, and P. M. Gresshoff (eds.) *Biology of Plant-Microbe Interactions*, International Society for Molecular Plant-Microbe Interactions, St. Paul, Minnesota, pp. 227-232.

Chen, B. and Nuss, D. L. (1999) Infectious cDNA clone of hypovirus CHV1-Euro7: a comparitive virology approach to investigate virus-mediated hypovirulence of the chestnut blight fungus *Cryphonectria parasitica*, *J. Virol.* **73**, 985-992.

Choi, G.H., Chen, B., and Nuss, D.L. (1995) Virus-mediated or transgenic suppression of a G-protein α subunit and attenuation of fungal virulence, *Proc. Natl. Acad. Sci. USA*, **92**, 305-309.

Choi, G.H. and Nuss, D.L. (1992) Hypovirulence of chestnut blight fungus conferred by an infectious viral cDNA, *Science*, **257**, 800-803.

Choi, G. H., Pawlyk, D. M., and Nuss, D. L. (1991a) The autocatalytic protease p29 encoded by a hypovirulence-associated virus of the chestnut blight fungus resembles the poty-virus-encoded protease HC-Pro, *Virology* **183**, 747-752.

Choi, G.H., Pawlyk, D.M., Rae, B., Shapira, R., and Nuss, D.L. (1993) Molecular analysis and over expression of the gene encoding endothiapepsin, an aspartic protease from *Cryphonectria parasitica*, *Gene,* **125**, 135-141.

Choi, G. H., Shapira, R., and Nuss, D. L. (1991b) Cotranslational autoproteolysis involved in gene expression from a double-stranded RNA genetic element associated with hypovirulence of the chestnut blight fungus, *Proc. Natl. Acad. Sci. U.S.A.* **88**, 1167-1171.

Churchill, A. C. L., Ciufetti, L. M., Hansen, D. R., Van Etten, H. D., and Van Alfen, N. K. (1990) Transformation of the fungal pathogen *Cryphonectria parasitica* with a variety of heterologous plasmids, *Curr. Genet.* **17**, 25-31.

Craven, M.G., Pawlyk, D.M., Choi, G.H., and Nuss, D.L. (1993) Papain-like protease p29 as a symptom determinant encoded by a hypovirulence-associated virus of the chestnut blight fungus, *J. Virology* **67**, 6513-6521.

Day, P. R., and Dodds, J. A. (1979) Viruses of plant pathogenic fungi, in P. A. Lemke (ed.), *Viruses and plasmids in fungi*, Marcel Dekker Inc., New York, pp. 201-238.

Day, P. R., Dodds, J. A., Elliston, J. E., Jaynes, R. A., and Anagnostakis, (1977) Double-stranded RNA in *Endothia parasitica*, *Phytopathology* **68**, 1391-1396.

Elliston, J.E. (1978) Pathogenicity and sporulation in normal and diseased strains of *Endothia parasitica* in American chestnut, in W. L. MacDonald, F. C. Cech, J. Luchok, and C. Smith (eds.), *Proc. Amer. Chestnut Symp.*, West Virginia University Press, Morgantown, pp95-100.

Elliston, J.E. (1985) Characterization of dsRNA-free and ds-RNA-containing strains of *Endothia parasitica* in relation to hypovirulence, *Phytopathology* **75**, 151-158.

Enebak, S.A., MacDonald, W.L., and Hillman, B.I. (1994) Effect of dsRNA associated with isolates of *Cryphonectria parasitica* from the central Appalachians and their relatedness to other dsRNAs from North America and Europe, *Phytopathology* **84**, 528-534.

Galbiati, M., Guzzi, F., Magee, A. I., Milligan, G., and Parenti, M. (1994) N-terminal fatty acylation of the α-subunit of the G-protein Gi1: only the myristoylated protein is a substrate for palmitoylation, *Biochem. Journal* **303**, 697-700.

Gallego, C., Gupta, S. K., Winitz, S., Eisfelder, B. J., and Johnson, G. L. (1992) Myristoylation of the Gαi2 polypeptide, a G protein α subunit, is required for its signaling and transformation functions. *Proc. Natl. Acad. Sci. USA* **89**, 9695-9699.

Gao, S., Choi, C.H., Shain, L. and Nuss, D.L. (1996) Cloning and targeted disruption of *epng-1*, encoding the major *in vitro* extracellular endopolygalacturonase of the chestnut blight fungus, *Cryphonectria parasitica. Appl. Environ. Microbiology* **62,** 1984-1990.

Gao, S. and Nuss, D.L. (1996) Distinct roles for two G protein α subunits in fungal virulence, morphology and reproduction revealed by targeted gene disruption, *Proc. Natl. Acad. Sci. USA* **93**, 14122-14127.

Gao, S. and Nuss, D. L. (1998) Mutagenesis of putative acylation sites alters function, localization and accumulation of a Giα subunit of the chestnut blight fungus *Cryphonectria parasitica, Mol. Plant-Microbe Interactions* **11**, 1130-1135.

Gao, S. and Shain, L. (1995) Activity of polygalacturonase produced by *Cryphonectria parasitica* in chestnut bark and its inhibition by extracts from American and Chinese chestnut, *Physiol. Mol. Plant Pathol.* **46**, 199-213.

Ghabrial, S. A. (1998) Origin, adaption and evolutionary pathways of fungal viruses, *Virus Genes* **16**,119-131.

Gilman, A.G. (1987) G proteins: Transducers of receptor-generated signals, *Ann. Rev. Biochem.* **56**, 615-649.

Heiniger, U. and Rigling, D. (1994) Biological control of chestnut blight in Europe, *Ann. Rev. Phytopathol.* **32**, 581-599.

Hillman, B. I., Fulbright, D. W., Nuss, D. L. and Van Alfen, N. K. (1995) *Hypoviridae*, in F. A. Murphy, C. M. Fauquet, D. H. L. Bishop, S. A. Ghabrial, A. W. Jarvis, G. P. Martelli, M. A. Mayo, and M. D. Summers, (eds), *Virus Taxonomy* , Springer Verlag, New York, pp.261-264.

Hillman, B.I., Halpern, B.T., and Brown, M.P. (1994) A viral dsRNA element of the chestnut blight fungus with a distinct genetic organization, *Virology* **201**, 241-250.

Hillman, B.I., Tian, Y., Bedker, P.J., and Brown, M.P. (1992) A North American hypovirulent isolate of the chestnut blight fungus with European isolate-related dsRNA, *J. Gen. Virology* **73**, 681-686.

Hong, Y., Cole, T. E., Brasier, C. M., and Buck, K. W. (1998a) Novel structures of two virus-like RNA elements from a diseased isolate of the Dutch elm disease fungus, *Ophiostoma novo-ulmi*, *Virology* **242**, 80-89.

Hong, Y., Cole, T. E., Brasier, C. M., and Buck, K. W. (1998b) Evolutionary relationships among putative RNA-dependent RNA polymerases encoded by a mitochondrial virus-like RNA in the Dutch Elm disease fungus, *Ophiostoma novo-ulmi*, by other viruses and virus-like RNAs and by the Arabidopsis mitochondrial genome, *Virology* **246**, 158-169.

Horwitz, B. A., Sharon, A., Shun-Wen, L., Ritter, V., Sandrock, T., Yoder, O. C. and Turgeon, B. G. (1999) A G protein alpha subunit from *Cochliobolus heterostrophus* involved in mating and appressorium formation, *Fungal Gen & Biol.* **16**, 19-32.

Ivey, F. D., Hodge, P. N., Turner, G. E., and Borkovich, K. A. (1996) The G_α homologue *gna-1* controls multiple differentiation pathways in *Neurospora crassa.*, *Mol. Biol. Cell* **7**, 1283-1297.

Jian, J., Lakshman, D. K. and Travantzis, S. M. (1997) Association of distinct double-stranded RNAs with enhanced or diminished virulence in *Rhizoctonia solani* infecting potato, *Mol. Plant-Microbe Interact* **10**, 1002-1009.

Jian, J., Lakshman, D. K. and Travantzis, S. M. (1998) A virulence-associated, 6.4-kb, double-stranded RNA from *Rhizoctonia solani* is phylogenetically related to plant bromoviruses and electron transport enzymes, *Mol. Plant-Microbe Interact* **11**, 601-609.

Kasahara, S., and Nuss, D. (1997) Targeted disruption of a fungal G-protein β subunit gene results in increased vegetative growth but reduced virulence, *Mol. Plant-Microbe Interact.* **8**, 984-993.

Kazmierczak, P., Pfeiffer, P., Zhang, L., and Van Alfen, N.K. (1996) Transcriptional repression of specific host genes by the mycovirus *Cryphonectria* hypovirus 1, *J. Virology* **70**, 1137-1142.

Kim, D.H., Rigling, D., Zhang, L., and Van Alfen, N.K. (1995) A new extracellular laccase of *Cryphonectria parasitica* is revealed by deletion of *Lac*1, *Mol. Plant-Microbe Interact.* **8**, 259-266.

Kim, D. U., Park, S. K., Chung, K. S., Choi, M. U., and Yoo, H. S. (1996) the G protein β subunit Gβp1 of *Schizosaccharomyces pombe* is a negative regulator of sexual development, *Mol. Gen. Genet.* **252**, 20-32.

Koonin, E. V., Choi, G. H., Nuss, D. L., Shapira, R., and Carrington, J. C. (1991) Evidence for common ancestry of a chestnut blight hypovirulence-associated double-stranded RNA and a group of positive-strand RNA plant viruses, *Proc. Natl. Acad. Sci. USA* **88**, 10647-10651.

Lakshman, D. K., Jain, J. and Travantzis, S. M. (1998) A double-stranded RNA element from a hypovirulent strain of *Rhizoctonia solani* occurs in DNA form and is genetically related to the pentafunctional AROM protein of the shikimate pathway, *Proc. Natl. Acad. Sci. USA* **95**, 6425-6429.

Lilly, P., Wu, L., Welker, D. L. and Deverotes, P. N. (1993) A G-protein β-subunit is essential for *Dictyostelium* development, *Genes Dev.* **7**, 986-995.

Linder, M. E., Pang, I., Duronio, R. J., Gordon, J. I., Sternweis, P. C., and Gilman A. G. (1991) Lipid modifications of G protein subunits: myristoylation of Goα increases its affinity for βγ, *J. Biol. Chem.* **266**, 4654-4659.

Liu, S. and Dean, R. A. (1997) G protein α subunit genes control growth, development and pathogenecity of *Magnaporthe grisea*, *Mol. Plant-Microbe Interact.* **9**, 1075-1086.

Liu, Y.-C. and Milgroom, M. G. (1996) Correlation between hypovirus transmission and the number of vegetative incompatibility (vic) gene different amoung isolates from natural populations of *Cryphonectria parasitica*, *Phytopathology* **86**, 79-86.

MacDonald, W. L. and Fulbright, D. W. (1991) Biological control of chestnut blight: use and limitation of transmissible hypovirulence, *Plant Disease* **75**, 656-661.

Mumby, S. M., Heukeroth, R. O., Gordon, J. I., and Gilman, A. G. (1990) G-protein α-subunit expression, myristoylation, and membrane association in COS cells, *Proc. Natl. Acad. Sci. USA* **87**, 728-732.

Nuss, D. L. (1992) Biological control of chestnut blight: an example of virus mediated attenuation of fungal pathogenesis, *Microbiol. Rev.* **56**, 561-576.

Nuss, D. L. (1996) Using hypoviruses to probe and perturb signal transduction processes underlying fungal pathogenesis, *The Plant Cell* **8**, 1845-1853.

Nuss, D. L. and Koltin, Y. (1990) Significance of dsRNA genetic elements in plant pathogenic fungi, *Annu. Rev. Phytopathol.* **28**, 37-58.

Polashoch, J. J. and Hillman, B. I. (1994) A small mitochondrial double-stranded (ds) RNA element associated with a hypovirulent strain of the chestnut blight fungus and ancestrally related to yeast cytoplasmic T and W dsRNAs, *Proc. Natl. Acad. Sci. USA* **91**, 8680-8684.

Powell, W.A., and Van Alfen, N.K. (1987a) Differential accumulation of poly(A)+ RNA between virulent and double-stranded RNA-induced hypovirulent strains of *Cryphonectria (Endothia) parasitica*, *Mol. Cell. Biol.* **7**, 3688-3693.

Powell, W.A., and Van Alfen, N.K. (1987b) Two nonhomologous viruses of *Cryphonectria (Endothia) parasitica* reduce accumulation of specific virulence-associated polypeptides, *J. Bacteriol.* **169**, 5324-5326.

Regenfelder, E., Spellig, T., Hartmann, A., Lauenstein, S., Bolker, M., and Kahmann, R. (1997) G proteins in *Ustilago maydis*: transmission of multiple signals ?, *EMBO J.* **16**, 1934-1942.

Rigling, D. and Van Alfen, N.K. (1991) Regulation of laccase biosynthesis in the plant pathogenic fungus *Cryphonectria parasitica* by double-stranded RNA, *J. Bacteriol.* **173**, 8000-8003.

Shapira, R., Choi, G., and Nuss, D. L. (1991) Virus-like genetic organization and expression strategy for a double-stranded RNA genetic element associated with biological control of chestnut blight, EMBO J. **10**, 731-739.

Shapira, R., and D. L. Nuss. (1991) Gene expression by a hypovirulence-associated virus of the chestnut blight fungus involves two papain-like protease activities, *J. Biol Chem.* **266,** 19419-19425.

Turner, G. E. and Borkovich, K. A. (1993) Identification of a G protein α subunit from *Neurospora crassa* that is similar to the Gi family, *J. Biol. Chem.* **268**, 14805-14811.

D.L. Nuss

Van Alfen, N.K. (1986) Hypovirulence of *Endothia (Cryphonectria) parasitica* and *Rhizoctonia solani*, in K.W. Buck (ed.), *Fungal Virology*, CRC Press, Boca Raton, Fla., pp. 143-162.

Varley, D.A., Podila, G.K., and Hiremath, S.T. (1992) Cutinase in *Cryphonectria parasitica*. the chestnut blight fungus: suppression of cutinase gene expression in isogenic hypovirulent strains containing double-stranded RNA, *Mol. Cell. Biol.* **12**, 4539-4544.

Wang, P., Larson, T. G., Chen, C-H., Pawlyk, D. M., Clark, J. A., and Nuss, D. L. (1998) Cloning and characterization of a general amino acid control transcriptional activator from the chestnut blight fungus *Cryphonectria parasitica*, *Fungal Genetics and Biology* **23**, 81-94.

Wang, P. and Nuss, D.L. (1995) Induction of a *Cryphonectria parasitica* cellobiohydrolase I gene is suppressed by hypovirus infection and regulation by a G-protein-linked signaling pathway involved in fungal pathogenesis, *Proc. Natl. Acad. Sci. USA* **92**, 11529-11533.

Yang, Q. and Borkovich, K. A. (1999) Mutational activation of a Gαi, causes uncontrolled proliferation of aerial hyphae and increased sensitivity to heat and oxidative stress in *Neurospora crassa*, *Genetics* **151**, 107-117.

Zhang, L., Churchill, A.C.L., Kazmierczak, P., Kim, D., and Van Alfen, N.K. (1993) Hypovirulence-associated traits induced by a mycovirus of *Cryphonectria parasitica* are mimicked by targeted inactivation of a host gene, *Mol. Cell. Biol.* **13**, 7782-7792.

Zhang, L., Villalon, Y., Sun, P., Kazmierczak, P., and Van Alfen, N.K. (1994) Virus-associated down-regulation of the gene encoding cryparin, an abundant cell-surface protein from the chestnut blight fungus, *Cryphonetria parasitica*, *Gene* **139**, 59-64.

INSECT PATHOGENIC FUNGI: FROM GENES TO POPULATIONS

MICHAEL J. BIDOCHKA, ANDRENA M. KAMP
AND J.N. AMRITHA DE CROOS

Department of Biology, Trent University, Peterborough, ON, Canada, K9J 7B8

1. Introduction

Despite many advances in pest control, including the transgenic expression of insect toxic proteins in crop plants, insects continue to represent serious competition for food globally. Furthermore, insect resistance to chemical insecticides (Charlesworth, 1998), to *Bacillus thuringiensis* (Bt; Tabashnik *et al.*, 1997) and to Bt-transgenic crops (Tabashnik, 1997) suggests that crop protection alternatives should be exploited. One of these alternatives is the utilization of entomopathogenic fungi, one of the most important factors regulating insect populations in nature. Spectacular epizootics produced by entomopathogenic fungi are known to decimate grasshoppers, aphids, leafhoppers, soil-dwelling larvae and other insects whose populations have the potential for explosive growth (Tanada and Kaya, 1993).

Approximately 1,000 species of fungi are pathogenic to insects, and fungal infection has been noted for insects in habitats as diverse as freshwater, soil, and above ground vegetation. A number of these fungi are presently being exploited or developed for commercial crop and forest protection; to date, successes achieved in insect control using entomopathogenic fungi are encouraging (see Roberts and Hajek, 1992). Although not generally reported in the literature, many failures in biocontrol trials suggest that the biology of entomopathogenic fungi must be more fully understood if their full potential is to be realized.

There are approximately ten fungal species currently being developed or utilized for insect control (Table 1). Of these, two species in particular, *Metarhizium anisopliae* and *Beauveria bassiana*, have received attention in a rigorous molecular and biochemical approach toward understanding the enzymes, structural genes and cell signaling mechanisms involved in infection processes. Concurrent with these studies, those directed at fungal epizootiology or utilizing computer modeling have advanced our understanding of disease transmission in the field.

Although we are developing a better understanding of the virulence factors and epizootiology of entomopathogenic fungi, certain other features of the biology of these fungi are less well represented in the literature. These include the analysis of fungal population structure, an understanding of their genetics, the relative importance of pathogenesis versus saprophytism in population dynamics, and an appreciation of fungal ecology. It is the purpose of this chapter to provide an overview of entomopathogenic fungi from the genetic to the population level. In this light, we hope

171

J. W. Kronstad (ed.), Fungal Pathology, 171–193.

TABLE 1. Examples of entomopathogenic fungi currently being used or developed for insect biocontrol.

Fungus	Taxon	Targetted insects
Beauveria bassiana	Deuteromycetes	Many, including European corn borer, Asiatic corn borer whitefly, thrips, aphids and grasshoppers.
Beauveria brongniartii	Deuteromycetes	Many, including scarab grubs, and cockchafers.
Metarhizium anisopliae	Deuteromycetes	Many, including a range of beetle and moth larvae.
Metarhizium flavoviride	Deuteromycetes	Grasshoppers and locusts.
Paecilomyces fumosoroseus	Deuteromycetes	Whitefly, thrips, spider mites.
Paecilomyces lilcanus	Deuteromycetes	Planthoppers.
Verticillium lecanii	Deuteromycetes	Whitefly, aphid and thrips.
Entomophaga maimaiga	Entomophthorales	Specific for gypsy moth larvae.
Entomophaga grylli	Entomophthorales	Specific for certain grasshopper families.
Entomophaga muscae	Entomophthorales	Specific for certain Diptera (flies).
Zoophthora radicans	Entomophthorales	Aphids, caterpillars, psyllids, leafhoppers.
Lagenidium gigantum	Oomycetes	Mosquitoe larvae.

to provide an incentive for the study of some of these other aspects of fungal biology which could contribute to the full application of entomopathogenic fungi as biocontrol agents. We hope to re-emphasize an insightful quote from Edward A. Steinhaus (1960), generally acknowledged as the father of insect pathology in North America, which states "Unfortunately, in much of insect pathology there is a tendency, in field observations as well as in laboratory research, to concentrate our attentions on the pathogen and on its insect host without fully appreciating the role of environmental factors."

There is a tremendous amount of genetic variation in deuteromycetous entomopathogenic fungi (e.g., Bidochka *et al.*, 1994). However, how this genetic variation translates to virulence, saprophytism or the ability to survive under various environmental conditions is still quite unclear. Within a given species, strains may be host and/or habitat specific or may be identified by geographical origin. The contribution of host or environment in driving genetic adaptation within entomopathogenic fungal populations is paramount for identifying isolates best suited for insect control in a particular habitat. Which is most important in driving natural selection and genetic adaptation in entomopathogenic fungi: host or habitat? Or both? The current paradigm in fungal pathology is that genes involved in insect virulence drive population genetic dynamics. Under this paradigm the insect host is the single most important selective factor in determining the genetics of the fungus, while the environment or habitat contributes to the distribution of the insect and has a minor role, perhaps in saprophytism or spore survivability (Figure 1A). In the following sections, we explore the relative contributions of host insect and environment in driving the population structure of entomopathogenic fungi.

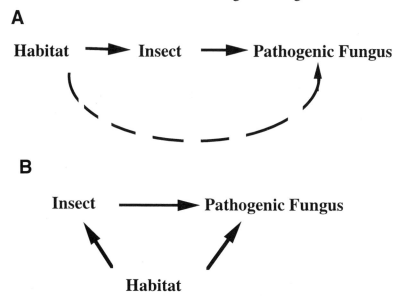

Figure 1. Two models showing the relative influences of habitat and insect host on the genetic structure of entomopathogenic fungi. Arrows indicated routes of influence of each factor on insect and fungus. A: current paradigm where insect host is the predominant factor and evidence suggests that obligate, entomophthoralean fungi are predominantly influenced in this manner with habitat playing a minor role. B: a revised model where habitat and insect host play a role in influencing population structure particularly in facultative, deuteromycetous entomopathogenic fungi.

2. Fungal Genetics, Insect Host and Environment

The direct interaction between insect host and fungus has largely occupied our study of fungal pathology - perhaps at the expense of the analysis of other factors contributing to the fungal biology of insect pathogens. With the advent of biochemical and molecular techniques, the disease process itself has received relatively more attention than the factors that contribute to disease in the field. Certainly an understanding of the genetic and molecular basis of pathogenicity has advanced the potential for genetic engineering of these fungi in an effort to improve control (St. Leger and Bidochka, 1996). But before proceeding further with genetic engineering, perhaps other areas of study are equally important, such as those involved in understanding the partitioning of genes involved in virulence and those implicated in saprophytism. For example, one can potentially elucidate most of the factors involved in virulence and theoretically engineer the most virulent fungus in the laboratory. But placement of the engineered genes into a fungal strain that survives poorly in agricultural habitats, - perhaps due to a low resistance to UV light, a sensitivity to desiccation, or poor saprophytic abilities, would fail to achieve the desired effect.

We also have a poor understanding of the non-pathogenic phase of the life cycle of many species of entomopathogenic fungi. The non-pathogenic phase of obligate fungal pathogens may simply be survival of the fungal conidium or spore until contact with an appropriate host. On the other hand, some of the facultative pathogens must survive as saprophytes until contact with an appropriate host. Certainly, this saprophytic phase in the case of facultative pathogens, or the "surviving" phase, in the case of obligate pathogens, is of considerable importance if we are to fully understand the life cycles of these fungi and utilize these fungi to their full potential as biocontrol agents. Transmission is also a key factor that must be understood if entomopathogens are to be genetically manipulated and released. Under some biocontrol strategies we will need to understand the requirements for efficient disease transmission in the field that are characteristic of epizootics.

Entomopathogenic fungi generally fall within two taxonomic groupings, the Entomophthorales (Zygomycetes) and the Deuteromycetes, with several ascomycetous and oomycetous examples. The Entomophthorales are generally regarded as more obligate pathogens, sometimes having host specificity at the insect family or subfamily level. For example, the pathotypes of the *Entomophthora grylli* complex have specificity in infection at the taxonomic level of grasshopper subfamily (Carruthers *et al.*, 1997). These fungi cannot grow saprophytically and survive as resting spores. On the other hand, many of the Deuteromycetes are considered to be facultative pathogens with larger host ranges and the ability to live, to some extent, saprophytically in the soil. For example, *Metarhizium anisopliae* has over 200 insect-host species, is cosmopolitan in distribution and is commonly isolated from soils. Figure 2 shows a continuum of fungal econutritional groupings from saprophyte to obligate pathogen. An appreciation of this continuum, which also applies to plant pathogenic fungi, is particularly important for understanding the pathogenic life style and for engineering effective biocontrol agents.

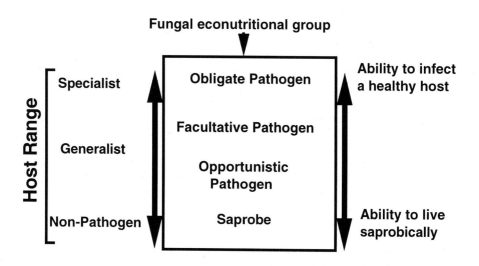

Figure 2. Econutritional groups of fungi according to nutritional mode and ecological behavior (after Cooke and Whipps, 1993). Note that this schematic indicates modes of nutrition for pathogenic and saprophytic fungi and omits the biotrophic fungi.

Here we will consider the molecular mechanisms of fungal pathogenesis of insects, the fungal population structure and the potential influences of host and environment on the population structure. While studies have shown that there is a genetic component of the fungus devoted to insect pathogenesis, the genetic component devoted to environmental survival, saprophytism and the potential for transmission has been seriously neglected. We hope to take a broad view of the molecular and environmental studies needed for the successfully exploitation of entomopathogenic fungi as biocontrol agents.

3. Fungal Pathogenesis

There are many recent comprehensive reviews on infection processes for entomopathogenic fungi (Hajek and St. Leger, 1994; St. Leger and Bidochka, 1996). Therefore, we will not attempt to provide a detailed review of the literature but instead will outline some of the main features and general themes which have emerged. In general, fungi infect their hosts by transgressing the host cuticle - the outermost integument. The early aspects of fungal pathogenesis may be subdivided into (1) conidial attachment, (2) germination and appressorium production, (3) invasion and transcuticular penetration, and (4) hemocoelic entry, and (5) host response to fungal invasion. The vast majority of research conducted on fungal infection processes of insects has been conducted with the deuteromycetous fungi, *M. anisopliae* and *B. bassiana*, principally because these fungi are easy to grow in the laboratory and are amenable to current molecular techniques.

In this section, fungal infection processes will be considered in deuteromycetous fungi; however, it is perceived that the general processes of insect infection also apply to entomophthoralean fungi. The following review may also be remarkably reminiscent of the infection processes in phytopathogenic fungi. The primary differences are related to the battery of extracellular enzymes required in order to penetrate the insect cuticle or plant cuticle. The general differences in composition for these host barriers are illustrated in Figure 3. For both the plant and insect pathogens, conidial attachment, germination and fungal perception of the host environment, appressorium formation, penetration and soft tissue invasion and evasion of host responses are critical to successful infection.

In the obligate, or more host specific, fungal pathogens host cues are particularly important in establishing infection. One of the challenging problems facing plant and insect pathology is to determine the molecular basis of the specificity between host and fungal pathogen. There are only several examples where the molecular basis of specificity in a phytopathogen is understood, and most of these involve production of a host-specific toxin by the fungus such as the cysteine-rich proteins (Templeton *et al.*, 1994). To our knowledge, no such examples are known for entomopathogenic fungi.

3.1. CONIDIAL ATTACHMENT

M. anisopliae and *B. bassiana* produce a vast number of conidia either on infected insects or after saprophytic growth on an organic substrate. When grown on complex media such as potato dextrose agar, these fungi can produce more than 10^9 conidia per Petri dish). However, conidial production can also be highly variable when different isolates are compared (Melzer and Bidochka, 1998). These conidia are about 5 -10 μm in size and

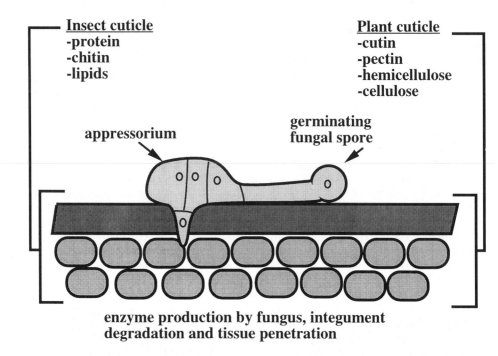

Figure 3. Similarities in the mechanisms of transcuticular penetration by plant and insect pathogenic fungi. General differences in the composition of host integument, which reflect differences in the battery of hydrolytic enzymes produced by insect pathogens versus plant pathogens, is shown.

are easily carried by the wind. In contrast, spores produced by many of the entomophthoralean fungi are significantly larger (up to 50 μm) and are generally deposited not much further than the infected insect. Some of these fungi such as the fly pathogen, *Entomophaga muscae* have the ability to ballistically propel their adhesive spores. In fact, this phenomenon is so striking that one can sometimes notice a white ring of ballistically projected spores around an infected fly cadaver adhered to a window. In contrast to the adhesive spores produced by many of the Entomophthorales, the deuteromycetous fungi have conidia that have a hydrophobic surface (Bidochka *et al.*, 1995a; Boucias *et al.*, 1988). The initial interactions between these conidia and the insect cuticle (which includes a lipid complement) is through non-specific hydrophobic interactions (Boucias *et al.*, 1988).

3.2. GERMINATION AND APPRESSORIUM FORMATION

The insect cuticle is the first and most formidable barrier that the fungus must transgress. Germination cues include both chemical and topographical signals. Utilizable cuticular nutrients include certain long chain fatty acids, some amino acids and sugars (Woods and Grula, 1984). However, the failure to germinate has been attributed to certain inhibitory compounds such as short chain fatty acids, quinone and phenols (Hsiao *et al.*, 1992). The presence or absence of specific cuticular compounds is highly variable among insect species and may be specific germination cues for some of the obligate fungal pathogens (i.e. the Entomophthorales). Micro-environments on the surface of the cuticle may also be essential for germination since these areas may form pockets having higher humidity levels.

There are many common themes on the general modes of pathogenesis in plant and insect pathogens. In the initial stages of infection, the conidia of entomopathogens such as *Metarhizium* spp. and plant pathogens such as *Colletotrichum trifolii* attach, swell, form a germ tube, and the germ tube differentiates into an appressorium (St. Leger *et al.*, 1991; Dickman, 1995). Appressoria are formed soon after germination and the length of the germination tube before appressorium formation is dependent on the surface topography of the insect cuticle (St. Leger *et al.*, 1991). Figure 4 shows cuticle invasion by *B. bassiana*. Note that this species of entomopathogenic fungus may or may not form an appressorium before the infection peg enters the cuticle.

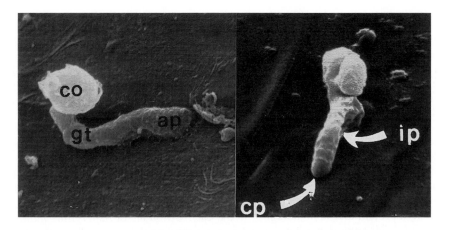

Figure 4. Penetration of insect cuticle by *B. bassiana*. The left panel shows the formation of an appressorium while the right panel shows direct penetration by the infection peg. co, conidia; gt, germination tube; ap, appressorium; ip, infection peg; cp, point of cuticle penetration.

Protein phosphorylation is almost universal in signal transduction and cell differentiation in mammalian systems. It seems likely that phosphorylation events also mediate conidial germination and appressorium formation in insect and plant pathogenic pathogen. In a series of studies coordinated primarily by R.J. St. Leger and reviewed in St. Leger and Bidochka (1996) it was shown that various protein kinases, calmodulin dependent and Ca^{2+}-activated protein kinases function in selective activation of signal transduction elements that are likely important in cell differentiation and appressorium formation. Furthermore, GTP-binding proteins and adenylate cyclase activity in the plasma membrane of *M. anisopliae* are seemingly involved in appressorium formation. It has also been shown that the disruption of Ca^{2+} gradients redirects apical growth to undifferentiated growth. A similar model is proposed for differentiation in the plant pathogen *C. trifolii* (Warwar and Dickman, 1996; Yang and Dickman, 1997)

One of the major problems in insect and plant pathology is the elucidation of the mechanisms by which the fungus senses an appropriate host and responds with an appropriate infection response. Knowledge of selective activation of specific signal transduction elements which lead to germination, appressorium formation and finally cuticle invasion would be important in determining how the fungus recognizes host specific cues, such as different cuticular nutrients or surface topologies. St. Leger *et al.* (1994) found that two isolates of *M. anisopliae*, one a homopteran pathogen and the other a coleopteran pathogen, germinated differently in the presence of glucose. The homopteran pathogen germinated and produced appressoria in glucose which is likely to be supplemented in the excretions of Homoptera such as aphids. The coleopteran pathogen germinated poorly in glucose. An understanding of glucose, or any other host nutrient, uptake and phosphorylation or modification and interaction with adenylate cyclase or other regulatory proteins in signal transduction may be particularly useful in understanding mechanisms of host specific cues for fungal germination.

3.3. CUTICLE INVASION

A penetration peg forms beneath the appressorium and breaches the insect cuticle. During this phase of infection there is a massive synthesis of a single subtilisin-like extracellular protease termed Pr1. The sequence of this enzyme has been determined from both *M. anisopliae* (St. Leger *et al.*, 1992) and *B. bassiana* (Joshi *et al.*, 1995). Pr1 gene disruption experiments and genetically enhanced Pr1 production through a constitutive promoter indicate that Pr1 is encoded by a pathogenicity gene (St. Leger *et al.*, 1996).

The *pr1* gene is transcriptionally regulated by levels of carbon and nitrogen in the environment, so that when grown on the nutrient poor insect cuticle, the fungus produces Pr1 in large amounts and has the ability to degrade insect cuticle allowing access to previously bound cuticular proteins. The putative transcription control elements which may play a role carbon catabolite repression of *pr1* have been identified (Screen *et al.*, 1997). The *pr1* promoter region had a carbon-response regulator gene, *crr1* encoding a protein that is homologous to the CREA proteins of *Aspergillus nidulans, A. niger, Trichoderma reesei* and *T. harzianum* . Recent work has also shown that *pr1* as well as several other proteases and a hydrophobin are transcriptionally regulated by ambient pH so that enzymes were synthesized only in the pH range at which they optimally operated (St. Leger *et al.*, 1998). The pH at the site of fungal penetration increases presumably due to the release of proteolytic byproducts such as ammonium. Pr1 transcription increased at pH 8 which is also the optimal pH for activity. Furthermore, at pH 8, in the absence of cuticle, Pr1 production was derepressed which suggests that ambient pH can override the inductive effects of cuticle on *pr1* transcription.

Other proteases produced by *M. anisopliae* are thought to be involved in hydrolyzing cuticular proteins and peptide products. These include trypsins, metalloprotease, aminopeptidases, dipeptidyl peptidase and carboxypeptidases (reviewed in St. Leger and Bidochka, 1996). Altogether, more than 20 different proteases or their isoforms have been described from *M. anisopliae*. Other extracellular enzymes, such as chitinases, N-acetyl-D-glucosaminidases, lipases and esterases also function to degrade insect cuticular components (St. Leger and Bidochka, 1996), but their involvement as virulence factors is still questionable.

3.4. HEMOCOELIC ENTRY AND HOST RESPONSE TO FUNGAL INVASION

M. anisopliae changes its morphology through cellular differentiation, presumably as a response to its immediate environment. Germ tubes produce appressoria, an infection peg penetrates the cuticle, and once the cuticle is transgressed, yeast-like bodies termed blastospores are produced. These cells disperse throughout the nutrient rich insect hemolymph. Blastospores have a different cell wall composition than mycelia and the alteration of carbohydrate moieties of fungal cell wall glycoproteins is thought to be at least one mechanism utilized by fungi to evade insect immune non-self recognition responses (Pendland and Boucias, 1993). This may also be a factor in host specificity and avoidance of immune responses in entomophthoralean fungi (Bidochka and Hajek, 1996).

Many extensive reviews have recently described insect immune responses and we refer the reader to these excellent articles (Gillespie *et al.*, 1997; Soderhall, 1996; Kanost and Jiang, 1996). In summary the immune responses are of two types; a cellular and a humoral response. The cellular responses involve insect hemocytic recognition and encapsulation of a non-self foreign body such as an invading fungus. The humoral responses involve the constitutive production or induction of proteins and peptides most of which that are known are antibacterial (Gillespie *et al.*, 1997). However, genetic analysis of the activation of an antifungal peptide, drosomycin, has recently been studied in *Drosophila* (Levashina *et al.*, 1999).

Once in the hemolymph, cytotoxic compounds may be produced by the blastospores and these include destruxins (Samuels *et al.*, 1988) and, potentially, a combination of other cyclic depsipeptides and hydrophobins (Gillespie and Claydon, 1989; St. Leger *et al.*, 1992c). Destruxins are cyclic depsipeptide toxins produced by *M. anisopliae* that cause insect paralysis and death. Destruxins have been shown to induce degranulation (exocystis) of insect hemocytes and perhaps impair the insect's cellular immune responses to the invading fungus (Cerenius *et al.*, 1990). Hydrophobins are a class of low molecular weight, hydrophobic, structural proteins that are also released extracellularly and resemble cerato-ulmin, a toxin involved in Dutch elm disease (Stringer and Timberlake, 1993). The insect succumbs possibly due to a combination of mechanical damage to internal organs, nutrient depletion and/or toxicosis (Gillespie and Claydon, 1989). The relative contribution of any of these factors depends on the host and fungal species or isolate under consideration.

3.5. PATHOGENICITY GENES

From the above descriptions, it seems likely that there are several classes of genes involved in pathogenicity and that all are potential targets for enhancement using various molecular techniques. These include (a) genes involved in detecting the presence of a host, and initiating the germination/differentiation process (e.g., phosphoprotein phosphatases, receptors, GTP-regulated adenylate cyclase, calmodulin and Ca^{2+}-dependent protein kinases); (b) genes coding a variety of proteases and other hydrolytic enzymes involved in breaching the insect cuticle (e.g. Pr1, other proteases, chitinases);

(c) genes that encode peptides that interfere with the host's immune system and cytotoxic peptides (e.g., cell-wall altering cyclic depsipeptides, hydrophobin); (d) genes that specify alterations in cell-wall glycoprotein and thus contribute to the evasion of the host immune responses (e.g. glycosyltransferases).

3.6. GENES NOT ASSOCIATED WITH PATHOGENICITY

Another category of genes related to the successful disease transmission of a fungal pathogen have no direct involvement with a fungal-insect interaction. This may be an heterogeneous group of genes involved in successful existence as a saprophyte, or through low humidity tolerance, UV resistance, temperature tolerance or conidial longevity. At the genetic level, this is a particularly neglected area of study. Pathogen infectivity and virulence on the one hand, and survival outside the host and capacity to disperse on the other, are crucial in epizootiology. In the following section we will review environmental factors affecting entomopathogenic fungi in order to define potential areas of research.

4. Environmental Factors Affecting the Distribution of Insect-Infecting Fungi.

As with all fungi, growth and survival are influenced by the habitat in which they occur. Successful survival can subsequently lead to insect epizootics and amplification of entomopathogenic fungi in the environment. Here, we examine six species of entomopathogenic fungi, the factors that determine their distribution and abundance and, when information is available, their variation in response to differing conditions.

4.1. DEUTEROMYCETES

4.1.1. *M. anisopliae*

M. anisopliae has a worldwide distribution. For example, this species has been isolated from South America (Tigano-Milani, *et al.*, 1995), Asia, Africa (Humber, 1992), the Arctic Circle (Finland) (Vanninen, 1996), Canada (Widden and Parkinson, 1979) and Russia, (Arkhipova, 1965) as well as from an Antarctic island (Roddam and Rath, 1997). Recently, we have seen that in Ontario, Canada alone, *M. anisopliae* was recovered from approximately 67% of the soil samples taken in areas ranging from temperate deciduous forests, agricultural lands to near northern boreal forests (Bidochka *et al.*, 1998).

      M. anisopliae typically grows at an optimum of 22-27°C; however, certain strains of *M. anisopliae,* such as those isolated from the Antarctic, can germinate and grow better in cold temperatures than can other strains (Roddam and Rath, 1997). Cold activity can be defined as the ability to grow and germinate at temperatures as low as 5°C (Roddam and Rath, 1997). We have also isolated several cold-active strains of *M. anisopliae* from Ontario, Canada (De Croos and Bidochka, 1999). Furthermore, using a PCR-based subtractive hybridization technique, we have been able to isolate several transcripts from *M. anisopliae* that are unique to cold active growth (De Croos and Bidochka, unpublished results).

High humidity (>75%) is also considered to be critical for fungal growth and conidiation (Millstein *et al.*, 1983). Milner *et al.* (1997) tested the pathogenicity of *M. anisopliae* towards termites at different humidity levels. Termite mounds are typically maintained at high humidity levels approaching 100%, but can drop to 86% without insect mortality (Milner *et al.*, 1997). Conidia of *M. anisopliae* showed an increase in germination frequency as the relative humidity approached 100% (Milner *et al.*, 1997). Significantly, they found no consistent effect on pathogenicity with a change in humidity, suggesting that humidity does not influence the ability of *M. anisopliae* to kill termites.

The intensity and duration of light during the sun's daily cycle may also effect the distribution of *M. anisopliae*. Germination was enhanced when conidia of *M. anisopliae* were placed in formulations conferring UV protection (Alves *et al.*, 1998). Control plates that were not exposed to simulated sunlight had improved conidia germination (Alves *et al.*, 1998). Canopy protection, as seen in forested habitats, could provide protection to fungal species that cannot withstand UV exposure. Agricultural habitats tend to be in open areas with full exposure to UV light and this can have a detrimental effect on non-tolerant fungi. Conidia of *M. anisopliae* exposed to an artificial light source had a maximum half-life of 165 minutes (Zimmerman, 1982).

4.1.2. *Beauveria bassiana*

B. bassiana, like *M. anisopliae,* has a broad host range and a worldwide soil distribution (Ogarkova and Ogarkov, 1986; Vanninen, 1996; Bidochka *et al.*, 1998). *B. bassiana* has similar physiological constraints to *M. anisopliae* with an optimum growth temperature of 25-30°C and relative humidity of 100% (Walstad, *et al.*, 1970). In Ontario, Canada, *B. bassiana* was recovered from 61% of sample sites (Bidochka *et al.*, 1998). *B. bassiana* was isolated more frequently from natural sites compared to cultivated sites. This is in contrast to *M. anisopliae*, which was found more frequently in the cultivated sites (Bidochka *et al.*, 1998). *B. bassiana* was isolated from soil samples, using a waxworm baiting technique (Zimmerman, 1986), and was recovered more frequently at 15°C than at 25°C (Bidochka *et al.*, 1998). This species is more psychrophylic than *M. anisopliae*. *B. bassiana* seems to require an insect host to persist over many generations and a heavily disturbed site like a cultivated area would preclude its occurrence. The distribution of these fungi may be, in part, related to their conidial coloration. *M. anisopliae* conidia are pigmented green while *B. bassiana* conidia are hyaline and are, perhaps, more susceptible to UV damage.

It is clear that the survival and distribution of fungi can be affected by exposure to visible light and UV light. However, in some instances light has been shown to also have a positive effect on infection. In *B. bassiana*, red-infrared light stimulated the release of conidia compared to darkness (Gottwald and Tedder, 1982). The same study showed that the production of conidia by *M. anisopliae* was inhibited by red-infrared light (Gottwald and Tedder, 1982).

Variation in ability to grow at different temperatures was also found for *B. bassiana*. Conidia of *B. bassiana* are remain viable when held at 15°C but were not viable when incubated at 55°C for ten days (Lingg and Donaldson, 1981). Such high temperatures can occur in subtropical and tropical habitats. Fargues *et al.*, (1997) looked at 65 isolates of *B. bassiana* and found that they were capable of growth over a broad temperature range (8° to 35°C). The isolates were from a variety of sources but there was no apparent relationship with geoclimatic origin and growth rate. However, isolates obtained from acridids grew better at higher temperatures (Fargues *et al.*, 1997). Acridids can raise their body temperature in response to *B. bassiana* infection and this may explain the presence of heat tolerant isolates (Inglis *et al.*, 1996). Acridids typically elevate their body temperature by intercepting as much solar radiation as possible through basking or habitat selection. Interestingly, high temperature exposure of grasshoppers lessened their chances of infection by *B. bassiana* (Inglis *et al.*, 1996).

4.1.3. *Verticillium lecanii*

V. lecanii is another deuteromycetous entomopathogen that is typically employed in greenhouses for the control of aphids (Quinlan, 1988). *V. lecanii* requires high humidity to cause infection and produce conidia. Infection of insects occurs optimally at 100% relative humidity with no transmission occurring below 80% (Milner and Lutton, 1986). The production of conidia by *V. lecanii* also decreased below 100% with no asexual sporulation occurring below 80% humidity (Milner and Lutton, 1986). It is possible that insects could avoid infection by staying in areas where humidity is lower. For example, aphid susceptibility to infection by *V. lecanii* in greenhouses was reduced near the woody stems of plants where humidity was lower as opposed to the apices which had a higher humidity (Quinlan, 1988).

An isolate of *V. lecanii* from the Antarctic was shown to be active at low temperatures (Fenice *et al.*, 1998). This extreme environment should include adaptable organisms that could cope with the colder temperatures. Assessment was made of the chitinolytic activity at optimal (40°C) and cold temperatures. The Antarctic isolate retained 50% of this enzyme activity despite the drop in incubation temperature to 5°C (Fenice *et al.*, 1998).

4.2. Entomophthorales

4.2.1. *Entomophaga grylli*

E. grylli is an obligate pathogen of grasshoppers (Carruthers *et al.*, 1997) and is one of the major causative agents of grasshopper epizootics worldwide (Roffrey, 1968; Pickford and Riegart, 1964, Milner, 1978). Within this species there are different pathotypes that show differences in their ability to infect grasshoppers at the subfamily level. Pathotype 1 infects grasshoppers in the subfamily Oedipodinae and is native to North America (Carruthers *et al.* 1997). Pathotype 2 is also native to North America and infects the subfamily Melanoplinae. Pathotype 3 is native to Australia but bioassays showed that it is pathogenic to the North American Oedipodinae and Melanoplinae as well as an Australian cyrtacanthracridine grasshoppers. Other pathotypes have been isolated from Japan and Indonesia and several others may exist in Africa and Europe (Carruthers *et al.*, 1997). Distribution of this entomopathogen appears wherever there are populations of acridids (Carruthers *et al.*, 1997).

A key feature of entomophthoralean insect infection, and infection by *E. grylli* in particular, is the tendency of infected insects to migrate to the apices of plants and die in a rigor mortis position while clasping the stems (Macleod, 1963). This position exposes the fungus at risk to extremely high temperatures by solar radiation and it has been shown that the viability of conidia produced by *E. grylli* is affected by both the duration and intensity of solar radiation (Carruthers *et al.*, 1988). Nevertheless, this elevated positioning also favours wind dispersal of spores produced on the cadaver.

Depending on the pathotype of *E. grylli*, a combination of conidia, hyphal bodies, resting spores and cryptoconidia may be produced. The resting spores can survive more extreme conditions and are involved in year to year persistence or vertical transmission. In contrast, the infective spores and cryptoconidia cannot withstand long term dormancy but germinate more quickly and are involved in horizontal disease transmission. The infective conidia of *E. grylli* are quickly killed by adverse environmental conditions such as low temperatures, extreme solar radiation, and low moisture levels (Carruthers et al, 1996).

4.2.2. *Entomophaga maimaiga*

E. maimaiga, like most other Entomophthorales, shows host preferences and preferentially attacks lepidopteran larvae in the family Lymantridae. *E. maimaiga* produces azygospores in late spring and early summer. These azygospores (resting spores) are typically double walled and thus have a greater ability to persist in the field. Bioassays have shown that once these azygospores germinated in the field, levels of insect infection were positively correlated with soil moisture (Hajek and Humber, 1997). *E. maimaiga* can germinate and produce conidia from 2 to 25°C with an optimal range of 20 to 25°C (Hajek *et al.*, 1990). The production of conidia occurs readily at a relative humidity of 95-100% and was minimal at 50% (Hajek *et al.*, 1990).

4.2.3. *Zoophthora radicans*

Z. radicans is an important biological control agent of aphids (Milner *et al.*, 1982; Shands *et al.*, 1962) and will infect a host insect through the formation of appressoria upon insect cuticle contact (Wraight *et al.*, 1990). Appressoria formation is necessary for infection and its environmental constraints differ from conidia formation (Magalhaes *et al.*, 1991). Conidial germination is significantly delayed at 10°C whereas the formation of appressoria is absent at that temperature (Magalhaes *et al.*, 1991). Appressoria were formed under low nutrient conditions. In contrast, the formation of germ tubes from conidia is not nutrient dependent (Magalhaes *et al.*, 1991). In fact, the precursors required for conidia germination are probably included in the spore because conidia can germinate in tap water. The formation of infective structures as opposed to reproductive structures appears to be more sensitive to environmental cues in this species.

      The major factor effecting the survival of *Z. radicans* conidia is solar radiation. *Z. radicans* was not viable after twenty-four hours of field exposure, (Furlong and Pell, 1997). When conidia were exposed to simulated tropical light for four hours, there was a significant decrease in infectivity as compared to a one hour exposure (Furlong and Pell, 1997). Rainfall and humidity were also investigated, but their influences did not have a great impact on the persistence or infectivity of *Z. radicans* .

5. Genetic Responses to the Environment in Fungi

Fungi have the capability to live in extreme environmental conditions including extremes of salinity, acidity, and temperature, as well as in the presence of toxic organic compounds and sewage sludge with high concentrations of heavy metal (Zibilske and Wagner, 1982).

One of the best studied environmental responses in eukaryotic and prokaryotic organisms is the synthesis of heat shock proteins (Lindquist, 1986). All, or a subset of, heat shock proteins can be induced in response to environmental stresses such as high salinity (Torzilli, 1997), temperature (Xavier and Khachatourians, 1996) and hypochlorous acid (Dukan *et al.* 1996). It is not yet clear what subset of "heat shock proteins" specifically induced during incubation at high temperatures may actually be "stress related proteins".

In fungi, other than yeast, the isolation and characterization of novel transcripts resulting from environmental stresses, in particular those not related to nutrient depletion, have not been well documented. *Aspergillus nidulans* grown under conditions of salt shock and salt adaptation was found to alter expression of proteins and showed *de novo* protein synthesis (Redkar et al, 1996). The same study also showed an increase in ribosome transcription, indicating alterations in gene expression under salt stress. In the plant pathogenic fungus *Fusarium oxysporum* f. sp. *cucumerinim*, a stress induced gene, *sti*35, was isolated and sequenced (Choi *et al.*, 1990). The gene was induced by treating the fungus with ethanol. RNA blot analysis showed that the gene was also induced in response to various other stresses such as copper (II) chloride and heat (37°C) (Choi *et al.*, 1990).

On the other hand, the expression of genes related to nutrient depletion in fungi is very well studied (e.g. Johnston, 1987; Hinnebusch, 1988). Nitrogen starvation in the plant pathogenic fungi, *Magnaporthe grisea*, resulted in the induction of genes for pathogenesis (Talbot *et al.*, 1997). Nitrogen limitation reflects conditions typical of fungal infection of rice leaves. Starvation stress in this case seems to influence fungal gene expression during plant infection (Talbot *et al.*, 1997). In the insect pathogen, *M. anisopliae*, starvation conditions analogous to those found on the insect cuticle result in the expression of starvation stress genes (*ssg*) such as the subtilisin-like protease (pr1; St. Leger *et al.*, 1992b) and hydrophobin (*ssg*A; St. Leger *et al.*, 1992c).

There has been little published data on induced gene expression in other entomopathogens. Physiological stress did induce expression of a chitinase gene, *ech*42, in the mycopathogen *Trichoderma atroviride* (Mach *et al.*, 1999). Transcription of *ech*42 was induced upon exposure to low temperatures, high osmotic pressure and ethanol (Mach *et al.*, 1999).

6. Population Structure and Genetic Variation

Screening for genetic variability within fungal populations is generally used to examine population structure and evolutionary origins, but it can also contribute to an understanding of the host range and the variability in environmental distribution within entomopathogenic fungi. These genetic variants may have potential as biocontrol agents themselves or certain genes may confer desirable traits in other fungal strains; they may also be transgenically introduced into other biocontrol agents such viruses, bacteria, nematodes or protozoa. Studies of genetic variability also provide diagnostic genetic markers that are useful to distinguish an introduced strain from background strains in field applications for epizootiological studies.

Genetic loci screened for allelic variation may be neutral or they may be diagnostic of specific pathogenicity genes. A variety of techniques, such as allozyme analysis, random amplification of polymorphic DNA by the polymerase chain reaction (RAPD-PCR), vegetative compatibility groups (VCG) and restriction fragment length polymorphisms (RFLP), are presently used to examine genetic variation in entomopathogenic fungi and to explore how this variation relates to host virulence. However, little has been done to relate variation in environmental tolerance to genetic variants among isolates. Many of the traits used to assess genetic variation such as allozymes, RFLP, RAPDs, minisatellites and microsatellites may not, in fact, be neutral. There is increasing evidence that perhaps there is no region of the genome that is altogether selectively neutral. For example, there is strong evidence that lactate dehydrogenase alleles in guppyfish have different enzyme kinetics and thermodynamic properties in different temperature regimes and this influences clinal distribution of these alleles (Schulte *et al.*, 1997). Fragile-X syndrome (mental retardation in humans) is associated with microsatellite repeat in the *fmr*1 gene (Hirst *et al.*, 1992).

6.1. DEUTEROMYCETES

Genetic variability in *M. anisopliae* has been examined fairly extensively (St. Leger *et al.* 1992a; Bidochka *et al.* 1994; Bridge *et al.* 1997), but the factors that influence population structure still remain unclear. Although population structure is presumed to be clonal, parasexuality has been observed in laboratory experiments (Tinline and Noviello 1971) although this may have no bearing on potential for recombination in the field.

The bulk of population level genetic investigations attempt to correlate genotypic variability with geographical origin or host specificity and the results are typically ambiguous. Again, the leading paradigm positing the insect host as the key factor driving genetic structure may be misleading (Figure 1A). Several examples of such misinterpretation follow.

Allozyme polymorphisms in 120 strains revealed 48 distinct genotypic classes indicating that, on a global level, genetic variability among *M. anisopliae* isolates is high (St. Leger *et al.* 1992a). RAPD-PCR profiles for *M. anisopliae* also revealed high divergence between strains but no associations with insect host or geographical origin (Bidochka *et al.* 1994). On the other hand, genetically similar Brazilian isolates of *M. anisopliae* showed no association with insect host but did show some geographic clustering (Tigano-Milani *et al.*, 1995). Equally as ambiguous are the results found by Leal *et al.* (1994) where *M. anisopliae* isolates from the same geographical areas may or may not be genetically related.

Allozyme analysis of *B. bassiana* also revealed a high degree of genetic diversity but no associations of the various genotypic classes to host or geographical origin were observed (St. Leger *et al.* 1992a). Chromosome polymorphisms have also been observed in *B. bassiana* and the authors suggested an association with host insect (Viaud *et al.* 1996). RFLP and RAPD analysis of isolates from different insect hosts and different geographical locations revealed a high degree of genetic variability in *B. bassiana* (Maurer *et al.* 1997). Here, it was suggested that clonal lineages were host specific and not associated with geographical origin (Maurer *et al.* 1997). Isolates were collected from insect hosts and cluster analysis of the RAPD and RFLP results revealed that in some instances isolates collected from a specific host grouped together in cluster analysis (Maurer *et al.* 1997). One genotype was associated with the genus *Ostrinia* (Lepidoptera; Pyralidae), another with *Diatraea,* and another group with insects of the genus *Sitona* (Curculionidae; Coleoptera); the isolates collected from the pyralid genus *Maliarpha* and the coleopteran Chrysomelidae did not cluster into distinct groups (Maurer *et al.* 1997). One factor which was not considered in this study was the influence of habitat on the fungal groups. It is well established that habitat influences distribution and abundance of insect species and our research with another deuteromycetous fungus, *M. anisopliae*, suggests that fungal genotype distribution is influenced by habitat (Kamp, 1999; see below). It would be noteworthy to establish whether the association of the moth or beetle larvae species with fungal genotype that was found by Maurer *et al.* was co-incidental to the habitat from which the insect species and fungal genotypes co-occurred.

One particular study by Leal *et al.*, (1997) identified RFLPs in a virulence factor, the *pr*1 gene, in isolates of *M. anisopliae*. Since this gene has been previously identified as critical for pathogenesis (St. Leger *et al.*, 1996), we could assume that there is strong selection pressure for maintaining nucleotide integrity at this locus. However, there was a correlation between RFLP profile and geographic origin for certain isolates. Examination of the predicted structure of Pr1 based on nucleotide substitutions suggested that some subtle differences may be found on the surface of the enzyme (Butt *et al.*, 1998), perhaps reflecting those involved with electrostatic charge and enzyme binding to insect cuticle (Bidochka and Khachatourians, 1994). It is also possible that subtle differences in Pr1 are associated with different selective constraints on this enzyme activity in various locals. This could be due to differences in climatic conditions or a different subset of insects found in different areas and their differing cuticular compositions. An alternative explanation is that these variations are selectively near neutral and reflect geographical patterns of genetic drift and/or founder effects.

Our laboratory has examined whether habitat plays a role in determining the genetic structure of *M. anisopliae*. Kamp (1999) assessed allelic variation at ten loci in 83 isolates of *M. anisopliae* from Ontario. There was significant linkage disequilibrium among the alleles suggesting little recombination between two distinctive clonal lineages (Figure 4). These clonal groups are geographically isolated except in one area in Southern Ontario where the two lineages overlap. We also found a strong association of *pr*1 RFLPs with habitat. The clonal groups had no association with insect host, but there was a strong association with habitat type.

These results suggest that, in this fungus, habitat, not insect host, drives population structure. One clonal lineage was isolated primarily from agricultural areas in more southern regions, while the other lineage was found predominantly in forested areas and had a more northerly distribution. The fungi from agricultural habitats were from crops as diverse as hay, beans, peaches, grapes, peas, wheat, corn, and barley. The fungi from forested areas were from coniferous, deciduous or mixed forests. The identifying feature of the agricultural habitat was that it was an exposed, open habitat while the forested habitat was a closed, cooler, moister habitat; there was no general association of an insect species with "forested" or "agricultural" habitats. This strong association of entomopathogenic fungal genotype with a specific "type" of habitat is not often reported in the literature, but our results indicate that habitat selection may be the principle driving force influencing the genetic structure of many facultative pathogens (Figure 1B). One of the potential selective forces that we examined was the ability of the isolates in the two clonal groups to grow at low temperatures ($8°C$). There was a strong, statistically significant association of cold-active growth found in isolates of the clonal group found predominantly in the forested habitats.

Considerable variation in dsRNA viruses within isolates of *M. anisopliae* has also been reported (Melzer and Bidochka, 1998). However, more stringent analysis of the results has shown that homologous dsRNA bands were shared only within a single clonal group and this clonal group was restricted to a single vegetative compatibility

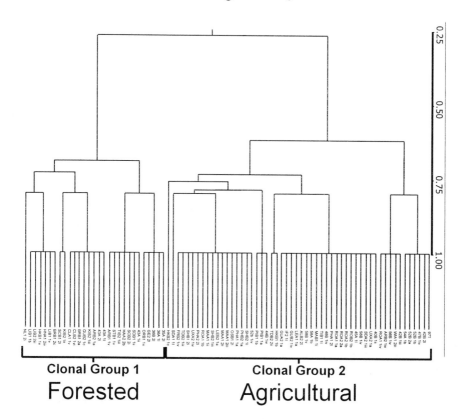

Figure 4. Dendrogram illustrating genetic relatedness among 83 Ontario isolates of the insect pathogenic fungus *M. anisopliae* . The cluster analysis was performed using allelic differences at five polymorphic allozyme loci. Two distinct, deep-rooted, clonal groups are indicated. Isolates in clonal group 1 were isolated primarily from forested habitats while isolates in clonal group 2 were isolated primarily from agricultural habitats.

group (VCG; Bidochka *et al.*, 2000). This suggests that dsRNA are either maintained and passed vertically within a clonal group or are passed horizontally through genetically related isolates within the same VCG. This evidence shows that recombination between genetically diverse isolates is very rare in nature.

V. lecanii, like other deuteromycetous insect pathogens, is a ubiquitous pathogen. Though it is of interest for biological control of agricultural insect pests, its population structure has been not been widely investigated in comparison with other species. Like other species discussed in this section, genetic variability is present within a presumably clonal population structure (Typas *et al.* 1998). Mitochondrial DNA (mtDNA) polymorphisms can be used to distinguish between isolates with greater success than with other species such as *B. bassiana* and *M. anisopliae* (Typas *et al.* 1998). The drawback of this method is the large amount of DNA that is required to do this type of analysis. On the other hand, it is a relatively easy procedure and could therefore be extremely useful in establishing molecular markers. Using molecular markers, released strains could be easily identified and distinguished from background strains and large numbers of isolates could be easily scanned.

6.2. ENTOMOPHTHORALES (ZYGOMYCETES)

The entomophthoralean insect pathogens are obligate pathogens, surviving only in a host or as resting spores. These fungi, in contrast with the deuteromycetes, are generally host specific and data to date suggests that these species have a more homogeneous genetic structure (St. Leger and Bidochka 1996). The entomophthorales reproduce through asexual conidia or produce resting spores through an asexual process called azygosporogenesis, depending on the age of the host cadaver (Hajek 1997; St. Leger and Bidochka 1996). The lack of recombination in the entomophthorales suggests a clonal population structure. However, the appropriate evaluation of population genetics/structure has not been accomplished for any of the entomophthorales.

E. maimaiga is an obligate pathogen of the gypsy moth (*Lymantria dispar*) and has few alternate hosts (Hajek 1997). It is specific at the lepidopteran level to the family Lymantridae and has been identified as the cause of major epizootics in gypsy moth populations (Hajek *et al.* 1990). Although samples sizes and the number of loci that were scored were low in fungal isolates, there appears to be little genetic variability at the species level (Hajek *et al.,* 1990; Walsh *et al.,* 1990). Therefore, molecular markers can effectively differentiate between closely related species and to determine fungal epidemiology (Walsh *et al.* 1990; Hajek *et al.* 1990). For example, the use of allozymes and RFLP analysis to differentiate between *E. maimaiga* and *E. aulicae* allowed Hajek *et al.* (1990) to establish that *E. maimaiga* was responsible for recent gypsy moth epizootics in the United States. *E. aulicae* shows host preferences in the lepidopteran family Geometridae. Molecular markers can also be used for tracking strains that are released for biological control. The lack of variability makes the identification of molecular markers of specific strains particularly useful in tracking released strains of *E. maimaiga*, and for monitoring fungal dispersal and range.

E. grylli is a pathogen of grasshoppers presently being considered both for augmentation of and introduction into biological control programs (Carruthers *et al.* 1997). At least four or five pathotypes, based on life cycles and host specificity, have been identified in this species complex (Carruthers *et al.* 1997). Allozyme analysis, as well as rDNA polymorphisms, have been used to differentiate between pathotypes. Although sample sizes were low, RAPD results suggest that there is little intrapathotype genetic variation (Bidochka et al., 1995b). An extensive screening program using cloned DNA probes for specific pathotypes proved to be quite useful in assessing the fate of pathotype 3 (Australian pathotype) released into the U.S.A. (Bidochka *et al.* 1996).

Z. radicans, a species complex similar to that of *E. grylli*, is composed of many species differing in host specificity (Hodge *et al.* 1995). Irrespective of geographical origin, groups associated with a particular host are largely genetically homogeneous (Hodge *et al.* 1995). One group with pathogenicity for the potato leafhopper (*Empoasca fabae*), has been investigated for its possible use as a biological control agent. In order to establish genetic strains for release, biological markers were required to differentiate between applied and background populations so that strains could be effectively monitored. Hodge *et al.* (1995) were able to use RAPD banding profiles to establish molecular markers for the released strain as well as subsequent identification of recovered isolates in further monitoring for the effectiveness of the applied strain. They were also able to establish RAPD banding profiles for other groups which were consistently different from the *E. fabae* pathogens but showed homogeneity within host specific groups.

The results on the population structures of any of the entomophthorales are, at this time, inconclusive partly due to small samples sizes and number of loci assessed for variation. However, the association of genetic profiles of the entomophthoralean fungi with insect host provides evidence that the insect host drives the population genetic structure of these fungi while habitat or geographic origin has little influence (Figure 1A).

The ability to track strains through the environment and investigate their effectiveness is an important factor in determining the economic feasibility of fungi as biological control agents. The use of molecular markers in zygomycetes has proven to be extremely effective for these purposes. However, the difficulty with most of the entomopathogenic zygomycetes is their inability to grow under mass produced fermentation conditions.

7. Conclusions

In order to effectively utilize and manipulate entomopathogenic fungi as biological control agents, a more complete understanding of the genetics of fungal interactions with host and environment is needed. Much of the recent focus has been directed at elucidating mechanisms of pathogenesis. There is no doubt that these fungi have evolved genetic and biochemical mechanisms well adapted for insect infection. But equally important is an understanding of the genetic mechanisms involved in saprophytic growth and environmental tolerances. Our research has shown that habitat selection plays a critical role in the population genetics of *M. anisopliae*. The influence of habitat on driving the population genetic structure and the resulting high degree of genetic variation in other deuteromycetous entomopathogenic fungi has not been critically evaluated. On the other hand, there is evidence that the host insect has a much greater influence on the population genetic structure in entomophthoralean fungi. The influence of habitat or host insect on the genetics of deuteromycetous or entomophthoralean fungi may be viewed from their econutritional perspective; is the fungal isolate a facultative or obligate insect pathogen? A better understanding of the genetic resources and genetic variability that are available in insect pathogenic fungi should prove useful in further developing our ability to use fungi, or their gene products, to their full potential as biocontrol agents.

Acknowledgments
The preparation of this article was supported by an operating grant from the National Sciences and Engineering Research Council of Canada (NSERC). We thank Kendra Adema for critically reviewing this manuscript.

References

Alves, R. T., Bateman, R.P., Prior, C., and Leather, S.R. (1998) Effects of simulated solar radiation on conidial germination of *Metarhizium anisopliae* in different formulations, *Crop Protection* **17**, 675-679.
Arkhipova, V.D. (1965) Fungus diseases of the codling moth, *Carpocapsa pomonella* L. (Lepidoptera: Tortricidae), *Entomol. Rev.* **44**, 432-435.
Bidochka, M.J. (in review) Monitoring the Fate of Biocontrol Fungi. CAB Press, UK.
Bidochka, M.J. and Hajek, A.E. (1996) Protoplast plasma membrane glycoproteins in two species of entomophthoralean fungi, *Mycol. Res.* **100**, 1094-1098.

Bidochka, M.J. and Khachatourians, G.G. (1994) Basic proteases of entomopathogenic fungi differ in their absorptive properties to insect cuticle, *J. Invertebr. Pathol.* **64**, 26-32.

Bidochka, M. J., Kasperski, J.E. and Wild, G.A.M. (1998) Occurrence of the entomopathogenic fungi *Metarhizium anisopliae* and *Beauveria bassiana* in soils from temperate and near-northern habitats, *Can. J. Bot.* **76**, 1198-1204.

Bidochka, M.J., Melzer, M.J., Lavender, T.M., and Kamp, A.M. (2000) Genetically related isolates of the entomopathogenic fungus *Metarhizium anisopliae* harbor homologous dsRNA viruses. *Mycological Research* (in press).

Bidochka, M.J., Walsh, S.R.A., Ramos, M.E., St.Leger, R.J., Silver, J.C. and Roberts, D.W. (1996) Fate of biological control introductions: Monitoring an Australian fungal pathogen of grasshoppers in North America, *Proceedings of the National Academy of Sciences of the USA* **93**, 918-921.

Bidochka, M.J., St. Leger, R.J., Joshi, L. and Roberts, D.W. (1995a) The rodlet layer from aerial and submerged conidia of the entomopathogenic fungus *Beauveria bassiana* contains hydrophobin, *Mycol. Res.* **99**, 403-406.

Bidochka, M.J., Walsh, S.R.A., Ramos, M.E., St.Leger, R.J., Silver, J.C. and Roberts, D.W. (1995b) Pathotypes in the *Entomophaga grylii* species complex of grasshopper pathogens differentiated with random amplified of polymorphic DNA and cloned-DNA probes, *Appl. Environ. Microbiol.* **61**,556-560.

Bidochka, M.J., McDonald, M.A., St Leger, R.J. and Roberts, D.W. (1994) Differentiation of species and strains of entomopathogenic fungi by random amplification of polymorphic DNA (RAPD), *Curr. Gen.*, **25**, 107-113.

Boucias, D.G., Pendland, J.C. and Latge, J.P. (1988) Nonspecific factors involved in attachment of entomopathogenic Deuteromycetes to host insect cuticle. *Appl. Environ. Microbiol.* **54**, 1795-1805.

Bridge, P.D., Prior, C., Sagbohan, J., Lomer, C.J., Carey, M. and Buddie, A. (1997) Molecular characterization of isolates of *Metarhizium* from locusts and grasshoppers. *Biodiver. Conserv.* **6**,177-189.

Butt, T.M., Segers, R., Leal, S.C. and Kerry, B.R. (1998) Variation in the subtilisins of fungal pathogens of insects and nematodes, in P. Bridge, Y. Couteaudier and J. Clarkson, (eds.), *Molecular Variability of Fungal Pathogens*, CAB International, New York, pp. 149-169.

Carruthers, R.I., Ramos, M.E., Larkin, T.S., Hostetter, D.L. and Soper, R.S. (1997) The *Entomophaga grylii* (Fresenius) Batko species complex: Its biology, ecology, and use for biological control of pest grasshoppers, *Mem. Entomol. Soc. Can.* **171**, 329-353.

Carruthers, R. I., Feng, Z., Ramos, M.E. and Soper, R.S. (1988) The effect of solar radiation on the survival of *Entomophaga grylli* (Entomophthorales: Entomophthoraceae) conidia, *J. Invertebr.Pathol.* **52**, 154-162.

Cerenius, L., Thornqvist, P-O., Vey, A., Johansson, M.W. and Soderhall, K. (1990) The effect of the fungal toxin destruxin E on isolated crayfish haemocytes, *J. Insect Physiol.* **36**, 785-789.

Charlesworth, B. (1998) Adaptive evolution: The struggle for dominance, *Curr. Biol.* **8**, 502-504.

Choi, G. H., Marek, E.T., Schardl, C.L., Richey, M.G., Chang, S. and Smith, D.A. (1990) *sti*35, a stress-responsive gene in *Fusarium* spp. *J. Bacteriol.* **179**, 4522-4528.

Cooke, R.C. and Whipps, J.M. (1993) *Ecophysiology of Fungi,* Blackwell Scientific Publ., Oxford, U.K. pp. 337.

De Croos, J. N. A. and Bidochka, M.J. (1999) Effects of low temperature on growth parameters in the entomopathogenic fungus, *Metarhizium anisopliae, Can. J. Microbiol.*, 45: 1055-1061.

Dickman, M.B., Buhr, T.L., Warwar, V., Truesdell, G.M., and Huang, C.X. (1995) Molecular signals during the early stages of alfalfa anthracnose, *Can. J. Bot.* **73**, Supplement 1, S1169-S1177.

Dukan, S., Dudon, S., Smulski, D.R. and Belkin, S. (1996) Hypochlorous acid activates the heat shock response and saxRS systems of *Escherichia coli., Appl. Environ. Microbiol.* **62**, 4004-4008.

Fargues, J., Goetel, M.S., Smits, N., Ouedraogo, A. and Rougier, M.. (1997) Effect of temperature on vegetative growth of *Beauveria bassiana* isolates from different origins, *Mycologia* **89**, 383-392.

Fenice, M., Selbmann, L., Di Giambattista, R. and Federici, F. (1998) Chitinolytic activity at low temperature of an Antarctic strain (A3) of *Verticillium lecanii, Res. Microbiol.* **149**, 289-300.

Furlong, M. J. and Pell, J.K. (1997) The influence of environmental factors on the persistence of *Zoophthora radicans* conidia, *J. Invertebr. Pathol.* **69**, 223-233.

Gillespie, A.T. and Claydon, N. (1989) The use of entomogenous fungi for pest control and the role of toxins in pathogenesis, *Pesticide Science* **27**, 203-215.

Gillespie, J.P., Kanost, M.R. and Trenczek, R. (1997) Biological mediators of insect immunity, *Ann. Rev. Entomol.* **42**, 611-643.

Gottwald, T. R. and Tedder, W.L. (1982) Studies on conidia release by the entomogenous fungi *Beauveria bassiana* and *Metarhizium anisopliae* (Deuteromycotina: Hyphomycetes) from adult pecan weevil (Coleoptera: Curiculionidae) cadavers, *Environ.Entomol.* **11**, 1274-

Hajek, A.E. (1997) *Entomophaga maimaiga* reproductive output is determined by the spore type initiating an infection, *Mycol. Res.* **101**, 971-974.

Hajek, A. E. and Humber, R.A. (1997) Formation and germination of *Entomophaga maimaiga* azygospores, *Can. J. Bot.* **75**, 1739-1747.

Hajek, A.E. and St. Leger, R.J. (1994) Interactions between fungal pathogens and insect hosts, *Ann. Rev. Entomol.* **39**, 293-322.

Hajek, A.E., Humber, R.A., Elkington, J.S., May, B., Walsh, S.R.A. and Silver, J.C. (1990) Allozyme and restriction length polymorphism analyses confirm *Entomophaga maimaiga* responsible for 1989 epizootics in North American gypsy moth populations, *Proc. Nat. Acad. Sci. USA* **87**, 6979-6982.

Hinnebusch, A.G. (1988) Mechanisms of gene regulation in the general control of amino acid biosynthesis in *Saccharomyces cerevisiae, Microbiol. Rev.* **52**, 248-273.

Hirst, M.C., Knight, S.J.L., Bell, M.V., Super, M., and Davies, K.E. (1992) The fragile X syndrome, *Clin. Sci.* **83**, 255-264.

Hodge, K.T., Sawyer, A.J. and Humber, R.A. (1995) RAPD-PCR for identification of *Zoopthora radicans* isolates in biological control of the potato leafhopper, *J. Invertebr.Pathol.* **65**, 1-9.

Hsiao, W.-F., Bidochka, M.J. and Khachatourians, G.G. (1992) Effects of diphenols on the growth of three entomopathogenic fungi, *Can. J. Microbiol.* **38**, 1000-1003.

Humber, R.A. (1992) Collection of entomopathogenic fungal cultures: Catalog of strains. U.S. Department of Agriculture, Agricultural Research Services. ARS-110, 177 pp.

Inglis, G.D., Johnson, D.L. and Goettel, M.S. (1996) Effects of temperature and thermoregulation on mycosis by *Beauveria bassiana* in grasshoppers, *Biol. Control* **7**, 131-139.

Johnston, M. (1987) A model fungal gene regulatory mechanism: the *GAL* genes of Saccharomyces cerevisiae, *Microbiol. Rev.* **51**, 458-476.

Joshi, L., St. Leger, R.J. and Bidochka, M.J. (1995) Cloning of a cuticle-degrading protease from the entomopathogenic fungus, *Beauveria bassiana FEMS Microbiol. Lett.* **125**, 211-218.

Kamp, A.M. (1999) Habitat associated clonal lineages in the insect pathogenic fungus *Metarhizium anisopliae* in Ontario. B.Sc. (hons) thesis., Trent University, 56 pp.

Kanost, M.R. and Jiang, H. (1996) Proteinase inhibitors in invertebrate immunity, in K. Soderhall, S. Iwanaga and G.R. Vasta (eds.), *New Directions in Invertebrate Immunology,* SOS Publications, Fair Haven, N.J.

Leal, S.C.M., Bertioli, D.J., Butt, T.M., Carder, J.H., Burrows, P.R. and Peberdy, J.F. (1997) Amplification and restriction endonuclease digestion of the Pr1 genes for the detection and characterization of *Metarhizium* strains, *Mycol. Res.* **101**, 257-265.

Leal, S.C.M., Bertioli, D.J., Butt, T.M. and Peberdy, J.F. (1994) Characterization of isolates of the entomopathogenic fungus *Metarhizium anisopliae* by RAPD-PCR, *Mycol. Res.* **98**, 1077-1081.

Levashina, E.A., Langley, E., Green, C., Gubb, D., Ashburner, M., Hoffmann, J.A. and Riechhart, J-M. (1999) Constitutive activation of Toll-mediated antifungal defense in serpin-deficient *Drosophila*, *Science* **285**, 1917-1919.

Lindquist, S. (1986) The heat shock response, *Ann. Rev. Biochem.* **55**, 1151-1191.

Lingg, A. J. and Donaldson, M.D. (1981) Biotic and abiotic factors affecting stability of *Beauveria bassiana* conidia in soil, *J. Invertebr. Pathol.* **38**, 191-200.

Mach, R. L., Peterbauer, C.K., Payer, K., Jaksits, S., Woo, S.L., Zeilinger, S., Kullnig, C.M., Lorito, M. and Kubicek, C.P. (1999) Expression of two major chitinase genes of *Trichoderma atroviride* (*T. harzianum* P1) is triggered by different regulatory signals. *Appl. Environ. Microbiol.* **y 65**, 1858-1863.

Macleod, D. M. (1963) Entomophthorales Infections, in E. A. Steinhaus (ed.), *Insect Pathology: An Advanced Treatise.* 189-231. Academic Press, London and New York.

Magalhaes, B. P., Humber, R.A., Shields, E.J. and Roberts, D.W. (1991) Effects of environment and nutrition on conidium germination and appressoria formation by *Zoophthora radicans* (Zygomycetes: Entomophthorales): a pathogen of the potato leafhopper (Homoptera: Cicadellidae), *Entomol. Soc. Am.* **20**, 1460-1468.

Maurer, P., Conteaudier, Y., Girard, P.A., Bridge, P.D. and Riba, G. (1997) Genetic diversity of *Beauveria bassiana* and relatedness to host range, *Mycol.Res.***101**, 159-164.

Melzer, M.J. and Bidochka, M.J. (1998) Diversity of double-stranded RNA viruses within populations of entomopathogenic fungi and potential implications for fungal growth and virulence, *Mycologia* **90**, 586-594.

Millstein, J. A., Brown G.C. and Nordin, G.L. (1983) Microclimate moisture and conidial production in *Erynia* sp. (Entomophthorales: Entomopthoraceae): In vivo production rate and duration under constant and fluctuating moisture regimes, *Environ. Entomol.* **12**, 1344- 1349 .

Milner, R. J. (1978) On the occurrence of *Entomophthora grylli*, a fungal pathogen of grasshoppers in Australia, *J. Australian Entomol. Soc.* **17**, 293-296.

Milner, R. J., Soper, R.S., Lutton, G.G. (1982) Field release of an Israeli strain of the fungus *Zoophthora radicans* (Brefeld) Batko for the biological control of *Therioaphis trifolii* (Monell) *f. maculata, J. Australian Entomol. Soc.* **21**, 113-118.

Milner, R. J. and G. G. Lutton, G.G. (1986) Dependence of *Verticillium lecanii* (fungi: Hyphomycetes) on high humidities and sporulation using *Myzus persicae* (Homoptera: Aphididae) as hosts, *Environ.Entomol.* **15**, 380-382.

Milner, R. J., Staples, J.A. and Lutton, G.G. (1997) The effect of humidity on germination and infection of termites by the hyphomycete, *Metarhizium anisopliae, J. Inverterbr. Pathol.***69**, 64-69.

Ogarkova, G.R., and Ogarkov, B.N. (1986) [Occurence and distribution of entomogenous fungi in forest and agricultural biocoenoses], *Mikologiya i Fitopatologiya* **20,** 170-175.

Pendland, J.C. and Boucias, D.G. (1993) Variations in the ability of galactose and mannose-specific lectins to bind to cell wall surfaces during growth of the insect pathogenic fungus *Paecilomyces farinosus, Eur. J. Cell Biol.* **60**, 322-330.

Pickford, R. and Riegart, P.W. (1964) The fungus disease caused by *Entomophthora grylli* Fres., and its effects on grasshopper populations in Saskatchewan in 1963, *Can.Entomol.* **96**, 1158-1166.

Quinlan, R. J. (1988) Use of fungi to control insects in glasshouses, in M. N. Burge (ed.), *Fungi in biological control systems* , Manchester University Press, UK, pp. 19-36.

Redkar, R. J., Lemke, P.A. and Singh, N.K. (1996) Altered gene expression in *Aspergillus nidulans* in response to salt stress, *Mycologia* **88**, 256-263.

Roberts, D. W. and Hajek, A.E. (1992) Entomopathogenic Fungi as Bioinsecticides, in Leatham, G.F. (ed.) *Frontiers in Industrial Mycology.* Chapman and Hall, New York, New York, pp. 144-159.

Roddam, L. F. and Rath, A.C. (1997) Isolation and characterisation of *Metarhizium anisopliae* and *Beauveria bassiana* from subantarctic Macquarie Island, *J.Invertebr.Pathol.***69**, 285-288.

Roffrey, J. (1968) The occurrence of the fungus *Entomophthora grylli* Fres., on lands and grasshoppers in Thailand, *J.Invertebr.Pathol.* **11**, 237- 241.

St. Leger R.J. and Bidochka, M.J. (1996) Insect fungal interactions, in K. Soderhall, S. Iwanaga and G.R. Vasta (eds.), *New Directions in Invertebrate Immunology,* SOS Publications, Fair Haven, N.J. pp.443-479.

St. Leger, R.J., Joshi, L., and Roberts, D.W (1998) Ambient pH is a major determinant in the expression of cuticle-degrading enzymes and hydrophobin by *Metarhizium anisopliae, Appl. Environ. Microbiol.* **64**, 709-713.

St. Leger, R.J., Joshi, L., Bidochka, M.J. and Roberts, D.W. (1996) Construction of an improved mycoinsecticide overexpressing a toxic protease, *Proc. Nat. Acad. Sci. USA* **93**, 6349-6354.

St. Leger, R.J., Bidochka, M.J. and Roberts, D.W. (1994) Germination triggers of *Metarhizium anisopliae* conidia are related to host species, *Microbiol.* **140**, 1651-1660.

St. Leger, R.J., May, B., Allee, L.L., Frank, D.C., Staples, R.C., Roberts, D.W. (1992a) Genetic differences in allozymes and in formation of infection structures among isolates of the entomopathogenic fungus *Metarhizium anisopliae.* *J. Invertebr. Pathol.* **60**, 89-101.

St. Leger, R.J., Frank, D.C., Roberts, D.W. and Staples, R.C. (1992b) Molecular cloning and regulatory analysis of the cuticle-degrading protease structural gene from the entomopathogenic fungus *Metarhizium anisopliae Eur. J. Biochem.* **204**, 991-1001.

St. Leger, R.J., Staples, R.C. and Roberts, D.W. (1992c) Cloning and regulatory analysis of starvation-stress gene, *ssg*A, encoding a hydrophobin-like protein from the entomopathogenic fungus, *Metarhizium anisopliae* , *Gene* **120**, 119-124.

St. Leger, R.J., Goettel, M. and Roberts, D.W. (1991) Pre-penetration events during infection of host cuticle by *Metarhizium anisopliae, J.Invertebr.Pathol.***58**, 168-179.

Schulte, P.M., Gomez-Chiarri, M., and Powers, D.A. (1997) Structural and functional differences in the promoter and 5' flanking region of Ldh-B within and between populations of the teleost *Fundulus heteroclitus. Genetics* **145**, 759-769.

Screen, S., Bailey, A., Charnley, K, Cooper, R., and Clarkson, J. (1997) Carbon regulation of the cuticle-degrading enzyme PR1 from *Metarhizium anisopliae* may involve a trans-acting DNA-binding protein CRR1, a functional equivalent of the *Aspergillus nidulans* CREA protein, *Curr. Genet.* **31**, 511-518.

Shands, W. A., Hall, I.M., and Simpson, G.W. (1962) Entomophthoraceous fungi attacking the potato aphid in Northeastern Maine in 1960, *J. Econ. Entomol..* **55**, 174-9.

Soderhall, K., Cerenius, L. and Johansson, M.W. (1996) The prophenoloxidase activating system in invertebrates, in K. Soderhall, S. Iwanaga and G.R. Vasta (eds.), *New Directions in Invertebrate Immunology,* SOS Publications, Fair Haven, N.J. pp.229-253.

Steinhaus, E.A. (1960) Symposium: Selected topics in microbial ecology. II. The importance of environmental factors in the insect-microbe ecosystem, *Bacteriol. Rev.* **24,** 365-373.

Tabashnik, B.E. (1997) Seeking the root of insect-resistance to transgenic plants, *Proc. Nat. Acad. Sci. USA* **94**, 3488-3490.

Tabashnik, B.E., Liu, Y-B., Malvar, T., Heckel, D.G., Masson, L., Ballester, V., Granero, F., Mensua, J.L. and Ferre, J. (1997) Global variation in the genetic and biochemical basis of diamondback moth resistance to *Bacillus thuringiensis, Proc. Nat. Acad. Sci. USA* **94**, 12780-12785.

Talbot, N. J., McCafferty, H. R. K., Ma, M., Moore, K. and Hamer, J.E. (1997) Nitrogen starvation of the rice blast fungus *Magnaporthe grisea* may act as an environmental cue for disease symptom expression, *Physiol. Mol. Plant Pathol.* **50**, 179-195.

Tanada, Y and Kaya, H.H. (1993) Insect Pathology. San Diego, Academic Press.

Templeton, M.D., Rikkerink, E.H.S. and Beever, R.E. (1994) Small, cysteine-rich proteins and recognition in fungal-plant interactions, *Mol. Plant-Microbe Interactions* **7**, 320-325.

Tigano-Milani, M.S., Gomes, A.C.M.M., and Sobral, B.W.S. (1995) Genetic variability among Brazilian isolates of the entomopathogenic fungus *Metarhizium anisopliae, J. Invertebr. Pathol.* **65**, 206-210.

Tinline, R.D. and Noviello, C. (1971) Heterokaryosis in the entomogenous fungus *Metarrhizium anisopliae, Mycologia* **63**, 701-712.

Torzilli, A. P. (1997) Tolerance to high temperature and salt stress by a salt marsh isolate of *Aureobasidium pullulans, Mycologia* **89**, 786-792.

Typas, M.A., Mavridou, A. and Kouvelis, V. (1998) Mitochondrial DNA differences provide maximum intraspecific polymorphism in the entomopathogenic fungi *Verticillium lecanii* and *Metarhizium anisopliae* and allow isolate detection and identification, in P. Bridge, Y. Couteaudier and J.M. Clarkson (eds.), *Molecular Variability of Fungal Pathogens* CAB International, New York.

Vainninen, I. (1996) Distribution and occurrence of four entomopathogenic fungi in Finland: effect of geographical location, habitat type and soil type, *Mycol. Res.* **100**, 93-101.

Viaud, M., Couteaudier, Y., Levis, C. and Riba, G. (1996) Genome organization in *Beauveria bassiana*: Electrophoretic karyotype, gene mapping, and telomeric fingerprint, *Fungal Genet. Biol.* **20**,175-183.

Walsh, S.R.A., Tyrrell, D., Humber, R.A. and Silver, J.C. (1990) DNA restriction fragment length polymorphisms in the rDNA repeat unit of *Entomophaga*, *Experi. Mycol.* **14**, 381-392.

Walstad, J.D., Anderson, R.F. and Stambaugh, W.J. (1970) Effects of environmental conditions on two species of Muscardine fungi (*Beauveria bassiana* and *Metarhizium anisopliae*), *J. Invertebr. Pathol.* **16**, 221-226.

Warwar, V., and Dickman, M.B. (1996) Effects of calcium and calmodulin on spore germination and appressorium development in *Colletotrichum trifolii, Appl. Environ. Microbiol.* **62**, 74-79.

Widden, P., and Parkinson, D. (1979) Populations of fungi in a high arctic ecosystem, *Can. J. Bot.* **57**, 1324-1431.

Woods, S.P. and Grula, E.A. (1984) Utilizable surface nutrients on *Heliothis zea* available for growth of *Beauveria bassiana, J. Invertebr. Pathol.* **43**, 259-269.

Wraight, S. P., Butt, T.M., Galaini-Wraight, S., Allee, L.L. and Roberts, D.W. (1990) Germination and infection process of the entomophthoralean fungus *Erynia radicans* on the potato leafhopper *Empoasca fabae, J. Invertebr. Pathol.* **56,** 157-174.

Xavier, I. J. and G. G. Khachatourians. 1996. Heat-shock response of the entomopathogenic fungus *Beauveria brongniartii, Can. J. Microbiol.* **42,**577-58.

Yang, Z., Dickman, M.B. (1997) Regulation of cAMP and cAMP dependent protein kinase during conidial germination and appressorium formation in *Colletotrichum trifolii, Physiol. Mol. Plant Pathol.* **50,** 117-127.

Zibilske, L. M. and Wagner, G.H. (1982) Bacterial growth and fungal genera distribution in soil amended with sewage sludge containing cadmium, chromium and copper, *Soil Science.* **134**, 364-370.

Zimmerman, G. (1986) The 'Galleria bait method' for detection of entomopathogenic fungi in soil, *J. Appl. Entomol.* **102**, 213-215.

Zimmerman, G. (1982) Effect of high temperature and sunlight on the viability of conidia of *Metarhizium anisopliae, J. Invertebr. Pathol.* **40**, 36-41.

PATHOGENESIS AND GENOME ORGANIZATION OF THE RICE BLAST FUNGUS

S. KANG AND E. MULLINS
Department of Plant Pathology, Pennsylvania State University
University Park, PA 16802

T. M. DEZWAAN
The DuPont Ag Enterprise, DuPont
Newark, DE 19714

M. J. ORBACH
Department of Plant Pathology, University of Arizona
Tucson, AZ 85721

1. Introduction

Plant diseases caused by fungi present a serious challenge to our efforts to feed the world's rapidly increasing population. Manipulation of plant genes through genetic engineering and traditional breeding provides an opportunity to improve the disease resistance in crops of agricultural significance. In addition, the application of target-based drug design will lead to the identification of various novel chemical agents that will possess significant potential as disease control agents. The success of such approaches will depend heavily on the biological knowledge of fungal pathogens in such areas as mechanisms of virulence and pathogenicity, population structure and dynamics, and the nature and mechanisms of genetic changes underpinning the evolution of new races. As a result of limited resources, coupled with the necessity to study many fungal diseases of economic significance, comprehensive research focused on these aspects of fungal biology has been restricted to a limited number of model systems. *Magnaporthe grisea*, the causal agent for rice blast disease, has become one such system. Worldwide, rice blast is one of the most economically devastating crop diseases. In addition to its agricultural significance, the rice blast system presents many advantages as an experimental model: (a) Extensive genetic and molecular analyses of host specificity and pathogenicity have been carried out; (b) Molecular and cellular bases of infection processes are well characterized; (c) Rice is a model genetic system for monocots; (d) Well-saturated genetic and physical maps of *M. grisea* are available; (e) Genome sequencing of both rice and *M. grisea* is in progress; (f) Genetic and race diversity of fungal populations in many rice-growing areas has been extensively surveyed. In this chapter we aim to provide an overview of the nature of the interaction between *M.*

195

J. W. Kronstad (ed.), Fungal Pathology, 195–235.

grisea and its hosts and the corresponding mechanisms employed at each level, from molecules to populations.

2. Reproduction and Host Range of *Magnaporthe grisea*

Magnaporthe grisea (Hebert) Barr. (anamorph, *Pyricularia grisea* Sacc.) is a haploid, filamentous, heterothallic ascomycete in the class pyrenomycetes. The fungus can infect all parts of the rice shoot, resulting in the formation of a lesion at the infected site. The size and appearance of lesions vary significantly depending on environmental conditions, developmental stages of rice and the compatibility between the fungus and the infected rice (Bonman, 1992). Neck blast caused by infection of the panicle neck node is the most damaging symptom. The sexual cycle of *M. grisea* is that of a typical pyrenomycete producing eight-spored asci inside a perithecium. Crosses can only be completed on certain types of media (Valent et al., 1986) and have not been observed in the field. The asexual cycle can occur on host plants as well as on various artificial substrates.

2.1. REPRODUCTION

Although the *M. grisea* sexual stage can be routinely produced in the laboratory, it has not been observed in nature, where the main mode of propagation is asexual growth and the production of conidia. Recent evidence based on apparent recombinant genotypes suggests that sexual reproduction in the field mat have occurred in some parts of the world (Kumar et al., 1999). The fungus reproduces asexually by conidiogenesis which is initiated by the formation of an aerial, bottle-shaped conidiophore (Howard, 1994). Expansion and swelling of the conidiophore gives rise to a spherical conidium initial that subsequently elongates to form a three-celled, pyriform conidium. Single nuclei in the three conidial cells are isogenic, having originated from a single progenitor nucleus through mitotic division. Additional conidia are formed in a sympodial manner, resulting in the production of three to five conidia on a mature conidiophore. Germ tube growth typically occurs from the apical and/or basal cell of a conidium although all three cells are competent for germination.

The sexual stage of the fungus was first demonstrated by crossing two crabgrass isolates (Hebert, 1971). Long beaked perithecia with a pigmented, globoid base develop at the intersection of two cultures when compatible strains are inoculated on media conducive to sexual development. Although the formation of perithecia in infected plants in the fields has not yet been reported, certain strains of opposite mating types produced perithecia containing viable ascospores following co-inoculation on the same rice plant (Hayashi et al., 1997; Silué and Nottegehem, 1990). Approximately two weeks after the pairing, asci containing eight crescent-shaped ascspores form in the perithecia. As with other ascomycetes, compatibility for mating is governed by alternate alleles in the mating-type locus, *MAT1* (Hebert, 1971). Analyses of the mating-type genes from a diverse group of ascomycetes revealed that the alternative mating-type alleles in each system contain non-homologous sequences embedded in essentially identical flanking DNA (Kronstad and Staben, 1997). The term idiomorph was introduced to designate unrelated sequences at the same locus (Metzenberg and Glass, 1990). Using genomic subtraction, the mating-type genes of *M. grisea* have been cloned and shown to be typical of other ascomycetes in containing idiomorphic alleles (Kang et al., 1994).

In contrast to the control of mating compatibility by a single locus, female fertility in *M. grisea* appears to be controlled by multiple loci (Kolmer and Ellingboe, 1988; Valent et al., 1991). Although the genetics of female fertility has not been studied in great detail, a spontaneous mutant was isolated that initiates development of maternal sexual tissue in the absence of a mate. This mutant, isolated from a finger millet (*Eleucine coracana*) pathogen, produces normal-looking perithecia but no asci or ascospores in the absence of a strain of opposite mating type (Tharreau et al., 1997). When paired with strains of opposite mating type this mutant, termed *sfp1*, mates normally. Mutants of similar phenotype were also identified in other strains (Namai et al., 1986; Yamanaka et al., 1984). The fertility of *M. grisea* field isolates, ranging from total sterility (inability to mate with any other strain) to full fertility (ability to mate as a male or as a female), generally correlates with their host of origin. Rice pathogen field isolates are almost all female sterile except for a limited number of hermaphroditic strains (Hayashi et al., 1997; Kumar et al., 1999; Leung et al., 1988). In contrast, field isolates that infect weeping lovegrass (*Eragrostis curvula*), goosegrass (*Eleusine indica*) or finger millet (*Eleusine coracana*) are usually hermaphrodites capable of mating to produce numerous viable ascospores. Hermaphroditic laboratory strains that infect various host plants including rice have been developed and used for different genetic analyses (Kolmer and Ellingboe, 1988; Leung et al., 1988; Valent et al., 1986; Valent et al., 1991)

2.3. HOST RANGE OF THE FUNGUS

As a collective species, *M. grisea* has a very broad host range on monocotolydenous species. Individual strains cause disease on crops such as rice, wheat, barley, finger millet, pearl millet, and maize, as well as on many grasses that grow near cultivated crops (Asuyama, 1963; Urashima et al., 1993). More than 50 grass species have been identified as potential hosts of *M. grisea*. Although most of these species belong to the family *Poaceae* in the order *Cyperales*, *M. grisea* has also been isolated from members of the four families (*Cannaceae*, *Musaceae*, *Zingiberaceae* and *Marantaceae*) in the order *Zingiberales* (Asuyama, 1963; Pappas, 1998).

3. The *M. grisea* Disease Cycle and the Molecular Basis of Pathogenicity

During the *M. grisea* disease cycle, at least five discrete developmental stages, essential for pathogenicity, have been defined: germination, appressorium differentiation, penetration, ramification and conidiation (Figure 1). The transition through each of these stages, during which the fungus grows on and then into its host, is coordinated with changes in external conditions. At each stage the fungus assimilates informational cues from its surrounding environment to elicit a specific cellular response.

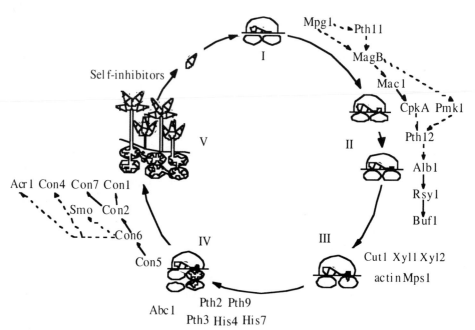

Figure 1. The *M. grisea* disease cycle and the molecular basis of pathogenicity. The five major stages of the disease cycle are shown (I. germination, II. appressorium differentiation, III. penetration, IV. ramification, and V. conidiation). The fungal components affecting pathogenicity at each stage are indicated. Arrows connecting components indicate order of function. The solid arrows represent an experimentally confirmed order of function whereas dashed arrows indicate a hypothetical order of function.

3.1. GERMINATION

Upon encountering a host, *M. grisea* conidia attach tightly to the surface via an adhesive called "spore tip mucilage" that is stored in an apical compartment (Hamer et al., 1989). Germination quickly follows attachment, occuring within the first hour of contact with the host (Bourett and Howard, 1990). Because *M. grisea* is a facultative saprobe, germination is not contingent on specific host cues, but it is dependent on the presence of sufficient ambient moisture. Hydrated conidia germinate efficiently in suspension and on a variety of surfaces (Xiao et al., 1994). However, conidia appear to produce self-inhibitors that prevent germination at high conidial densities (Hegde and Kolattukudy, 1997). Self-inhibitors of germination are reversible, static-type compounds that are produced by many fungi to prevent germination in sub-optimal conditions (Trione, 1981). Because the density of *M. grisea* conidia is highest at the site of conidiation (i.e., the surface of diseased leaf tissue) the self-inhibitors prevent germination on an exhausted nutrient supply.

3.2. APPRESSORIUM DIFFERENTIATION

Between 4 and 8 hours after the onset of germination, the *M. grisea* germ tube differentiates into a hemispherical infection structure, termed the appressorium. Appressoria are deployed by many pathogenic fungi as a means of breaching the host cuticle to access the susceptible internal tissue (Mendgen and Deising, 1993). Although the function of appressoria in different fungal species is similar, some fungi penetrate their hosts indirectly through openings such as stomata, whereas other fungi, including *M. grisea*, penetrate directly through the cuticle and underlying epidermal cell wall (Howard, 1997). In addition, the environmental cues that trigger appressorium differentiation vary among fungi, and often reflect an intimate co-evolution of host and parasite. For example, some rust species only form appressoria on substrates with a surface topography that mimics that of their hosts (Allen et al., 1991).

Appressorium differentiation in *M. grisea* is also dependent on specific substrate features, but these features are not peculiar to the host plant. A hard contact surface is necessary for appressorium differentiation. Although conidia are able to germinate and begin hyphal growth on a soft surface such as agar, or at the interface of water and a hydrophobic liquid, they are unable to begin appressorium development (Xiao et al., 1994). The hydrophobicity and chemical composition of the contact surface are also important determinants of appressorium differentiation, and these surface features appear to provide the primary inductive cues that activate differentiation. The extent to which surface hydrophobicity contributes to appressorium differentiation in *M. grisea* is currently a source of debate. Lee and Dean described a tight correlation between the efficiency of appressorium differentiation and the hydrophobicity of the substrate (Lee and Dean, 1994), whereas Jelitto et al. observed no such correlation (Jelitto et al., 1994). These results likely reflect variations among *M. grisea* strains, and it has recently been shown that there is plasticity in the way that inductive substrate cues are processed into a pathogenic response (DeZwaan et al., 1999). The majority of *M. grisea* strains form appressoria more efficiently on hydrophobic surfaces, such as the plant leaf and Teflon, than on hydrophilic surfaces, such as glass and the hydrophilic side of Gelbond (agarose-coated polyester) (DeZwaan et al., 1999).

A clue to the role of hydrophobicity in appressorium differentiation comes from studies with the *M. grisea* fungal hydrophobin protein, MPG1. Fungal hydrophobins are secreted proteins that self assemble into a layer at the interface of the fungal cell wall and the surrounding microenvironment (Kershaw and Talbot, 1998). *mpg1* mutants form appressoria at a reduced frequency and demonstrate impaired pathogenicity (Beckerman and Ebbole, 1996; Talbot et al., 1996). This suggests that MPG1 may facilitate interaction with the host surface, possibly by assembling at the interface of germlings and the hydrophobic surface of the host plant. It has been proposed that the MPG1 layer may interact with the host surface and provide a conformational cue that promotes efficient appressorium differentiation (Kershaw and Talbot, 1998). An intriguing possibility is that MPG1 provides conformational information that results in the activation of a pathogenicity-specific mechanosensor at the *M. grisea* plasma membrane. Although it is currently unknown whether receptors of this type contribute to fungal pathogenicity, the rust fungus *Uromyces appendiculatus* carries a mechanosensitive cation channel at the plasma membrane that has been proposed to play a role in sensing surface topography and activating appressorium differentiation (Zhou et al., 1991).

Lee and Dean examined the contribution of substrate chemical composition to appressorium formation by assaying compounds that commonly occur on leaf surfaces for their ability to induce differentiation (Lee and Dean, 1994). Appressorium differentiation was very efficient in the presence of nanomolar concentrations of plant cutin monomers, and a stringent structure-activity relationship was observed with these and related compounds. Although there was specificity in the action of these compounds, the cutin monomers that were most effective at inducing differentiation were common to many plant species, and the *M. grisea* strains examined in this study were able to form appressoria on both host and non-host plants. Thus, the pre-penetration events of the *M. grisea* disease cycle are not host-specific (Gilbert et al., 1996).

Appressorium development in *M. grisea* occurs in two phases, a surface recognition phase and a differentiation phase. During the first phase, in which the fitness of the surface is determined, the tip of the germ tube enlarges and becomes hook-shaped, and apical vesicles polarize toward the plant surface. Also during this phase, the germ tube tip has a "nose down" appearance that suggests an intimate interaction between the fungal and plant surfaces. This "recognition phase" appears to be prior to the commitment to form appressoria (Bourett and Howard, 1990). Consistent with this hypothesis, conidia that are germinated on non-inductive surfaces (i.e., hydrophilic surfaces lacking the appropriate surface chemistry) execute these early developmental events but fail to complete appressorium differentiation (DeZwaan et al., 1999). During the second phase of appressorium differentiation, the germ tube tip continues to enlarge to form a symmetrical appressorium. The appressorium becomes deeply melanized over its entire surface except at the appressorial pore, where it interfaces with the host (Howard and Ferrari, 1989). A septum then forms between the appressorium and the remainder of the germ tube to yield a unicellular structure with a single nucleus (Bourett and Howard, 1990). As the appressorium matures, an "O"-ring encircles and seals the appressorial pore, which tightly adheres the appressorium to the underlying substrate (Howard and Ferrari, 1989).

A number of studies have begun to elucidate the molecular basis of substrate recognition that occurs during the first stage of appressorium differentiation. Exogenous cyclic AMP (cAMP) suppresses the appressorium deficiency of germlings inoculated on non-inductive surfaces, suggesting that inductive substrates activate a cAMP-mediated kinase cascade that controls differentiation (Lee and Dean, 1993). Consistent with this, mutations in the MAC1 adenylate cyclase, which converts ATP to cAMP, prevent appressorium development beyond the first phase of differentiation, and this defect is suppressed by the addition of exogenous cAMP (Adachi and Hamer, 1998; Choi and Dean, 1997).

Cyclic AMP signaling in many eukaryotes is mediated via receptor-coupled heterotrimeric G proteins that govern the activity of adenylate cyclase (Simonds, 1999). A recent study of the heterotrimeric G protein α subunit encoded by the *MAGB* gene suggests that cAMP levels may be modulated in a similar manner in *M. grisea* (Liu and Dean, 1997). *magB* mutants produce reduced numbers of appressoria, similar to *mac1* mutants (Choi and Dean, 1997), and this *magB* defect is also suppressed by exogenous cAMP (Choi and Dean, 1997; Liu and Dean, 1997). This suggests that MAGB may activate MAC1, resulting in elevated intracellular cAMP levels and activation of appressorium differentiation (Liu and Dean, 1997). Interestingly, *magB* mutants fail to form appressoria specifically on hydrophobic surfaces but continue to form appressoria in response to soluble cutin monomers (Liu and Dean, 1997), whereas *mac1* mutants fail to form appressoria in response to both inductive cues (Choi and Dean, 1997). This suggests that MAGB may specifically mediate differentiation in response to surface hydrophobicity, and may function in parallel with a cutin monomer-activated pathway. These parallel signaling pathways then appear to converge on MAC1.

A novel membrane protein encoded by the *PTH11* gene, which plays an upstream role in substrate signaling, has recently been identified (DeZwaan et al., 1999). Similar to *mac1* and *magB* mutants, *pth11* mutants fail to differentiate beyond the first stage of appressorium morphogenesis, and the *pth11* mutant defect is suppressed by exogenous cAMP. In contrast to MAGB and MAC1, which, in addition to their roles in appressorium development, function during mating and vegetative growth, PTH11 only functions in appressorium differentiation. *pth11* mutants exhibit defects in appressorium development but are unimpaired for all other stages of the disease or life cycle. The signaling role of PTH11 and its functional specificity suggest that it may act as a receptor for inductive substrate cues. However, PTH11 lacks the prototypical seven transmembrane regions of eukaryotic serpentine receptors (Bockaert and Pin, 1999) and is predicted to have as many as nine transmembrane regions. PTH11 is required for appressorium differentiation in response to both hydrophobicity and cutin monomers, thus it is not clear how it may fit into the putative MAGB-MAC1 signaling pathway. PTH11 has also been found to repress differentiation on non-inductive surfaces in some *M. grisea* strains, indicating that it can function as both a positive and negative regulator of differentiation (DeZwaan et al., 1999).

Protein kinases in fungi have been shown to be important mediators in developmental responses to extracellular signals. At least two kinases in *M. grisea* are required for appressorium morphogenesis. The first is a cAMP-dependent protein kinase A homolog encoded by the *CPKA* gene. The appressoria of *cpkA* mutants are smaller than those of wild type strains and fail to penetrate the host surface (Xu et al., 1997). *cpkA* mutants remain responsive to cAMP (Xu et al., 1997), suggesting that cyclic AMP-dependent protein kinases in addition to CPKA may exist in *M. grisea*. Also, *cpka* mutants are able to differentiate beyond the first stage of appressorium morphogenesis, albeit incompletely, whereas *mac1* mutants do not. This suggests that CPKA is only required for a subset of MAC1/cAMP-dependent events.

The *PMK1* gene encodes the second kinase that is required for appressorium differentiation. PMK1 is a homolog of the FUS3 and KSS1 mitogen-activated protein kinases of *Saccharomyces cerevisiae* (Xu and Hamer, 1996). In *S. cerevisiae*, the FUS3 and KSS1 are required for morphological changes of the cell that occur during mating and pseudohyphal growth (Madhani and Fink, 1998). The PMK1 kinase of *M. grisea* is also required for cell morphological changes. *pmk1* mutants are able to complete the early "recognition" stage of appressorium differentiation, as evidenced by their ability to undergo germ tube hooking and apical enlargement. However, *pmk1* mutants fail to complete appressorium differentiation and are, thus, unable to penetrate the host plant (Xu and Hamer, 1996). In addition, PMK1 is also required for invasive growth, suggesting that it may perform multiple functions during the course of pathogenesis (Xu and Hamer, 1996). It is currently unclear whether PMK1 is a downstream component of cAMP-mediated signaling or acts in a parallel-signaling pathway.

Signal transduction in fungi as in other eukaryotic cells culminates in the deployment of specific transcription factors that regulate gene expression to elicit a specific cellular response (Banuett, 1998). This also appears to be the case with appressorium signaling in *M. grisea*. The *PTH12* gene encodes a homeodomain transcription factor that is essential for appressorium differentiation (A.M. Carroll and J.A. Sweigard, personal communication). The phenotype of *pth12* mutants is similar to that of *cpkA* mutants; appressoria form but they fail to completely mature and are unable to penetrate the host surface. At present the kinase pathway that regulates PTH12 function is unknown. Also, additional transcriptional regulators of appressorium differentiation are likely to exist because *pth12* mutants are able to partially complete appressorium differentiation.

3.3. PENETRATION

Following the formation of a mature appressorium, *M. grisea* physically punctures the host cuticle and underlying epidermal cell layer with a narrow penetration peg (Howard and Valent, 1996). Penetration through the host surface is a turgor-driven process that is dependent on the formation of a cell wall layer composed of fungal melanin (Chumley and Valent, 1990; Howard and Ferrari, 1989; Howard et al., 1991). Appressorial turgor is generated by an increase in the internal glycerol concentration (de Jong et al., 1997), and it is believed that the melanin layer mediates turgor build-up by preventing the free diffusion of glycerol out of the cell (de Jong et al., 1997; Money, 1997). Pressure levels as high as 8 Mpa (over 1000 psi) can be reached (Howard et al., 1991). Because melanin is essential for appressorial penetration, the genes responsible for melanin production are required for pathogenicity. Mutational analysis has identified three unlinked genes controlling melanin biosynthesis; the *ALB1* gene encoding a polyketide synthase, the *RSY1* gene encoding scytalone dehydratase, and the *BUF1* gene encoding polyhydroxynaphthalene reductase, an NADPH-dependent dehydrogenase with similarity to a ketoreductase involved in *Aspergillus parasiticus* aflatoxin biosynthesis (Howard and Valent, 1996). It will be interesting to determine whether the expression of these genes is governed by any of the signaling pathways that mediate appressorium morphogenesis.

A period of quiescence lasting between 24 and 48 hours precedes penetration peg formation. How the formation of this structure is activated or how the delocalized force of appressorial turgor is focused to mediate host penetration is still unknown. One model predicts that penetration occurs through a weakened region of the host surface that underlies the appressorial pore. Enzymes that digest the host cuticle and cell wall have been proposed to weaken the host surface to facilitate penetration. Consistent with this model, penetration occurs more rapidly on rice leaves, which are likely to be degraded by enzymatic digestion, than it does on mylar, a non-biodegradable substrate (Howard, 1994). Three genes encoding cuticle and cell wall-degrading enzymes have been identified in *M. grisea*; two *endo*-β-1,4-xylanase genes, *XYL1* and *XYL2* (Wu et al., 1995), and the cutinase gene, *CUT1* (Sweigard et al., 1992). Deletion mutants of these genes are fully pathogenic as is a *xyl1 xyl2* double mutant, indicating that none of these enzymes are essential for infection (Sweigard et al., 1992; Wu et al., 1997). However, additional endoxylanase activity was detected in the supernatant of a *xyl1 xyl2* mutant strain, and additional cutinase activity was detected in the supernatant of a *cut1* mutant strain (Sweigard et al., 1992; Wu et al., 1997). Thus, additional cell wall degrading enzymes may function in conjunction with CUT1, XYL1 and XYL2 to facilitate host penetration.

Polarized growth of the penetration peg through the weakened surface would likely be accompanied by reorganization of the fungal cell wall and cytoskeleton. This would parallel the dynamic nature of these structures at other points where polarized growth occurs, such as at hyphal tips and hyphal branch initiation. The penetration peg is an actin rich structure, thus the actin cytoskeleton may aid in the generation of force or perform a stabilizing role during the course of penetration (Bourett and Howard, 1992). A mitogen activated protein kinase homolog, encoded by the *MPS1* gene, is required for penetration, and *mps1* mutants form appressoria that appear morphologically normal but fail to penetrate the host surface (Xu et al., 1998). MPS1 is a homolog of the yeast SLT2 kinase that is required for reorganization of the actin cytoskeleton and polarized cell wall deposition during bud emergence and apical bud growth (Mazzoni et al., 1993; Xu et al., 1998). It will be interesting to see if MPS1 functions in a similar manner, mediating cytoskeletal and cell wall reorganization during penetration peg formation.

3.4. RAMIFICATION

Following penetration of the host cuticle and the epidermal cell layer, the penetration peg gives rise to invasive hyphae that spread cell to cell throughout the host tissue. The two primary environmental stresses that the fungus encounters during this stage appear to be nutrient deprivation and host fungitoxins that may be deployed to defend against infection. Two lines of experimental evidence have revealed a link between the nutrient poor environment the pathogen encounters within the host and pathogenicity. First, nitrogen starvation of vegetative *M. grisea* cultures elicited the production of secreted proteins that caused senescence of rice leaves (Talbot et al., 1997). This suggests that nutrient limitation may induce the expression of genes that cause symptom development in the host. Second, mutational analyses have shown that genes controlling nutrient metabolism are essential for pathogenicity (Sweigard et al., 1998). These genes include three that were isolated through a pathogenicity mutant search: *PTH2*, an acyltransferase gene homolog, *PTH9*, a neutral trehalase gene that may be required for utilization of trehalose as a carbon source and *PTH3*, an imidazole glycerol phosphate dehydratase gene homolog. Also, the *M. grisea* homologs of the *S. cerevisiae HIS4* and *HIS7* genes (Sweigard et al., 1998; A.M. Carroll and J.A. Sweigard, personal communication) are required for pathogenicity. *HIS4* encodes three enzymatic functions in histidine biosynthesis, phosphoribosyl-AMP cyclohydrolase, phosphoribosyl-ATP pyrophosphohydrolase, and histidinol dehydrogenase, and *HIS7* encodes the enzymatic functions glutamine amidotransferase/cyclase. Interestingly, *PTH3*, *HIS4* and *HIS7* all encode components of the histidine biosynthetic pathway, indicating that the histidine concentration in the plant is insufficient to complement histidine auxotrophy in *M. grisea*. Mutations in these metabolic genes do not dramatically affect the ability to form appressoria but the size of the lesions produced is greatly reduced compared to wild type (Sweigard et al., 1998; A.M. Carroll and J.A. Sweigard, personal communication). This suggests that these strains are specifically deficient in their ability to grow invasively. This may be due to the combined effect of the metabolic deficiency caused by the mutations and the nutrient limited environment of the host.

Although no host fungitoxins have been identified that affect *M. grisea* pathogenicity, recent evidence suggests that a drug efflux pump is necessary for growth within host tissue. The ATP-driven drug efflux pump ABC1 was identified by REMI mutagenesis as a pathogenicity factor of *M. grisea* (Urban et al., 1999). *abc1* mutants form appressoria and penetrate efficiently but fail to grow invasively throughout the host tissue. Thus, this pump is thought to provide resistance to antifungal compounds present in host cells (Urban et al., 1999).

3.5. CONIDIATION

Because conidia are both the infectious propagules responsible for initiating infection and the dispersal units for spreading disease, conidiation is essential for pathogenicity. Mutational analysis has revealed two classes of genetic loci that affect conidiation. Mutations in the first class fail to produce conidia because they are defective in conidiophore differentiation. The loci *CON5* and *CON6* fall into this class. *con5* mutants are completely blocked for conidiophore formation whereas *con6* mutants produce abundant conidiophores that fail to form conidia (Shi et al., 1997). These results suggest that *CON5* is epistatic to *CON6*.

Mutations in the second class of genetic loci affecting conidiation result in altered conidium morphology. The *SMO*, *CON1*, *CON2*, *CON4*, *CON7* and *ACR1* loci belong to this class. The *SMO* locus was identified in a screen for mutants that are defective for appressorium formation (Hamer et al., 1989). Analysis of *smo* mutants revealed defects in multiple specialized structures including appressoria, conidia, and asci. Thus, *SMO* may encode a global controller of cell morphogenesis that is required for multiple developmental events. The *CON1*, *2*, *4* and *7* loci were identified in a genetic screen for mutants with defects in conidium morphogenesis. The epistatic relationship among these loci was then tested by crossing two *con* mutants to produce the double mutant (Shi and Leung, 1995). In general, the epistatic relationships among these loci reflect the phenotype of each of the single mutants. *con2* mutants form very underdeveloped conidia that have only one or two cells whereas *con1* and *con7* mutants form hyper-elongated three-celled conidia, and the epistatic relationship places *con2* upstream of both *con1* and *con7* (Shi et al., 1997). This type of epistasis analysis has revealed the existence of a core pathway governing conidiophore formation and conidiation wherein the order of gene action is *con5* > *con6* > *con2* > *con1* and *con7*. However, conidiation also involves other loci that function in parallel with this core pathway. For example, *con4* mutants form ellipsoidal three-celled conidia and thus, *con2* would be predicted to be epistatic to *con4*. However, when the *con2/con4* double mutant was made the phenotype of the spores was intermediate between the two mutants, suggesting that they function in parallel pathways (Shi et al., 1997). A common feature among the four conidium morphology mutants (*CON1, 2, 4* and *7*) is a reduction in pathogenicity, suggesting that normal spore morphogenesis is a prerequisite for downstream events in the disease cycle. It is possible that this is because abnormal spores are less able to adhere properly to host tissue.

The final genetic locus that is known to be required for normal conidium morphology is defined by *ACR1*. Like the *con* mutants, *acr1* mutants fail to produce multiple bottle-shaped three-celled conidia in a sympodial manner on each conidiophore, instead producing a single long chain of conidiogenous cells at the conidiophore apex (Lau and Hamer, 1998) in an acropetal mode. The mutant also exhibits reduced pathogenicity on rice, most likely due to both a reduced ability to attach to plant surfaces and a reduction in appressorium formation. *ACR1* is the only conidiation gene that has been cloned and sequenced, and it encodes a novel glutamine-rich protein that has weak similarity to some transcriptional repressors. Thus, it is thought that ACR1 may be a negative regulator of conidiogenesis that arrests further growth beyond the three-cell stage of development (Lau and Hamer, 1998).

4. Genome Organization

Because of the high purported level of genetic instability in *M. grisea*, there has been a great deal of interest in the genome organization and chromosome structure of *M. grisea*. Physical characterization of the *M. grisea* genome has been analyzed using a variety of approaches. Overall genome structure has been analyzed using molecular karyotype methodology and genetic mapping. Two components of the genome that may play a role in genetic instability, telomeres and repetitive DNA elements, are discussed.

4.1. KARYOTYPES

The karyotype of *M. grisea* isolates from various hosts has been analyzed using pulsed-field gel electrophoresis (Orbach et al., 1996; Talbot et al., 1993). The number of "A" chromosomes, ranging in size from 2 Mb to greater than 10 Mb, is largely uniform although the sizes vary among strains due to apparent translocations. However, the number and size of "B" chromosomes varied considerably among the strains used in these surveys, where a correlation between the presence of B chromosomes and low levels of sexual fertility appeared to exist (Orbach et al., 1996).

It was estimated that the genome size of *M. grisea* falls in the range of 35-49 Mb (Skinner et al., 1993). To test whether karyotype variation correlates with race variation, the electrophoretic karyotypes of a number of rice isolate races from the United States and China were studied (Talbot et al., 1993). The degree of karyotype variation, which is most likely due to chromosomal deletions and rearrangements, was high even among strains in the same genetic lineage, obscuring their close genetic and race relatedness. It was suggested that the apparent lack of sexual recombination among the rice pathogen populations underpins the accumulation of chromosome polymorphisms (Talbot et al., 1993). Infertile field isolates, including most rice pathogens, have a high degree of chromosome length polymorphisms. In contrast, fertile strains from diverse hosts around the world have a relatively uniform karyotype (Orbach et al., 1996). While the occurrence of karyotype changes was observed in a rice strain after prolonged serial transfer in culture, such polymorphisms did not appear to alter the race of the isolate (Talbot et al., 1993).

4.2. TELOMERES

The ends of chromosomes in *M. grisea* terminate with a hexanucleotide repeat, (TTAGGG)$_n$ (Farman and Leong, 1995; Sweigard et al., 1993), which is also present in the telomeres of vertebrates, slime molds and other fungi, including *Neurospora crassa* (Schechtman, 1990), *Cladosporium fulvum* (Coleman et al., 1993) and *Fusarium oxysporum* (Powell and Kistler, 1990). Because of evidence that two avirulence genes, which exhibit genetic instability, were located near telomeres in the rice pathogen O-137, three chromosomal ends from this strain were cloned and characterized. One of them contains the *AVR2-YAMO* gene (renamed *AVR-Pita* based on evidence of its interaction with the rice resistance gene *Pi-ta*) which prevents the fungus from infecting rice cultivar Yashiro-mochi. The *AVR-Pita* open reading frame ends just 48 bp upstream of the 20 copies of the hexameric telomere repeat (Valent and Chumley, 1994). This gene has been found to have three different types of mutation: deletions, point mutations and an insertion (Valent and Chumley, 1994). Sequencing of a second telomeric region from O-137 revealed the presence of transposable elements and a novel gene family found specifically in rice pathogens in sub-telomeric positions (S. Kang, unpublished). In a 13 kb telomeric fragment, which corresponds to a chromosome end different from the end containing *AVR-Pita*, part of a Pot3 transposable element (Farman et al., 1996) is located at the distal end away from the 156 bp telomeric repeat sequence (Figure 2). A different type of transposon, Mg-SINE (Kachroo et al., 1995), is also present 800 bp downstream from Pot3. A novel gene, designated *TLH1* (*T*elomere-*L*inked *H*elicase *1*), is present between the telomeric repeat and Pot3 and encodes a protein of 818 amino acids that shows significant similarity to helicases from several organisms (Kang, unpublished). The *TLH1* gene appears to be a member of a gene family specific to rice pathogens. From a survey of 18 field isolates of *M. grisea* from various hosts the *TLH1* probe did not hybridize to any of the 16 isolates from hosts other than cultivated rice (*Oryza sativa*) with the exception of an isolate from *Panicum maximum* that contains a single copy of *TLH1*. In contrast, two isolates from *O. sativa*, Guy11 and O-137, had approximately 10 copies each of this gene. The sequences homologous to the *TLH1* gene segregated with at least six different telomeres in the mapping progeny from a cross between parental strains derived from Guy11 and O-137 (Sweigard et al., 1993), suggesting that many subtelomeric regions of these rice pathogens have sequences homologous to the *TLH1* gene. Sequences flanking *TLH1* have also been amplified during the course of evolution of this gene family. Genetic and physical mapping of telomeres in Guy11, 2539 and their progeny also showed that at least one telomere-associated sequence was duplicated (Farman and Leong, 1995).

A third chromosome end of O-137, isolated as a 19 kb fragment, has also been characterized (S. Kang, unpublished). Similar to the telomere linked to *AVR-Pita*, this telomere was altered in a group of spontaneous mutants of O-137 that gained virulence on rice cultivar Tsuyuake (see section 7.2 for more information about these mutants). Three different types of transposons were identified on this telomeric fragment (Figure 2). The 114 bp-long telomeric repeat was immediately followed by an approximately 6 kb repetitive DNA element designated STR for Sub-Telomeric Repeat. The STR sequence revealed that the element is a novel retrotransposon containing Long Terminal Repeats (LTRs). The telomeric repeat is attached to one of the LTRs of STR. A truncated copy of Pot2 flanks the other LTR, suggesting that this telomere of O-137 was generated by a transposition event where STR inserted into a Pot2 element in an ancestral strain of O-137, causing the breakage of a chromosome. The subsequent addition of a telomeric repeat to one of the LTRs of STR most probably repaired the broken end of the chromosome. An Mg-SINE element immediately follows the truncated Pot2.

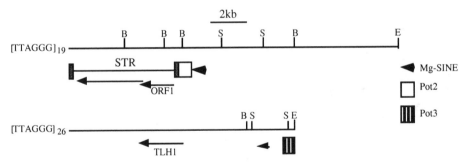

Figure 2. Restriction map and organization of two telomeres in rice pathogen O-137. Restriction sites for *Bam*HI (B), *Eco*RI (E) and *Sac*I (S) are marked. Positions of the repetitive DNA elements were determined by sequencing the two telomeric clones. The [TTAGGG]ₙ represents the telomeric repeat of *M. grisea*.

4.3. REPETITIVE DNA ELEMENTS

Many different types of repetitive DNA elements have been identified in *M. grisea* (Dobinson et al., 1993; Farman et al., 1996; Farman et al., 1996; Hamer et al., 1989; Kachroo et al., 1995; Kachroo et al., 1994; Kang et al., 1995; Sone et al., 1993; Valent and Chumley, 1991). Most of them are transposable elements, including both retrotransposons that transpose through an RNA intermediate and inverted terminal repeat (ITR) transposons that directly transpose (see Figure 3 for their structural organizations). Three different families of LTR-retrotransposons, *grasshopper* (*grh*), MAGGY and STR, belonging to the *Gypsy* class have been identified to date (Dobinson et al., 1993; Farman et al., 1996; S. Kang, unpublished). *Grh* contains 198 bp LTRs that are flanked by 5 bp inverted repeats (IR), and its transposition appears to generate a 5-bp target site duplication (Dobinson et al., 1993). *Grh* is present only in a subgroup of *Eleusine* pathogens isolated from broad geographic locations, suggesting that the element has been horizontally acquired subsequent to the evolution of *Eleusine*-pathogenic isolates. While MAGGY also generates a 5-bp target site duplication, its LTR (253 bp) is longer than that of *grh* and it is flanked by a 6-bp IR. Transposition events of MAGGY have been detected in two *M. grisea* strains (Shull and Hamer, 1996). The copy number of MAGGY is high (50-100) among rice pathogens but varies significantly (0-50) among isolates from other hosts (Farman et al., 1996). A recent report suggests one way in which the MAGGY copy number may be controlled. A degenerate copy of MAGGY was cloned from a strain isolated from common millet (*Panicum miliaceum*), which showed that an RIP (Repeat Induced Point mutation)-like process might have been involved in inactivating the invading MAGGY (Nakayashiki et al., 1999). While MGR583 in the same strain was not methylated, certain cytosine residues of this degenerate copy appeared selectively methylated in an isolate from *P. miliaceum*. While the second ORF of STR exhibited significant similarity to the corresponding proteins encoded by *grh* and MAGGY (S. Kang, unpublished), its distribution among strains was quite different. In contrast to the mostly host-specific occurrence of *grh* and MAGGY, isolates from diverse hosts have similar copy numbers of STR (10-20 copies) with only a very few exceptions.

Two different types of non-LTR retrotransposons, one LINE (Long Interspersed Nuclear Element) and two SINEs (Short Interspersed Nuclear Element), have been identified in *M. grisea* (Hamer et al., 1989; Kachroo et al., 1995; Sone et al., 1993; Valent and Chumley, 1991). The copy number of the 6 kb-long LINE MGR583 (or MGL) is high in rice pathogens (> 50 copies) and is of variable number in strains from hosts other than rice. Some strains have as many copies as in rice pathogens and others have fewer than ten copies (Hamer et al., 1989; M.A. Meyn and M.J. Orbach, unpublished). Although four mRNAs, 7.5, 2.6, 2.4 and 0.5 kb in length, hybridize to MGL (Valent and Chumley, 1991), its transposition has not yet been observed (Meyn et al., 1998). The first SINE identified in *M. grisea* was MGSR1 (Magnaporthe grisea short repeat 1), which was isolated as a 0.8 kb repeated sequence for RFLP analysis. It has two features of SINEs, an RNA polymerase III internal promoter region and a poly-T 3'-end (Sone et al., 1993) and it is amplified primarily in rice pathogens (~40 copies), while non-rice pathogens contain few MGSR1 copies. A second SINE, Mg-SINE, was identified as an insertion element within Pot2 (Kachroo et al., 1995). This element is approximately 500 bp long and has three features conserved in SINEs, the presence of two promoter sequences (A box and B box) for RNA polymerase III, a target site duplication and an A/T-rich region at the 3'-end. The copy number of Mg-SINE was estimated to be ~100 in both rice and non-rice isolates (Kachroo et al., 1995). Interestingly, the last 240 bp of Mg-SINE are 90% identical to MGL (Meyn et al., 1998), suggesting that Mg-SINE was derived from MGL and may utilize the reverse transcriptase encoded by MGL for its transposition. This homology most likely led to an overestimation of the MG-SINE copy number because the MG-SINE probe detects both MG-SINE and MGL. Using a probe specific for the unique region of MG-SINE, rice pathogenic strains appeared to have about 50 copies per genome (M.A. Meyn and M.J. Orbach, unpublished). Most non-rice isolates had similar numbers of MGL and MG-SINE copies. For example, those strains with low numbers of MGL also had fewer MG-SINE copies. However three strains were found to have high MGL copy numbers and few MG-SINE copies (M.A. Meyn and M.J. Orbach, unpublished). The homology between MGL and MG-SINE also explains the high copy number 0.5 kb-RNA that hybridizes to the 3' end probe of MGL. This RNA is probably the MG-SINE RNA and not derived from MGL. Unlike the other MGL transcripts, the 0.5 kb transcript is not enriched in poly(A) RNA preparations suggesting that it is not polyadenylated. One would not expect a SINE to contain a poly(A) tail because they are transcribed by RNA polymerase III, not RNA polymerase II which produces polyadenylated RNAs.

Two different families of ITR transposons, Pot2 and Pot3, have been identified in *M. grisea* (Farman et al., 1996; Kachroo et al., 1994). Pot2 consists of an ORF encoding a putative transposase flanked by 43-bp ITRs (Kachroo et al., 1994). A comparison of sequences flanking 12 Pot2 elements revealed a conserved insertion site sequence similar to that of Tc1, an ITR transposon of *Caenorhabditis elegans* that is the type element of the Tc1 family of transposons. The target sequence (TA) was duplicated upon the transposition of Pot2. A high copy number of Pot2 (~100) was present in both rice and non-rice pathogens. MGR586, a repetitive DNA probe for population analysis (see Section 6.1.), contains part of a novel ITR transposon termed Pot3 (Farman et al., 1996). The structure of Pot3 is similar to that of Pot2: 42-bp ITRs and an ORF for transposase (Farman et al., 1996). However, in contrast to Pot2, sequences flanking nine independent Pot3 elements did not show significant similarity except in the duplicated sequence (TA) also found at both sides of Pot3 (S. Kang and B. Valent, unpublished). Most rice pathogens have approximately 50 copies of Pot3, whereas isolates from other grass species possess 0-10 copies. However, three *Pennisetum* isolates have a copy number similar to that of the rice pathogens, suggesting that amplification of Pot3 has independently occurred at least twice since the introduction of Pot3 into *M. grisea* (S. Kang and B. Valent, unpublished). The nearly ubiquitous presence of Pot3 among various host specific forms supports the hypothesis that the introduction of Pot3 to the *M. grisea* genome predates the diversification of host specificity (Shull and Hamer, 1994). The presence of sequences homologous to Pot3 in ginger pathogen G-228 and *Cyperus* pathogen G-231 suggests that timing of the Pot3 introduction may even predate the speciation of *M. grisea*. Both G-228 and G-231 are genetically distant from *M. grisea* strains isolated from other hosts, suggesting that they are probably separate species (Borromeo et al., 1993; Shull and Hamer, 1994).

Members of MGR586, Mg-SINE, Pot2, *grh* and MAGGY appear scattered throughout the genome (Dobinson et al., 1993; Kachroo et al., 1995; Kachroo et al., 1994; Nitta et al., 1997; Romao and Hamer, 1992). However, each family of transposable elements does not appear randomly distributed relative to other families of transposons (Nishimura et al., 1998; Nitta et al., 1997). Instead, some of them tend to cluster in the genome. In strain 2539, MAGGY was frequently associated with MGR586, Pot2 and Mg-SINE (Nitta et al., 1997). Furthermore, the clustering of transposable elements has also been observed in a field isolate pathogenic to rice (Nishimura et al., 1998).

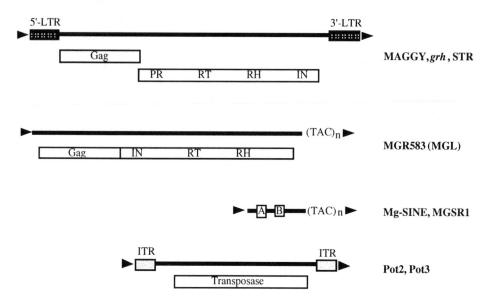

Figure 3. Structural features of the transposons identified in *M. grisea.* The drawings do not accurately reflect their relative sizes. Arrowheads flanking individual transposons indicate the target site duplication generated during their transposition. The labeled boxes underneath the transposons denote open reading frames: Gag (proteins showing similarity to retroviral core proteins), PR (protease), RT (reverse transcriptase), RH (RNase H), IN (integrase). The third ORF that is present in certain LTR-type retrotransposons appears missing in MAGGY, *grh* and STR. The two boxes, labeled A and B, in Mg-SINE and MGSR1 indicate the internal RNA polymerase III promoter sequences. LTR and ITR indicate Long Terminal Repeat and Inverted Terminal Repeat, respectively.

4.4. GENETIC MAPS

Four genetic maps of *M. grisea* based on different pairs of parental strains have been reported to date (Dioh et al., 1996; Nitta et al., 1997; Romao and Hamer, 1992; Skinner et al., 1993; Sweigard et al., 1993). Two of these maps contain more than 200 markers, including phenotypic markers, cloned genes, random cosmid clones, telomeres and repetitive DNA elements (Nitta et al., 1997; Sweigard et al., 1993). An integrated genetic map between these two maps spans approximately 840 cM with an average resolution of 4.5 cM (Nitta et al., 1997)

5. Molecular Basis of Host Specificity

A group of genes called avirulence (*AVR*) genes have been identified in many plant pathogens (Laugé and de Wit, 1998; Leach and White, 1996). A pathogen expressing one or more of these *AVR* genes is unable to infect those hosts that express any of the matching disease resistance (*R*) genes, due to the induction of host defense responses. Such a mechanism, termed a gene-for-gene interaction (Flor, 1971), governs the compatibility of plant-pathogen interactions in many cases, and thus the question of how this mechanism operates has been a subject of extensive investigations (Bent, 1996; Crute and Pink, 1996). Accumulated evidence to date indicates that the cultivar specificity of rice pathogens follows this gene-for-gene model. Numerous *AVR* genes have been genetically identified in *M. grisea* (Chao and Ellingboe, 1997; Ellingboe, 1992; Ellingboe et al., 1990; Lau et al., 1993; Lau and Ellingboe, 1993; Leung et al., 1988; Silué et al., 1992a; Silué et al., 1992b; Valent and Chumley, 1987; Valent et al., 1991). By convention, most *AVR* genes in *M. grisea* are named according to the rice cultivar upon which they are avirulent. For example, *AVR1-CO39* is the first *AVR* gene identified, which prevents infection of cultivar CO-39. In addition to *AVR* genes, two different classes of genes (*S* and *M* for suppressor and modifier) have been shown to control compatibility toward certain cultivars (Ellingboe, 1992; Lau et al., 1993; Lau and Ellingboe, 1993). The function of the *S* gene is to suppress the expression or function of specific *AVR* gene(s), whereas the *M* gene is required for the expression or function of one or more *AVR* genes. The question as to whether all the *AVR* genes in *M. grisea* have corresponding *S* and *M* genes remains unanswered.

The interfertility of *M. grisea* strains that infect different grass species has allowed for the genetic analyses of host specificity at the species level. Single genes that determine the compatibility to weeping lovegrass have been identified in independent crosses (Valent and Chumley, 1987; Valent and Chumley, 1991; Valent et al., 1986; Yaegashi, 1978). The *PWL1* gene, which prevents the fungus from infecting weeping lovegrass, was identified in a genetic cross between two grass pathogens (Valent et al., 1986). Another cross between two laboratory strains pathogenic on rice and weeping lovegrass, respectively, identified a second gene, *PWL2*, with a similar phenotype to *PWL1* (Valent and Chumley, 1991). Browning or auto-fluorescence of weeping lovegrass cells around developing colony margins correlated with the presence of the *PWL1* gene in an invading fungus (Heath et al., 1990). The fungus was unable to grow beyond these brown cells, which suggests that weeping lovegrass recognizes the product or by-product of *PWL1* and subsequently initiates a successful defense response. Numerous *R* genes against *M. grisea* have been identified in rice (Marchetti et al., 1987; Silué et al., 1992a; Valent et al., 1998; Wang et al., 1994; Yamada et al., 1976). The cloning of several *AVR* genes in *M. grisea* provides an opportunity to study the molecular basis of their avirulence function. These are discussed below. Cloning of additional *AVR* genes is also in progress: *AVR1-MARA* (Mandel et al., 1997), *AVR2-MARA* (T. Harper and M.J. Orbach, unpublished) *AVR1-TSUY* (S. Kang and B. Valent, unpublished), and *AVR1-MedNo*, *AVR1-Ku86*, and *Avr1-Irat7* (Dioh et al., 1996; Percet et al., 1998).

5.1. THE *AVR-Pita* GENE

The *AVR-Pita* gene (formerly known as *AVR2-YAMO*), which prevents *M. grisea* from infecting rice cultivar Yashiro-mochi, is located within 1.5 kb of the tip of a chromosome in rice pathogen O-137 and encodes a protein of 223 amino acids (Valent, 1997). The protein contains a putative signal peptide for secretion and a sequence motif characteristic of neutral zinc metalloproteases. The mutation of specific residues in this motif renders *AVR-Pita* nonfunctional as an *AVR* gene, suggesting that its enzymatic activity might be necessary for triggering the host defense response in cultivar Yashiro-mochi. However, recent data suggest a physical interaction between the *AVR-Pita* gene product and the product encoded by its corresponding resistance gene, *Pi-ta* (Valent et al., 1998). Some *M. grisea* isolates from *Digitaria* or *Pennisetum*, which are unable to infect rice, possess a functional gene homologous to *AVR-Pita*. Expression of *AVR-Pita* is induced *in planta*, suggesting a role as a virulence factor in those hosts that lack the *Pi-ta* gene (Valent et al., 1998).

5.2. THE *AVR-CO39* GENE

The *AVR-CO39* gene, which prevents infection of rice cultivar CO39, was cloned by chromosome walking (Farman and Leong, 1998). The gene was initially identified in a series of backcrosses between O-135, a rice pathogen isolate from China, and 4091-5-8, a hermaphroditic laboratory strain that is pathogenic on weeping lovegrass and goosegrass (Valent et al., 1991). Since O-135 is virulent on cultivar CO39, the gene appears to have been inherited from 4091-5-8, a progeny of two grass pathogen isolates that are not pathogenic on rice (Valent et al., 1986). Three additional *AVR* genes, *AVR1-M201*, *AVR1-YAMO* and *AVR2-MARA* were also inherited from 4091-5-8, suggesting that certain grass pathogens nonpathogenic on rice contain genes that act as functional *AVR* genes corresponding to certain *R* genes in rice (Valent et al., 1991). The chromosome walk to *AVR-CO39* covered an approximately 610 kb region of chromosome 1 and encountered three gaps that were caused by the presence of repetitive or unclonable DNA around the *AVR1-CO39* locus. Cleavage of the genome at defined positions using a technique termed RecA-assisted Achilles' Cleavage (RecA-AC) was employed to jump over the gaps (Koob et al., 1993). There existed a 14-fold variation in the relationship between genetic and physical distance (16 kb/cM to 218 kb/cM) over this 610 kb region. Two overlapping cosmid clones identified during the chromosome walk conferred avirulence to Guy11, a strain virulent on cultivar CO39. Subcloning of these cosmid clones and subsequent functional complementation tests with the resulting subclones localized the *AVR-CO39* gene on a 1.05 kb region. Several small open reading frames (ORFs) were present on this 1.05 kb region (Leong et al., 1998). Mutation of the translational start codon of individual ORFs showed that two of them, ORF1 and ORF3, are necessary for the avirulence phenotype. The ORF3 encodes a putative protein of 89 amino acids and contains a putative signal peptide.

5.3. THE *PWL* GENE FAMILY

The *PWL2* gene, responsible for inducing resistance on weeping lovegrass, was cloned by chromosome walking and used as a probe to investigate the distribution of genes homologous to *PWL2* in more than 100 *M. grisea* strains isolated from diverse host species (Kang et al., 1995; Sweigard et al., 1995). The number of genes homologous to *PWL2* and the degree of their sequence homology was highly variable even among isolates from the same host species, suggesting that the *PWL2* gene is a member of a multigene family. Several *PWL* genes and their alleles, including *PWL1*, have been cloned from strains isolated from diverse hosts (Kang et al., 1995). The *PWL1* gene encodes a protein that exhibits 75% identity to the *PWL2* gene product. The *PWL3* gene, cloned from a finger millet pathogen, is allelic to the *PWL4* gene isolated from a weeping lovegrass pathogen. Both alleles are nonfunctional in triggering the defense response in weeping lovegrass. However, the *PWL4* ORF is functional as demonstrated by placing it under the control of the promoter of either *PWL1* or *PWL2* (Kang et al., 1995). The products encoded by members of the *PWL* gene family have the following common characteristics: (a) Their amino terminus has features conserved among eukaryotic signal peptides for secretion. (b) Several glycine residues are well conserved and evenly distributed throughout the protein. (c) They are highly hydrophilic with many charged residues. The genomic organization and sequence of the *PWL* gene family suggests possible modes of evolution of the gene family.

5.3.1. *Accelerated Evolution of the PWL3/4 Gene*

A member of the *PWL* gene family appears to have experienced a much faster sequence evolution than the rest of the genome. The DNA sequence of the ORFs of the allelic *PWL3* and *PWL4* show only 88% identity, while the sequences flanking these ORFs are more than 98% identical. This result suggests that sequence divergence among these members of the *PWL* gene family is not the consequence of early gene divergence, but rather resulted from an accelerated evolution of the gene family. The timing and mechanisms of accelerated sequence evolution in the *PWL3/4* locus have been investigated (S. Kang and J. Hamer, unpublished). More than 90 *M. grisea* strains isolated from diverse hosts and geographic locations were screened for the presence of these alleles. In contrast to *PWL2*, which is nearly ubiquitous and often exists in multiple copies, a single copy of *PWL3* or *PWL4* is present only in certain groups of strains, including some isolates from rice, *Eleusine* species, and wheat. None of the strains isolated from *Digitaria* or *Pennisetum* species carry this gene, suggesting that the ancestral gene for *PWL3* and *PWL4* was introduced into *M. grisea* after the separation of a genetic lineage leading to *Digitaria* and *Pennisetum* pathogens from the lineage including rice, *Eleusine*, and wheat pathogens. Twenty-five *PWL3* and *PWL4* alleles were amplified by PCR and sequenced. These new alleles are highly similar (>98% identity) to either *PWL3* or *PWL4,* and the rate of sequence divergence between the ORF and the flanking sequences was about the same within each group. This result suggests that *PWL3* and *PWL4* have been stabilized and fixed in the *M. grisea* population after an episode of rapid sequence evolution.

5.3.2. *Roles of Transposable Elements in the Evolution of the PWL Gene Family*

All of the *PWL* genes characterized to date are closely associated with repetitive DNA element(s), many of which are transposons (Figure 4). At least four different types of repetitive DNA elements have been identified near members of the gene family. These repetitive DNA elements appear to have played an important role in the amplification of the *PWL* gene family. Recombination between homologous repetitive DNA elements amplified the *PWL2* gene in certain groups of *M. grisea* (Figure 4). Reverse transcriptase encoded by various retrotransposons in *M. grisea* (Dobinson et al., 1993; Farman et al., 1996; Valent and Chumley, 1991) may also have mediated the amplification of the gene family through retrotransposition. Sequences flanking the coding regions of the *PWL1*, *PWL2* and *PWL3/PWL4* (allelic) genes are completely different among themselves, and these genes are not linked. It seems unlikely that DNA-mediated translocation is responsible for such an organization. However, none of the sequenced *PWL* genes have poly(A) tracts at the 3'-end of the gene, a characteristic commonly associated with pseudogenes (Weiner et al., 1986). Retrotransposons at the 5'end of an ancestral *PWL* gene could also have mobilized the gene to different locations of the genome through "exon shuffling" (Eickbush, 1999; Moran et al., 1999).

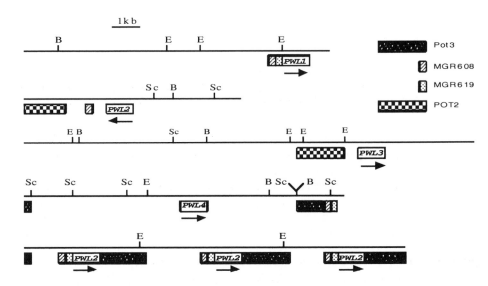

Figure 4. Restriction maps of the *PWL* genes and the repetitive DNA elements associated with these *PWL* genes. Restriction sites for *Bam*HI (B), *Eco*RI (E), and *Sac*I (Sc) are marked. Positions of the repetitive DNA elements were determined either by sequencing or by hybridizing individual restriction fragments to a blot that contains previously cloned repetitive DNA elements. The arrows denote the direction of transcription.

5.4 OTHER *AVR* GENES

Several *AVR* genes have been identified following a cross of two rice pathogen field isolates, Guy11 and ML25 (Silué et al., 1992a; Silué et al., 1992b). RFLP (restriction fragment length polymorphism) and RAPD (random amplified polymorphic DNA) markers were used to identify linked DNAs to three independent genes from this cross, *AVR1-Irat7*, *AVR1-Mednoï* and *AVR1-Ku86*. Markers linked to *AVR1-Irat7* were used to perform a chromosome walk which resulted in the identification of a cosmid clone that when introduced into a strain virulent on rice cultivar Irat7, conferred avirulence (W. Dioh and M.-H. Lebrun, personal communication). A candidate gene for *AVR1-Irat7* in the cosmid appears different from other fungal avirulence genes identified to date as it is not a secreted small protein or peptide (M.-H. Lebrun, personal communication).

The *AVR1-MARA* gene of strain 4224-7-8 has also been localized between linked RFLP markers to a region of at least 90 kb on chromosome 2 of the consolidated RFLP map (Nitta et al., 1997). Most of this region is not represented in various chromosomal DNA libraries and parts are unclonable in *Escherichia coli*. To localize the *AVR1-MARA* within the region, a transformation-mediated locus replacement approach was taken (U. Gunawardena and M.J. Orbach, unpublished). A hygromycin phosphotransferase (hph) selectable marker was cloned between two fragments of the locus and used to transform avirulent strain 4224-7-8. Drug resistant transformants were screened for replacement of portions of the locus by the hph marker. This localized *AVR1-MARA* to a region of the locus of about 60 kb, which is currently being isolated using a combined long range and inverse PCR approach. Interestingly, the DNA content of this region is very AT rich (65-70% AT) in contrast to the *M. grisea* genome average of 45% AT. The virulent locus, *avr1-MARA*, has been isolated and contains two deletions that cover most of the locus, suggesting that virulence is due to the absence of some of these sequences (Mandel et al., 1997).

6. Population Structure and Dynamics

Management of rice blast through the breeding of blast-resistant varieties has had only limited success due to the frequent breakdown of resistance under field conditions (Bonman et al., 1992; Kiyosawa, 1982). The frequent variation of race in pathogen populations has been proposed as the principal mechanism involved in the loss of resistance (Ou, 1980). The race of a given *M. grisea* isolate is defined by its pattern of compatibility to a set of eight tester cultivars known as differential varieties (Ling and Ou, 1969). Although it is generally accepted that race change in *M. grisea* occurs in nature, the degree of its variability has been a controversial subject. A number of studies reported extremely high rates of appearance of new races (Giatgong and Frederiksen, 1968; Ou and Ayad, 1968; Ou et al., 1970; Ou et al., 1971). Various potential mechanisms, including heterokaryosis (Suzuki, 1965), parasexual recombination (Genovesi and Magill, 1976), and aneuploidy (Kameswar Row et al., 1985; Ou, 1980), have been proposed to explain frequent race changes. In contrast, other studies have shown that although race change could occur, its frequency was much lower than that predicted by earlier studies (Bonman et al., 1987; Latterell and Rossi, 1986; Marchetti et al., 1976). The appearance of new races could be a result of the selection of rare individuals in a population through pressure from resistant cultivars rather than from modification of an avirulent strain. Analysis of the population structure of rice pathogens using genetic markers in addition to their virulence spectrum on a reference set of rice cultivars has provided many new insights into the evolutionary dynamics of *M. grisea* in the field.

6.1. GENETIC MARKERS FOR THE ANALYSIS OF POPULATION STRUCTURE

Race analysis alone, based on the compatibility to a set of rice cultivars, is not sufficient to allow for a comprehensive understanding of the genetic relationship among rice pathogens. Considering that more than one gene can control the compatibility/incompatibility to a specific cultivar, a similar pattern of compatibility to a set of test cultivars does not necessarily indicate a close genetic relationship. For instance, two of the races identified in the United States, IB-49 and IG-1, each consist of two genetically distinct groups of strains that share less than 40% similarity in their multi-locus fingerprints. In contrast, there are some strains that share more than 80% similarity in their DNA fingerprinting patterns but belong to different races (Levy et al., 1991). A variety of markers have been used to determine the population structure at the genetic level (Shull and Hamer, 1994). These markers include isozyme polymorphisms (Leung and Williams, 1986), karyotypes (Orbach et al., 1996; Talbot et al., 1993), and RFLPs. The RFLPs analyzed include those in the mitochondrial genome (Borromeo et al., 1993), in the ribosomal DNA (Borromeo et al., 1993; Lebrun et al., 1991), those detected by single copy genomic probes (Borromeo et al., 1993; Lebrun et al., 1991), those generated by a repetitive element-based polymerase chain reaction (George et al., 1998) and RFLPs associated with a transposable element, MGR586, as a probe (Chen et al., 1995; Correa-Victoria and Zeigler, 1993; Correa-Victoria et al., 1994; Han et al., 1993; Kumar et al., 1999; Levy et al., 1993; Levy et al., 1991; Roumen et al., 1997; Xia et al., 1993; Zeigler et al., 1995; Zeigler et al., 1997; Zeigler et al., 1994). The RFLPs generated with MGR586, known as MGR fingerprints, have been the most informative in elucidating the genetic diversity and variation of rice pathogens both within and between geographic regions. This is primarily because MGR586 is randomly distributed throughout the genome (Romao and Hamer, 1992) and RFLPs associated with MGR586 were sufficiently variable to distinguish strains but not so hypervariable that DNA fingerprint patterns were not generally maintained within an asexually reproducing population (Levy et al., 1991).

6.2. POPULATION STRUCTURE OF RICE PATHOGENS

A large number of strains from many countries have been subjected to MGR fingerprinting analysis. Detailed studies of strains from the USA (Levy et al., 1991; Xia et al., 1993), Columbia (Correa-Victoria and Zeigler, 1993; Correa-Victoria et al., 1994; Levy et al., 1993), Europe (Roumen et al., 1997), India (Kumar et al., 1999),

Korea (Han et al., 1993), and the Philippines (Chen et al., 1995; Zeigler et al., 1995) have been performed. Individual strains were grouped based on their genetic relatedness, which was estimated from pairwise comparisons between their fingerprints. These studies showed that in most countries rice pathogens mainly consist of a few clonally derived lineages that are genetically distinct. In the United States eight distinct clonal lineages were identified among 42 isolates representing the eight races collected over 30 years (Levy et al., 1991). The average similarity of fingerprinting patterns among isolates within individual lineages exceeded 90%, while the similarity between lineages ranged from 30-80%, suggesting that sexual or parasexual recombinations between rice pathogen isolates have occurred rarely, if at all, in the United States. If sexual or parasexual recombination had been occurring frequently, one would expect to see continuous variation in MGR profiles rather than the presence of a small number of discrete lineages. The relationship between race and genetic lineage within these USA field isolates exhibited a one-to-one relationship in six lineages (Levy et al., 1991). The remaining two lineages contain two races. However, a larger number of isolates collected from two rice fields in Arkansas revealed a slightly more complex picture than the previous study (Xia et al., 1993). Although no new lineage different from the previously identified eight lineages was identified, three of the four USA lineages detected in this study contain more than one race. A study with 41 isolates from five European countries showed the presence of five lineages (Roumen et al., 1997), and surveys of isolates from Colombia and the Philippines have shown that more lineages (17 and 10, respectively) are present there. However, races within individual lineages typically differed by only one or a few compatible/incompatible interactions on the international differential cultivar set, suggesting that genetic lineage can still serve as a good indicator for the spectrum of virulence among rice pathogens in these countries (Chen et al., 1995; Levy et al., 1993). The appearance of multiple races within a lineage is not surprising based on the frequent mutation of certain *AVR* genes, such as *AVR-Pita* and *AVR1-TSUY* (see Sections 7.1. and 7.2.).

6.2.1. *Recombination vs. Clonal*

Several lines of evidence suggest that the role of sexual recombination in the population dynamics of rice pathogens in many regions has, if at all, been minimal (Zeigler, 1998). These data include the presence of distinct genetic lineages within a single geographic region, the predominance of a single mating type in most areas, very low fertility and a high degree of karyotype diversity. However, it was pointed out that the existence of small scale parasexual recombinations between individuals could not be completely ruled out based on the results from MGR fingerprinting analyses (Zeigler et al., 1994). Subsequently, evidence for parasexual recombination between field isolates under unselected conditions has been reported (Zeigler et al., 1997). Another piece of evidence challenging the exclusive clonality of rice pathogens came from a survey of strains in the Indian Himalayas, a center of rice diversity (Kumar et al., 1999). Based on a high degree of diversity, dynamic fungal populations and the presence of hermaphroditic isolates, it was suggested that sexual recombination may have affected, to some extent, the structure of some populations in this region. The test for gametic disequilibrium using single- or low copy-RFLP markers failed to reject the null hypothesis of gametic equilibrium (i.e., random mating), hence supporting the possibility of sexual recombinations in this location (Kumar et al., 1999).

6.2.2. *Lineage Exclusion: A New Strategy for Resistance Breeding*

Conventional breeding efforts for blast resistance have focused on introducing those *R* genes that are incompatible to the races identified in a given geographic area, to appropriate cultivars which will then exclude the detected races from infecting the resulting new cultivars. However, such an approach has had only limited success in providing durable resistance mainly due to the frequent appearance of new races in the field. An alternative approach to breeding resistant cultivars, termed lineage exclusion, was developed to produce more durable levels of resistance (Zeigler et al., 1994). The principle of lineage exclusion is to develop a cultivar that contains *R* genes not only to the detected races in an area, but also the *R* genes to all of the races that could potentially evolve within the region based on knowledge of the local genetic lineages. Lineage exclusion is based on the following hypotheses: (a) Individual lineages have specific virulence potentials (i.e., the number and types of races that are present and those that could be potentially produced); (b) Race changes may occur among members in a given lineage only within its virulence potential; (c) Certain *R* gene(s) may be difficult to overcome by all isolates in a lineage due to fitness constraint, the requirement of multiple genetic changes, or both; (d) A given *R* gene may be highly durable against all isolates in certain lineages, but may be ineffective against members of other lineages. Since members of a particular lineage cannot easily overcome certain *R* genes, combining those *R* genes conferring durable resistance against one or more lineages can be used to breed a cultivar resistant to all lineages in a given area. Cultivars bred for lineage exclusion are currently being tested for their resistance in southern India (R. Sivaraj, S. S. Gnanamanickam and M. Levy, personal communication) and in Columbia (M. Levy, personal communication). The success of the lineage-exclusion based breeding strategy as a durable blast control measure will not be fully realized until we have a better understanding of the mechanisms underlying the emergence of new races. A clearer view of those mechanisms that are operating and how frequently they contribute to the breakdown of resistance will minimize the incidence of rapid breakdown of disease resistance by assisting the identification of proper *R* genes for breeding and/or engineering. Such information will also facilitate the deployment of disease resistance in a way that could suppress the appearance of newly emerging races.

6.3. THE POPULATION STRUCTURE OF *M. GRISEA* ISOLATES FROM OTHER HOSTS

In contrast to the wealth of information available with respect to the population structure and dynamics of rice pathogens, only limited data have been obtained for *M. grisea* isolates originating from other hosts. This has not been due to the lack of tools to develop such information. With the variety of repetitive DNAs that are present in *M. grisea* it should be possible to develop DNA fingerprint information for populations on other host plants. Two examples where the population structure of non-rice pathogens has been studied are illustrated. Both examples are cases where the pathogen caused epidemics on new hosts of economic significance.

6.3.1. *Blast Disease on Wheat in Brazil*

Blast disease on wheat was first detected in 1985 in the state of Paraná in Brazil. By 1991, all wheat growing states of the country were affected by the disease (Urashima et al., 1993). Because rice blast was common in the state of Paraná, a fingerprinting analysis, using MGR586 as a probe, was used to test the hypothesis that wheat blast epidemics were induced by rice pathogens (Valent and Chumley, 1994). The MGR586 probe in rice pathogens typically hybridized to 50 or more resolvable *Eco*RI fragments. However, isolates from other grass species contained only a few hybridizing *Eco*RI fragments, suggesting that the rice pathogens are genetically isolated from these isolates (Borromeo et al., 1993; Hamer et al., 1989). This disparity between the MGR fingerprints of rice and other grass pathogens was used to demonstrate that wheat blast in Brazil had not been caused by the rice pathogens indigenous to the area. In contrast to the presence of a large copy number of MGR586 in the Brazilian rice pathogens, the wheat pathogens in Brazil only contain one to two copies of MGR586 (Valent and Chumley, 1994). The uniformity in the wheat pathogens' karyotype coupled with their high level of sexual fertility suggest that the wheat pathogens in Brazil have been derived from a fertile subpopulation of *M. grisea* (Orbach et al., 1996). Candidates include a group of strains infecting *Eleusine* spp. (Urashima et al., 1993).

6.3.2. *Blast Epidemics on Perennial Ryegrass in the United States*

Severe outbreaks of blast disease on perennial ryegrass (*Lolium perenne*) in golf course fairways have been reported around several eastern states (Uddin, 1999). Loss of turf exceeded 90% in numerous golf courses. Although the disease on annual ryegrass (*Lolium multiflorum* Lam.) was first reported in Louisiana and Mississippi in 1971 (Bain et al., 1972; Carver et al., 1972), blast disease has not been considered a major threat to perennial ryegrass until very recently. To determine the origin of the causal agent for the recent episode of blast epidemics, isolates from perennial ryegrass were characterized using both molecular markers and infection assays on an array of hosts (Viji et al., 1999). The isolates from perennial ryegrass contained 20-30 copies of MGR586, suggesting that they are genetically distinct from the rice pathogen population in the United States. None of the perennial ryegrass isolates caused lesions on rice. A comparison of genetic similarity among the isolates from perennial ryegrass, using the fingerprints generated by the Pot2 transposable element as a probe, suggests that these isolates recently originated from one or two related strains. The presence of a single mating type (MAT1-2) among all the isolates also supports this supposition. Only those isolates from wheat showed significant similarity to these isolates in the Pot2 fingerprinting analysis. The perennial ryegrass isolates had the same sequence in the internal transcribed sequence of the genes encoding ribosomal RNA as that of the wheat isolates. The ability of the perennial ryegrass isolates to infect various wheat varieties also supports their close genetic relationship. Isolates from wheat have also been shown to be pathogenic on perennial ryegrass (Urashima et al., 1993).

7. Mechanisms Underpinning the Evolution of New Races

Although questions about the frequency of race changes in *M. grisea* remain open, molecular genetic analyses of the fungus have started providing new insights into the nature and mechanisms responsible for race change in some strains. The practical significance of investigating the question as to how virulence changes arise, resulting in the generation of new races, is not limited to the rice blast system, since the breakdown of disease resistance seems to be a common problem among many crops. A better understanding of such mechanisms could theoretically allow a survey of pathogen populations for their potential for the evolution of new races. Since the compatibility of interactions between *M. grisea* and rice follows the gene-for-gene model, the mutation of *AVR* genes or genes controlling the expression and function of *AVR* genes is likely to be a major mechanism for virulence change. Mutations in the *S*- and *M*-type genes could account for simultaneous, multiple virulence changes. Some of the *AVR* genes, including *AVR-Pita*, *AVR1-TSUY*, and *PWL2*, have been shown to be unstable (Valent and Chumley, 1994). Strains carrying one or more of these unstable *AVR* genes frequently produce spontaneous gain-of-virulence mutants. Molecular genetic analysis on the nature of genetic changes in these mutants has provided an insight into the mechanisms underpinning the evolution of new races.

7.1. MECHANISMS OF MUTATIONS UNDERPINNING THE GAIN OF VIRULENCE ON RICE CULTIVAR YASHIRO-MOCHI

Gain of virulence on cultivar Yashiro-mochi was often associated with the alteration or disappearance of telomeric restriction fragments linked to *AVR-Pita* (Valent and Chumley, 1994). Analysis of eight spontaneous virulent mutants of a laboratory strain, 4375-R-6 (see Figure 5 for its pedigree), revealed that three different types of mutation were responsible for the gain of virulence on Yashiro-mochi (Valent and Chumley, 1994). Five of the mutants had a deletion, ranging in size from 100 bp to over 12.5 kb, that spans either a part of *AVR-Pita* or the whole gene plus its flanking sequences. Two mutants had a point mutation within the ORF of *AVR-Pita*. Insertion of Pot3 in the putative promoter region of *AVR-Pita* was responsible for the gain of virulence in the remaining mutant.

7.2. MULTIPLE GENETIC CHANGES ASSOCIATED WITH THE GAIN OF VIRULENCE ON CULTIVAR TSUYUAKE

Rice pathogen O-137, an isolate from China, is unable to infect a number of rice cultivars, including Tsuyuake and Yashiro-mochi, due to the presence of the *AVR* genes *AVR1-TSUY* and *AVR-Pita*, respectively (Valent and Chumley, 1991). However, this strain frequently produces spontaneous mutants that have become virulent on one or both of these rice cultivars. Similar to *AVR-Pita*, *AVR1-TSUY* is very closely linked to a telomere and mutates frequently (S. Kang and B. Valent, unpublished). Six independent spontaneous mutants of O-137 (designated as CP714, CP726, CP819,

CP820, CP821 and CP822) that had become virulent on cultivar Tsuyuake were isolated following infection of Tusyuake with monoconidial cultures of O-137. All six mutants remained virulent on Sariceltik, a cultivar susceptible to O-137, and avirulent on weeping lovegrass, a host that is resistant to O-137. Thus, spontaneous genetic change(s) in O-137 specifically altered its compatibility to cultivar Tsuyuake without changing its compatibility to weeping lovegrass and cultivar Sariceltik. Specific telomeric restriction fragments of O-137 were missing in all six mutants (see Section 4.2.). The nature of mutation in this telomere appears complex, involving deletion and/or rearrangement. DNA fragments corresponding to this telomeric region, as well as any other loci that might be deleted in these mutants, were cloned by genomic subtraction. This technique specifically enriches for DNA fragments corresponding to loci present in one genome, but absent in the other (Straus and Ausubel, 1990). Genomic DNAs of two of the virulent mutants, CP821 and CP822, were used individually for subtracting DNA of O-137 to isolate DNA fragments deleted in each mutant. Southern analyses using these fragments as probes showed that in addition to a change at a specific telomere, multiple deletion mutations were present at specific areas of the genome in these mutants. Some of the mutations were present in all six mutants, and both CP821 and CP822 contained at least one deletion mutation unique to each strain. Cloning and characterization of the sequences spanning the deletion breakpoints suggests that recombination between homologous transposons may be responsible for some of the deletions (S. Kang, unpublished).

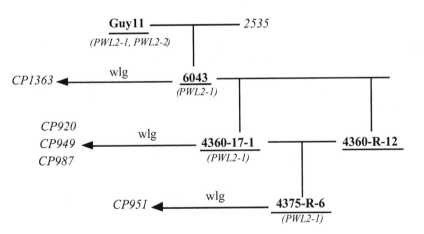

Figure 5. Segregation of the *pwl2* genes in Guy11. The horizontal lines connecting pairs of strains indicate genetic crosses. Strains with underlined names fail to infect weeping lovegrass due to the presence of one or both of the *PWL2* genes. Arrows labeled "wlg" indicate spontaneous mutants (CP strains) that have gained the ability to infect weeping lovegrass. The strains virulent on weeping lovegrass are italicized.

7.3. GAIN OF VIRULENCE TOWARD WEEPING LOVEGRASS

The descendants of Guy11, including 6043, 4360-17-1 and 4375-R-6, that are avirulent on weeping lovegrass, frequently produce spontaneous virulent mutants on weeping lovegrass (Sweigard et al., 1995). Analysis of the spontaneous mutants derived from these strains showed that the *PWL2* gene and more than 30 kb of flanking DNA sequences had been deleted (Sweigard et al., 1995). However, Guy11 failed to produce such mutants. The segregation pattern of *PWL2* in a cross between Guy11 and 2539 suggested that its stability in Guy11 is due to the presence of an additional gene conferring avirulence toward weeping lovegrass. Low-stringency hybridization showed that Guy11 contains two copies of *PWL2* (*PWL2-1* and *PWL2-2*) and a single copy of *PWL3* (S. Kang and J.E. Hamer, unpublished). Both *PWL2* genes are functional in conferring avirulence to weeping lovegrass, and 6043 inherited *PWL2-1*, but not *PWL2-2*, from GUY11. These *PWL2* genes and their flanking regions (> 2 kb to both directions) are identical in sequence, but are located on different chromosomes.

The genetic stability of the two identical, but unlinked, *PWL2* genes in Guy11 has been investigated using nine progeny from a cross between Guy11 and 2539 (S. Kang and J. Hamer, unpublished): three of the progeny carried only *PWL2-1*, three progeny contained only *PWL2-2*, and three progeny contained both of them. Two monoconidial isolates from each progeny were individually cultured to produce conidia for an infection assay. Five pots of weeping lovegrass were sprayed with 1×10^6 conidia per pot. The total number of lesions in each pot was counted as a way of estimating the mutational frequency. Both *PWL2* genes were unstable, frequently producing spontaneous virulent mutants. Strains carrying *PWL2-1* produced 40-100 lesions per pot (4×10^{-5} to 1×10^{-4}), and those carrying *PWL2-2* produced 2-10 lesions per pot (2×10^{-6} to 1×10^{-5}). Strains containing both *PWL2* genes also produced lesions (one or two lesions per pot). The mutational frequency of each gene was not significantly different among progeny carrying the same gene, suggesting that the genetic background of the rest of the genome does not significantly affect the mutational frequency. However, since *PWL2-2* is more stable than *PWL2-1* by an order of magnitude, local chromosomal contexts around these *PWL2* genes appear to significantly influence their genetic stability. In these spray inoculum assays, fewer than half the spores sprayed onto the plants actually land on plant tissue. In addition, experiments with low inoculum densities suggested that only 1% of the pathogenic spores that actually land on plant tissue produce lesions (Sweigard et al., 1995). Considering these factors, the actual mutational frequency was probably two orders of magnitude higher than those listed above.

7.4. EVOLUTION OF NEW RACES IN ARKANSAS

Analysis of new races that appeared in Arkansas in the 1980's provides some support for the hypothesis that the breakdown of blast resistance in the field may be caused by spontaneous mutations of *AVR* genes (Correll et al., 1998). As described above, it has been demonstrated in the laboratory that some *AVR* genes undergo frequent spontaneous mutations in certain strains, resulting in the generation of new races. To demonstrate that this mechanism occurs in the field, where the population structure is mainly clonal, one must show that new races are gain of virulence mutants of an existing genetic lineage. The *M. grisea* population in the USA is primarily composed of eight genetic lineages (Levy et al., 1991), making it relatively easy to trace the origins of any new races. In Arkansas four genetic lineages (A, B, C, and D) have been identified in the contemporary *M. grisea* population. Three lines of evidence suggest that the new races IC-17 and IC-1k capable of infecting cultivars Newbonnet (released in 1983) and Katy (released in 1989), respectively, originated from race IG-1 (avirulent on both Newbonnet and Katy) in the resident population. Historical evidence (Lee, 1994) indicated that there had been a progression of races in Arkansas from IG-1 to IC-17 (virulent on Newbonnet), and then to IC-1k (virulent on both Newbonnet and Katy). The population data suggest that all three races (IG-1, IC-17, and IC-1k) belong to genetic lineage B. Also, it has been demonstrated that, using certain IG-1 isolates in genetic lineage B, the observed changes of race in the field can be experimentally reproduced. Spontaneous mutations affecting certain *AVR* genes occur more readily among isolates in this lineage. The mechanisms of the mutations responsible for the observed race changes in Arkansas are currently being investigated (J. Correll, personal communication).

7.5. MECHANISMS OF GENETIC INSTABILITY IN GENES OTHER THAN *AVR* GENES

Genetic instability is not unique to *AVR* genes. The *BUF1* gene, which encodes an enzyme involved in the melanin biosynthetic pathway, spontaneously mutates at a high frequency in some rice pathogens, but the frequency of mutation differs significantly from strain to strain (Chumley and Valent, 1990). The *SMO1* gene, whose mutation causes abnormal spore morphology and reduces pathogenicity, also mutates frequently (Hamer et al., 1989). In contrast to these unstable genes, other *M. grisea* genes exhibit normal mutation frequencies (Valent, 1997), indicating that genetic instability may affect only certain parts of the genome and that the affected parts can vary from strain to strain. It has been proposed that recombination between homologous repetitive DNA elements might be responsible for frequent mutations and chromosomal rearrangements (Talbot et al., 1993; Valent and Chumley, 1991). Depending on the relative orientation and position of the repetitive elements that are recombined, homologous recombination between these elements could cause deletion, inversion, duplication, or translocation. The *AVR-Pita* gene and many members of the *PWL* gene family are closely associated with certain types of repetitive DNA. No repetitive DNA element is present within 20 kb of a stable *BUF1* allele in WGG-FA40, but a rice pathogen carrying an unstable allele appears to have several types of repetitive DNA elements around the locus (M. Farman, personal communication).

8. Genomics and Rice Blast Disease

One of the goals of the Human Genome Project has been to sequence the genomes of several model organisms, which possess a rich background of genetics and biology. Complete genomic sequences of *S. cerevisiae, C. elegans* and *E. coli,* are already available and those of *Drosophila melanogaster* and *Arabidopsis thaliana* are rapidly being completed. In addition, the genomes of many other bacterial species and archaea have been sequenced and progress is currently being made on the genomes of many additional species (http://www.tigr.org/tdb/mdb/mdb.html). Comparative structural and functional genomic analysis across species will significantly speed up our efforts to understand these model organisms as well as their less studied but equally important relatives. The rice blast system is well positioned to serve as a model for understanding the mechanisms of fungal pathogenesis using "genome-based biology" due to ongoing efforts to sequence the genomes of both *M. grisea* and its host, rice.

8.1. CURRENT STATUS OF RICE GENOME SEQUENCING

The 430 Mb genome size of *Oryza sativa* (rice) is significantly smaller than that of other agriculturally significant cereals (Arumuganathan and Earle, 1991): barley (4900 Mb), corn (2,500 Mb) and wheat (16,000 Mb). However, gene content and orders are well conserved within these genomes (Devos and Gale, 1997), hence a completed rice genome sequence will significantly further our understanding of the biology and evolution of all these cereals. Furthermore, comparative genome analyses between rice and *Arabidopsis* will also provide a tremendous insight into the evolution and genetic basis of many important biological functions in plants.

A major effort to analyze the rice genome was initiated in Japan in 1991. The main objectives of this effort were to identify rice genes using EST (Expressed sequence Tag) analysis, to construct and integrate genetic and physical maps of rice and to develop bioinformatic tools (Sasaki, 1999; Sasaki, 1998). Partial sequences of 40,000 cDNA clones, representing one-third to one-half of the genes in rice, have been obtained. Using 2,300 genetic markers, mainly consisting of rice ESTs, an RFLP map of 1521 cM with 12 linkage groups has been constructed. A physical map that covers approximately 65% of the rice genome has also been assembled using YAC (Yeast Artificial Chromosome) clones. Based on the success of this earlier effort, in 1998 an international consortium, including Japan, USA, China and Korea along with the European Union, was organized to initiate the sequencing of the entire genome (Sasaki, 1999). The goal of this scientific coalition is to have more than 170 Mb of the genome sequenced by 2003. However, if Celera, a newly formed genomics company, succeeds in meeting its goals, the rice genome sequence will be completely deciphered by 2001(Marshall, 1999). The rice genome sequence data will be of tremendous utility in gaining an understanding of the response to blast infection. First, all "*R*-like" genes can be identified by sequence comparisons to known *R* gene motifs. Subsequently, all of the rice genes can be placed in microarrays in order to identify those that are up and down regulated during infection and to compare gene responses to infection by avirulent versus virulent *M. grisea* strains.

8.2. CURRENT STATUS OF SEQUENCING THE *M. GRISEA* GENOME

A concerted effort toward the goal of completely sequencing the *M. grisea* genome has been initiated through the formation of an international consortium (http://www.genome.clemson.edu/blast_frame.html). The proposed strategy is to initially generate a physical contig map of the genome in a sequence ready format through end-sequencing and DNA fingerprinting of BAC (Bacterial Artificial Chromosome) clones (Venter et al., 1996). The sequence of seed BAC clones across the genome will be determined by sequencing random clones of small fragments (3-10 kb fragments in plasmid vectors and/or 1kb fragments in M13 vectors) derived from the seed BAC clones. The framework of BAC clones will assist in identifying BAC clones that minimally overlap to the seed clones. Three BAC libraries of *M. grisea* have already been constructed (Diaz-Perez et al., 1996; Nishimura et al., 1998; Zhu et al., 1997). One of these libraries consists of 9216 clones (25-fold genome coverage) with an average insert size of 130 kb (Zhu et al., 1997) and has been fingerprinted and end-sequenced (Dean, 1999; Zhu et al., 1999). A sequence-ready framework of chromosome 7 (4.2 Mb), consisting of five contigs that cover more than 95% of the chromosome, has been completed (Zhu et al., 1999), and the assembly of a framework for the rest of the genome is in progress. These resources will significantly facilitate the whole genome sequencing. A private member of this consortium (DuPont) has already generated about 3 Mb of genomic sequence by sequencing random clones of 1-2 kb genomic DNA fragments (J. Sweigard, personal communication).

In parallel to the whole genome sequencing effort, EST (Expressed Sequence Tag) analysis of *M. grisea* genes has been initiated in both public and private sectors. To date, several thousand ESTs have been identified by sequencing random clones in cDNA libraries specific for various growth and developmental conditions (Choi et al., 1999). To identify genes involved in appressorium formation, 2717 clones from an appressorium-stage specific cDNA library have been sequenced (Choi et al., 1999). For about 45% of the sequences, significant homology with sequences in the GenBank database have been detected. A large-scale expression profiling of genes on a specific chromosome has also been carried out (Zhu et al., 1999). Using the sequence-ready framework of chromosome 7 (Zhu et al., 1999), 466 cDNA clones, corresponding to 310 genes located on chromosome 7, have been isolated from an appressorium-stage specific cDNA library. These clones have been sequenced and mapped to a 100 kb resolution on the chromosome. Expression profiles of these genes under four developmental stages (appressoria, mycelia, conidia and germinating conidia) have been studied. DuPont has also generated about 5,500 ESTs from a cDNA library that was made from a mixture of mRNA obtained from cultures in 12 different growth and developmental conditions (J. Sweigard, personal communication).

Considering the large number of novel genes that remain to be identified through the genome sequencing and EST analysis, the development of an efficient gene knockout technique for *M. grisea* is critical in order to systematically characterize the function of these genes. In *M. grisea*, as in many filamentous fungi, homologous integration of transforming gene knockout constructs is rather infrequent, making it necessary to screen a large number of transformants for a gene knockout mutant. Although increasing the size of flanking DNA fragments can enhance the frequency of homologous recombination, construction of such a vector can be laborious. A novel strategy for the efficient mutagenesis of a large number of target genes has been proposed (Hamer et al., 1999). This strategy is based on *in vitro* insertional mutagenesis of a target gene on a cosmid or BAC clone using purified transposase and a modified transposon that contains a selectable marker for fungal transformants (e.g., hygromycin B resistance gene). Clones that contain the target gene tagged by a transposon can be rapidly identified by PCR using one primer based on the target gene and the other based on the transposon. Due to the presence of long flanking sequences, the tagged clones should significantly increase the frequency of gene replacement in fungal transformants. Because *in vitro* transposition requires only a short sequence at the ends of the transposon (Goryshin and Reznikoff, 1998), the transposon could be easily modified to include a different marker for selecting fungal transformants.

8.3. BEYOND THE GENOME SEQUENCING

Information and technology resources derived from the ongoing genome sequencing efforts will undoubtedly transform the way we study both rice and *M. grisea*. These resources will also permit investigators to approach a wide range of problems associated with rice blast that seemed impractical or impossible to study in the not-so-distant past. A DNA microarray-based assay makes it possible to profile the gene expression patterns of tens of thousands of genes in a single experiment (Schena et al., 1995). This technique will allow investigators to monitor whole genome changes in gene expression in both *M. grisea* and rice during the various stages of infection. Since cellular activities are controlled not only at the transcriptional level but also at both the translational and post-translational levels, it will also be necessary to monitor the pattern of proteins synthesized under a given condition. Such a pattern could be determined using a combination of two-dimensional electrophoresis and mass spectrometry (Kahn, 1995). Judicious applications of these techniques will assist investigators in systematically identifying a large set of genes from both *M. grisea* and rice that might play an important role for pathogenesis and defense response, respectively. The genome of a completely sequenced *M. grisea* strain could also be used to study the nature and mechanisms of genome evolution among various field isolates from diverse hosts, which will increase our understanding as to how different host-specific forms of *M. grisea* have evolved. Considering the importance of resistance breeding in controlling rice blast disease, such an understanding is critical for sustainable rice cultivation, especially in many developing countries in which chemical control is often economically impractical.

Using current approaches employed for the functional analysis of the yeast genome (Oliver et al., 1998) as a crystal ball, one could easily make a long list of other potential experiments to comprehend the molecular and cellular basis of rice blast disease. It is highly probable that the length of such a list will continue to increase, as more novel techniques for utilizing the genome sequence data become available. The potential impacts of a completed genome sequence of *M. grisea* will reach beyond the rice blast community. Information and technology resources derived from it will serve as a stepping stone in furthering our understanding of the biology of other fungal pathogens of agricultural significance. Comparative genome analyses among various fungal pathogens will also provide an insight into the evolution and genetic basis of pathogenesis.

9. References

Adachi, K. and Hamer, J. E. (1998). Divergent cAMP signaling pathways regulate growth and pathogenesis in the rice blast fungus *Magnaporthe grisea*. Plant Cell **10**, 1361-1373.

Allen, E. A., Hazen, B. E., Hoch, H. C., Kwon, Y., Leinhos, G. M. E., Staples, R. C., Stumpf, M. A., and Terhune, B. T. (1991). Appressorium formation in response to topographical signals by 27 rust species. Phytopathology **81**, 323-331.

Arumuganathan, K. and Earle, E. D. (1991). Nuclear DNA content of some important plant species. Plant. Mol. Biol. Rep. **9**, 208-218.

Asuyama, H. (1963). Morphology, taxonomy, host range, and life cycle of *Piricularia oryzae*, in S. H. Ou (ed.) *The Rice Blast Disease*, The Johns Hopkins Press, Baltimore, pp. 9-22.

Bain, D. C., Patel, B. N., and Patel, M. V. (1972). Blast of ryegrass in Mississippi. Plant Dis. Rep. **56**, 210.

Banuett, F. (1998). Signaling in the yeasts: An informational cascade with links to the filamentous fungi. Microbiol. Mol. Biol. Rev. **62**, 249-274.

Beckerman, J. L. and Ebbole, D. J. (1996). *MPG1*, a gene encoding a fungal hydrophobin of *Magnaporthe grisea*, is involved in surface recognition. Mol. Plant-Microbe Interact. **9**, 450-456.

Bent, A. F. (1996). Plant disease resistance genes: function meets structure. Plant Cell **8**, 1757-1771.

Bockaert, J. and Pin, J. P. (1999). Molecular tinkering of G protein-coupled receptors: an evolutionary success. EMBO J. **18**, 1723-1729.

Bonman, J. M. (1992). Blast, in R. K. Webster and P. S. Gunnell (eds.), *Compendium of Rice Disease* The American Phytopathological Society Press, St. Paul. pp. 14-17.

Bonman, J. M., Khush, G. S., and Nelson, R. J. (1992). Breeding rice for resistance to pests. Annu. Rev. Phytopathol. **30**, 507-528.

Bonman, J. M., Vergel De Dios, T. I., Bandong, J. M., and Lee, E. J. (1987). Pathogenic variability of monoconidial isolates of *Pyricularia oryzae* in Korea and in the Philippines. Plant Dis. **71**, 127-130.

Borromeo, E. S., Nelson, R. J., Bonman, J. M., and Leung, H. (1993). Genetic differentiation among isolates of *Pyricularia* infecting rice and weed hosts. Phytopathology **83**, 393-399.

Bourett, T. and Howard, R. (1990). *In vitro* development of penetration structures in the rice blast fungus *Magnaporthe grisea*. Can. J. Botany **69**, 329-342.

Bourett, T. M. and Howard, R. J. (1992). Actin in penetration pegs of the fungal rice blast pathogen, *Magnaporthe grisea*. Protoplasma **168**, 20-26.

Carver, R. B., Rush, M. C., and Lindberg, G. D. (1972). An epiphytotic of ryegrass blast in Louisiana. Plant Dis. Rep. **56**, 157-159.

Chao, C. T. and Ellingboe, A. H. (1997). Genetic analysis of avirulence/virulence of an isolate of *Magnaporthe grisea* from a rice field in Texas. Phytopathology **87**, 71-76.

Chen, D., Zeigler, R. S., Leung, H., and Nelson, R. (1995). Population structure of *Pyricularia grisea* at two screening sites in Philippines. Phytopathology **85**, 1011-1020.

Choi, W. and Dean, R. A. (1997). The adenylate cyclase gene *MAC1* of *Magnaporthe grisea* controls appressorium formation and other aspects of growth and development. Plant Cell **9**, 1973-1983.

Choi, W., Fang, E. G. C., Sasinowski, M., and Dean, R. A. (1999). Sequence analysis and characterization of genes expressed during appressorium formation in *Magnaporthe grisea*. Fungal Genet. Newsl. **46S**, 40.

Chumley, F. G. and Valent, B. (1990). Genetic analysis of melanin-deficient, nonpathogenic mutants of *Magnaporthe grisea*. Mol. Plant-Microbe Interact. **3**, 135-143.

Coleman, M. J., McHale, M. T., Arnau, J., and Oliver, R. P. (1993). Cloning and characterization of telomeric DNA from *Cladosporium fulvum*. Gene **132**, 67-73.

Correa-Victoria, F. J., and Zeigler, R. S. (1993). Pathogenic variability in *Pyricularia grisea* at a rice blast "hot spot" breeding site in Eastern Colombia. Plant Dis. **77**, 1029-1035.

Correa-Victoria, F. J., Zeigler, R. S., and Levy, M. (1994). Virulence characteristics of genetic families of *Pyricularia grisea* in Colombia. in R. S. Zeigler, P. S. Teng and S. A. Leong (eds.), *Rice Blast Disease*, CAB International, Wallingford, pp. 211-229.

Correll, J. C., Harp, T. L., and Lee, F. L. (1998). Differential changes in host specificity among MGR586 DNA fingerprint groups of the rice blast pathogen. in 2nd International Rice Blast Conference Program Abstracts (Montpellier, France), pp. 34.

Crute, I. R. and Pink, D. A. (1996). Genetics and utilization of pathogen resistance in plants. Plant Cell **10**, 1747-1755.

de Jong, J. K., McCormack, B. J., Smirnoff, N., and Talbot, N. J. (1997). Glycerol generates turgor in rice blast. Nature **389**, 244-245.

Dean, R. (1999). Construction of a sequence-ready framework for the rice blast fungus based on fingerprinting contigs and BAC-end sequencing. Fungal Genet. Newsl. **46S**, 42.

Devos, K. M. and Gale, M. D. (1997). Comparative genetics in the grasses. Plant Mol. Biol. **35**, 3-15.

DeZwaan, T. M., Carroll, A. M., Valent, B., and Sweigard, J. A. (1999). *Magnaporthe grisea* Pth11p is a novel plasma membrane protein that mediates appressorium differentiation in response to inductive substrate cues. Plant Cell **11**, 2013-2030.

Diaz-Perez, S. V., Crouch, V. W., and Orbach, M. J. (1996). Construction and characterization of a *Magnaporthe grisea* bacterial artificial chromosome library. Fungal Genet. Biol. **20**, 280-288.

Dioh, W., Tharreau, D., Gomez, R., Roumen, E., Orbach, M. J., Notteghem, J. L., and Lebrun, M. H. (1996). Mapping avirulence genes in in the rice blast fungus *Magnaporthe grisea*. in G. S. Khush (ed.), *Rice Genetics III*, IRRI Press, Los Bonos, pp. 916-920.

Dobinson, K. F., Harris, R. E., and Hamer, J. E. (1993). *Grasshopper*, a long terminal repeat (LTR) retroelement in the phytopathogenic fungus *Magnaporthe grisea*. Mol. Plant-Microbe Interact. **6**, 114-126.

Eickbush, T. (1999). Exon shuffling in retrospect. Science **283**, 1465-1467.

Ellingboe, A. H. (1992). Segregation of avirulence/virulence on three rice cultivars in 16 crosses of *Magnaporthe grisea*. Phytopathology **82**, 597-601.

Ellingboe, A. H., Wu, B. C., and Robertson, W. (1990). Inheritance of avirulence/virulence in a cross of two isolates of *Magnaporthe grisea* pathogenic to rice. Phytopathology **80**, 108-111.

Farman, M. L. and Leong, S. A. (1998). Chromosome walking to the *AVR1-CO39* avirulence gene of *Magnaporthe grisea*: Discrepancy between the physical and genetic maps. Genetics **150**, 1049-1058.

Farman, M. L. and Leong, S. A. (1995). Genetic and physical mapping of telomeres in the rice blast fungus, *Magnaporthe grisea*. Genetics **140**, 479-492.

Farman, M. L., Taura, S., and Leong, S. A. (1996). The *Magnaporthe grisea* DNA fingerprinting probe MGR586 contains the 3' end of an inverted repeat transposon. Mol. Gen. Genet. **251**, 675-681.

Farman, M. L., Tosa, Y., Nitta, N., and Leong, S. A. (1996). MAGGY, a retrotransposon in the genome of the rice blast fungus *Magnaporthe grisea*. Mol. Gen. Genet. **251**, 665-674.

Flor, H. H. (1971). Current status of the gene-for-gene concept. Annu. Rev. Phytopathol. **9**, 275-296.

Genovesi, A. D. and Magill, C. W. (1976). Heterokaryosis and parasexuality in *Pyricularia oryzae* Cavara. Can. J. Microbiol. **22**, 531-536.

George, M. L. C., Nelson, R. J., Zeigler, R. S., and Leung, H. (1998). Rapid population analysis of *Magnaporthe grisea* by using rep-PCR and endogenous repetitive DNA sequences. Phytopathology **88**, 223-229.

Giatgong, P. and Frederiksen, R. A. (1968). Pathogenic variability and cytology of monoconidial subcultures of *Pyricularia oryzae*. Phytopathology **59**, 1152-1157.

Gilbert, R. D., Johnson, A. M., and Dean, R. A. (1996). Chemical signals responsible for appressorium formation in the rice blast fungus *Magnaporthe grisea*. Physiol. Mol. Plant Pathol. **48**, 335-346.

Goryshin, I. Y. and Reznikoff, W. S. (1998). Tn5 *in vitro* transposition. J. Biol. Chem. **273**, 7367-7374.

Hamer, J. E., Farrall, L., Orbach, M. J., Valent, B., and Chumley, F. G. (1989). Host species-specific conservation of a family of repeated DNA sequences in the genome of a fungal plant pathogen. Proc. Natl. Acad. Sci. USA **86**, 9981-9985.

Hamer, J. E., Valent, B., and Chumley, F. G. (1989). Mutations at the *SMO* genetic locus affect the shape of diverse cell types in the rice blast fungus. Genetics **122**, 351-361.

Hamer, L., Page, A., Woessner, J., Tanzer, M., Adachi, K., Uknes, S., and Hamer, J. E. (1999). Genome wide mutagenesis and sequencing for filamentous fungi. Fungal Genet. Newsl. **46S**, 104.

Han, S. S., Ra, D. S., and Nelson, R. J. (1993). Comparison of RFLP-based phylogenetic trees and pathotypes of *Pyricularia oryzae* in Korea. RDA J. Agric. Sci. **35**, 315-323.

Hayashi, N., Li, C. Y., Li, J. L., and Naito, H. (1997). *In vitro* production on rice plants of perithecia of *Magnaporthe grisea* from Yunnan, China. Mycol. Res. **101**, 1308-1310.

Heath, M. C., Valent, B., Howard, R. J., and Chumley, F. G. (1990). Correlations between cytologically detected plant-fungal interactions and pathogenicity of *Magnaporthe grisea* toward weeping lovegrass. Phytopathology **80**, 1382-1386.

Hebert, T. T. (1971). The perfect stage of Pyricularia grisea. Phytopathology **61**, 83-87.

Hegde, Y. and Kolattukudy, P. E. (1997). Cuticular waxes relieve self-inhibition of germination and appressorium formation by the conidia of *Magnaporthe grisea*. Physiol. Mol. Plant Path. **51**, 75-84.

Howard, R. J. (1997). Breaching the outer barriers - Cuticle and cell wall penetration. in G. C. Carroll and P. Tuzynski (eds.), *The Mycota V*, Springer-Verlag, Berlin, pp. 43-60.

Howard, R. J. (1994). Cell biology of pathogenesis, in R. S. Zeigler, P. S. Teng and S. A. Leong (eds.), *Rice Blast Disease*, CAB International, Wallingford, pp. 3-22.

Howard, R. J. and Ferrari, M. A. (1989). Role of melanin in appressorium function. Exp. Mycol. **13**, 403-418.

Howard, R. J., Ferrari, M. A., Roach, D. H., and Money, N. P. (1991). Penetration of hard substrates by a fungus employing enormous turgor pressure. Proc. Natl. Acad. Sci. USA **88**, 11281-11284.

Howard, R. J. and Valent, B. (1996). Breaking and entering: Host penetration by the fungal rice blast pathogen *Magnaporthe grisea*. Annu. Rev. Microbiol. **50**, 491-512.

Jelitto, T. C., Page, H. A., and Read, N. D. (1994). Role of external signals in regulating the pre-penetration phase of infection by the rice blast fungus, *Magnaporthe grisea*. Planta **194**, 471-477.

Kachroo, P., Leong, S. A., and Chattoo, B. B. (1995). Mg-SINE: A short interspersed nuclear element from the rice blast fungus, *Magnaporthe grisea*. Proc. Natl. Acad. Sci. USA **92**, 11125-11129.

Kachroo, P. K., Chattoo, B. B., and Leong, S. A. (1994). Pot2, an inverted repeat transposon from *Magnaporthe grisea*. Mol. Gen. Genet. **245**, 339-348.

Kahn, P. (1995). From genome to proteome: Looking at a cell's proteins. Science **270**, 369-370.

Kameswar Row, K. V. S. R., Aist, J. R., and Crill, J. P. (1985). Mitosis in the rice blast fungus and its possible implications for pathogenic variability. Can. J. Bot. **63**, 1129-1134.

Kang, S., Chumley, F. G., and Valent, B. (1994). Isolation of the mating-type genes of the phytopathogenic fungus *Magnaporthe grisea* using genomic subtraction. Genetics **138**, 289-296.

Kang, S., Sweigard, J. A., and Valent, B. (1995). The *PWL* host specificity gene family in the rice blast fungus *Magnaporthe grisea*. Mol. Plant-Microbe Interact. **8**, 939-948.

Kershaw, M. J. and Talbot, N. J. (1998). Hydrophobins and repellents: proteins with fundamental roles in fungal morphogenesis. Fungal Gen. Biol. **23**, 18-33.

Kiyosawa, S. (1982). Genetic and epidemiological modeling of breakdown of plant disease resistance. Annu. Rev. Phytopathol. **20**, 93-117.

Kolmer, J. A. and Ellingboe, A. H. (1988). Genetic relationships between fertility and pathogenicity and virulence to rice in *Magnaporthe grisea*. Can. J. Bot. **66**, 891-897.

Koob, M., Burkiewicz, A., Kur, J., and Szybalski, W. (1993). RecA-AC: single-site cleavage of plasmids and chromosomes at any predetermined site. Nucleic Acids Res. **20**, 5831-5836.

Kronstad, J. W. and Staben, C. (1997). Mating in filamentous fungi. Annu. Rev. Genet. **31**, 245-276.

Kumar, J., Nelson, R. J., and Zeigler, R. S. (1999). Population structure and dynamics of *Magnaporthe grisea* in the Indian Himalayas. Genetics **152**, 971-984.

Latterell, F. M. and Rossi, A. E. (1986). Longevity and pathogenic stability of *Pyricularia oryzae*. Phytopathology **76**, 231-235.

Lau, G. and Hamer, J. E. (1998). Acropetal: a genetic locus required for conidiophore architecture and pathogenicity in the rice blast fungus. Fungal Genet. Biol. **24**, 228-239.

Lau, G. W., Chao, C. T., and Ellingboe, A. H. (1993). Interaction of genes controlling avirulence/virulence of *Magnaporthe grisea* on rice cultivar Katy. Phytopathology **83**, 375-382.

Lau, G. W., and Ellingboe, A. H. (1993). Genetic analysis of mutations to increased virulence in *Magnaporthe grisea*. Phytopathology **83**, 1093-1096.

Laugé, R. and de Wit, P. J. G. M. (1998). Fungal avirulence genes: structure and possible functions. Fungal Genet. Biol. **24**, 285-297.

Leach, J. E. and White, F. F. (1996). Bacterial avirulence genes. Annu. Rev. Phytopathol. **34**, 153-179.

Lebrun, M.-H., Capy, M. P., Garcia, N., Dutertre, M., Brygoo, Y., Notteghem, J. L., and Vales, M. (1991). Biology and genetics of *Pyricularia oryzae* and *pyricularia grisea* populations: Current situation and development of RFLP markers. in *Rice Genetics II*, IRRI, Los Banos.

Lee, F. N. (1994). Rice breeding programs, blast epidemics and blast management in the United States, in R. S. Zeigler, P. S. Teng and S. A. Leong (eds.), *Rice Blast Disease*, CAB International, Wallingford, pp. 489-500.

Lee, Y.-H. and Dean, R. (1994). Hydrophobicity of contact surface induces appressorium formation in *Magnaporthe grisea*. FEMS Microbiol. Lett. **115**, 71-76.

Lee, Y.-H. and Dean, R. A. (1993). cAMP regulates infection structure formation in the plant pathogenic fungus *Magnaporthe grisea*. Plant Cell **5**, 693-700.

Leong, S. A., Farman, M. L., Punekar, N., Mayama, S., Nakayashi, H., Eto, Y., and Tosa, Y. (1998). Molecular characterization of the *AVR-CO39* from *Magnaporthe grisea*. in 2nd International Rice Blast Conference Program Abstracts (Montpellier, France), pp. 50.

Leung, H., Borromeo, E. S., Bernardo, M. A., and Notteghem, J. L. (1988). Genetic analysis of virulence in the rice blast fungus *Magnaporthe grisea*. Phytopathology **78**, 1227-1233.

Leung, H. and Williams, P. H. (1986). Enzyme polymorphism and genetic differentiation among geographic isolates of the rice blast fungus. Phytopathology **76**, 778-783.

Levy, M., Correa-Victoria, F. J., Zeigler, R. S., Xu, S., and Hamer, J. E. (1993). Genetic diversity of the rice blast fungus in a disease nursery in Colombia. Phytopathology **83**, 1427-1433.

Levy, M., Romao, J., Marchetti, M. A., and Hamer, J. E. (1991). DNA fingerprinting with a dispersed repeated sequence resolves pathotype diversity in the rice blast fungus. Plant Cell **3**, 95-102.

Ling, K. C. and Ou, S. H. (1969). Standardization of the international race numbers of *Pyricularia oryzae*. Phytopathology **59**, 339-342.

Liu, S. and Dean, R. A. (1997). G protein α subunit genes control growth, development, and pathogenicity of *Magnaporthe grisea*. Mol. Plant-Microbe Interact. **10**, 1075-1086.

Madhani, H. D. and Fink, G. R. (1998). The riddle of MAP kinase signaling specificity. Trends Genet. **14**, 151-155.

Mandel, M. A., Crouch, V. W., Gunawardena, U. P., Harper, T. M., and Orbach, M. J. (1997). Physical mapping of the *Magnaporthe grisea AVR1-MARA* locus reveals the virulent allele contains two deletions. Mol. Plant-Microbe Interact. **10**, 1102-1105.

Marchetti, M. A., Lai, X., and Bollich, C. N. (1987). Inheritance of resistance to *Pyricularia oryzae* in rice cultivars grown in the United States. Phytopathology **77**, 799-804.

Marchetti, M. A., Rush, M. C., and Hunter, W. E. (1976). Current status of rice blast in the Southern United States. Plant Dis. Rep. **60**, 721-725.

Marshall, E. (1999). A high-stakes gamble on genome sequencing. Science **284**, 1906-1909.

Mazzoni, C., Zarov, P., Rambourg, A., and Mann, C. (1993). The *SLT2* (*MPK1*) MAP kinase homolog is involved in polarized growth in *Saccharomyces cerevisiae*. J. Cell Biol. **123**, 1821-1833.

Mendgen, K. and Deising, H. (1993). Tansley Review No. 48: Infection structures of fungal plant pathogens - a cytological and physiological evaluation. New Phytol. **124**, 193-213.

Metzenberg, R. L. and Glass, N. L. (1990). Mating type and mating strategies in *Neurospora*. BioEssays **12**, 53-59.

Meyn, M. A., Farrall, L., Chumley, F. G., Valent, B., and Orbach, M. J. (1998). LINES and SINES in *Magnaporthe grisea*. in 2nd International Rice Blast Conference Program Abstracts (Montpellier, France), pp. 53.

Money, N. P. (1997). Mechanism linking cellular pigmentation and pathogenicity in rice blast disease. Fungal Genet. Biol. **22**, 151-152.

Moran, J. V., DeBerardinis, R. J., and Kazazian Jr., H. H. (1999). Exon shuffling by L1 retrotransposition. Science **283**, 1530-1534.

Nakayashiki, H., Nishimoto, N., Ikeda, K., Tosa, Y., and Mayama, S. (1999). Degenerate MAGGY elements in a subgroup of *Pyricularia grisea*: a possible example of successful capture of a genetic invader by a fungal genome. Mol. Gen. Genet. **261**, 958-966.

Namai, T., Endo, M., Yusa, M., Yamanaka, S., and Ehara, Y. (1986). *Pyricularia* spp. produced perithecium-like structure in single culture and its productive process. Ann. Phytopath. Soc. Japan **52**, 312-319.

Nishimura, M., Nakamura, S., Hayashi, N., Asakawa, S., Shimizu, N., Kaku, H., Hasebe, A., and Kawasaki, S. (1998). Construction of a BAC library of the rice blast fungus *Magnaporthe grisea* and finding specific genome regions in which its transposons tend to cluster. Biosci. Biotechnol. Biochem. **62**, 1515-1521.

Nitta, N., Farman, M. L., and Leong, S. A. (1997). Genome organization of *Magnaporthe grisea*: integration of genetic maps, clustering of transposable elements and identification of genome duplications and rearrangements. Theor. Appl. Genet. **95**, 20-32.

Oliver, S. G., Winson, M. K., Kell, D. B., and Baganz, F. (1998). Systematic functional analysis of the yeast genome. Trends Biotech. **16**, 373-378.

Orbach, M. J., Chumley, F. G., and Valent, B. (1996). Electrophoretic karyotypes of *Magnaporthe grisea* pathogens of diverse grasses. Mol. Plant-Microbe Interact. **9**, 261-271.

Ou, S. H. (1980). Pathogen variability and host resistance in rice blast disease. Annu. Rev. Phytopathol. **18**, 167-187.

Ou, S. H. and Ayad, M. R. (1968). Pathogenic races of *Pyricularia oryzae* derived from monoconidial cultures. Phytopathology **58**, 179-182.

Ou, S. H., Nuque, F. L., Ebron, T. T., and Awoderu, V. A. (1970). Pathogenic races of *Pyricularia oryzae* derived from monoconidial cultures. Plant Dis. Rep. **54**, 1045-1049.

Ou, S. H., Nuque, F. L., Ebron, T. T., and Awoderu, V. A. (1971). A type of stable resistance to blast disease of rice. Phytopathology **61**, 703-706.

Pappas, A. C. (1998). Pyricularia leaf spot: a new disease of ornamental plants of the family Marantaceae. Plant Dis. **82**, 465-469.

Percet, K., Sibuet, S., Dioh, W., Tharreau, D., Notteghem, J. L., and Lebrun, M. H. (1998). Avirulence genes cloning in the rice blast fungus *Magnaporthe grisea*. in 2nd International Rice Blast Conference Program Abstracts (Montpellier, France), pp. 49.

Powell, W. A. and Kistler, H. C. (1990). In vivo rearrangement of foreign DNA by *Fusarium oxysporum* produces linear self-replicating plasmids. J. Bacteriol. **172**, 3163-3171.

Romao, J. and Hamer, J. E. (1992). Genetic organization of a repeated DNA sequence family in the rice blast fungus. Proc. Natl. Acad. Sci. USA **89**, 5316-5320.

Roumen, E., Levy, M., and Notteghem (1997). Characterization of the European pathogen population of *Magnaporthe grisea* by DNA fingerprinting and pathotype analysis. Europ. J. Plant Pathol. **103**, 363-371.

Sasaki, T. (1999). Current status of and future prospects for genome analysis in rice, in K. Shimamoto (ed.), *Molecular Biology of Rice*, Springer-Verlag, Tokyo, pp. 3-16.

Sasaki, T. (1998). The rice genome project in Japan. Proc. Natl. Acad. Sci. USA **95**, 2027-2028.

Schechtman, M. G. (1990). Characterization of telomere DNA from *Neurospora crassa*. Gene **88**, 159-165.

Schena, M., Shalon, D., Davis, R. W., and Brown, P. O. (1995). Quantitative monitoring of gene expression patterns with a complementary DNA microarray. Science **270**, 467-470.

Shi, Z., Christian, D., and Leung, H. (1997). Interactions between spore morphogenetic mutations affect cell types, sporulation, and pathogenesis in *Magnaporthe grisea*. Mol. Plant-Microbe Interact. **11**, 199-207.

Shi, Z. and Leung, H. (1995). Genetic analysis of sporulation in *Magnaporthe grisea* by chemical and insertional mutagenesis. Mol. Plant-Microbe Interact. **8**, 949-959.

Shull, V. and Hamer, J. E. (1994). Genomic structure and variability in *Pyricularia grisea*. in R. S. Zeigler, P. S. Teng and S. A. Leong (eds.), *Rice Blast Disease*, CAB International, Wallingford, pp. 65-86.

Shull, V. and Hamer, J. E. (1996). Rearrangement at a DNA-fingerprint locus in the rice blast fungus. Curr. Genet. **30**, 263-271.

Silué, D. and Notteghem, J. L. (1990). Production of perithecia of *Magnaporthe grisea* on rice plants. Mycol. Res. **94**, 1151-1152.

Silué, D., Notteghem, J. L., and Tharrea, D. (1992a). Evidence for a gene-for-gene relationship in the *Oryza sativa-Magnaporthe grisea* pathosystem. Phytopathology **82**, 577-580.

Silué, D., Tharrea, D., and Notteghem, J. L. (1992b). Identification of *Magnaporthe grisea* avirulence genes to seven rice cultivars. Phytopathology **82**, 1462-1467.

Simonds, W. F. (1999). G protein regulation of adenylate cyclase. Trends Pharmacol. Sci. **20**, 66-73.

Skinner, D. Z., Budde, A. D., Farman, M. L., Smith, J. R., Leung, H., and Leong, S. A. (1993). Genome organization of *Magnaporthe grisea*: genetic map, electrophoretic karyotype and occurrence of repeated DNAs. Theor. Appl. Genet. **87**, 545-557.

Sone, T., Suto, M., and Tomita, F. (1993). Host species-specific repetitive DNA sequence in the genome of *Magnaporthe grisea*, the rice blast fungus. Biosci. Biotech. Biochem. **57**, 1228-1230.

Straus, D. and Ausubel, F. M. (1990). Genomic subtraction for DNA corresponding to deletion mutations. Proc. Natl. Acad. Sci. USA **87**, 1889-1893.

Suzuki, H. (1965). Origin of variation in *Pyricularia oryzae*, in S. H. Ou (ed.), *The Rice Blast Disease*, Johns Hopkins Press, Baltimore, pp. 111-149.

Sweigard, J. A., Carroll, A. M., Farrall, L., Chumley, F. G., and Valent, B. (1998). *Magnaporthe grisea* pathogenicity genes obtained through insertional mutagenesis. Mol. Plant-Microbe Interact. **11**, 404-412.

Sweigard, J. A., Carroll, A. M., Kang, S., Farrall, L., Chumley, F. G., and Valent, B. (1995). Identification, cloning, characterization of *PWL2*, a gene for host-species specificity in the rice blast fungus. Plant Cell **7**, 1221-1233.

Sweigard, J. A., Chumley, F. G., and Valent, B. (1992). Cloning and analysis of *CUT1*, a cutinase gene from *Magnaporthe grisea*. Mol. Gen. Genet. **232**, 174-182.

Sweigard, J. A., Valent, B., Orbach, M. J., Walter, A. M., Rafalski, A., and Chumley, F. G. (1993). Genetic map of the rice blast fungus *Magnaporthe grisea*. in S. J. O'Brien (ed.) *Genetics Maps*, Cold Spring Harbor Laboratory, New York, pp. 3.112-3.117.

Talbot, N. J., Kershaw, M. J., Wakley, G. E., and de Vries, O. M. (1996). *MPG1* encodes a fungal hydrophobin involved in surface interactions during infection-related development by *Magnaporthe grisea*. Plant Cell **8**, 985-989.

Talbot, N. J., McCafferty, H. R. K., Ma, M., Moore, K., and Hamer, J. E. (1997). Nitrogen starvation of the rice blast fungus *Magnaporthe grisea* may act as an environmental cue for disease symptom expression. Physiol. Mol. Plant Path. **50**, 179-195.

Talbot, N. J., Salch, Y. P., Ma, M., and Hamer, J. E. (1993). Karyotypic variation within clonal lineages of the rice blast fungus, *Magnaporthe grisea*. Appl. Environ. Microbiol. **59**, 585-593.

Tharreau, D., Notteghem, J. L., and Lebrun, M. H. (1997). Mutations affecting perithecium development and sporulation in *Magnaporthe grisea*. Fungal Genet. Biol. **21**, 206-213.

Trione, E. J. (1981). Natural regulators of fungal development, in R. Staples and G. Toenniessen (eds.), *Plant Disease Control: Resistance and Susceptibility*, John Wiley and Sons, New York. pp.85 - 102.

Uddin, W. (1999). Gray leaf spot 'blasts' U.S. golf course turf. Golf Course Management **April**, 52-56.

Urashima, A. S., Igarashi, S., and Kato, H. (1993). Host range, mating type, and fertility of *Pyricularia grisea* from wheat in Brazil. Plant Dis. **77**, 1211-1216.

Urban, M., Bhargava, T., and Hamer, J. E. (1999). An ATP-driven drug efflux pump is a novel pathogenicity factor in rice blast disease. EMBO J. **18**, 512-521.

Valent, B. (1997). The rice blast fungus, *Magnaporthe grisea*, in G. C. Carroll and P. Tuzynski (eds.), *The Mycota V*, Springer-Verlag, Berlin, pp. 37-53.

Valent, B., Bryan, G. T., Jia, Y., Farrall, L., Wu, K., Orbach, M. J., Donaldson, G. K., Hershey, H. P., and McAdams, S. A. (1998). Molecular characterization of the *Magnaporthe grisea* avirulence gene *AVR2-YAMO* and corresponding resistance gene *Pi-ta*. in 2nd International Rice Blast Conference Program Abstracts (Montpellier, France), pp. 50.

Valent, B. and Chumley, F. G. (1994). Avirulence genes and mechanisms of genetic instability in the rice blast fungus, in R. S. Zeigler, P. S. Teng and S. A. Leong (eds.), *Rice Blast Disease*, CAB International, Wallingford, pp. 111-153.

Valent, B. and Chumley, F. G. (1987). Genetic analysis of host species specificity in *Magnaporthe grisea*, in C. J. Arntzen and C. Ryan (eds.), *Molecular Strategies for Crop Protection*, Alan R. Liss, New York, pp. 83-93.

Valent, B. and Chumley, F. G. (1991). Molecular genetic analysis of the rice blast fungus, *Magnaporthe grisea*. Annu. Rev. Phytopathol. **29**, 443-467.

Valent, B., Crawford, M. S., Weaver, C. G., and Chumley, F. G. (1986). Genetic studies of fertility and pathogenicity in *Magnaporthe grisea* (*Pyricularia oryzae*). Iowa State J. Res. **60**, 569-594.

Valent, B., Farrall, L., and Chumley, F. G. (1991). *Magnaporthe grisea* genes for pathogenicity and virulence identified through a series of backcrosses. Genetics **127**, 87-101.

Venter, J. C., Smith, H. O., and Hood, L. (1996). A new strategy for genome sequencing. Nature **381**, 364-366.

Viji, G., Wu, B., Uddin, W., and Kang, S. (1999). Characterization of *Pyricularia grisea* isolates from perennial ryegrass from various geographic regions in the United States. Phytopathology **89**, S80.

Wang, G.-L., Mackill, D. J., Bonman, J. M., McCouch, S. R., Champox, M. C., and Nelson, R. J. (1994). RFLP mapping of genes conferring complete and partial resistance to blast in a durably resistant rice cultivar. Genetics **136**, 1421-1434.

Weiner, A. M., Deininger, P. L., and A., E. (1986). Nonviral retroposons: genes, pseudogenes, and transposable elements generated by the reverse flow of genetic information. Annu. Rev. Biochem. **55**, 631-661.

Wu, S.-C., Ham, K.-S., Darvill, A. G., and Albersheim, P. (1997). Deletion of two *endo*-β-1,4-xylanase genes reveals additional isozymes secreted by the rice blast fungus. Mol. Plant-Microbe Interact. **10**, 700-708.

Wu, S.-C., Kauffmann, S., Darvill, A. G., and Albersheim, P. (1995). Purification, cloning and characterization of two xylanases from *Magnaporthe grisea*, the rice blast fungus. Mol. Plant-Microbe Interact. **8**, 506-514.

Xia, J. Q., Correll, J. C., Lee, F. N., Marchetti, M. A., and Rhoads, D. D. (1993). DNA fingerprinting to examine microgeographic variation in the *Magnaporthe grisea* (*Pyricularia grisea*) population in two rice fields in Alkansas. Phytopathology **83**, 1029-1035.

Xiao, J.-Z., Watanabe, T., Kamakura, T., Ohshima, A., and Yamaguchi, I. (1994). Studies on the cellular differentiation of *Magnaporthe grisea*: Physicochemical aspects of substratum surfaces in relation to appressorium formation. Physiol. Mol. Plant Path. **44**, 227-236.

Xu, J., and Hamer, J. E. (1996). MAP kinase and cAMP signaling regulate infection structure formation and pathogenic growth in the rice blast fungus *Magnaporthe grisea*. Genes Dev. **10**, 2696-2706.

Xu, J., Urban, M., Sweigard, J. A., and Hamer, J. E. (1997). The *CPKA* gene of *Magnaporthe grisea* is essential for appressorial penetration. Mol. Plant-Microbe Interact. **10**, 187-194.

Xu, J.-R., Staiger, C. J., and Hamer, J. E. (1998). Inactivation of the mitogen-activated protein kinase Mps1 from the rice blast fungus prevents penetration of host cells but allows activation of plant defense responses. Proc. Natl. Acad. Sci. USA **95**, 12713-12718.

Yaegashi, H. (1978). Inheritance of pathogenicity in crosses of Pyricularia isolates from weeping lovegrass and finger millet. Ann. Phytopath. Soc. Japan **44**, 626-632.

Yamada, M., Kiyosawa, S., Yamaguchi, T., Hirano, T., Kobayashi, T., Kushibuchi, K., and Watanabe, S. (1976). Proposal of a new method for differentiating races of *Pyricularia oryzae* Cavara in Japan. Ann. Phytopath. Soc. Japan **42**, 216-219.

Yamanaka, S., Nakayama, S., and Namai, T. (1984). A perithecium-like structure produced on agar by a *Pyricularia* isolate from goosegrass. Ann. Phytopath. Soc. Japan **50**, 545-548.

Zeigler, R. S. (1998). Recombination in *Magnaporthe grisea*. Annu. Rev. Phytopathol. **36**, 249-275.

Zeigler, R. S., Cuoc, L. X., Scott, R. P., Bernardo, M. A., Chen, D. H., Valent, B., and Nelson, R. J. (1995). The relationship between lineage and virulence in *Pyricularia grisea* in the Philippines. Phytopathology **85**, 443-451.

Zeigler, R. S., Scott, R. P., Leung, H., Bordeos, A. A., Kmar, J., and Nelson, R. J. (1997). Evidence of parasexual exchange of DNA in the rice blast fungus challenges its exclusive clonality. Phytopathology **87**, 284-294.

Zeigler, R. S., Thome, J., Nelson, R., Levy, M., and Correa-Victoria, F. J. (1994). Lineage exclusion: A proposal for linking blast population analysis to resistance breeding, in R. S. Zeigler, P. S. Teng and S. A. Leong (eds.), *Rice Blast Disease*, CAB International, Wallingford, pp. 287-292.

Zhou, X.-L., Stumpf, M. A., Hoch, H. C., and Kung, C. (1991). A mechanosensitive channel in whole cells and in membrane patches of the fungus *Uromyces*. Science **253**, 1415-1417.

Zhu, H., Blackmon, B. P., Sasinowski, M., and Dean, R. A. (1999). Physical map and organization of chromosome 7 in the rice blast fungus, *Magnaporthe grisea*. Genome Res. **9**, 739-750.

Zhu, H., Choi, S., Johnston, A. K., Wing, R. A., and Dean, R. A. (1997). A large-insert (130 kbp) bacterial artificial chromosome library of the rice blast fungus *Magnaporthe grisea*: Genome analysis, contig assembly, and gene cloning. Fungal Genet. Biol. **21**, 337-347.

Zhu, H., Choi, W., and Dean, R. A. (1999). Localization and expression profiling of genes expressed during appressorium formation on chromosome 7 in *Magnaporthe grisea*. Fungal Genet. Newsl. **46S**, 72.

PHYTOPHTHORA

SOPHIEN KAMOUN
Department of Plant Pathology
The Ohio State University-Ohio Agricultural Research and Development Center
1680 Madison Avenue
Wooster, Ohio 44691

Ever since the potato late blight epidemics of the mid-nineteenth century, members of the oomycete genus *Phytophthora* have emerged as major pathogens of innumerable crops. Nowadays, with more than fifty species recognized and with destructive diseases caused on thousands of plant species, *Phytophthora* remains an active subject of research and a nagging problem to farmers and growers. Renewed interest in *Phytophthora* diseases occurred in recent years with the reemergence of *Phytophthora infestans,* the Irish potato famine fungus. Late blight has turned into a global threat to potato and tomato production following a series of severe late blight epidemics that coincided with the migration to Europe and North America of aggressive A2 mating type strains (Fry and Goodwin, 1997a; Fry and Goodwin, 1997b). The International Potato Center estimated that worldwide losses in potato production caused by late blight have recently exceeded $3 billion annually, making *P. infestans* the single most important biotic constraint to global food production (Anonymous, 1996).

In parallel to the resurgence of *Phytophthora* in the field and in the news, oomycete research has entered an exciting phase. Recent technical developments, such as routine DNA transformation (Judelson, 1996a), use of reporter genes (Judelson, 1997b; Kamoun et al., 1998b; van West et al., 1998), genetic manipulation using gene silencing (van West et al., 1998; Kamoun et al., 1998c), and the development of detailed genetic maps (van der Lee et al., 1997; Whisson et al., 1995), will facilitate cloning and allow functional analyses of numerous candidate genes involved in interactions with the plant and other important phenomena. In addition, genomics approaches, or the wholesale study of *Phytophthora* genes, promises to bring research to yet another level (Kamoun et al., 1999b).

In this chapter, I summarize basic information on the biology of *Phytophthora* and review recent advances in molecular study and genetic manipulation of these pathogens. For additional detail, readers should refer to other recent reviews. Two

J. W. Kronstad (ed.), Fungal Pathology, 237–265.

review articles that discuss the genetics of the genus were published in the last three years (Judelson, 1996a; Judelson, 1997a). For more general but detailed information on *Phytophthora*, the monumental treatise by Erwin and Ribeiro (1996) describes many aspects of the biology of *Phytophthora*, provides detailed protocols and methods of study, separate and detailed descriptions of all known species, and an exhaustive list of references.

1. Oomycetes: A unique group of eukaryotic plant pathogens

Oomycetes include diverse organisms, some of which are pathogenic on fish, crayfish, and animals. Plant pathogenic oomycetes are usually classified in the Peronosporales and include in addition to *Phytophthora*, numerous genera of the biotrophic downy mildews (such as *Peronospora* and *Bremia*), the white rust pathogens *Albugo*, and more than 100 species of the genus *Pythium*. Altogether, these pathogens cause several billions of dollars of damages on crop, ornamental, and native plants. Some of the economically important *Phytophthora* pathogens and their major host plants are listed in Table 1.

Table 1. Some economically important *Phytophthora* species

Phytophthora species	Major host plant(s)
Phytophthora capsici	Pepper
Phytophthora cinnamomi	Avocado and numerous forest species in the Australasian region
Phytophthora fragariae var. fragariae	Strawberry
Phytophthora infestans	Potato and tomato
Phytophthora palmivora	Cocoa, papaya, coconut and palm species
Phytophthora sojae	Soybean

Oomycetes exhibit filamentous growth habit, and are often inaccurately referred to as fungi. Modern biochemical analyses as well as phylogenetic analyses based on sequences of ribosomal and mitochondrial genes suggest that oomycetes share little taxonomic affinity to filamentous fungi, but are more closely related to golden-brown algae and heterokont algae in the Kingdom Stramenopila (Kumar and Rzhetsky, 1996; Paquin et al., 1997; Van de Peer and De Wachter, 1997). Therefore, oomycetes include a unique group of eukaryotic plant pathogens, which evolved the ability to infect plants independently from true fungi. This suggests that oomycetes may have distinct genetic and biochemical mechanisms for interacting with plants. For example, in contrast to filamentous fungi (Osbourn, 1996a; Osbourn, 1996b), oomycetes contain little or no membrane sterols, the target of toxic saponins, and are therefore unaffected by these compounds which are abundantly present in plants.

2. Biology of *Phytophthora*

2.1. GENERAL CHARACTERSTICS

In the vegetative stage, *Phytophthora* exhibits filamentous mycelial growth (Erwin and Ribeiro, 1996). Mycelium is generally coenocytic with no or a few septa. The multiple diploid nuclei can differ genetically to form so called heterokaryotic strains. *Phytophthora* can produce both asexual and sexual spores. Asexual sporangia (or more accurately zoosporangia, singular zoosporangiospore) emerge directly from the hyphae through structures known as sporangiophores. Under optimal conditions of temperature and moisture, sporangia release swimming spores, the biflagellate zoospores. Sexual spores, or oospores, are formed when the male structure, antheridium, associates with the female, egg bearing oogonium. Some species of *Phytophthora* are self-fertile or homothallic, whereas others are self-sterile or heterothallic. Heterothallic species are divided into A1 and A2 mating types and crossing occurs when these two types of strains contact each other.

2.2. LIFE AND DISEASE HISTORY OF A FOLIAR *PHYTOPHTHORA*

A typical aerial species is the late blight pathogen *P. infestans*. *P. infestans* is generally considered a specialized pathogen causing disease on leaves and fruits of potato and tomato crops. In potato, infection of tubers may also occur. Only sporadic reports of natural infection of plants outside the genera *Solanum* and *Lycopersicon* have been reported (Erwin and Ribeiro, 1996). The life cycle and infection process of *P. infestans* is well known (Coffey and Wilson, 1983; Hohl and Suter, 1976; Judelson, 1997a; Kamoun et al., 1999c; Pristou and Gallegly, 1954). Infection generally starts when motile zoospores, that swim on the leaf surface, encyst and germinate. Germ tubes form an appressorium and a penetration peg, which pierces the cuticle and penetrates an epidermal cell to form an infection vesicle. Branching hyphae with narrow, digit-like haustoria expand from the site of penetration to neighboring cells through the intercellular space. Later on, infected tissue necrotizes and the mycelium develops sporangiophores that emerge through the stomata to produce numerous asexual spores called sporangia. Penetration of an epidermal cell by *P. infestans* was noted in all examined interactions including those with plant species unrelated to the solanaceous hosts (Gross et al., 1993; Kamoun et al., 1999c; Kamoun et al., 1998c; Naton et al., 1996; Schmelzer et al., 1995). Fully resistant plants, such as some of the potato lines bearing *R* genes or the nonhosts *Solanum nigrum*, parsley, and *Nicotiana* spp. display a typical localized hypersensitive response (HR), a programmed cell death of plants, at all infection sites (Colon et al., 1992; Freytag et al., 1994; Gees and Hohl, 1988; Gross et al., 1993; Kamoun et al., 1999c; Kamoun et al., 1998c; Naton et al., 1996; Schmelzer et al., 1995). The HR can be highly localized to a single epidermal cell or can affect a group of cells surrounding the penetrating hyphae (Kamoun et al., 1999c; Kamoun et al., 1998c). It appears that the HR is associated with all known forms of genetic resistance to *P. infestans* and may be involved in nonhost resistance and partial resistance phenotypes (Kamoun et al., 1999c).

2.3. LIFE AND DISEASE HISTORY OF A SOIL *PHYTOPHTHORA*

A typical soil species is the root and stem rot pathogen *Phytophthora sojae* (formerly known as *Phytophthora megasperma* f. sp. *glycinea*). *P. sojae* is a specialized pathogen causing disease on soybean and some lupine species (Erwin and Ribeiro, 1996; Schmitthenner, 1989). This pathogen generally infects roots and sometimes spreads to the stem of soybean plants. However, leaves may also be infected if they get in contact with soil particles containing the pathogen (Schmitthenner, 1985; Schmitthenner, 1989). The life history of *P. sojae* is well known, but contradictory reports on the cytology of infection have been published (Enkerli et al., 1997; Ward et al., 1989). In the field, infection generally starts when motile zoospores, swimming in the water interface in the soil, are attracted to host plant root exudates. Once on the root surface, zoospores encyst and germinate. In a study that mimicked natural infection of soybean roots, Enkerli et al. (1997) examined the ultrastructure of early infection events on both resistant and susceptible plants. On all plant genotypes, penetration by germ tubes occurred directly between two epidermal cells without the formation of an appressorium. On susceptible soybean roots, *P. sojae* established a short biotrophic phase by growing in the intercellular space, forming numerous haustoria and inducing little response from the plant. Later on, starting at about 15 hours after inoculation, the root turned necrotic as the pathogen reached the vascular tissue and started growing intracellularly. In roots of resistant soybean plants containing the root rot resistance genes *Rps1a* or *Rps1b*, the timing of events was different. Rapid host cell necrosis occurred and the pathogen rarely established haustoria. Penetration of root tissue was limited as *P. sojae* rarely reached the vascular tissue.

Even though *P. sojae* infects roots, many researchers use hypocotyl infection assays for convenience and reproducibility. Interestingly, cytological examination of hypocotyl infection revealed the formation of appressoria by *P. sojae* germinating cysts prior to penetration unlike the direct intercellular penetration observed on roots (Ward et al., 1989). It remains unclear at this stage whether this discrepancy truly reflects different modes of infection of soybean tissues by the pathogen or simply the different experimental procedures used.

2.4. *P. INFESTANS* VS. *P. SOJAE*: TWO CONTRASTING PATHOGENS

The genus *Phytophthora* is diverse. Table 2 illustrates contrasting features in the biology and genetics of two economically important *Phytophthora* pathogens, *P. infestans* and *P. sojae*. The noted differences in such basic aspects as mode of sexual reproduction or genome size suggest that genetic studies on *Phytophthora* should consider this diversity of the genus and must incorporate comparative analyses. This will prove essential in order to achieve a thorough understanding of these organisms.

Table 2. General characteristics of *Phytophthora infestans* and *Phytophthora sojae*

Feature	P. infestans	P. sojae
Host plants	Potato, tomato, and related species	Soybean and lupines
Infected tissue	Leaves, stems, and tubers	Roots and stems, occasionally leaves
Penetration of plant tissue	Intracellular penetration of an epidermal cell through appressoria formation	Intercellular penetration between two root epidermal cells, no appressoria formation
Infection process	Early biotrophic phase with formation of haustoria followed by saprophytic growth	Early biotrophic phase with formation of haustoria followed by saprophytic growth
Sporangia	Deciduous, dislodge easily from the sporangiophores	Persistent, remain attached to the sporangiophores
Main habitat	Aerial plant parts	Soil
Sexual behavior	Self-sterile or heterothallic	Self-fertile or homothallic
Genome size	250 Mb	62 Mb
Haploid chromosome count	8-10	10-13

3. Main themes of molecular research

3.1. VIRULENCE AND PATHOGENICITY

Due to the lack of random mutagenesis systems in *Phytophthora*, indirect approaches have been taken to identify genes that may play a role in virulence and pathogenicity. Pieterse et al. (1993a) showed that a number of *P. infestans* genes are activated during interaction with potato. Using a differential hybridization approach, several in planta-induced genes (so called *ipi* genes) were isolated (Pieterse et al., 1994a; Pieterse et al., 1993a; Pieterse et al., 1991; Pieterse et al., 1993b). Based on their typical pattern of expression, it was proposed that these genes might play a role in pathogenicity. Two such *in planta*-induced genes are the closely related *ipiO1* and *ipiO2* (Pieterse et al., 1994a; Pieterse et al., 1994b). The putative 152 amino-acid

IPIO1 protein bears a signal peptide at the N-terminus and is presumably extracellular. A careful analysis of the expression of *ipiO in vitro* and during various stages of infection of potato using RNA blot hybridizations and histochemical localization of GUS activity from a *P. infestans* transformant carrying a fusion between the *ipiO1* promoter and the β-glucuronidase reporter gene, was recently reported by van West et al. (1998). These experiments showed that *ipiO* expression is limited to particular stages of the life cycle of *P. infestans*. *IpiO* mRNA was detected in zoospores, cysts, germinating cysts, and younger mycelium, but not in sporangia and mature mycelium. In diseased potato leaves, *ipiO* expression was limited to the outer zones in invading biotrophic hyphae, but did not occur in the necrotic areas of lesions. This suggests that the IPIO protein could be located at the interface between the invading hyphae and plant cells (van West et al., 1998). Other *ipi* genes identified through the differential hybridization procedure of Pieterse et al. (1993a) encode homologs of ubiquitin (encoded by *ubi3R*) and calmodulin (*calA*), and a novel glycine-rich protein family (*ipiB1*, *ipiB2*, and *ipiB3*) (Pieterse et al., 1991; Pieterse et al., 1994b; Pieterse et al., 1993b). However, future functional genetic analyses are required to unequivocally determine whether these *ipi* genes play a role in pathogenicity.

Penetration of plant cells by *P. infestans* germinating cysts requires the formation of specialized infection structures, such as appressoria. Kraemer et al. (1997) used a biochemical approach to identify polypeptides associated with these infection structures. Two-dimensional SDS-polyacrylamide gel electrophoresis of total and newly synthesized proteins in hyphae, cysts, germinating cysts and appressoria induced *in vitro* on artificial membranes were compared. Several polypeptides that showed specific developmental changes were detected but their identity remains unknown.

Degradation of plant cell wall is likely to be a component of virulence of *Phytophthora*, particularly during penetration of plant epidermis and establishment of haustoria. However, little is known about genes encoding degrading enzymes. Munoz and Bailey (1998) described a *Phytophthora capsici* gene encoding a 218 amino-acid cutinase that may be essential for infection of pepper. Following a pilot sequencing project of *P. infestans* cDNAs, a number of sequences encoding homologs of degradative enzymes, such as polygalacturonases, were identified (Kamoun et al., 1999b). Future functional assays should help determine whether these degradative enzymes are important virulence factors for *Phytophthora*.

3.1. SPECIFICITY IN INTERACTION WITH PLANTS

3.1.1. Race-specific avirulence genes

Perception by the plant of signal molecules, elicitors, produced by the avirulent pathogen, leads to the induction of effective defense responses including a programmed cell death response termed hypersensitive response (HR) (Baker et al., 1997; Dangl et al., 1996; Lamb et al., 1989). This model has been genetically defined by Flor's gene-for-gene hypothesis (Flor, 1956; Flor, 1971). According to this hypothesis, a resistance reaction is determined by the simultaneous expression of a pathogen avirulence or *Avr* gene with the corresponding plant resistance or *R* gene (Staskawicz et al., 1995). No plant *R* gene targeted against *Phytophthora* or race-specific *Avr* gene of *Phytophthora* has been described yet (Judelson, 1996a; Kamoun et al., 1999c). Subsequently, the mechanisms underlying the evolution of new virulence traits in *Phytophthora* remain unknown even though races of *P. infestans* virulent on a wider range of plants have been known for fifty years (Black and Gallegy, 1957). However, preliminary genetic analyses of *P. infestans* and *P. sojae* indicate that race/cultivar specificity usually follows the gene-for-gene model.

Through classical breeding, a total of eleven dominant late blight *R* genes have been introgressed into potato from the Mexican wild species *Solanum demissum*. The genetics of *P. infestans* virulence/avirulence on these potato lines has been examined. In a first study describing genetic crosses between *P. infestans* isolates with different virulence patterns, it was shown that avirulence on potato plants carrying the *R-2* and *R-4* resistance genes is dominant (Al-Kherb et al., 1995). In a second study using a different cross, van der Lee et al. (1998) reported that avirulence on potato plants carrying one of the *R-3*, *R-4*, *R-10*, and *R-11* resistance genes is dominant. Interestingly, *Avr3*, *Avr10*, and *Avr11* appeared closely linked in this cross and formed a tight cluster on linkage group VIII of an AFLP linkage map generated with the same progeny (van der Lee et al., 1997). van der Lee et al. (1998) then screened pooled DNA from virulent and avirulent progenies using the bulked segregant analysis (BSA) approach. This led to the identification of a total of 18 AFLP markers linked to the *Avr* cluster. These tightly linked markers should greatly facilitate the molecular cloning of this *P. infestans Avr* gene cluster.

Genetically resistant soybean cultivars containing one or combinations of thirteen known *Rps* resistance genes introduced from related *Glycine* germplasm have resulted in significant protection in the field against *P. sojae*. However, new races of *P. sojae* that can infect *Rps* cultivars are rapidly evolving (Forster et al., 1994). Populations of *P. sojae* that can defeat the most commonly displayed *Rps* genes, including the successful *Rps1-k*, have recently increased in number and distribution (Schmitthenner et al., 1994; A.F. Schmitthenner and A. Dorrance, pers. comm.). More than forty races of the pathogen have been described and dozens of new races are being characterized countrywide. Genetic analyses of crosses between *P. sojae* strains differing in virulence patterns have shown that avirulence to *Rps1b*, *Rps1d*, *Rps1k*, *Rps3b*, *Rps4*, and *Rps6* is dominant (Gijzen et al., 1996; Tyler et al., 1995; Whisson et al., 1994; Whisson et al., 1995). Interestingly, a number of *P. sojae* avirulence genes, such as *Avr1b* and *Avr1k*, and *Avr4* and *Avr6*, appear very tightly linked (Gijzen et al., 1996; Whisson et al., 1995). Progress has been made toward cloning these two *Avr* clusters. A Bacterial Artificial Chromosome (BAC) clone that contains both the *Avr1b* and *Avr1k* genes was identified and fine mapping of the two genes was performed (Arredondo et al., 1998). In addition, three overlapping cosmids containing *Avr4* and *Avr6* and covering 67.3 kb and 10.1 cM were identified (Whisson et al., 1998). Currently, both of these *Avr* clusters are being characterized using subcloning and complementation experiments.

A summary of all known clusters of *Avr* genes in *Phytophthora* is shown in Table 3.

It should be noted that in both *P. infestans* and *P. sojae*, avirulence did not always segregate as a dominant trait (Al-Kherb et al., 1995; Whisson et al., 1998). In some cases, avirulence was recessive in some but not all crosses, and in other cases aberrant segregation ratios were observed. It remains unclear at this stage whether this departure from Flor's hypothesis reflects a genetic basis of cultivar/race specificity in *Phytophthora* that sometimes does not fit the gene-for-gene model or whether these genes are subject to aberrant segregation in some crosses.

Table 3. Clusters of race-specific *Avr* genes in *Phytophthora*

Phytophthora spp.	*Avr* cluster	Plant spp./*R* gene	Reference
P. infestans	*Avr3/Avr10/Avr11*	*Solanum demissum* *R3/R10/R11*	Van der Lee et al. (1998)
P. sojae	*Avr1b/Avr1k*	*Glycine* *Rps1b/Rps1k*	Whisson et al. (1995), Arredondo et al. (1998)
P. sojae	*Avr4/Avr6*	*Glycine Rps4/Rps6*	Whisson et al. (1995), Gijzen et al. (1996)

3.1.2. Elicitins

A family of extracellular protein elicitors, termed elicitins, has been identified in *P. infestans* and other *Phytophthora* species and evidence has accumulated for a role of these molecules in delimiting the host-range of *Phytophthora* (Grant et al., 1996; Kamoun et al., 1998c; Yu, 1995). Elicitins are highly conserved 10-kDa proteins that are secreted by all *Phytophthora* species and several *Pythium* species (Huet et al., 1995; Kamoun et al., 1993b; Pernollet et al., 1993). A list of all known elicitins and elicitin-like proteins from *Phytophthora* is shown in Table 4. Elicitins induce defense responses, including the HR, on a restricted number of plants, specifically *Nicotiana* species within the Solanaceae family, and radish and turnips within the Cruciferae family (Bonnet et al., 1996; Kamoun et al., 1994). Recognition by plant cells is thought to be determined by the interaction of elicitins with a high-affinity binding site in the tobacco plasma membrane (Wendehenne et al., 1995; Yu, 1995).

Table 4. The elicitin family of *Phytophthora*

Phytophthora spp.	Elicitin	Class	Predicted cellular localization
P. cactorum	CAC-A	I-A (acidic)	Secreted
P. capsici	CAP-A (capsicein)	I-A (acidic)	Secreted
P. cinnamomi	CIN-B	I-B (basic)	Secreted
P. cryptogea	CRY-A1	I-A (basic)	Secreted
P. cryptogea	CRY-B (cryptogein)	I-B (basic)	Secreted
P. cryptogea	CRY-HAE20, CRY-HAE26	II (highly acidic)	Secreted
P. drechsleri	DRE-A, DRE-B	I-B (basic)	Secreted
P. infestans	INF1	I-A (acidic)	Secreted
P. infestans	INF2A, INF2B	III	Surface protein
P. infestans	INF4	Not yet classified	Secreted
P. infestans	INF5, INF6	Not yet classified	Surface protein
P. infestans	INF7	Not yet classified	Surface protein
P. megasperma	MGM-A	I-A (acidic)	Secreted
P. megasperma	MGM-B	I-B (basic)	Secreted
P. parasitica	PARA1 (parasiticein)	I-A (acidic)	Secreted
P. parasitica	PARA2	II (highly acidic)	Secreted
P. parasitica	PARA3	Not yet classified	Surface protein
P. sojae	SOJ-1, SOJ-2, SOJ-3, SOJ-4 (sojein)	I-A (acidic)	Secreted

The three-dimensional structure of cryptogein, the major basic elicitin of *Phytophthora cryptogea*, was determined using crystallography and nuclear magnetic resonance spectroscopy (Boissy et al., 1996; Fefeu et al., 1997; Gooley et al., 1998). Cryptogein displays on one side five loosely conserved alpha-helices and on the other side, a highly conserved beak structure formed by two antiparallel beta sheets and a Ω-loop. Based on the high amino-acid sequence similarity between different members of the elicitin family, this overall structure is likely to be conserved.

There are a number of experiments that suggest that elicitins function as avirulence factors in *Phytophthora*-plant interactions (Table 5). Treatment of tobacco plants with purified elicitins or expression of an elicitin gene in transgenic tobacco induces resistance to *Phytophthora parasitica* var. *nicotiana* (Ricci et al., 19989; Kamoun et al., 1993b; Keller et al., 1999; Tepfer et al., 1998). In *P.*

parasitica, the absence of elicitin production correlates with virulence on tobacco, a plant species that strongly responds to elicitins (Bonnet et al., 1994; Kamoun et al., 1994; Kamoun et al., 1993b; Ricci et al., 1989). Moreover, in sexual progeny of *P. parasitica*, elicitin production segregated with low virulence (Kamoun et al., 1994), suggesting that elicitins function as avirulence factors in *P. parasitica*-tobacco interactions (Kamoun et al., 1993a; Yu, 1995). Elicitin recognition has also been proposed to be a component of nonhost resistance of *Nicotiana* species to *P. infestans* (Kamoun et al., 1997b; Kamoun et al., 1998c). Conclusive direct evidence of the role of elicitins as avirulence factors came from functional analyses of the *P. infestans* elicitin gene *inf1*. Using a single step transformation procedure with an antisense construct of the *inf1* elicitin gene, stable *P. infestans* strains deficient in the production of INF1 elicitin were engineered (Kamoun et al., 1998c). Two of these strains showed increased virulence on the plant species *N. benthamiana*, indicating that the recognition of INF1 is a major determinant of the resistance response of *N. benthamiana* to *P. infestans*.

Elicitins can also induce resistance to pathogens other than *Phytophthora* (Table 5). Keller et al. (1999) generated transgenic tobacco plants containing a fusion between the pathogen inducible tobacco promoter hsr203J and the *P. cryptogea* cryptogein (*cry-B*) gene. These transgenic plants exhibited the HR under inducing conditions and showed enhanced resistance to the fungal pathogens *Thielaviopsis basicola*, *Erysiphe cichoracearum*, and *Botrytis cinerea*. Kamoun et al. (1993b) showed that coinfiltration of radish leaves with purified elicitin solutions and the bacterial pathogen *Xanthomonas campestris* pv. *armoraciae* resulted in reduced leaf spot symptoms and decreased growth of the pathogen. Functional expression of the *inf1* avirulence gene from engineered potato virus X (PVX) genome resulted in localized HR lesions on tobacco plants and inhibited spread of the engineered virus (Kamoun et al., 1999a). In contrast, a PVX construct producing an INF1 mutant form with reduced elicitor activity caused systemic necrotic symptoms, and was unable to inhibit PVX spread (Kamoun et al., 1999a). Taken together, these results demonstrate that the HR induced by elicitins is a highly versatile defense mechanism active against a number of unrelated pathogens.

Table 5. Experimental evidence for a role of elicitins as avirulence factors

elicitin	Plant tested	Experiment	Reference
CAP-A, cryptogein (CRY-B), PARA1	Tobacco	Application of purified elicitins to tobacco plants induced resistance to *Phytophthora parasitica* var. *nicotiana*	Ricci et al. (1989), Kamoun et al. (1993b)
CRY-B	Tobacco	Transgenic tobacco	Keller et al. (1999),

elicitin	Plant tested	Experiment	Reference
		plants producing cryptogein displayed enhanced resistance to *P. p.* var. *nicotiana*, as well as to the fungal pathogens *Thielaviopsis basicola*, *Erysiphe cichoracearum*, and *Botrytis cinerea*.	Tepfer et al. (1998)
CRY-B, PARA1	Radish	Application of purified elicitins to radish plants induced resistance to the bacterial pathogen *Xanthomonas campestris* pv. *armoraciae*	Kamoun et al. (1993b)
INF1	*Nicotiana benthamiana*	*P. infestans* strains silenced for the *inf1* gene were able to infect *N. benthamiana*	Kamoun et al. (1998c)
INF1	Tobacco	A recombinant potato virus X (PVX) expressing the *inf1* gene became avirulent on tobacco	Kamoun et al. (1999a)

Elicitin-like genes were isolated from *P. infestans* using PCR amplification with degenerate primers, low stringency hybridization techniques, and random sequencing of cDNAs (Kamoun et al., 1997a; Kamoun et al., 1999b; unpublished data). In total, seven elicitin and elicitin-like genes have now been identified in *P. infestans* (Table 4). All these genes encode putative extracellular proteins that share the 98 amino-acid elicitin domain that corresponds to the mature INF1. Major structural features of the elicitin domain (six cysteins, predicted secondary structure etc...) are conserved in the *inf* gene family. Five *inf* genes (*inf2A*, *inf2B*, *inf5*, *inf6*, and *inf7*) encode predicted proteins with a C-terminal domain in addition to the N-terminal elicitin domain. Sequence analysis of these C-terminal domains shows a high frequency of serine, threonine, alanine, and proline. The amino-acid composition and the distribution of these four residues suggest the presence of clusters of *O*-linked glycosylation sites (Kamoun et al., 1997a; Wilson et al., 1991). Interestingly, numerous surface and cell wall associated proteins consist of a signal peptide and a functional extracellular domain followed by a serine-threonine rich *O*-

glycosylated domain (Jentoft, 1990). Such proteins were shown to have a 'lollipop on a stick' structure in which the O-glycosylated domain forms an extended rod that anchors the protein to the cell wall leaving the extracellular N-terminal domain exposed on the cell surface (Jentoft, 1990). Therefore, these atypical INF proteins may be surface or cell wall associated glycoproteins that interact with plant cells during infection. Whether these elicitin-like genes encode active elicitor proteins and function as avirulence determinants remains to be determined.

The *P. infestans-Nicotiana* system is proving useful as a model system for dissecting the molecular components that determine nonhost resistance. In addition to the use of gene silencing technology for functional analysis of elicitin genes in *Phytophthora* (Kamoun et al., 1998c), the use of viral vectors, such as potato virus X (PVX), for expression of elicitin genes in *Nicotiana* (Kamoun et al., 1999a) should help in determining patterns of elicitor activity and specificity of the individual members of this diverse gene family and will allow the dissection of *Nicotiana* resistance into specific components (Kamoun, 1998).

The intrinsic biological function of elicitins in *Phytophthora* has long remained a mystery. Due to their high expression level in sporulating mycelium and abundant secretion, elicitins were thought to be structural proteins (Yu 1995; Kamoun et al., 1997b). However, more conclusive evidence came recently with the demonstration that these proteins can bind a number of sterols, such as dehydroergosterol, and may function as sterol-carrier proteins (Mikes et al., 1998; Mikes et al., 1997), a biological function of essential importance to *Phytophthora* spp. since they cannot synthesize sterols and must assimilate them from external sources (Hendrix, 1970). Functional genetic analyses and *in vivo* studies should help determine whether elicitins function indeed as sterol carriers in *Phytophthora*. In addition, it would be interesting to determine by structure-function mutagenesis studies whether or not the residues involved in sterol binding are distinct from those involved in interactions with tobacco receptor(s).

3.1.3. Other elicitors

Parsley is a nonhost of *P. sojae* and *P. infestans*. Following inoculation with *Phytophthora*, parsley cells exhibit a complex and coordinated series of morphological and biochemical defense responses that culminate into HR cell death (Hahlbrock et al., 1995; Naton et al., 1996; Somssich and Hahlbrock, 1998). An extracellular 42 kD glycoprotein elicitor from *P. sojae* or a 13 amino-acid oligopeptide (Pep-13) derived from this protein are sufficient to induce changes in plasma membrane permeability, an oxidative burst, activation of defense genes, and accumulation of defense compounds (Nurnberger et al., 1994). In addition to molecular signals, local mechanical stimulations, perhaps similar to those caused by the invading pathogen, induce some of the early morphological reactions and potentiate the response to the elicitor (Gus-Mayer et al., 1998). However, the signal(s) that lead to the HR in the parsley system remain unknown.

A 34-kDa glycoprotein elicitor (known as CBEL) was identified from *Phytophthora parasitica* var. *nicotianae*, a pathogen of tobacco (Sejalon-Delmas et al., 1997; Villalba Mateos et al., 1997). CBEL induced lipoxygenase (LOX) activity and accumulation of hydroxyproline-rich glycoproteins in tobacco (Sejalon-Delmas et al., 1997). A cDNA encoding CBEL was described (Villalba Mateos et al., 1997). The deduced amino acid sequence contained two direct repeats of a cysteine-rich domain, joined by a threonine/proline-rich region reminiscent of the cellulose-binding domain of fungal glycanases. However, CBEL did not show hydrolytic activity on a variety of glycans. Instead, CBEL showed lectin-like activity and bound purified fibrous cellulose as well as plant cell wall extracts suggesting a dual function for this protein.

3.1.4. P. mirabilis vs. P. infestans

Phytophthora mirabilis, a host-specific species closely related to *P. infestans*, infects *Mirabilis jalapa* (four-o'clock) but is unable to infect potato and tomato. Interspecific hybrids between these two *Phytophthora* species were essentially unable to infect the original host plants suggesting that avirulence on the nonhosts is dominant (Goodwin and Fry, 1994). In addition, in contrast to the parental strains, large HR-like necrotic lesions were induced by several of the hybrids on tomato indicating an alteration of the extent of the HR. Future genetic work could help identify the components of host-specificity in these interactions. Since both *P. infestans* and *P. mirabilis* naturally occur in central Mexico on different hosts and are reproductively isolated, Goodwin (1998) proposed that this is an example of sympatric speciation in *Phytophthora*.

3.2. CHEMOTAXIS

Zoospores of *Phytophthora* exhibit a number of tactic responses that allow them to reach infection sites in roots of host plants (Erwin et al., 1983). Electrotactic swimming of zoospores of *Phytophthora palmivora* has been reported and is thought to reflect attraction to weak electrical fields generated by roots (Morris and Gow, 1993). In addition, positive chemotaxis of zoospores towards plant derived compounds is well known. For example, zoospores of *P. sojae* are attracted to the isoflavones daidzein and genistein, which are exuded from the roots of soybean plants into the rhizosphere (Morris and Ward, 1992; Tyler et al., 1996). Morris and Ward (1992) observed that zoospores of other *Phytophthora* species are not attracted to soybean isoflavones suggesting that selective chemotaxis might play a role in host-range determination. Tyler et al. (1996) identified differences in the ability of a series of *P. sojae* strains to sense various compounds with structural similarity to soybean isoflavones. This led to the prospect of genetic analysis of chemotactic response in *P. sojae*.

In addition to zoospore chemotaxis, chemotropic and contact-induced responses have been noted for germinating cysts *of P. sojae* (Morris et al., 1998). The hyphal tips of germinating cysts detected and penetrated pores in artificial membranes and produced multiple appressoria on smooth, impenetrable surfaces (Morris et al., 1998). These processes may help the hyphae identify appropriate penetration sites on the root surface.

3.3. MATING TYPE DETERMINANTS

Heterothallic species of *Phytophthora* display two types of strains with either the A1 or A2 mating type. A genetic model of mating type inheritance was developed based on genetic analyses of mating type and molecular markers linked to the mating type locus of *P. infestans* and *P. parasitica* (Fabritius and Judelson, 1997; Judelson, 1996b; Judelson, 1996c). Apparently, heterozygosity (A/a) determines the A1 type and homozygosity (a/a) the A2 type. A number of abnormal genetic events, such as distorted segregation, nonrandom assortment of alleles, translocations, and duplications, occur at the mating type locus of several *P. infestans* strains (Judelson, 1996b; Judelson, 1996c; Judelson et al., 1995). Cosmid and BAC contigs containing the mating type locus were assembled (Judelson et al., 1998). Fine genetic mapping, complementation assays, and identification of expressed sequences in this region are under way and will help characterize the mating type locus.

3.4. GENOME STRUCTURE

The genome size of different *Phytophthora* species shows great divergence. For example, whereas the *P. sojae* genome was estimated at 62 Mb (Mao and Tyler, 1991), the *P. infestans* genome is thought to be 250 Mb (Tooley and Therrien, 1987). The haploid chromosome count of many *Phytophthora* species is not exactly known due to the occurrence of several small chromosomes that are difficult to resolve under light microscopy. The haploid chromosome number of *P. infestans* was estimated at n=8-10 by microscopy (Sansome and Brasier, 1973), and that of *P. sojae* at n=10-13 (Sansome and Brasier, 1974). Using pulse field gel electrophoresis, a total of eight chromosome-sized DNA bands were identified (Judelson et al., 1992b). However, based on intense ethidium bromide staining, at least two of these bands were suspected to be doublets or triplets (Judelson et al., 1992b).

One frequently ascribed feature to *Phytophthora* species is a high abundance of repetitive sequences. In *P. sojae*, five families of tandemly repeated sequences were identified following genomic substraction of chromosomal DNA from different isolates of the pathogen (Mao and Tyler, 1996). These sequences varied in copy number between the isolates and were all localized on single chromosomes of *P. sojae*. In *P. infestans*, repetitive DNA was estimated to cover at least 50% of the genome (Judelson and Randall, 1998). Following screening of a library of *P. infestans* for repetitive sequences, a total of 33 distinct families of repetitive DNA were discovered (Judelson and Randall, 1998). These elements were either tandemly repeated or dispersed throughout the *P. infestans* genome. Copy numbers varied from 70 to 8,400 per haploid genome. A number of these repeated sequences occurred in other distantly related *Phytophthora* species. However, some elements were specific to *P. infestans* and the closely related *P. mirabilis*.

Based on sequencing of a 60 kb BAC clone of *P. sojae*, a high density of genes was uncovered (Arredondo et al., 1998). Several pairs of genes are less than 300 bp apart. This led to the suggestion that in *P. sojae*, functional genes are located in high-density gene islands separated by clusters of repetitive sequences, which may constitute around 50% of the genome (Arredondo et al., 1998).

3.5. POPULATION GENETICS

Modern analyses of population genetics of *Phytophthora* using variation in isozymes and RFLP markers resulted in great insight into the population structures and macroevolution of several species (Fry et al., 1993; Goodwin et al., 1994; Drenth et al., 1994; Drenth et al., 1996; Goodwin, 1997). For example, based on DNA and isozyme fingerprint analyses, it was suggested that before the 1980s a single clonal lineage of the A1 mating type, termed US-1, dominated most populations of *P. infestans* worldwide (Goodwin et al., 1994). US-1 isolates are thought to have propagated by asexual reproduction from *P. infestans* strains introduced in the 1840s from Mexico to North America and later to Europe and the rest of the world (Fry et al., 1993; Goodwin et al., 1994; Fry and Goodwin, 1995; Fry and Goodwin, 1997a; Fry and Goodwin, 1997b; Goodwin, 1997). However, recent migration events, probably from Mexico, to Europe and North America of populations of *P. infestans* that include aggressive and A2 mating type strains led to the displacement of US-1 populations and the establishment of a sexual cycle in some localities (Fry et al., 1993; Drenth et al., 1994; Fry and Goodwin, 1995; Fry and Goodwin, 1997a; Fry and Goodwin, 1997b). Particularly devastating epidemics caused by the migrant strains were recently observed worldwide as late blight re-emerged as a serious threat to potato production worldwide (Anonymous, 1996; Fry and Goodwin, 1997a; Fry and Goodwin, 1997b).

Similar to *P. infestans*, the use of RFLP markers helped dissect the distribution of genetic variation in *P. sojae*. *P. sojae* appears to show a moderate degree of diversity (Forster et al., 1994; Drenth et al., 1996). A number of isolates with different virulence patterns on soybean plants containing *Rps* genes have identical or near identical RFLP patterns, suggesting that some race types may have arisen in clonal lineages (Forster et al., 1994). In addition, and based on the distribution of alleles among isolates within each group, some races of *P. sojae* may also have arisen by rare outcrosses in this homothallic species (Forster et al., 1994). Drenth et al. (1996) compared US and Australian populations of *P. sojae* using RFLP markers. Genotypic diversity was much lower in Australian populations of *P. sojae* (ranging from 2.5 to 14.3%) compared to US populations (60%), suggesting that *P. sojae* may have been established in Australia following a single introduction of the pathogen. In addition, all five races that occur in Australia appear to have emerged in clonal lineages.

The levels of analysis obtained using RFLP and isozyme markers remain limited particularly in populations showing little polymorphism, such as clonal lineages of *P. infestans* and *P. sojae*. The AFLP DNA fingerprinting technique, which was used to generate a high-resolution genetic map of *P. infestans* (van der Lee et al., 1997), could also be useful in achieving fine levels of genetic analysis of *Phytophthora* populations. For example, Kamoun et al. (1998a) identified a total of 49 polymorphic AFLP bands in a set of 18 US-1 isolates of *P. infestans* with no known isozyme or RFLP polymorphisms.

3.6. MOLECULAR PHYLOGENY

Classical taxonomy of the genus *Phytophthora* has been based on often-inconsistent morphological markers. This classification has been confusing and unsatisfactory. Recent studies re-examined the evolution of the genus using molecular markers, such as the Internal Transcribed Spacer (ITS) regions of rDNA. Cooke (1998) reported sequencing the ITS region of over 180 isolates representing 50 *Phytophthora* species. There was sufficient polymorphism in the ITS sequences to differentiate most species and to generate a coherent phylogenetic tree. The genus appeared monophyletic and showed less overall diversity than *Pythium*. However, the downy-mildews *Peronospora* showed strong affinity with *Phytophthora* raising the possibility that this group evolved from a biotrophic stock of *Phytophthora*. A good understanding of the phylogenetic relationships between *Phytophthora* species is essential for comparative biological studies and for designing sound disease management approaches.

A more detailed level of phylogenetic analysis can also be achieved in asexual populations of *Phytophthora*. For example, since *P. infestans* isolates of the US-1 genotype have propagated clonally by asexual reproduction, it is possible to reconstruct the phylogeny of these isolates given enough polymorphic molecular markers. Kamoun et al. (1998a) conducted a phylogenetic analysis on a set of 18 European US-1 isolates using markers generated by AFLP DNA fingerprinting. A phylogenetic tree was generated using parsimony analysis and 14 of the tested isolates were found to belong to three statistically significant branches within the US-1 lineage. Since five of the examined isolates were deficient in the production of the extracellular protein INF1 elicitin, this analysis helped determine whether this phenotype arose from one common mutant ancestor or whether loss of INF1 production occurred repeatedly in *P. infestans* (Kamoun et al., 1998a).

4. Molecular tools for genetic manipulation

4.1. DNA TRANSFORMATION

Stable DNA transformation is a prerequisite for genetic manipulation and functional analysis. Protocols and plasmids for transformation of oomycetes had to be developed from scratch as vectors available for transformation of filamentous fungi did not work in *Phytophthora* probably due to different sequence requirements for the transcriptional machinery (Judelson and Michelmore, 1991; Judelson et al., 1992a). A number of promoters, such as *ham34* and *hsp70*, from the oomycete downy-mildew *Bremia lactucae*, showed strong activity in transient transformation assays of *P. infestans* protoplasts (Judelson and Michelmore, 1991). These were then linked to genes encoding antibiotic resistance (*nptII* and hygromycin resistance genes) and successfully used as selection vectors for stable DNA transformation of various *Phytophthora* species. The standard transformation protocol is based on liposome-PEG mediated transformation of protoplasts, followed by regeneration of the protoplasts and antibiotic selection on agar medium (Judelson et al., 1991; Judelson et al., 1992b). High frequency rates of co-transformation (up to 50%) were observed especially if the two plasmids are linearized with restriction enzymes with compatible ends (Judelson, 1993). This finding turned out to be quite useful as the gene of interest can be rapidly cloned in convenient expression cassettes and co-transformed with the selection plasmid. So far, homologous recombination has not been detected and integration of the introduced DNA into the genome is thought to be through heterologous recombination. High rates of tandemly integrated plasmids are often observed. Overall, the rates of transformation remain limited for some applications, however, a single transformation experiment with *P. infestans* can now lead up to 200 independent transformants.

Several species, such as *P. infestans*, *P. sojae*, and *P. palmivora*, have been transformed using the PEG/liposome protoplast protocol (Judelson et al., 1991; Judelson et al., 1992b; P. van West, pers. comm.). The limiting step in transformation appears to be in the heterologous integration of the introduced plasmids and the low regeneration rate of the protoplasts. Recently, transformation of *P. sojae* via electroporation of zoospores, which naturally lack a cell wall, has been achieved (B. Tyler, pers. comm.). This could help in avoiding problems with low frequencies of protoplast regeneration and may ultimately result in improved frequencies of transformation.

4.2. REPORTER GENES

Several reporter genes including those encoding β-glucuronidase (GUS), luciferase, and the green fluorescent protein (GFP) have been used successfully in *Phytophthora* (Judelson, 1997b; van West et al., 1998; P. van West, pers. comm.). *Phytophthora* transformants expressing the GUS reporter gene have been used to monitor disease progression *in planta*, to evaluate disease resistance, to study promoter expression and to visualize morphological structures during mating (Judelson, 1997b; Kamoun et al., 1998b; van West et al., 1998). A transgenic *P. infestans* strain containing a transcriptional fusion between the promoter of the plant-induced *ipiO* gene and GUS proved useful in determining spatial patterns of expression of the *ipiO* promoter during infection of potato (van West et al., 1998). GUS staining was limited to the biotrophic stage of infection, particularly the edge of the invading hyphae suggesting that *ipiO* is highly expressed at that stage. Kamoun et al. (1998b) described the use of a *P. infestans* strain constitutively expressing high levels of GUS to measure fungal biomass and to estimate levels of general resistance of potato. However, due to the cost of GUS assays and the biosafety requirements for a transgenic strain, it is unclear whether using such GUS-tagged strains would confer significant advantages in routine evaluation of disease resistance in *P. infestans*-potato studies over using traditional methods for disease evaluation.

P. infestans strains expressing GUS were also used in studies on the mating process. By pairing strains containing a GUS transgene with nontransformed strains, Judelson (1997b) easily determined whether oospores resulted from hybridization or from self-interactions. Depending on the cross examined, 5 to 99% of the total oospores formed resulted from outcrossing. The use of the GUS marker also allowed determining levels of sexual preference, whether a strain is likely to act as a male or a female in a particular cross. A1 and A2 isolates behaved mainly as females and males, respectively. However, the sexual preference of a particular strain varied depending on its mating counterpart.

4.3. GENE SILENCING

Since no homologous recombination was detected following transformation of *Phytophthora*, a classical gene disruption approach was considered to have a low chance of success in this diploid organism. Therefore, attempts at targeted gene knockout followed the gene silencing approach that proved successful in plants. Following transformation of *P. infestans* with sense, antisense, and promoterless constructs of the endogenous single locus *inf1* gene, silencing of *inf1* was observed in up to 20% of the transformants (van West et al., 1999). Silencing was accompanied by the complete absence of *inf1* mRNA and INF1 protein and proved stable over repeated vegetative culture of the pathogen both *in vitro* and *in planta* (Kamoun et al., 1998c; van West et al., 1999). Nuclear run-on assays indicated that silencing is regulated at the transcriptional level (van West et al., 1999). No hypermethylation was observed in both transgenic and endogenous sequences of *inf1*. Due to the stability and total efficacy of this phenomenon, functional analyses could be performed with the silenced strains.

Efficient silencing of the *inf1* gene was also manifested in heterokaryotic mycelia, obtained after protoplast fusion of a transgenic-silenced strain and a non-silenced strain (van West et al., 1999). This observation suggests the involvement of a *trans*-nuclear silencing factor and rules out DNA-DNA interactions as the basis for gene silencing. Furthermore, homokaryotic wild-type strains, obtained following nuclear separation of silenced heterokaryotic strains, displayed stable gene silencing indicating that the presence of nuclear transgenic sequences was not essential to ensure and maintain silencing of the endogenous *inf1* gene. Apparently an inter-nuclear process, perhaps based on a *trans*-acting silencing factor, is responsible for the gene silencing phenomenon in the heterokaryotic *P. infestans* strains (van West et al., 1999). The facile transfer of gene silencing from one genetic background to another by passage through a heterokaryon should prove a very useful technology for constructing strains silenced for multiple genes.

Spontaneous silencing of transgenic GUS sequences was described in *P. infestans* (Judelson and Whittaker, 1995) and could be mechanistically related to the transcriptional silencing of the *inf1* gene. Following culturing, *P. infestans* strains transformed with GUS spontaneously lost all detectable GUS mRNA and activity. No correlation with changes in transgene structure or hypermethylation was detected. There is a tremendous variation in incidence of this spontaneous gene silencing phenomenon in *P. infestans*. For example an independent set of transgenic *P. infestans* strains containing the GUS gene failed to display spontaneous gene silencing (S. Kamoun, unpublished data). Thus it remains unclear whether spontaneous gene silencing is a major problem that can hamper stable expression of genes in *Phytophthora*.

4.4. GENETIC MAPPING

The development of comprehensive genetic and physical maps should greatly facilitate gene isolation from *Phytophthora* species. A detailed genetic map of *P. infestans* based on AFLP markers has been published (van der Lee et al., 1997). The data was generated from 73 F1 progeny from a cross between two homokaryotic and diploid isolates of *P. infestans*. A total of 183 AFLP markers, 7 RFLP markers and the mating type locus were mapped into 10 major and 7 minor linkage groups covering a total of 827 cM. More recently, a tight cluster of three avirulence genes, *Avr3*, *Avr10*, and *Avr11*, were placed on linkage group VIII of the AFLP map (van der Lee et al., 1998). A linkage map of *P. sojae* was constructed using 106 F2 individuals from two crosses (Whisson et al., 1995). The map was based on 22 RFLP markers, 228 RAPD, and 7 avirulence genes and covered 10 major and 12 minor linkage groups for a total of 830.5 cM. Progress toward a full physical map is under way. Arredondo et al. (1998) reported progress toward the construction of a BAC contig covering the entire 62 Mb genome of *P. sojae*.

4.5. GENOMICS

It is obvious that cDNA and genomic sequencing approaches can accelerate the genetic characterization of *Phytophthora* and other oomycetes. Understanding the genetic make-up of economically important species, such as *P. infestans* and *P. sojae*, promises to lead to novel approaches for disease control and management. Based on these premises, a number of *Phytophthora* geneticists initiated a collaborative genomics effort, known as the *Phytophthora* Genome Initiative (PGI; http://www.ncgr.org/pgi/index.html). Within, the framework of PGI, a pilot cDNA sequencing project was performed (Kamoun et al., 1999b). A total of 1,000 Expressed Sequence Tags (ESTs) corresponding to 760 unique sets of sequences were identified using random sequencing of clones from a cDNA library constructed from mycelial RNA of *P. infestans*. A number of software programs, represented by a relational database and an analysis pipeline, were developed for the automated analysis and storage of the EST sequence data. A set of 419 non-redundant sequences, which correspond to a total of 632 ESTs (63.2%), were identified as showing significant matches to sequences deposited in public databases. A putative cellular identity and role was assigned to all 419 sequence sets. All major functional categories were represented by at least several ESTs. Four novel cDNAs containing sequences related to elicitins were among the most notable genes identified. Two of these elicitin-like cDNAs were among the most abundant cDNAs examined (1.7% of all cDNAs).

A number of high throughput genomics projects are being initiated and include large scale EST sequencing of particular developmental and infection stages, BAC-end sequencing for the construction of BAC contigs, and targeted sequencing of BAC contigs of particular interest. There is no doubt that as with other organisms, the genetic study of *Phytophthora* will be totally transformed by the anticipated overflow of DNA sequence data.

4.6. FUNCTIONAL GENOMICS

Once a sizable amount of DNA sequence data is available for *Phytophthora*, there will be an essential need for robust assays to perform functional genetic analyses. Currently, gene silencing is the only proven technology to generate *Phytophthora* strains deficient in particular gene products (Kamoun et al., 1998c; van West et al., 1999). However, this approach has only been proven for a single gene so far, and it remains to be determined whether it will prove successful for a reasonable number of genes to allow its systematic application. Another disadvantage of gene silencing is a potential lack of specificity if a family of closely related genes is involved. In addition, depending on the gene examined, screening for silenced transformants can prove tedious and should be done preferably at the protein level, in addition to the RNA level. There are however a number of advantages for using gene silencing to generate strains with altered phenotype. For example, no specific plasmid needs to be constructed for the transformation experiment. van West et al. (1999) showed that a promoterless full length cDNA clone of the *inf1* gene could be used without modifications to transform *P. infestans* and generate silenced strains. However, the most exciting prospect of gene silencing for functional genomic analyses is the observed spread of the silenced state from silenced transgenic nuclei to non-transformed nuclei, when mixed together in a heterokaryotic strain (van West et al., 1999). This suggests that it might be sufficient to engineer silencing in a limited number of nuclei of a hyphae to ultimately silence the entire hyphae. One can easily devise rapid or transient transformation assays to test such an approach. Similar approaches are being contemplated in *Caenorhabditis elegans*, since it was observed that localized treatments of the worms with double-stranded RNA resulted in systemic gene silencing and resulted in a progeny silenced for the target gene (Fire et al., 1998).

Other methods for targeted gene knockout need also to be investigated for development of functional analysis tools for *Phytophthora*. Gene disruption through homologous recombination has only been examined superficially in *Phytophthora* and needs to be revisited with perhaps constructs containing large pieces of flanking DNA. Transposon mutagenesis, with either endogenous or ubiquitous transposons, is also a potentially useful technique and may also result in increased transformation frequencies, particularly if it turns out that frequency of transposition is higher than the low frequency of heterologous recombination. For both of these techniques, and considering that *Phytophthora* is a diploid at the vegetative stage, it will be more judicious to use homothallic species, such as *P. sojae* instead of a heterothallic species, such as *P. infestans*. This will allow facile selfing of the transformed strains to recover progeny that are homozygous at the mutated locus.

5. Conclusion

Research on *Phytophthora* has entered an exciting phase, with a noticeable increased discovery pace in recent years. In the coming years, continuous technological improvements as well as the expected impact of genomics should strengthen this trend and hopefully attract fresh talent to the field. Even though the classical biology of many *Phytophthora* species is well known (Erwin et al., 1983; Erwin and Ribeiro, 1996), there remains an urgent need for additional molecular and genetic studies. Hopefully, this will lead to a better understanding of the molecular genetics of these exceptional organisms and to novel approaches for management of the devastating diseases they cause.

6. Acknowledgments

I would like to thank my colleagues in the *Phytophthora* field for numerous discussions that undoubtedly sharpened my thoughts. Salaries and research support were provided by State and Federal Funds appropriated to the Ohio Agricultural Research and Development Center, the Ohio State University.

7. References

Al-Kherb, S. M., Fininsa, C., Shattock, R. C. and Shaw, D. S. (1995) The inheritance of virulence of *Phytophthora infestans* to potato, *Plant Pathol.* **44**, 552-562.

Anonymous (1996) *Late blight: A global initiative*, International Potato Center (CIP), Lima, Peru.

Arredondo, F., Shan, W. X., Chan, A., Hraber, P., Waugh, M., Sobral, B. and Tyler, B. M. (1998) Construction and DNA sequencing of a BAC contig spanning the *Phytophthora sojae* genome, *Abstract from "Advances in Phytophthora Molecular Genetics" Symposium*, British Society for Plant Pathology, Heriot-Watt University, Edinburgh.

Baker, B., Zambryski, P., Staskawicz, B. and Dinesh-Kumar, S. P. (1997) Signaling in plant-microbe interactions, *Science* **276**, 726-733.

Black, W. and Gallegy, M. E. (1957) Screening of *Solanum* species for resistance to *Phytophthora infestans*, *American Potato J.* **34**, 273-281.

Boissy, G., de La Fortelle, E., Kahn, R., Huet, J. C., Bricogne, G., Pernollet, J. C. and Brunie, S. (1996) Crystal structure of a fungal elicitor secreted by *Phytophthora cryptogea*, a member of a novel class of plant necrotic proteins, *Structure* **4**, 1429-39.

Bonnet, P., Bourdon, E., Ponchet, M., Blein, J.-P. and Ricci, P. (1996) Acquired resistance triggered by elicitins in tobacco and other plants, *Eur. J. Plant Pathol.* **102**, 181-192.

Bonnet, P., Lacourt, I., Venard, P. and Ricci, P. (1994) Diversity in pathogenicity of tobacco and in elicitin production among isolates of *Phytophthora parasitica*, *J. Phytopathology* **141**, 25-37.

Coffey, M. D. and Wilson, U. E. (1983) Histology and cytology of infection and disease caused by *Phytophthora*, in D. C. Erwin, S. Bartnicki-Garcia and P. H. Tsao (eds.), *Phytophthora*, Am.

Phytopathol. Soc., St. Paul, pp. 289-301.

Colon, L. T., Eijlander, R., Budding, D. J., van Ijzendoorn, M. T., Pieters, M. M. J. and Hoogendoorn, J. (1992) Resistance to potato late blight (*Phytophthora infestans* (Mont.) de Bary) in *Solanum nigrum*, *Solanum villosum* and their sexual hybrids with *Solanum tuberosum* and *Solanum demissum*, *Euphytica* **66**, 55-64.

Cooke, D. (1998) Whither *Phytophthora*? A revised classification of *Phytophthora* and other oomycetes on the basis of ITS analysis, *Abstract from "Advances in Phytophthora Molecular Genetics" Symposium* , British Society for Plant Pathology, Heriot-Watt University, Edinburgh.

Dangl, J. L., Dietrich, R. A. and Richberg, M. H. (1996) Death don't have no mercy: Cell death programs in plant-microbe interactions, *Plant Cell* **8**, 1793-1807.

Drenth, A., Tas, I. C. Q. and Govers, F. (1994) DNA fingerprinting uncovers a new sexually reproducing population of *Phytophthora infestans* in the Netherlands, *Eur. J. Plant Pathol.* **100**, 97-107.

Drenth, A., Whisson, S. C., Maclean, D. J., Irwin, J. A. G., Obst, N. R. and Ryley, M. J. (1996) The evolution of races of *Phytophthora sojae* in Australia, *Phytopathology* **86**, 163-169.

Enkerli, K., Hahn, M. G. and Mims, C. W. (1997) Ultrastructure of compatible and incompatible interactions of soybean roots infected with the plant pathogenic oomycete *Phytophthora sojae*, *Can. J. Bot.* **75**, 1493-1508.

Erwin, D. C. and Ribeiro, O. K. (1996) *Phytophthora Diseases Worldwide*, APS Press, St. Paul, Minnesota.

Fabritius, A. L. and Judelson, H. S. (1997) Mating-type loci segregate aberrantly in *Phytophthora infestans* but normally in *Phytophthora parasitica*: implications for models of mating- type determination, *Curr. Genet.* **32**, 60-65.

Fefeu, S., Bouaziz, S., Huet, J. C., Pernollet, J. C. and Guittet, E. (1997) Three-dimensional solution structure of beta cryptogein, a beta elicitin secreted by a phytopathogenic fungus *Phytophthora cryptogea*, *Protein Sci.* **6**, 2279-84.

Fire, A., Xu, S., Montgomery, M. K., Kostas, S. A., Driver, S. E. and Mello, C. C. (1998) Potent and specific genetic interference by double-stranded RNA in *Caenorhabditis elegans*, *Nature* **391**, 806-11.

Flor, H. H. (1956) The complementary genetic systems in flax and flax rust, *Adv. Genet.* **8**, 29-54.

Flor, H. H. (1971) Current status of the gene-for-gene concept, *Annu. Rev. Phytopathology* **9**, 275-296.

Forster, H., Tyler, B. M. and Coffey, M. D. (1994) *Phytophthora sojae* races have arisen by clonal evolution and by rare outcrosses, *Mol. Plant-Microbe Interact.* **7**, 780-791.

Freytag, S., Arabatzis, N., Hahlbrock, K. and Schmelzer, E. (1994) Reversible cytoplasmic rearrangements precede wall apposition, hypersensitive cell death and defense-related gene activation in potato/*Phytophthora infestans* interactions, *Planta* **194**, 123-135.

Fry, W. E. and Goodwin, S. B. (1995) Recent migrations of *Phytophthora infestans*, in L. J. Dowley, E. Bannon, L. R. Cooke, T. Keane and E. O'Sullivan (eds.), *Phytophthora infestans 150* , Boole Press Ltd., Dublin, pp. 89-95.

Fry, W. E. and Goodwin, S. B. (1997a) Re-emergence of potato and tomato late blight in the United States, *Plant Dis.* **81**, 1349-1357.

Fry, W. E. and Goodwin, S. B. (1997b) Resurgence of the Irish potato famine fungus, *Bioscience* **47**,

363-371.

Fry, W. E., Goodwin, S. B., Dyer, A. T., Matsuzak, J. M., Drenth, A., Tooley, P. W., Sujkowski, L. S., Koh, Y. J., Cohen, B. A., Spielman, L. J., Deahl, K. L., Inglis, D. A. and Sandlan, K. P. (1993) Historical and recent migrations of *Phytophthora infestans*: chronology, pathways, and implications, *Plant Dis.* **77**, 653-661.

Gees, R. and Hohl, H. R. (1988) Cytological comparison of specific (*R3*) and general resistance to late blight in potato leaf tissue, *Phytopathology* **78**, 350-357.

Gijzen, M., Forster, H., Coffey, M. D. and Tyler, B. (1996) Cosegregation of *Avr4* and *Avr6* in *Phytophthora sojae*, *Can. J. Bot.* **74**, 800-802.

Goodwin, S. B. (1997) The population genetics of *Phytophthora*, *Phytopathology* **87**, 462-473.

Goodwin, S. B. (1998) Probable sympatric speciation in *Phytophthora* mediated by changes in host specificity, *Abstract from "Advances in Phytophthora Molecular Genetics" Symposium* , British Society for Plant Pathology, Heriot-Watt University, Edinburgh.

Goodwin, S. B., Cohen, B. A. and Fry, W. E. (1994) Panglobal distribution of a single clonal lineage of the Irish potato famine fungus, *Proc. Natl. Acad. Sci, USA* **91**, 11591-11595.

Goodwin, S. B. and Fry, W. E. (1994) Genetic analyses of interspecific hybrids between *Phytophthora infestans* and *Phytophthora mirabilis*, *Exp. Mycol.* **18**, 20-32.

Gooley, P. R., Keniry, M. A., Dimitrov, R. A., Marsh, D. E., Keizer, D. W., Gayler, K. R. and Grant, B. R. (1998) The NMR solution structure and characterization of pH dependent chemical shifts of the beta-elicitin, cryptogein, *J. Biomol. NMR* **12**, 523-34.

Grant, B. R., Ebert, D. and Gayler, K. R. (1996) Elicitins- proteins in search of a role, *Australas. Plant Pathol.* **25**, 148-157.

Gross, P., Julius, C., Schmelzer, E. and Hahlbrock, K. (1993) Translocation of cytoplasm and nucleus to fungal penetration sites is associated with depolymerization of microtubules and defence gene activation in infected, cultured parsley cells, *EMBO J.* **12**, 1735-1744.

Gus-Mayer, S., Naton, B., Hahlbrock, K. and Schmelzer, E. (1998) Local mechanical stimulation induces components of the pathogen defense response in parsley, *Proc. Natl. Acad. Sci. U S A* **95**, 8398-403.

Hahlbrock, K., Scheel, D., Logemann, E., Nurnberger, T., Parniske, M., Reinold, S., Sacks, W. R. and Schmelzer, E. (1995) Oligopeptide elicitor-mediated defense gene activation in cultured parsley cells, *Proc. Natl. Acad. Sci. USA* **92**, 4150-4157.

Hendrix, J. W. (1970) Sterols in growth and reproduction of fungi, *Ann. Rev. Phytopathol.* 8, 111-130.

Hohl, H. R. and Suter, E. (1976) Host-parasite interfaces in a resistant and susceptible cultivar of *Solanum tuberosum* inoculated with *Phytophthora infestans*: leaf tissue, *Can. J. Bot.* **54**, 1956-1970.

Huet, J. C., Le Caer, J. P., Nespoulous, C. and Pernollet, J. C. (1995) The relationships between the toxicity and the primary and secondary structures of elicitinlike protein elicitors secreted by the phytopathogenic fungus *Pythium vexans*, *Mol. Plant Microbe Interact.* **8**, 302-310.

Jentoft, N. (1990) Why are proteins *O*-glycosylated?, *Trends Biochem. Sci.* **15**, 291-294.

Judelson, H. J. (1993) Intermolecular ligation mediates efficient cotransformation in *Phytophthora infestans*, *Mol. Gen. Genet.* **239**, 241-250.

Judelson, H. J. (1996a) Recent advances in the genetics of oomycete plant-pathogens, *Mol. Plant-Microbe Interact.* **9**, 443-449.

Judelson, H. J. (1997a) The genetics and biology of *Phytophthora infestans*: Modern approaches to a historical challenge, *Fun. Gen. Biol.* **22**, 65-76.

Judelson, H. J. and Michelmore, R. W. (1991) Transient expression of genes in the oomycete *Phytophthora infestans* using *Bremia lactucae* regulatory sequences, *Curr. Genet.* **19**, 453-459.

Judelson, H. J., Tyler, B. M. and Michelmore, R. W. (1991) Transformation of the oomycete pathogen, *Phytophthora infestans*, *Mol. Plant-Microbe Interact.* **4**, 602-607.

Judelson, H. J., Tyler, B. M. and Michelmore, R. W. (1992a) Regulatory sequences for expressing genes in oomycete fungi, *Mol. Gen. Genet.* **234**, 138-146.

Judelson, H. J. and Whittaker, S. L. (1995) Inactivation of transgenes in *Phytophthora infestans* is not associated with their deletion, methylation, or mutation, *Curr. Genet.* **28**, 571-579.

Judelson, H. S. (1996b) Chromosomal heteromorphism linked to the mating type locus of the oomycete *Phytophthora infestans*, *Mol. Gen. Genet.* **252**, 155-61.

Judelson, H. S. (1996c) Genetic and physical variability at the mating type locus of the oomycete, *Phytophthora infestans*, *Genetics* **144**, 1005-1013.

Judelson, H. S. (1997b) Expression and inheritance of sexual preference and selfing potential in *Phytophthora infestans*, *Fun. Genet. Biol.* **21**, 188-197.

Judelson, H. S., Coffey, M. D., Arredondo, F. R. and Tyler, B. M. (1992b) Transformation of the oomycete pathogen *Phytophthora megasperma* f. sp. *glycinea* occurs by DNA integration into single or multiple chromosomes, *Current Genetics* **23**,211-218

Judelson, H. S. and Randall, T. A. (1998) Families of repeated DNA in the oomycete *Phytophthora infestans* and their distribution within the genus, *Genome* **41**, 605-15.

Judelson, H. S., Randall, T. A. and Fabritius, A.-L. (1998) Classical and molecular genetics of mating in *Phytophthora*, *Abstract from "Advances in Phytophthora Molecular Genetics" Symposium* , British Society for Plant Pathology, Heriot-Watt University, Edinburgh.

Judelson, H. S., Spielman, L. J. and Shattock, R. C. (1995) Genetic mapping and non-Mendelian segregation of mating type loci in the oomycete, *Phytophthora infestans*, *Genetics* **141**, 503-12.

Kamoun, S. (1998) Dissection of nonhost resistance of *Nicotiana* to *Phytophthora infestans* using a potato virus X vector, *Phytopathology* **88**, S45.

Kamoun, S., Honee, G., Weide, R., Lauge, R., Kooman-Gersmann, M., de Groot, K., Govers, F. and de Wit, P. J. G. M. (1999a) The fungal gene *Avr9* and the oomycete gene *inf1* confer avirulence to potato virus X on tobacco, *Mol. Plant-Microbe Interact.* **12**,459-462.

Kamoun, S., Hraber, P., Sobral, B., Nuss, D. and Govers, F. (1999b) Initial assessement of gene diversity for the oomycete plant pathogen *Phytophthora infestans*, *Fun. Genet. Biol.* **28**,94-106.

Kamoun, S., Huitema, E. and Vleeshouwers, V. G. A. A. (1999c) Resistance to oomycetes: A general role for the hypersensitive response?, *Trends Plant Sci.* **4**,196-200.

Kamoun, S., Klucher, K. M., Coffey, M. D. and Tyler, B. M. (1993a) A gene encoding a host-specific elicitor protein of *Phytophthora parasitica*, *Mol. Plant-Microbe Interact.* **6**, 573-581.

Kamoun, S., Lindqvist, H. and Govers, F. (1997a) A novel class of elicitin-like genes from *Phytophthora infestans*, *Mol. Plant-Microbe Interact.* **10**, 1028-1030.

Kamoun, S., van der Lee, T., van den Berg, G., de Groot, K. E. and Govers, F. (1998a) Loss of production of the elicitor protein INF1 in the clonal lineage US-1 of *Phytophthora infestans*, *Phytopathology* **88**, 1315-1323.

Kamoun, S., van West, P., de Jong, A. J., de Groot, K., Vleeshouwers, V. and Govers, F. (1997b) A gene encoding a protein elicitor of *Phytophthora infestans* is down-regulated during infection of potato, *Mol. Plant-Microbe Interact.* **10**, 13-20.

Kamoun, S., van West, P. and Govers, F. (1998b) Quantification of late blight resistance of potato using transgenic *Phytophthora infestans* expressing beta-glucuronidase, *Eur. J. Plant Pathol.* **104**, 521-525.

Kamoun, S., van West, P., Vleeshouwers, V. G., de Groot, K. E. and Govers, F. (1998c) Resistance of *Nicotiana benthamiana* to *Phytophthora infestans* is mediated by the recognition of the elicitor protein INF1, *Plant Cell* **10**, 1413-26.

Kamoun, S., Young, M., Forster, H., Coffey, M. D. and Tyler, B. M. (1994) Potential role of elicitins in the interaction between *Phytophthora* species and tobacco, *App. Env. Microbiol.* **60**, 1593-1598.

Kamoun, S., Young, M., Glascock, C. and Tyler, B. M. (1993b) Extracellular protein elicitors from *Phytophthora*: Host-specificity and induction of resistance to fungal and bacterial phytopathogens, *Mol. Plant-Microbe Interact.* **6**, 15-25.

Keller, H., Pamboukdjian, N., Ponchet, M., Poupet, A., Delon, R., Verrier, J.L., Roby, D., and Ricci, P. (1999) Pathogen-induced elicitin production in transgenic tobacco genrates a hypersensitive response and nonspecific disease resistance, *Plant Cell* **11**, 223-236

Kraemer, R., Freytag, S. and Schmelzer, E. (1997) In vitro formation of infection structures of *Phytophthora infestans* in associated with synthesis of stage specific polypeptides, *Eur. J. Plant Pathol.* **103**, 43-53.

Kumar, S. and Rzhetsky, A. (1996) Evolutionary relationships of eukaryotic kingdoms, *J. Mol. Evol.* **42**, 183-93.

Lamb, C. J., Lawton, M. A., Dron, M. and Dixon, R. A. (1989) Signals and transduction mechanisms for activation of plant defenses against microbial attack, *Cell* **56**, 215-224.

Mao, Y. and Tyler, B. M. (1991) Genome organization of *Phytophthora megasperma* f.sp. *glycinea*, *Exp. Mycol.* **15**, 283-291.

Mao, Y. and Tyler, B. M. (1996) The *Phytophthora sojae* genome contains tandem repeat sequences which vary from strain to strain, *Fun. Gen. Biol.* **20**, 43-51.

Mikes, V., Milat, M. L., Ponchet, M., Panabieres, F., Ricci, P. and Blein, J. P. (1998) Elicitins, proteinaceous elicitors of plant defense, are a new class of sterol carrier proteins, *Biochem. Biophys. Res. Commun.* **245**, 133-9.

Mikes, V., Milat, M. L., Ponchet, M., Ricci, P. and Blein, J. P. (1997) The fungal elicitor cryptogein is a sterol carrier protein, *FEBS Lett.* **416**, 190-2.

Morris, B. M. and Gow, N. A. R. (1993) Mechanism of electrotaxis of zoospores of phytopathogenic fungi, *Phytopathology* **83**, 877-882.

Morris, P. F., Bone, E. and Tyler, B. M. (1998) Chemotropic and contact responses of *Phytophthora sojae* hyphae to soybean isoflavonoids and artificial substrates, *Plant Physiol.* **117**, 1171-8.

Morris, P. F. and Ward, E. W. B. (1992) Chemoattraction of zoospores of the soybean pathogen, *Phytophthora sojae*, by isoflavones, *Physiol. Mol. Plant. Pathol.* **40**, 17-22.

Munoz, C. I. and Bailey, A. M. (1998) A cutinase-encoding gene from *Phytophthora capsici* isolated by differential-display RT-PCR, *Curr. Genet.* **33**, 225-30.

Naton, B., Hahlbrock, K. and Schmelzer, E. (1996) Correlation of rapid cell death with metabolic changes in fungus-infected, cultured parsley cells, *Plant Physiol.* **112**, 433-444.

Nurnberger, T., Nennstiel, D., Jabs, T., Sacks, W. R., Hahlbrock, K. and Scheel, D. (1994) High affinity binding of a fungal oligopeptide elicitor to parsley plasma membranes triggers multiple defense responses, *Cell* **78**, 449-60.

Osbourn, A. (1996a) Preformed antimicrobial compounds and plant defense against fungal attack, *Plant Cell* **8**, 1821-1831.

Osbourn, A. (1996b) Saponins and plant defence- a soap story, *Trends Plant Sci.* **1**, 4-9.

Paquin, B., Laforest, M. J., Forget, L., Roewer, I., Wang, Z., Longcore, J. and Lang, B. F. (1997) The fungal mitochondrial genome project: evolution of fungal mitochondrial genomes and their gene expression, *Curr. Genet.* **31**, 380-95.

Pernollet, J.-C., Sallantin, M., Salle-Tourne, M. and Huet, J.-C. (1993) Elicitin isoforms from seven *Phytophthora* species: Comparison of their physico-chemical properties and toxicity to tobacco and other plant species, *Physiol. Mol. Plant Pathol.* **42**, 53-67.

Pieterse, C. M. J., Derksen, A. M. C. E., Folders, J. and Govers, F. (1994a) Expression of the *Phytophthora infestans ipiB* and *ipiO* genes *in planta* and *in vitro, Mol. Gen. Genet.* **244**, 269-277.

Pieterse, C. M. J., Riach, M. R., Bleker, T., van den Berg Velthuis, G. C. M. and Govers, F. (1993a) Isolation of putative pathogenicity genes of the potato late blight fungus *Phytophthora infestans* by differential screening of a genomic library, *Physiol. Mol. Plant Pathol.* **43**, 69-79.

Pieterse, C. M. J., Risseeuw, E. P. and Davidse, L. C. (1991) An *in planta* induced gene of *Phytophthora infestans* codes for ubiquitin, *Plant Mol. Biol.* **17**, 799-811.

Pieterse, C. M. J., van West, P., Verbakel, H. M., Brasse, P. W. H. M., van den Berg Velthuis, G. C. M. and Govers, F. (1994b) Structure and genomic organization of the *ipiB* and *ipiO* gene clusters of *Phytophthora infestans, Gene* **138**, 67-77.

Pieterse, C. M. J., Verbakel, H. M., Spaans, J. H., Davidse, L. C. and Govers, F. (1993b) Increased expression of the calmodulin gene of the late blight fungus *Phytophthora infestans* during pathogenesis on potato, *Mol. Plant-Microbe Interact.* **6**, 164-172.

Pristou, R. and Gallegly, M. E. (1954) Leaf penetration by *Phytophthora infestans, Phytopathology* **44**, 81-86.

Ricci, P., Bonnet, P., Huet, J.-C., Sallantin, M., Beauvais-Cante, F., Bruneteau, M., Billard, V., Michel, G. and Pernollet, J.-C. (1989) Structure and activity of proteins from pathogenic fungi *Phytophthora* eliciting necrosis and acquired resistance in tobacco, *Eur. J. Biochem.* **183**, 555-563.

Sansome, E. and Brasier, C. M. (1973) Diploidy and chromosomal structural hybridity in *Phytophthora infestans, Nature* **241**, 344-345.

Sansome, E. and Brasier, C. M. (1974) Polyploidy associated with varietal differentiation in the *megasperma* complex of *Phytophthora, Trans. Br. Mycol. Soc.* **63**, 461-467.

Schmelzer, E., Naton, B., Freytag, S., Rouhara, I., Kuester, B. and Hahlbrock, K. (1995) Infection-induced rapid cell death in plants: A means of efficient pathogen defense, *Can. J. Bot.* **73** (Suppl. 1), S426-S434.

Schmitthenner, A. F. (1985) Problems and progress toward control of Phytophthora root rot of soybean,

Plant Dis. **69**, 362-368.

Schmitthenner, A. F. (1989) Phytophthora rot, in J. B. Sinclair and P. A. Backman (eds.), *Compendium of soybean diseases* , APS Press, St. Paul, MN, pp. 35-38.

Schmitthenner, A. F., Hobe, M. and Bhat, R. G. (1994) *Phytophthora sojae* races in Ohio over a 10-year interval, *Plant Dis.* **78**, 269-276.

Sejalon-Delmas, N., Villalba Mateos, F., Bottin, A., Rickauer, M., Dargent, R. and Esquerre-Tugaye, M. T. (1997) Purification, elicitor activity, and cell wall localization of a glycoprotein from *Phytophthora parasitica* var. *nicotianae*, a fungal pathogen of tobacco, *Phytopathology* **87**, 899-909.

Somssich, I. E. and Hahlbrock, K. (1998) Pathogen defence in plants - a paradigm of biological complexity, *Trends Plant Sci.* **3**, 86-90.

Staskawicz, B. J., Ausubel, F. M., Baker, B. J., Ellis, J. G. and Jones, J. D. G. (1995) Molecular genetics of plant disease resistance, *Science* **268**, 661-667.

Tepfer, D., Boutteaux, C., Vigon, C., Aymes, S., Perez, V., O'Donohue, M.J., Huet, J.-C., Pernollet, J.-C. (1998) Phytophthora resistance through production of a fungal protein elicitor (beta-cryptogein) in tobacco, *Mol. Plant-Microbe Interact.* **11**,64-67.

Tooley, P. W. and Therrien, C. D. (1987) Cytophotometric determination of the nuclear DNA content of 23 Mexican and 18 non-Mexican isolates of *Phytophthora infestans*, *Exp. Mycol.* **11**, 19-26.

Tyler, B. M., Forster, H. and Coffey, M. D. (1995) Inheritance of avirulence factors and restriction fragment length polymorphism markers in outcrosses of the oomycete *Phytophthora sojae*, *Mol. Plant-Microbe Interact.* **8**, 515-523.

Tyler, B. M., Wu, M.-H., Wang, J.-M., Cheung, W. and Morris, P. F. (1996) Chemotactic preferences and strain variation in the response of *Phytophthora sojae* zoospores to host isoflavones, *Appl. Env. Microbiol.* **62**, 2811-2817.

Van de Peer, Y. and De Wachter, R. (1997) Evolutionary relationships among the eukaryotic crown taxa taking into account site-to-site rate variation in 18S rRNA, *J. Mol. Evol.* **45**, 619-30.

van der Lee, T., De Witte, I., Drenth, A., Alfonso, C. and Govers, F. (1997) AFLP linkage map of the oomycete *Phytophthora infestans*, *Fun. Gen. Biol.* **21**, 278-291.

van der Lee, T., Testa, A. and Govers, F. (1998) A high density map of an *Avr*-gene cluster in *Phytophthora infestans*, *Abstract from "Advances in Phytophthora Molecular Genetics" Symposium* , British Society for Plant Pathology, Heriot-Watt University, Edinburgh.

van West, P., de Jong, A. J., Judelson, H. S., Emons, A. M. C. and Govers, F. (1998) The *ipiO* gene of *Phytophthora infestans* is highly expressed in invading hyphae during infection, *Fun. Gen. Biol.* **23**, 126-138.

van West, P., Kamoun, S., van't Klooster, J. W. and Govers, F. (1999) Internuclear gene silencing in *Phytophthora*, *Mol. Cell* **3**,339-348.

Villalba Mateos, F., Rickauer, M. and Esquerre-Tugaye, M. T. (1997) Cloning and characterization of a cDNA encoding an elicitor of *Phytophthora parasitica* var. *nicotianae* that shows cellulose-binding and lectin-like activities, *Mol. Plant-Microbe Interact.* **10**, 1045-53.

Ward, E. W. B., Cahill, D. M. and Bhattacharyya, M. K. (1989) Early cytological differences between compatible and incompatible interactions of soybeans with *Phytophthora megasperma* f. sp. *glycinea*, *Physiol. Mol. Plant Pathol.* **34**, 267-283.

Wendehenne, D., Binet, M. N., Blein, J. P., Ricci, P. and Pugin, A. (1995) Evidence for specific, high-affinity binding sites for a proteinaceous elicitor in tobacco plasma membrane, *FEBS Lett.* **374**, 203-7.

Whisson, S. C., Drenth, A., Maclean, D. J. and Irwin, J. A. (1994) Evidence for outcrossing in *Phytophthora sojae* and linkage of a DNA marker to two avirulence genes, *Curr. Genet.* **27**, 77-82.

Whisson, S. C., Drenth, A., Maclean, D. J. and Irwin, J. A. (1995) *Phytophthora sojae* avirulence genes, RAPD, and RFLP markers used to construct a detailed genetic linkage map, *Mol. Plant-Microbe Interact.* **8**, 988-95.

Whisson, S. C., May, K. J., Drenth, A., Maclean, D. J. and Irwin, J. A. G. (1998) Genetics and cloning of avirulence genes from *Phytophthora sojae*, *Abstract from "Advances in Phytophthora Molecular Genetics" Symposium* , British Society for Plant Pathology, Heriot-Watt University, Edinburgh.

Wilson, I. B. H., Gavel, Y. and von Heijne, G. (1991) Amino acid distributions around *O*-linked glycosylation sites, *Biochem. J.* **275**, 529-534.

Yu, L. M. (1995) Elicitins from *Phytophthora* and basic resistance in tobacco, *Proc. Natl. Acad. Sci. USA* **92**, 4088-4094.

The Rust Fungi

Cytology, Physiology and Molecular Biology of Infection

Matthias Hahn

Department of Biology, Phytopathology, University of Konstanz

Universitätsstr. 10, D-78457 Konstanz, Germany

1. Introduction

Rust fungi (Basidiomycota; Order Uredinales) belong to the largest groups of fungi, with more than 6000 recognized species. Rust diseases have accompanied human history from the beginning of agriculture in the Fertile Crescent and Central America, by threatening the health and productivity of the earliest cereal (wheat, barley, maize) and legume (pea, bean) crops. In antiquity, the Romans have practiced the ceremony of the Robigalia, to sacrifice to and to appease Robigo, the god of cereal rust diseases. In the 17th century, French laws ordered the eradication of barberry bushes surrounding fields of wheat to avoid stem rust epidemics, before the causal agent of the disease (*P. graminis* f. sp. *tritici*) had even been recognized.

The life cycle of wheat stem rust on both plants was discovered by DeBary (1865), and finally clarified by Craigie (1927) who discovered the role of pycnia as sexual organs. The hypersensitive response, a form of programmed cell death that leads to the most effective resistance in plants, was described for the first time in 1915 by Stakman for the wheat-wheat stem rust interaction (Stakman, 1915). In 1905, resistance to a plant disease, yellow rust (*P. striiformis*) of wheat, was shown for the first time to be a single locus Mendelian character (Biffen, 1905). This observation, and the subsequent discovery of physiological races of wheat stem rust (Stakman and Piemeisal, 1917) initiated the era of science-based resistance breeding in cultivated crops.

Much of the early use of the electron microscope in plant pathology dealt with the cellular interaction between the haustoria-forming rust mycelium and infected host cells (Harder and Chong, 1984). Until today, few fungi have been studied in similar ultrastructural detail as the rusts. Flor's fundamental concept of gene-for-gene interactions governing race-cultivar specificity was developed using the interaction between flax and flax rust (Flor, 1956). This pathosystem has been successfully used recently for the cloning and molecular analysis of the host resistance genes (Luck *et al.*, 1998). Rusts are serious pathogens to many wild and commercially grown plants. In a recent report, a new epidemic caused by *P. distincta* was described that is highly destructive to cultivated and wild daisy (*Bellis perennis*) (Weber *et al.*, 1998). The

J. W. Kronstad (ed.), Fungal Pathology, 267–306.

diversity of parasitic strategies of rust fungi is beautifully exemplified by a recent analysis of *P. monoica*. Members of this rust species have been shown to mimic flower formation on their cruciferous host plants not only by appearance but also by the production of nectar rewards and the emission of floral fragrances (Raguso and Roy, 1998).

Despite detailed cytological knowledge, our understanding of the molecular mechanisms of rust infection lags considerably behind that of other plant pathogenic fungi, several of which are also treated in this book. Due to the parasitic specialization of rust fungi, we are unable to apply to them some of the most efficient tools of molecular biology. While few reports have described transient expression of DNA introduced by particle bombardment (Bhairi and Staples, 1992; Li *et al.*, 1993) or microinjection (Barja *et al.*, 1998), no stable DNA transformation system for rust fungi is in sight that would allow one to test gene function by targeted mutagenesis. Nevertheless, the application of modern techniques in molecular biology has started to deepen our understanding of rust biology.

This review will discuss the state of our knowledge on morphological, cytological, biochemical and molecular aspects of rust infection. It starts out by describing the early phase of infection by dikaryotic and monokaryotic rust spores, and then moves towards the events that lead to the establishment of a parasitic, biotrophic mycelium. Current ideas of the mechanism of host-parasite nutrient transfer are then discussed, including the role of haustoria in this process. At the end of this chapter, the physiology and the responses of susceptible and resistant plants to rust infections are described, and the possible relevance of fungal elicitors and suppressors. At several occasions, in particular when little information on important issues is available for rust fungi, I will mention examples from other, better characterized pathosystems that might shed light on similar mechanisms to be discovered in rust pathogenesis. A number of reviews on several aspects treated in this review are recommended for further reading: On contact sensing and early differentiation (Hoch and Staples, 1987; Read *et al.*, 1992), enzyme secretion (Deising *et al.*, 1995b), penetration (Mendgen *et al.*, 1996), cytology of infection (Mims, 1991; Mendgen and Deising, 1993; Heath and Skalamera, 1997), haustorium ultrastructure (Harder and Chong, 1991) and function (Heath and Skalamera, 1997), signal exchange (Heath, 1997b) and nutritional relationships (Farrar and Lewis, 1987). Beyond the scope of this review are taxonomical, diagnostic, epidemiological and phytomedical aspects related to rust fungi.

2. Infectious rust spores

Rusts have among the most complex life cycles of all fungi, with up to five spore types that are successively produced in macrocyclic species. While these are considered to represent the ancient forms, there are also many species that have various kinds of reduced life cycles in which one or several spore types are missing (Petersen, 1974; Littlefield, 1981).

Three of the five types of rust spores can infect host plants, namely uredo- and aecidiospores (dikaryotic) and basidiospores (monokaryotic, haploid). In the heteroecious rusts that need two different host plants to complete their life cycle, dikaryotic spores infect one and monokaryotic spores the other species. Although biogenesis and morphology of uredo- and aecidiospores are different, the infection behaviour of these spores was found to be very similar in the case of *Uromyces vignae* (Stark-Urnau and Mendgen, 1993). With few exceptions, they perform stomatal penetration, followed by formation of an intercellular mycelium with haustoria, and finally, sporulation (Fig. 1).

penetration   parasitic   sporulation
growth

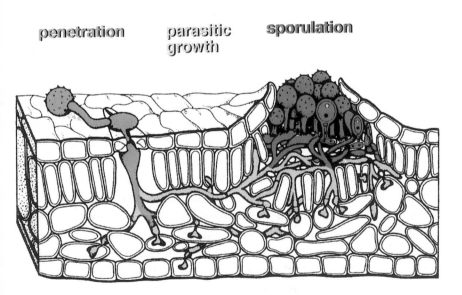

Figure 1: Schematic illustration of the infection cycle of dikaryotic rust spores.

In contrast, basidiospores usually penetrate directly, and the resulting mycelium appears to be morphologically less differentiated (see below). It can be expected that the differences in the developmental programs of dikaryotic and haploid spores are under the control of mating type genes, a situation that is similar to that in the other major group of plant pathogenic Basidiomycetes, the smut fungi, except that the latter can infect their host plant only in the dikaryotic phase (see chapter by Kahmann).

3. Spore and germ tube adhesion

Attachment and adhesion of fungal propagules to the host tissue is a prerequisite for successful infection. The first contact of the uredospore with the cuticular surface is stabilized by hydrophobic interactions (Clement *et al.*, 1993b; Terhune and Hoch, 1993; Clement *et al.*, 1994). The spines are usually the only parts of the uredospores that are involved in the early contact, leading sometimes to remarkable scanning electron microscopic pictures showing a cluster of spores being supported on the substratum by nothing but a few spines (Fig. 2; Clement *et al.*, 1994). Thereafter, humidity is required, for the hydration and swelling of the desiccated spore and the activation of metabolism, as well as for the release of adhesive spore surface components that greatly increase the area of contact to the substratum (Clement *et al.*, 1993a; Moloshok and Leinhos, 1993; Clement *et al.*, 1997). While mostly capillary water pads were observed underneath *U. fabae* spores 3 hours after incubation at saturating humidity (Clement *et al.*, 1994), rapid formation of adhesion pads containing extracellular material occured within 1 hour when spores were misted with water (Deising *et al.*, 1992). Adhesion pads were formed between neighbouring spores and between spores and hydrophobic surfaces such as teflon or broad bean leaves When formed on a leaf surface, adhesion pads also contain exudates from the host (Clement *et al.*, 1997). Adhesive material was also released from the germ pore shortly after germ tube emergence, and by the germ tube when it came in contact with the substratum (Clement *et al.*, 1994).

   Adhesive properties of uredospore germlings were analysed by treatments with hydrolytic enzymes and alkali. Proteases and ß-1,3-glucanases were shown to remove germlings of *U. appendiculatus* (Epstein *et al.*, 1987) and of *P. sorghi* (Chaubal *et al.*, 1991) from artificial surfaces, indicating that glycoproteins and ß-1,3-glucans are involved in adhesion. The composition of the extracellular material of spores of *U. fabae* was further studied by biochemical analyses. In spore washing fluids, proteins, neutral sugars and uronic acids were found, and significant activities of acid phosphatases, ß-glucosidases, proteases, and esterases including cutinase were detected (Deising *et al.*, 1992; Clement *et al.*, 1993a). Using hydrophobic interaction chromatography and polyacrylamide gel electrophoresis, a fractionation of various extracellular matrix proteins was achieved (Moloshok *et al.*, 1994). Treatment of spores with a serine esterase inhibitor, diisopropyl fluorophosphate, or autoclaving resulted in a strongly impaired adhesion to broad bean leaves. Addition of either complete spore washing fluids or esterase and cutinase fractions separated by native gel electrophoresis to autoclaved spores led to a significant increase in their adhesiveness (Deising *et al.*, 1992). While these data clearly support the involvement of esterases and cutinase in spore adhesion to the host cuticle, it is still unknown how this is achieved. As shown for powdery mildew conidia, cutinase activity might alter the properties of the leaf surface in a way that increases spore attachment (Nicholson *et al.*, 1993).

Figure 2: Attachment of a cluster of *Uromyces fabae* uredospores to the substratum by means of few spines (from Clement *et al.*, 1994, with permission).

A family of abundant cell wall proteins called hydrophobins has been identified in many higher fungi. By means of their ability to form amphipathic layers in fungal cell walls, they contribute to adhesion of hyphae to hydrophobic surfaces (Kershaw and Talbot, 1998). An important role of a hydrophobin in appressorium formation has been demonstrated for *Magnaporthe grisea* (Talbot *et al.*, 1996). A gene family (*esp*) that encodes small hydrophobic polypeptides was found to be highly expressed during germination of *P. graminis*, and was suggested to perform similar functions as hydrophobins (Liu *et al.*, 1993).

4. Factors affecting spore germination

It is well known that germination of uredospores is controlled not only by temperature and humidity (Kramer and Eversmeyer, 1992; Vallavielle-Pope *et al.*, 1995), but also by various chemicals, some of which are volatile. Derivatives of cinnamic and ferulic acid are present in the spores and prevent their premature germination in the uredosori (Macko, 1981). They are believed to inhibit dissolution of the germination pore(s) in the wall of the uredospore (Langen *et al.*, 1992). Stimulators of germination are also known, for example nonanal and ß-ionone, which are able to overcome the autoinhibitors and to induce precocious germination of uredospores within pustules on the host plant (French, 1992). Germination stimulators have been identified in rust uredospores, and are found in the oils and volatiles from host plants (French, 1992). Since teliospores are usually formed for the purpose of overwintering or sustaining of unfavorable periods, their germination often occurs slowly and only after a period of dormancy (Gold and Mendgen, 1983). Volatile stimulators of teliospore germination have also been identified (French et al., 1993). The mode of action of germination modulators remains to be investigated, and our general knowledge on the mechanisms and regulation of dormancy and germination of fungal spores is still very scarce.

5. Germ tube growth and infection structure formation

Basidiospores of U. vignae and P. xanthii penetrate within 8 hours after germination through the intact cuticle, with an appressorium that is morphologically not clearly distinguished from the germ tube (Gold and Mendgen, 1991; Morin et al., 1992). The stimulus for appressorium formation and infection structure differentiation seems to be the hardness of the underlying substratum (Freytag et al., 1988). Speed and efficiency of penetration are influenced by the thickness of the cuticle: U. vignae penetrates epidermal cell walls of the non-host Vicia faba, which are covered with a thinner cuticular layer than those of the host Vigna sinensis, more quickly and efficiently (Xu and Mendgen, 1991). Penetration is accompanied by modifications in the host cell wall, indicating the involvement of fungal cell wall degrading enzymes (see section 7). The penetration hypha is filled with microfilaments that seem to contribute to the shape of the invading hypha (Xu and Mendgen, 1994).

Rust uredospores and aecidiospores are unique in their sensing mechanisms governing the growth orientation of the germ tube, the finding of an stomatal opening and the subsequent differentation events leading to invasion of host tissue. Numerous studies have demonstrated that the major signals determining these events are provided by host leaf topography (Hoch et al., 1987a; Allen et al., 1991). Uredospore germlings are growing in close contact to the substratum, the hyphae being somewhat flattened (Kwon et al., 1991a). An accumulation of vesicles in the apex is responsible for continuous supply of membranes and cell wall precursors to the advancing germ tube tip. Actin filaments are in close association with the apical vesicles (Hoch et al., 1987b; Kwon et al., 1991b). Using a fluorescent dye that integrates into membranes, endocytosis and membrane turnover involving the apical vesicle cloud was observed in germ tubes of U. fabae (Hoffmann and Mendgen, 1998).

Within 2 to 4 min after apical contact with an inductive topography (a stoma or a ridge of about 0.5 μm height), the germ tube of U. appendiculatus responds by cessation of polarized growth and tip swelling (Kwon and Hoch, 1991). Microtubules and actin microfilaments that were mostly oriented parallel to the longitudinal axis of the non-induced germ tube, become progressively reoriented in the cell apex, with some filaments finally being aligned parallel to the ridge (Kwon *et al.*, 1991b). The apical vesicle cluster in the developing appressorium, together with this population of microtubules, is found in close proximity to the substratum (Kwon *et al.*, 1991a). Probably, the reorientation of the cytoskeleton is responsible for the subsequent changes in cell morphogenesis. Using mechanical stimulation with glass micropipettes of germ tubes at different sites, the most thigmo-responsive region was found within 10 μm from the cell apex (Corrêa and Hoch, 1995). Within one hour following the initial encounter with the topographical stimulus, the appressorium is fully differentiated and delineated from the cytoplasmatically emptied germ tube by a septum, and the first round of divisions of the two nuclei has been completed (Kwon *et al.*, 1991b).

While single artificial ridges of similar dimensions as French bean stomatal guard cell lips are highly inductive for appressorium formation of *U. appendiculatus*, the situation is different for the cereal rusts. Some of them, notably *P. graminis* f. sp. *tritici*, do not respond to single ridges but require a series of closely spaced artificial ridges on polystyrene replicas for efficient appressorium induction (Read *et al.*, 1997). This behaviour corresponds to the different architecture of the stomatal guard cells in grass leaves compared to dicot leaves. As a consequence, rust fungi specialized on dicots do not recognize grass stomata. For the cereal rusts, the consistently lower efficiency of artificial substrata to induce appressorium formation as compared to host leaves, led to the suggestion that additional, chemical factors (e.g. volatiles) are involved in the induction process (Grambow, 1977; Collins and Read, 1997).

Reduced efficiency of appressorium formation of *P. hordei* germ tubes was observed over the stomata of some genotypes of *Hordeum chilense*. The up to tenfold lower efficiency of stomatal penetration was shown to be due to a prominent wax layer over the guard cells of the 'avoidance' lines (Rubiales and Niks, 1996). Another example of reduced stomatal penetration as a component of quantitative resistance was described in the pathosystem wheat / *P. striiformis* (Broers and López-Atilano, 1996). These data demonstrate the potential feasibility of the breeding of crops that avoid rust infection by means of non-inductive surface topographies.

After induction of appressorium formation on artificial substrata, several *Uromyces* spp. sequentially form further infection structures such as penetration pegs, infection hyphae and haustorial mother cells, without further stimuli (Heath and Perumalla, 1988; Deising *et al.*, 1991). Under special conditions, even rudimentary haustoria can be induced in the absence of the host plant (Heath, 1990). For the formation of a fully differentiated parasitic mycelium with functional haustoria, however, yet unknown signals provided by the host plant are required.

6. Molecular events related to infection structure differentiation

A mechanosensitive cation channel was detected by patch clamp studies in protoplasts of *U. appendiculatus* germ tubes (Zhou *et al.*, 1991). It was proposed that this channel, which has a large conductance for monovalent and divalent cations including Ca^{2+}, transmits the topographical signal into the fungal cytoplasm, leading to oriented germ tube growth and appressorium differentiation. The nature of the signal transduction pathway is unknown in rust fungi. Hoch and Staples (1984) have shown that, by externally manipulating the cAMP levels of *U. appendiculatus* germ tubes, nuclear divisions and infection structure formation could be induced. This work is of interest in view of novel data obtained with several fungi pathogenic to plants, insects and man, that demonstrate a central role of cAMP in appressorium formation and pathogenic development (Kronstad, 1997; Alspaugh *et al.*, 1998). The involvement in appressorium induction of extracellular rust proteins similar to integrins (extracellular matrix proteins of mammalian cells) was suggested by studies performed with germ tubes of *U. appendiculatus*. Addition of peptides containing the sequence Arg-Gly-Asp, a motif that is bound by integrins, prevented their differentiation (Corrêa *et al.*, 1996).

Using highly synchronized uredospore germlings most of which were in the early stages of appressorium development, a differentiation-induced shift in the isoelectric point of a 21 kDa protein subsequently identified as superoxide dismutase was detected (Lamboy *et al.*, 1995). It is possible that post-translational modification (e.g. phosphorylation) of this enzyme is part of the series of events leading to appressorium formation.

Formation of uredospore-derived infection structures was found to be accompanied by the differential synthesis and degradation of specific proteins. In *U. appendiculatus*, 15 newly synthesized, differentiation-related proteins were identified (Staples and Hoch, 1988). In *U. fabae*, high resolution two-dimensional gel electrophoresis and silver staining revealed changes in 55 out of a total of 733 detectable protein spots during the course of germination and infection structure formation (Deising *et al.*, 1991). Genes that are activated during appressorium and infection structure formation have been isolated from *U. appendiculatus* (Bhairi *et al.*, 1989; Xuei *et al.*, 1992) and *U. fabae* (Deising *et al.*, 1995b). Comparison of their deduced amino acid sequences with current protein databases did not reveal significant similarity to known proteins. To analyse the role of one of these differentiation-specific genes (*Inf24*), a 24 bp antisense oligonucleotide towards *Inf24* was synthesized and microinjected into germ tubes of *U. appendiculatus*. Injection of this oligonucleotide was found to effectively block appressorium formation, whereas the sense oligonucleotide had no effect, indicating an essential role of *Inf24* in appressorium differentiation (Barja *et al.*, 1998). This is a rare example of the successful application of antisense oligonucleotides in fungi, but it is doubtful whether they are useful for studying gene function except in special cases.

Carbohydrate surface components of rust infection structures were analysed by staining with fluorescence-labeled lectins in combination with enzymatic and chemical treatments. Most obvious was a differential binding of several lectins to germ tubes and appressoria on one hand, and to penetration hyphae and infection vesicles (monokaryon) or infection hyphae and haustorial mother cells (dikaryon) on the other hand (Freytag and Mendgen, 1991). In contrast to the intense staining of germ tubes and appressoria by wheat germ agglutinin, this lectin showed a strongly reduced binding to those infection structures that grow within host tissue. This indicated that elicitor-active chitin in the fungal wall is either reduced or masked after penetration, possibly to avoid defense reactions of the host plant. Further information was provided by the observation that chitin deacetylase was released by infection structures of *U. fabae* in a similar time frame, beginning after appressorium formation (Deising and Siegrist, 1995). Whether deacetylation of wall chitin by this enzyme is responsible for reduced elicitor activity of the invading fungus and required for successful infection cannot be tested by a mutagenic approach with rust fungi. However, a lack of chitin on surfaces of infection hyphae has been observed in *Colletotrichum lagenarium* (O'Connell and Ride, 1990) and *C. graminicola* (Deising, personal communication). It is expected that the role of chitin deacetylase in fungal pathogenesis will be unraveled by cloning the corresponding gene(s) and the construction of disruption mutants with these fungi.

7. Host penetration and secretion of cell wall degrading enzymes

Like other plant pathogenic fungi, the rusts have to overcome the barrier of the host cell wall, but in a highly localized and nondestructive fashion in order to maintain the biotrophic relationship. Basidiospores penetrate epidermal cells directly and grow within and in between of host cells. In the case of dikaryotic rust hyphae, the only local cell wall penetration events occur during haustorium formation, with haustorium mother cells taking on the role of appressoria. The role of turgor pressure has been thoroughly documented for melanized appressoria formed by *Magnaporthe* and *Colletotrichum* spp. (see chapters by Kang et al. and Dickman). Dikaryotic rust appressoria develop turgor pressures that are one order of magnitude lower (approx. 0.35 MPa) than in appressoria of *Magnaporthe*. Nevertheless, this is still sufficient to deform stomatal guard cell lips and artificial membranes, and possibly to penetrate through closed stomata (Terhune *et al.*, 1993).

While secretion of extracellular enzymes occurs in all phytopathogenic fungi, the amounts released by rust fungi are very low and appear to be controlled mainly by development. The role of lytic enzymes in basidiospore penetration was studied by electron microscopy. Evidence for cell wall degrading activity was obtained by immunocytological studies using antibodies against pectin and xyloglucan (Xu and Mendgen, 1997). Within the epidermal wall, the zone surrounding the penetration hypha of *U. fabae* showed a significant reduction of epitopes representing methylated pectin and xyloglucan, whereas an increase in the number of polygalacturonic acid (pectate) epitopes was observed in the inner wall close to the penetration site. These results demonstrate that pectin esterase is released by the fungus, possibly to demethylate the wall pectin for subsequent depolymerization by pectate lyase (see below).

Secretion of host cell wall degrading enzymes during early differentiation of *U. fabae* uredosporelings was analysed by Deising and colleagues, using infection structures formed on scratched polyethylene membranes. In these studies, a remarkable correlation between differentiation and enzyme release was found. Whereas little exoenzyme activity was observed in germ tubes of any age, secretion started with the onset of appressorium formation and continued in a sequential and highly regulated fashion. Concomitantly with the appearance of appressoria (after 4 h), gelatin- and azocollagen-degrading proteases (Rauscher *et al.*, 1995) and various isoforms of cellulases (Heiler *et al.*, 1993) were detected. These enzymes increased in their activities during development of substomatal vesicles (>8 h), infection hyphae (>10 h) and haustorial mother cells (>12 h). Several pectin esterase isoenzymes started to rise during formation of substomatal vesicles (Frittrang *et al.*, 1992). Finally, a pectate lyase was detected at the time when haustorial mother cells were formed (Deising *et al.*, 1995a). In contrast to the situation in many other fungi, synthesis and secretion of cellulases and pectin esterases were not subject to substrate induction or glucose repression. Only in the case of pectate lyase, polygalacturonate, in addition to morphological differentiation, was required for enzyme induction. Based on their expression, and on the catalytic pH optima and the charges of these enzymes, a model for the mechanism of penetration of mesophyll cells during haustorium formation was developed (Deising *et al.*, 1995b). Briefly, the model states that during rust infection, which induces an alkalization of the apoplast (see section 12), the following events occur: At the penetration site, the activities of most of the lytic rust enzymes, as well as their ionic bonds to the negatively charged host cell wall, are progressively reduced. On the other hand, pectate lyase, with an isoelectric point above 10.5 and a similarly high pH optimum, remains firmly bound to the cell wall and becomes activated, which results in a highly localized breaching of the host cell wall (Deising *et al.*, 1995b). In addition, enzyme secretion might contribute to the intimate contact of intercellular hyphae to host cell walls. This could enable the fungus to compete more strongly for apoplastic host metabolites and to increase the efficiency of signal exchange.

8. Cytology of the biotrophic mycelium and haustoria

By using various light and electron microscopic techniques, including cytochemical stainings, lectin-gold and immunogold labeling, the ultrastructure and molecular components of parasitically growing rust hyphae and haustoria have been studied in great detail (Harder, 1984; Harder, 1989; Gold and Mendgen, 1991; Mendgen et al., 1991). Most obvious is a markedly different appearance of monokaryotic and dikaryotic mycelia. In contrast to dikaryotic mycelia, hyphae in the monokaryotic state appear less differentiated, and haustoria are morphologically similar to hyphae (Gold and Mendgen, 1991; Morin et al., 1992). Distinguishing features of various types of hyphae were often detected in the architecture of their walls, such as thickness, number and staining properties of layers, and molecular composition (Harder, 1989; Harder et al., 1989; Baka and Loesel, 1998). Further molecular differentiation of hyphal cells was recently achieved by immuno-localization of phase- and cell-specifically expressed proteins and carbohydrates (see below).

Haustoria, particularly in the dikaryotic phase, have a number of structural and molecular characteristics that are related to nutrient uptake (Harder and Chong, 1991). The extrahaustorial matrix represents a unique extracellular compartment that is separated from the apoplast by the haustorial neckband, a mineral-rich structure that connects host and fungal plasma membrane. The contributions of fungus and host to neckband synthesis are unknown. By preventing the flux of host metabolites from the matrix to the large volume of the apoplast, the neckband is believed to increase the efficiency of nutrient transfer from the infected host cell into the haustorium (see sections 10, 11). The extrahaustorial membrane (ehm), formed by invagination and resynthesis of the host plasma membrane during haustorium formation, differs from the non-invaginated domain of the plasma membrane in several respects. It has unique staining properties, increased thickness, high carbohydrate content, and it lacks intramembrane particles (Harder, 1989; Knauf et al., 1989; Harder and Chong, 1991). This lack might be correlated with the absence of ATPase activity (see section 11). Endocytosis via clathrin-coated pits was observed in the ehm of cowpea cells infected with monokaryotic haustoria of U. vignae (Stark-Urnau and Mendgen, 1995). Further evidence for the molecular differentiation of the ehm was obtained in the pea / powdery mildew system. Using a monoclonal antibody, a high molecular weight glycoprotein was found to be exclusively located in the ehm (Roberts et al., 1993). It is not yet known whether this glycoprotein is of plant or fungal origin. The cloning of genes encoding this or other ehm-specific proteins in interactions with haustoria-forming fungi would represent a major step forward towards an understanding of their targeting signals and of the unique molecular properties of the ehm.

In contrast to the plant plasma membrane, the ehm surrounding monokaryotic and dikaryotic haustoria does not form a regular cell wall, but seems to contribute to the adjacent extrahaustorial matrix (ema). In the ema, a variety of carbohydrates and glycoproteins similar to plant cell wall components were detected, such as cellulose, arabinogalactan proteins and hydroxyproline-rich glycoproteins (Harder and Chong, 1991; Stark-Urnau and Mendgen, 1995). However, some contribution by the pathogen to the ema, e.g. by secretion of extracellular proteins, appears to be likely. A technical problem associated with the cytological analysis of the host-parasite interface is a tendency of unspecific binding of antibodies to haustorial walls and the ema (Heller, 1995; Mendgen, unpublished observations).

In dikaryotic intercellular hyphae and haustoria, but not in germinated spores, the endoplasmic reticulum (ER) shows proliferations of partially inflated cisternae, called tubular-vesicular complexes (Welter *et al.*, 1988). Immunogold labeling revealed an increased concentration of proteins carrying the C-terminal ER retention signal –HDEL, such as BiP, a chaperone protein of the heat shock protein family (Bachem and Mendgen, 1995). It was suggested that the accumulation of ER in the parasitic mycelium represents an adaptation of the fungus to stress conditions experienced within the host plant. This hypothesis is supported by the observation that several plant-induced rust genes seem to encode stress defense-related proteins (see next section).

In monokaryotic infections, the rust mycelium frequently penetrates into the vascular system, a behaviour that is rarely observed in the dikaryon. Such infections are often systemic and can lead to abnormal growth or distortion of infected host tissue (Larous and Loesel, 1993). Apart from the regular hyphal morphologies known from most rust species, non-typical examples have been observed in colonies of tropical rust fungi: *P. gouaniae* produces haustorium mother cells that develop within wall-bound compartments of intercellular hyphae (Berndt, 1995), and several *Ravenelia* species form intracellular hyphae and haustoria in their dikaryophase (Berndt, 1997).

9. Differentiation and gene expression during biotrophic growth

The formation of rust infection structures in the absence of exogenously added nutrients suggests that metabolism in the initial phase of infection is fueled largely by the metabolic reserves of the uredospore. Transition to the biotrophic mode of nutrition is accomplished with the formation of the first haustorium within a host cell. While it is possible to study early infection structure development in the absence of the plant, analysis of the physiology, biochemistry and molecular biology of the parasitic phase is a very difficult task. Isolation of fungal haustoria from infected leaves as a tool for investigation of parasitic hyphae has been achieved first with pea powdery mildew (Gil and Gay, 1977). By raising monoclonal antibodies against isolated haustoria, haustorium-specific plasma membrane glycoproteins were identified (Mackie *et al.*, 1991; Mackie *et al.*, 1993). Taking this work as a model, methods for the isolation of rust haustoria have been published by Tiburzy *et al.* (1992), Cantrell and Deverall (1993), and Hahn and Mendgen (1992). Whereas the two former methods, similar to that described for powdery mildew haustoria, involve density gradient centrifugation of homogenates from rust-infected leaves, the method developed in our lab is based on the affinity of rust haustorial walls to the lectin concanavalin A. Using concanavalin A as a ligand coupled to large Sepharose beads, a highly efficient enrichment of haustoria was obtained. This method worked equally well for the isolation of haustoria from various rust fungi, but not from powdery mildews (Hahn and Mendgen, 1992; see below). By using a co-immunization protocol that increased the specificity of the immune response in mice against isolated haustoria of *Melampsora lini*, monoclonal antibodies were isolated that recognized three epitopes restricted to the haustorial cell wall (Hardham and Mitchell, 1998; Murdoch *et al.*, 1998). The loss of antibody binding after treatment with periodate, but not with pronase, indicated that the epitopes in the haustorial walls are carbohydrates.

Isolated haustoria from *U. fabae* were used for the isolation of mRNA and the construction of a haustorium-specific cDNA library (Hahn and Mendgen, 1997). Filter-bound clones from this library were sequentially hybridized to labeled cDNA probes from various stages of rust development. By this means, approximately 20% of the cDNAs were found to be preferentially or exclusively expressed in haustoria and rust-infected leaves. Clones representing 33 different *in planta*-induced rust genes (*PIGs*) were obtained. Some of the *PIGs* are highly expressed in haustoria, each representing more than 0.5% of the clones in the library (Hahn and Mendgen, 1997; Hahn, unpublished data; Table 1). DNA sequencing and database searching revealed similarities for less than half of these cDNAs to proteins encoded by known genes. The *PIGs* encoding putative nutrient transporters (*PIG2*, *PIG27* and *PIG33*) are discussed in the next section. *PIG1* and *PIG4*, two highly expressed genes which together comprise about 5% of all haustorial cDNAs, are homologs of yeast genes that are involved in thiamine biosynthesis. Their function was confirmed by complementation of the corresponding mutants of *Schizosaccharomyces pombe* (Sohn et al., submitted). This indicates that during parasitic growth, rust fungi have a large requirement for vitamin B1 which is apparently not supplied by the host plant. Interestingly, expression of *PIG1* and *PIG4* was not affected when thiamine was supplied to rust-infected leaves by infiltration. This is in striking contrast to the strong repression of their homologs in yeasts and *Aspergillus* by thiamine (Manetti *et al.*, 1994). The significance of most of the other *PIGs* remains speculative. It is intriguing that several of them encode proteins that have been shown to serve anti-stress functions in other organisms, namely *PIG11*, *PIG13*, *PIG32* (against oxidative stress), *PIG15* (against osmotic stress), and *PIG16* (against stress by host toxins; Hahn and Mendgen, 1997). These data are in line

M. Hahn

gene	% of haust. cDNA	postulated function	probability	copy number
PIG1	2.8	vitamin B1 synthesis	verified	1 − 2
PIG2	0.7	amino acid transporter	$1 \times e^{-111}$	1
PIG3	1.3	?	---	2 − 4
PIG4	2.4	vitamin B1 synthesis	verified	1
PIG5	2.0	?	---	1 - 3
PIG6	1.0	?	---	1 - 2
PIG7	0.7	?	---	?
PIG8	2.2	mannitol dehydrogenase	verified	1
PIG9	?	intermed. filament protein	$6 \times e^{-07}$	1 − 2
PIG10	0.6	?	---	2 - 3
PIG11	0.5	metallothionein	$2 \times e^{-02}$	1 − 2
PIG12	0.8	?	---	?
PIG13	?	metallothionein	$3 \times e^{-01}$	?
PIG14	0.8	?	---	?
PIG15	?	trehalose-6-P phosphatase	$2 \times e^{-08}$	1 − 2
PIG16	?	cyt P450 monooxygenase	$2 \times e^{-42}$	?
PIG17	?	?	---	?
PIG18	0.8	chitinase	$8 \times e^{-20}$	?
PIG19	?	?	---	?
PIG20	0.8	?	---	?
PIG21	?	?	---	?
PIG22	2.0	?	---	?
PIG23	?	?	---	?
PIG24	?	?	---	?
PIG25	?	?	---	?
PIG26	?	?	---	?
PIG27	?	amino acid transporter	verified	?
PIG28	?	peptidyl-prolyl isomerase	$5 \times e^{-61}$	1
PIG29	?	?	---	?
PIG30	?	?	---	?
PIG31	?	?	---	?
PIG32	?	anti-oxidant protein	$8 \times e^{-42}$	1
PIG33	0.8	glucose transporter	verified	1

Table 1: *In planta*-induced genes (*PIGs*), isolated from a haustorium-specific cDNA library of *Uromyces fabae* (Hahn et al. , 1997; Hahn, Sohn and Voegele, unpublished data). The functions were postulated based on sequence similarity to known genes. Probabilities were estimated using to the Blastx algorithm (Altschul et al., 1997); the lower the values, the more significant are the similarities. Functional verification was obtained by heterologous complementation of yeast mutants and *Xenopus* oocytes.

with the observed accumulation of stress-related ER proteins in parasitic rust hyphae (see previous section). However, application of various kinds of stresses to *in vitro* infection structures (e.g. addition of H_2O_2, NaCl or heavy metals) did not affect the expression of these genes (Köhler and Hahn, unpublished results).

Based on the results described above it was concluded that the transition of *U. fabae* from the early stages of infection to the biotrophic growth phase is accompanied by drastic changes in gene expression, and that this transition represents a major checkpoint in the developmental program of this fungus (Hahn and Mendgen, 1997). It is still unknown which signals from the plant, in addition to those that induce appressorium formation, are required to allow the fungus to perform and finish its development until sporulation. We hypothesize that, if all signals are provided in the correct spatial and temporal pattern, and if appropriate nutrients are available, development proceeds according to a largely predetermined program that is little affected by further environmental stimuli. This program includes the final stages of rust development, when sporulation occurs (Fig. 1).

10. Host-parasite nutrient transfer

In the following two sections, key questions relating to biotrophic nutrient transfer will be discussed, namely i) the nature of the transfered metabolites and ions, ii) nutrient concentrations and fluxes at the host-parasite interface, iii) the roles of haustoria and intercellular hyphae in nutrient uptake, and iv) mechanisms and components of trans-membrane transport.

Radioisotope labeling techniques have been used to identify metabolites that are transferred from host to parasite (reviewed by Mendgen, 1981). For instance, incubation of stem rust-infected wheat leaves with [3]H- and ^{14}C-labeled compounds, followed by analysis of labeled metabolites that had accumulated in uredospores, led to the conclusion that glucose (Pfeiffer *et al.*, 1969), several amino acids (Jäger and Reisener, 1969) and nucleosides (Bhattacharya and Shaw, 1967) are taken up by the rust fungus. Lysine and arginine appeared to be exclusively obtained from the plant and used without further metabolism. Instead, glutamate, alanine and glycine were taken up from the plant but also synthesized by the fungus (Jäger and Reisener, 1969; Reisener *et al.*, 1970). Feeding of infected bean leaves with radiolabeled amino acids and nucleosides, followed by autoradiography and electron microscopic analysis of the spatial distribution of radioactivity, demonstrated translocation of these metabolites to the rust mycelium. With this technique, ^{3}H-labeled lysine was found to be predominantly taken up by haustoria (Mendgen, 1979).

The establishment of axenic cultures from several rust species provided a tool for studying their metabolic capabilities in the absence of the host plant (Coffey and Shaw,

1972; Maclean, 1982; Williams, 1984). Although these cultures grow very slowly and usually only for a limited period, they can be maintained on defined media containing salts, sugars, amino acids and vitamins (Fasters *et al.*, 1993). This was the most convincing evidence against metabolic deficiencies being responsible for the obligate biotrophy of rust fungi, in contrast to bacterial parasites such as mycoplasmas that contain greatly reduced genomes lacking genes for most biosynthetic pathways (Razin *et al.*, 1998). ^{14}C-labeled sucrose was found to be extracellularly cleaved by invertase before being taken up. Within the mycelium, the pathways of conversion of ^{14}C-glucose into storage compounds, such as sorbitol, mannitol and trehalose was analysed by following the kinetics of movement of label into metabolites (Manners *et al.*, 1982; Manners *et al.*, 1984; Manners *et al.*, 1988). Sulfur had to be supplied in a reduced organic form, such as methionine and cysteine (Maclean, 1982). It remains to be shown to which extent axenic cultures provide an adequate system for the analysis of nutrient uptake and metabolism in the parasitic mycelium.

In leaves, apoplastic concentrations of sucrose, hexoses and amino acids are in the millimolar range, both in non-infected and rust-infected plants (Tetlow and Farrar, 1993; Lohaus *et al.*, 1995). These concentrations correspond to a 'very dilute culture medium' (Durbin, 1984) and are able to support growth of intercellularly growing fungal endophytes or parasites such as *Cladosporium fulvum* (see chapter by Oliver et al.). Unfortunately, data based on measurements with intercellular washing fluids cannot distinguish local differences in apoplastic metabolite concentrations, and they provide no data for the extrahaustorial matrix. Metabolite concentrations (or fluxes) are expected to be higher in the extrahaustorial matrix than in the apoplast, as suggested by the apparent sealing function of the neckband and the properties of the extrahaustorial membrane (see section 8).

For the powdery mildews that form a superficial mycelium from which haustoria penetrate into epidermal cell, the indispensable role of haustoria in nutrient uptake has been experimentally verified (Aist and Bushnell, 1991). In contrast, rust fungi (and downy mildews) could absorb nutrients either via intercellular hyphae from the apoplast, or via haustoria from the extrahaustorial matrix. In colonies of *P. hordei* on barley leaves, the surface area of the biotrophic mycelium was measured on serial sections. The contribution of haustoria to the total mycelial surface was found to be less than 20%. It was concluded that all the sugars required for fungal growth could be taken up either by haustoria or by intercellular hyphae (Kneale and Farrar, 1985). Molecular data supporting a nutritional role of haustoria are discussed in the next section.

11. Mechanisms of nutrient transfer

It is believed that the plasma membrane H^+-ATPase of parasite and host plays a key role in nutrient transfer. This enzyme is chiefly responsible for the generation of a trans-membrane electrochemical gradient, that is essential for the maintenance and regulation of ion homoeostasis and secondary active metabolite uptake (Serrano, 1989). A cytochemical method was used to assay ATPase activity at the fungus-plant interface. Fixed tissue sections were incubated with ATP and lead or cerium salts, and enzymatic release of inorganic phosphate was visualized by electron microscopy of heavy metal phosphate precipitates (Wachstein and Meisel, 1957). The reliability of this method has been questioned (Katz et al., 1988; Chauhan et al., 1991). It requires rigorous controls to exclude other phosphatases and non-specific precipitations, and care to avoid inactivation of the H^+-ATPase by fixatives and reagents. There is nevertheless no alternative method available yet with a comparable level of sensitivity and spatial resolution. Based on this method, an ATPase domain hypothesis has been proposed for plant cells invaded by rust and powdery mildew haustoria (Spencer-Phillips and Gay, 1981). It states that the cell wall-lined plant plasma membrane has mainly import functions, whereas the extrahaustorial membrane is destined for export of metabolites. The hypothesis was derived from the observation that, in contrast to the plasma membrane, the ehm lacked ATPase activity. Similar observations were made with dikaryotic haustoria of several rust fungi; in contrast, the ehm surrounding monokaryotic haustoria retained ATPase activity (except sometimes for the apical portion of the ehm), indicating a lower degree of cellular specialization in the monokaryon. Regarding the ATPase activity of the haustorial plasma membranes, variable results for different rust species were obtained (Baka et al., 1995). Solid immunocytological localization experiments are needed that confirm the postulated absence of host H^+-ATPase in the ehm, in order to get support for the long-held ATPase domain hypothesis by a different approach.

To study the role of the rust H^+-ATPase, enzyme assays were performed with membrane vesicles isolated from various stages of rust development. The activity of the H^+-ATPase was found to be several-fold higher in haustoria than in spores and germlings of U. fabae (Struck et al., 1996). This result supported the idea that the H^+-ATPase plays a special role in haustorial metabolite uptake. The enzyme was biochemically characterized in membrane fractions isolated from germlings (Struck et al., 1996), and from transgenic yeast cells in which both endogenous PMA genes had been replaced by the rust PMA1 gene (Struck et al., 1998). Northern analysis showed that the increased enzyme activity in haustoria was not correlated with increased transcript levels of PMA1. Instead, expression of a PMA1 derivative carrying a C-terminal deletion of 76 amino acids in yeast indicated that, similar to the plasma membrane ATPases of yeast and plants, the hydrophilic C-terminus of PMA1p could be involved in post-translational activation of the enzyme during biotrophic growth (Struck et al., 1998).

To further investigate the mechanism of nutrient uptake, molecular studies of the PIGs encoding putative metabolite transporters were performed. Northern analysis revealed that transcripts of PIG2 (amino acid transporter) and PIG33 (hexose transporter) were observed only in haustoria and in rust-infected leaves, but not in spores, germlings or infection structures formed in vitro (Hahn et al., 1997; Voegele and Hahn, unpublished). In contrast, PIG27 (amino acid transporter) mRNA was present in moderate amounts in germlings and infection structures, reaching their highest levels in haustoria (Fig. 3; Hahn et al., submitted). The cell-specific expression of the putative amino acid transporter PIG2p was confirmed by Western blot and immunofluorescence analysis, showing labeling only of haustorial plasma membranes (Fig. 4; Hahn et al., 1997). The functions of PIG27p and PIG33p were studied by heterologous expression of their cDNAs in yeast cells. PIG27 was shown to encode a transporter for amino acids by means of its ability to complement yeast strains defective in histidine and lysine uptake. Accumulation studies with ^{14}C-labeled histidine and the use of inhibitors provided evidence that PIG27p mediates active, proton symport-dependent transport (Hahn et al., submitted). Using a yeast mutant deficient in the uptake of hexoses, the function of PIG33p as a hexose transporter was also confirmed (Voegele and Hahn, unpublished). The yeast expression system now allows to study in detail the substrate specificities and transport properties of PIG27p and PIG33p.

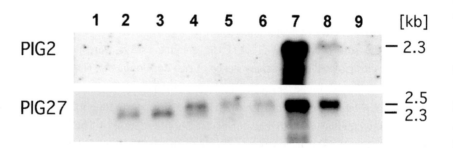

Figure 3: Northern hybridizations showing mRNA levels of amino acid transporter genes in U. fabae spores (1), germlings (2), in vitro grown infection structures (3: 6 h; 4: 12 h; 5: 18 h; 6: 24 h old), haustoria (7), rust-infected leaves (8) and non-infected leaves (9). DIG-labeled probes were used from the genes indicated on the left. Ten µg total RNA were loaded on each lane.

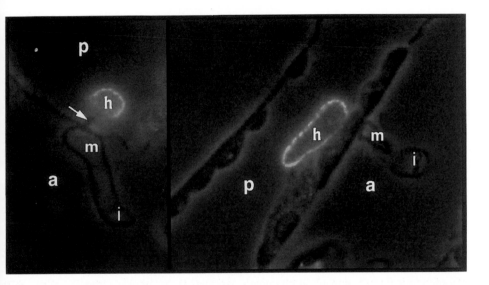

Figure 4: Localization of the PIG2p putative amino acid transporter in haustoria of *U. fabae* by immuno-fluorescence microscopy (from Hahn *et al.*, 1997, with permission). p: plant cell; a: apoplastic space; h: haustorium; m: haustorial mother cell; i: intercellular hypha.

From these results, the following conclusions can be drawn: i) Within the biotrophic rust mycelium, metabolite transporters are predominantly expressed in haustoria. This demonstrates a high degree of cellular specialization of the dikaryotic rust mycelium and supports the idea that haustoria are essential for efficient nutrient uptake. ii) The competence of the rust fungus for nutrient uptake is fully achieved not until the parasitic growth stage has been reached. iii) Nutrient uptake is active and dependent on the proton gradient formed by the plasma membrane ATPase. To confirm and refine this picture, further studies are required, in particular regarding the substrate specificity, regulation and localization of the transporters.

12. Physiology of rust-infected plants

Infections by rust fungi - as well as by other biotrophic pathogens - lead to profound changes in photosynthesis, respiration, metabolism, and metabolite movements in colonized host plants. The heterogeneity of infected tissue, consisting not only of invaded and non-invaded cell types, but also of neighbouring cells that are affected to different degrees by the presence of the parasite, considerably complicates analysis and interpretation of these changes in host physiology (Farrar and Lewis, 1987).

In barley leaves infected with *P. hordei*, transient increases in gross photosynthesis (oxygen released per unit chlorophyll) were observed (Owera *et al.*, 1983). This was accompanied by increased respiration (by host and parasite) in the diseased leaf tissue, resulting in reduced net photosynthesis compared to uninfected control leaves. Photosynthesis was considerably higher in pustules containing infected cells than in the leaf tissue between the pustules. Leaves of bluebell infected by *U. muscari* also showed a decrease in photosynthesis, but here, the decrease was confined to infected leaf areas (Scholes and Farrar, 1985). Another effect of rust infection, a progressive loss of chlorophyll, occured mainly in regions between fungal colonies in *P. hordei*-infected barley leaves (Scholes and Farrar, 1986). Within colonies, little changes were observed, and in those parts of the tissue that were invaded by the spreading mycelium, a regreening occured, leading to the formation of green islands (see below). Again, an inverse situation was found in rust-infected bluebell (Scholes and Farrar, 1985) and leek (Roberts and Walters, 1988), where loss of chlorophyll was found mainly within fungal pustules. In these studies, rates of photosynthesis were determined by the incorporation of $^{14}CO_2$. More recently, video-based quantitative imaging of chlorophyll fluorescence was introduced as a sensitive, non-destructive, and high-resolution measurement of photochemical and non-photochemical parameters of photosynthesis. With this technique, changes in photosynthesis of rust-infected French bean (Peterson and Aylor, 1995) and crown rust-infected oat (Scholes and Rolfe, 1996) were found to follow a complex spatial and temporal pattern during progression of these diseases.

Green islands, regions with increased chlorophyll surrounding infection sites, often appear in rust- or mildew-infected tissue at late stages of the diseases. They retain some photosynthetic activity, while in the surrounding non-infected regions, premature senescence occurs and photosynthesis drops to very low levels (Scholes and Rolfe, 1996). Green islands can be artificially induced by local application of cytokinins to senescing leaf tissue. Therefore, and since elevated levels of cytokinins and auxins have been measured in rust-infected leaves (Bushnell, 1984), their involvement in green island formation has been suggested. Until now, however, a causal relationship between phytohormones levels, green island formation and disease development has not been demonstrated.

Rust-infection sites show increased starch deposition, and they become powerful sinks for host assimilates (Bushnell, 1984). Infected leaves show a strongly reduced metabolite export, and an increasing proportion of assimilates is diverted to the pathogen. There is evidence that apoplastic invertases play a role in pathogenesis-induced starch accumulation and changes in source-sink relationships. By hydrolysing sucrose to glucose and fructose, they prevent assimilate export, since only sucrose but not hexoses can be loaded into the phloem. In addition, the regulation and tissue-specific expression of extracellular invertases in tomato indicate that they play a crucial role in sink induction (Godt and Roitsch, 1997). Fungi are known to possess uptake systems for hexoses, but do not seem to take up sucrose. In rust-infected plants, strongly increased invertase activities were measured (Long *et al.*, 1975; Greenland and Lewis, 1983). A crucial question is whether they are due to host or fungal invertases.

In stem-rust infected wheat, the increase in invertase activity correlated with the degree of colonization by the pathogen, and was therefore suggested to be due to the rust enzyme (Heisterüber *et al.*, 1994).

Following rust-infection, the pH's of leaf apoplastic fluids were found to be increased from 6.6 to 7.3 (barley / *P. hordei*; Tetlow and Farrar, 1993) and from 6.3 to 6.7 (broad bean / *U. fabae*; Deising *et al.*, 1995b), respectively. The pH increase could affect the activities of fungal cell wall degrading enzymes (see section 7) and promote redirection of sugar fluxes towards the fungus. A lower proton transmembrane gradient induced by the alkaline shift was proposed to decrease phloem loading of sucrose, and to activate fungal invertases (Tetlow and Farrar, 1993). The availability of plant mutants defective in apoplastic invertases (Tang *et al.*, 1999; Scholes, personal communication), and the purification and cloning of invertases from rust fungi (Moerschbacher, personal communication; Hahn and Voegele, unpublished results), will help to reveal origin and role of invertases during pathogenesis.

The accumulation of apoplastic hexoses at the infection site can have profound effects on host physiology, due to sugar sensing mechanisms of plants (Jang and Sheen, 1997). The suppression of photosynthetic genes by hexoses (Jang and Sheen, 1994) could be responsible for the decrease in photosynthesis following rust infection (see above). On the other hand, increased sugar concentrations can lead to the induction of host defense mechanisms. It has been oberserved that disease susceptibility of plants is dependent on the sugar content in leaf tissue. In so-called low-sugar diseases, a heightened resistance to pathogens was observed when the leaves contained elevated sugar levels (Horsefall and Dimond, 1957). In transgenic tobacco plants that overexpress yeast invertase either in the apoplast or the vacuole, systemic defense responses were found to be induced, in correlation to increased concentration of hexoses in the leaf tissue (Herbers *et al.*, 1996). Further evidence for the role of carbohydrate metabolism in pathogenesis comes from the observation that a gene of *Arabidopsis thaliana* encoding a monosaccharide transporter was induced by wounding, elicitor treatment and fungal infection (Truernit *et al.*, 1996). How these observations are related to the mechanisms of rust-induced changes in host physiology is a challenging question for future research.

13. Responses of susceptible and resistant plants to rust infection

For obligate biotrophic pathogens, a particular requirement for success is the ability to keep host cells alive. Lack of this ability leads to the hypersensitive reaction (HR), the most effective resistance response of plants that occurs in response to stress, elicitor treatment and infection by rust fungi and other pathogens (Heath, 1997a). Genetic and molecular analyses of *Arabidopsis* mutants have provided evidence for a highly regulated signal transduction pathway leading to HR (Richberg *et al.*, 1998). In compatible interactions, biotrophic pathogens have developed mechanisms to prevent triggering of this pathway, either by evading host detection or by suppressing signal transduction (Johal *et al.*, 1995).

The responses of plant cells to penetration by rust hyphae determine the outcome of the infection, resulting either in support of fungal growth or in rejection of the pathogen. Some responses are similar in compatible and incompatible rust infections, and therefore considered not to be crucial, whereas others occur only in resistant plants. In a cytological case study, the penetration of *U. vignae* basidiospores into epidermal cells of host (cowpea) and non-host (broad bean) plants was analysed (Xu and Mendgen, 1991). In the early phase of penetration (up to 8 h after inoculation), few differences were observed. The first visible response of epidermal cells occuring 4 to 6 hours after inoculation (during appressorium formation) was the formation of cytoplasmic aggregates. In both types of interactions, the plant nucleus migrated towards the penetration site but moved away shortly later. Papillae containing ß-1,3-glucans were observed at most infection sites in host and non-host cells. Close to the penetration site, endocytosis of ß-1,3-glucans by clathrin-coated vesicles was detected in non-host cells (Xu and Mendgen, 1994). After completion of intracellular hypha formation, the nucleus returned to the site of infection in the compatible but not in the incompatible interaction (Xu and Mendgen, 1991; Heath, 1997b).

From these and other studies, it can be concluded that the following host responses are unlikely to be determinative for the outcome of rust infections: i) formation of cytoplasmic aggregates (Xu and Mendgen, 1991), ii) migration of the nucleus towards the penetration site (Heath *et al.*, 1997), iii) deposition of callose (Skalamera and Heath, 1996; Skalamera *et al.*, 1997), iv) induction of apoplastic ß-1,3-glucanases and chitinases (Fink *et al.*, 1988; Münch-Garthoff *et al.*, 1997; Kemp et al., 1999), and v) accumulation and rearrangement of the ER around penetration hyphae and haustoria (Skalamera and Heath, 1995; Heath and Skalamera, 1997). Papillae, i.e. cell wall depositions composed mainly of ß-1,3-glucans, are also induced by purely mechanical stresses such as piercing with needles (Russo and Bushnell, 1989). They are effective barriers against powdery mildew penetration in barley carrying the *mlo* resistance gene, but do not seem to be important in rust infections (Skalamera and Heath, 1996). The physiological role of the other responses are not clear. With immunogold labeling, the ultrastructural localization of apoplastic ß-1,3-glucanase was studied in leaf rust-infected wheat leaves. The enzyme was detected in the host plasmalemma and in cell walls of host and fungi both in compatible and incompatible interactions, but differences in its subcellular locations were observed (Hu and Rijkenberg, 1998).

Some of the host responses seem to occur under the control and to the advantage of the parasite, such as the formation of the ehm and of ER-derived membranes surrounding haustoria. Evidence for parasite control of host responses comes from the observation that, in the same host plant, haustoria from different species of *Puccinia* trigger the formation of morphologically different membrane complexes (Harder and Chong, 1991). A gene that was tenfold induced in flax plants after compatible but not incompatible infection with flax rust was isolated (Roberts and Pryor, 1995). The gene encodes a protein with sequence motifs characteristic for aldehyde dehydrogenase. The authors suggest that it is induced by the rust fungus in the context of redirecting host metabolism in order to facilitate the biotrophic relationship.

Ageing of haustoria in compatible interactions is often accompanied by widening and more densely staining of the extrahaustorial matrix (Harder and Chong, 1991); see section 8). For monokaryotic haustoria of *U. vignae*, this was shown to occur by means of sequential secretion and deposition of additional layers from the ehm into the ema, containing callose, pectic components and fucose residues on xyloglucans (Mendgen *et al.*, 1995; Stark-Urnau and Mendgen, 1995). Encasement appears to represent a defense-related host response, resulting in a progressive loss of haustorial cytoplasmic integrity.

Plant defense responses that occur more rapidly and more pronounced in incompatible cultivars are candidates for being effective determinants of resistance. Most detailed studies on cytological changes correlated with host resistance have been performed in the laboratory of M. Heath, with the pathosystem *U. vignae* (monokaryon) / cowpea. As for other pathogenic interactions, strong evidence for a role of calcium in rust resistance-related signal transduction has been obtained. Using a calcium-sensitive fluorescent dye and confocal laser scanning microscopy, a rise in intracellular $[Ca^{2+}]$ was observed in living epidermal cells of an incompatible cowpea cultivar during penetration of the fungus through the cell wall. No such changes were observed in susceptible plants. Calcium channel inhibitors prevented the increase in $[Ca^{2+}]$ and delayed the onset of HR, indicating that elevated $[Ca^{2+}]$ is part of the signal transduction chain leading to hypersensitive cell death (Xu and Heath, 1998). Other early responses of resistant, but not susceptible, cowpea cells were increased numbers of nuclear pores and polyribosomes close to the penetration site. Since nuclear pore and polyribosome densities have been correlated with transcriptional and translational activity, respectively, in some aminal and plant systems, it was concluded that increased gene expression leading to HR is triggered during fungal penetration through the outer epidermal wall (Mould and Heath, 1999). In the same system, changes in the cytoskeletal organization of infected plant cells were similar in resistant and susceptible interactions, but treatments with anticytoskeletal drugs indicated that microtubules play a role in the deposition of phenolics, whereas microfilaments could be involved in the execution of HR (Skalamera and Heath, 1998). Later cellular events that accompanied the HR in cowpea cells were the cessation of cytoplasmic streaming and the cleavage of plant DNA into oligonucleosomal fragments (Ryerson and Heath, 1996). A similar type of DNA cleavage is a characteristic event during apoptosis of animal cells, indicating related mechanisms of the programmed cell death responses in plants and animals (Gilchrist, 1998).

In resistant wheat plants infected with *P. graminis*, the HR of mesophyll cells is frequently initiated shortly after the penetration peg of the haustorial mother cell has started to invaginate the host plasma membrane. Membrane damage occurs quickly, and the hypersensitively reacting cells display a bright yellow autofluorescence under UV light. Histochemical staining techniques and radiotracer experiments indicated that the fluorescence is due to the deposition of lignin-like material in these cells (Moerschbacher and Reisener, 1997). Enzymes involved in lignin biosynthesis (phenylalanine ammonium lyase, 4-coumarate: coenzyme A ligase, cinnamyl alcohol dehydrogenase, peroxidases) showed increased activities after infection in resistant plants. Treatments with inhibitors of these enzymes decreased the frequency of hypersensitive cell death, and led to increased growth of the incompatible fungi (Moerschbacher *et al.*, 1990a; Tiburzy and Reisener, 1990). Similar increases in lignification-associated enzymes have been described during defense responses of cowpea against non-pathogenic rust fungi (Fink *et al.*, 1991). In oat leaves, accumulation of diferulic acid was observed concomitantly with an increase in cell wall-bound peroxidase as a HR-associated response. At the same time, the cell walls of infected leaves became undigestible by cell wall degrading enzymes, presumably by peroxidase-mediated cross-linking of ferulic acids in the cell walls (Ikegawa *et al.*, 1996). In summary, these data strongly indicate that lignification-related processes are important components of the resistance response.

In oat leaves, HR was found to be associated with the accumulation of the phytoalexin, avenalumin. Crosses segregating for resistance to a specific race of *P. coronata* f. sp. *avenae*, accumulation of avenalumin and susceptibility against a host-specific toxin of the fungus *Cochliobolus victoriae*, victorin, indicated that all these responses are controlled by the same gene, Pc-2 (Mayama *et al.*, 1995; see chapter by Yoder and Turgeon). The recent discovery of the targets and the modes of action of victorin (Navarre and Wolpert, 1999) could lead to a molecular understanding of the unusual properties of a gene that seems to confer at the same time resistance to one fungus (crown rust) and susceptibility to another (*C. victoriae*).

In oat cells infected with incompatible isolates of *P. graminis*, the temporal patterns of accumulation of defence-related transcripts were compared to the occurrence of HR and the inhibition of fungal growth (Lin *et al.*, 1998). Defense-related genes were found to be activated in a sequential manner, but HR was observed only in a minority of infection sites. Since inhibition of fungal growth was found to be independent of plant cell death, it was concluded that HR was not required for successful defense in these interactions. Similarly, non-host resistance of various grasses against cereal rusts that resulted in the cessation of intercellular fungal growth was not always associated with death and autofluorescence of mesophyll cells (Luke *et al.*, 1987).

Taken together, HR of infected cells is a common and highly effective mechanism of resistance against rust fungi, but in addition, a variety of other resistance-related responses exist. These include the already mentioned avoidance of stomatal penetration (section 5; Broers and López-Atilano, 1996), encasement of haustoria, secretion of pathogenesis-related proteins (see above), silicon deposition in cell walls, and abnormal development of the extrahaustorial membrane (Heath and Skalamera, 1997). The role of pathogenesis-related proteins, such as ß-1,3 glucanases and chitinases (Münch-Garthoff et al., 1997) or thaumatin-like proteins (Lin et al., 1996) is often obscured by the fact that several isoforms exist that are differentially expressed, some being constitutively expressed, and others being induced to different degrees, either both in compatible and incompatible or only in incompatible interactions.

14. Induced resistance against rust fungi

Inoculations with virulent or avirulent pathogens or treatments with certain chemicals can lead to increased and often systemic resistance of genetically susceptible plants to a variety of pathogens (Stichter et al., 1997). While the phenomenon has been intensively studied in the last years with model plants such as tobacco and *Arabidopsis*, little is known about induced resistance against rust fungi, because these are no common pathogens of solanaceous and cruciferous plants. Recently, a study has been performed in which the mechanism of chemically induced resistance of broad bean against *Uromyces fabae* was investigated. Treatment of leaves with salicylic acid for one day led to the inhibition of fungal growth immediately after appressorial penetration through the stomatal pore. Consistent with this result was the observation that intercellular washing fluid prepared from induced resistant leaves inhibited rust differentiation *in vitro* after appressorium formation. Induced resistance correlated with the appearance of the extracellular pathogenesis-related protein PR-1, and PR-1 purified from resistant broad bean leaves strongly inhibited differentiation of *U. fabae* infection structures (Rauscher et al., 1999). These data strongly indicate that PR-1 is a major player in the induced defense of broad beans against rust fungi.

15. Elicitors and suppressors

Usually, plant defense responses leading to resistance are dependent on the recognition of the pathogen. Recognition occurs via elicitors, which are often surface components of the pathogen, for instance cell wall fragments. In compatible interactions, host defense responses are generally weak and do not inhibit fungal growth. Nevertheless, elicitor-active compounds appear to be universally present in plant pathogens, and have also been identified from a variety of rust fungi (Yamamoto, 1995).

The best characterized elicitor from rusts, the so-called Pgt elicitor, was isolated from germ tube walls of *P. graminis* f. sp. *tritici*, and shown to induce HR associated with lignification in wheat leaves (Kogel *et al.*, 1985). Treatment of wheat leaves with Pgt elicitor led to a different pattern of induction of lipoxygenase isoforms than treatment with other elicitors such as chitin and methyl jasmonate, indicating that independent elicitor-responsive signaling pathways exist in wheat (Bohland *et al.*, 1997). The Pgt elicitor was characterized as a 67 kDa peptidoglycan (Kogel *et al.*, 1988). Treatments with proteases, periodate and deglycosylating enzymes revealed that the elicitor-active component resided in the O-linked galactosyl side chains (Kogel and Beissmann, 1992). Elicitor activity was not cultivar specific, but dependent on the presence of a dominant gene on wheat chromosome 5A (Deverall *et al.*, 1994). Specific binding of purified, radiolabeled Pgt elicitor to wheat plasma membranes was demonstrated (Kogel *et al.*, 1991). If the gene residing on chromosome 5A encodes a receptor as suggested (Deverall *et al.*, 1994), binding of the elicitor should occur in a genotype-specific manner. By using monoclonal antibodies, the Pgt elicitor was localized in cell walls of intercellular hyphae and haustoria of *P. graminis* f. sp. *tritici*. (Tiburzy *et al.*, 1991; Marticke *et al.*, 1998). Elicitors with similar specificity as the Pgt elicitor were detected in germ tube extracts of *P. recondita* f. sp. *tritici*, and in intercellular washing fluids from rust-infected wheat leaves (Sutherland *et al.*, 1989). The physiological role of the Pgt elicitor remains unclear. Because it is permanently present during rust development, its failure to induce host defence responses in compatible interactions can be explained either by the lack of accessibility to the host receptor, or by negation of its activity in some way, e.g. by suppressors.

In contrast to the rather non-specific Pgt elicitor, a race-specific elicitor of necrosis, effective against the resistant cowpea cultivar Dixie Cream, was shown to be released by basidiospores of *U. vignae* race 1 (Chen and Heath, 1990). As judged by cytological and biochemical assays, the resistance response of Dixie Cream to elicitor treatment was similar to the response observed after inoculation with race 1 basidiospores (Chen and Heath, 1994). The elicitor was made by differentiated (containing intraepidermal vesicles) but not undifferentiated (containing germ tubes only) basidiospores, and it was also present in intercellular washing fluids obtained from basidiospore-infected cowpea leaves. In contrast, neither uredospore-derived infection structures nor intercellular washing fluids from race 1 uredospore-infected leaves released the elicitor (Chen and Heath, 1992). It is therefore likely that the elicitor is produced by the whole monokaryotic mycelium, but only by haustoria of the dikaryotic rust mycelium. This conclusion is in line with observations showing that resistance against dikaryon infection does not become effective until haustorium formation (Heath, 1989). In a remarkable effort, two elicitor-active peptides of 5600 and

5800 Da were purified from apoplastic fluids of race 1-infected cowpea leaves. Partial sequencing of the two peptides revealed that they are similar but encoded by different genes (D'Silva and Heath, 1997). Hopefully, this work will lead to the isolation of the fungal avirulence gene(s) and eventually to the identification of the corresponding host resistance gene. By using a microscopic assay for determination of leaf cell necrosis, evidence for the presence of race-specific elicitors in extracts from wheat leaves infected with *P. recondita* (dikaryon) was also obtained (Saverimuttu and Deverall, 1998).

Suppressors have been postulated to be involved in maintaining basic compatibility between biotrophic fungi and their host plants (Bushnell and Rowell, 1981). When extracts from rust-infected French bean leaves were injected into non-inoculated bean leaves, subsequent inoculation with the incompatible cowpea rust (*U. vignae*) resulted in an increased frequency of haustorium formation (Heath, 1981). Further evidence for the existence of suppressors is a phenomenon called induced susceptibility. It occurs when infection and haustorium formation by compatible powdery mildew or rust fungi permits penetration and development of avirulent pathogens in the immediate cellular vicinity (Fernandez and Heath, 1991; Adhikar and McIntosh, 1998). However, explanations other than the involvement of suppressors have been suggested for induced susceptibility (Yamaoka *et al.*, 1994), and experimental evidence for suppressors in rust infections is still preliminary (Moerschbacher *et al.*, 1990b; Beissmann and Kogel, 1991; Knogge, 1997).

Detailed molecular characterization of suppressors from fungi have been performed so far only with those from the non-biotrophic pea pathogen, *Microsphaerella pinodes*. These molecules, called supprescins, are glycopeptides that effectively suppress defense responses in host plants, probably by affecting the signal transduction pathway leading to defense gene activation (Wada *et al.*, 1995). Evidence for the molecular nature of a defense-suppressing mechanism has recently been described for the biotroph *Cladosporium fulvum*. Disruption mutants defective in the synthesis of extracellular proteins were shown to be severely impaired in pathogenicity, because of increased host defense responses that are possibly suppressed by these proteins in the wild type strain (Laugé *et al.*, 1997).

16. Concluding remarks

As described above, great progress has been made in the last decade in many aspects of rust pathogenesis. Nevertheless, due to our inability to genetically manipulate rust fungi, the knowledge about their interactions with plants has remained largely descriptive, except for the metabolically autonomous stages of early development. However, the use of transgenic plants offers a novel approach in various aspects of plant-fungus interactions, in particular with regard to our understanding of source-sink relationships, sugar sensing and metabolism, and resistance mechanisms. These are areas of rapidly advancing research in plant physiology that will significantly expand our understanding of fungal pathogenesis. For instance, the role of lignification during HR in the resistance against rust infection (see section 13) could be analysed by the construction of antisense mutants defective in lignin biosynthesis. A drawback is the fact that those plants that are easily transformable such as *Arabidopsis*, tobacco and tomato are not hosts to common rust fungi. To overcome this limitation, we have established in our laboratory a new pathosystem consisting of *Uromyces striatus* and *Medicago truncatula*, a legume that has been proposed as a new model plant to study plant-microbe interactions that do not occur with *Arabidopsis* (Cook, 1999; Hahn and Mendgen, unpublished data).

The remarkable progress made in the cloning of plant resistance genes have made it possible to study the molecular basis of gene-for-gene resistance to rust fungi. From the flax resistance gene L, thirteen alleles have been cloned. By comparing their sequences and their phenotypes, as well as those of recombinant alleles in transgenic plants, regions of the encoded proteins that determine race-specificity could be localized (Ellis *et al.*, 1999). If they interact directly with the products of the corresponding fungal avirulence genes, the latter could possibly be cloned by yeast two-hybrid screening using appropriate rust cDNA libraries. Progress has also been made in the physical mapping of rust resistance genes from other plants, e.g. wheat (Feuillet *et al.*, 1997), poplar (Villar *et al.*, 1996), pine (Wilcox *et al.*, 1996), and barley (Kilian *et al.*, 1997). Since there is evidence that avirulence genes do have significant roles for the fungal pathogen as well (Knogge, 1997; Laugé *et al.*, 1998), their cloning would be of great benefit for the understanding not only of the mechanisms of plant resistance but also of rust pathogenicity.

Our success in getting novel insights into the special ways rusts cause disease will depend on the effective use of modern experimental techniques, such as *in vivo* fluorescence imaging, immunocytology, electrophysiology, and comparative and functional genomics. In particular, genomic approaches bear a great potential in the near future. By sequencing large numbers of cDNAs or even complete genomes, and determining transcript levels from many or all genes simultaneously, metabolic activities can be monitored comprehensively. This approach has been demonstrated to be extremely powerful in yeast (DeRisi *et al.*, 1997) and other organisms including man, and is also highly promising for biotrophic plant / fungus interactions. With a haustorium-specific cDNA library of *U. fabae* (see section 9), an EST project has been initiated, with the goal of obtaining a large number of sequence-tagged clones, determining their putative functions by database searching, and analyzing gene expression during rust development by microarray hybridization (Hahn et al., unpublished data). While the majority of rust genes cloned so far have no similar counterparts in the current databases, the situation is likely to change within a few years when genomic sequences from other filamentous and plant pathogenic fungi become available (Dunn-Colemann and Prade, 1998). Although the function of rust genes cannot yet be proven by mutagenesis, targeted disruption of similar genes in related fungi (e.g. *Ustilago maydis*, see chapter by Kahmann), and complementation of such mutants with rust genes, could provide correlative evidence for their role in pathogenicity.

The most detailed knowledge on rust infection exists on the series of events associated with the induction and formation of appressoria (section 5). It would be of great interest to further understand the molecular components and the signaling mechanisms that are involved in the unique thigmosensing behaviour of dikaryotic rust sporelings. Regarding the role of haustoria, cytological and molecular data indicate that they are more efficient in the manipulation of host metabolism and absorption of host cellular pools of nutrients than intercellular hyphae. The abundant yields of spores produced by dikaryotic rust mycelia compared to the less differentiated monokaryotic rust mycelia and to strictly intercellularly growing fungi support the view that haustoria are required to support the very high nutritional demands during sporulation. Whether or not they are able to absorb specific metabolites from the extrahaustorial matrix that are not available in sufficient amounts in the apoplast (Heath and Skalamera, 1997) remains to be proven. While further studies on haustorium-specific gene expression will provide more answers to these questions, *in vivo* techniques are urgently required that are able to describe metabolite fluxes at a cellular level. Other roles of haustoria, such as intermediary metabolism and maintenance of the biotrophic relationship with the host plant, remain to be substantiated by molecular data. For instance, genes involved in the biosynthesis of vitamin B1 are strongly induced in haustoria (section 9), and haustorial cDNAs have been identified that encode proteins highly similar to enzymes involved in the synthesis of several amino acids (Hahn et al., unpublished data). This could mean that haustoria play an essential role not only in the uptake of host-derived nutrients but also in their transformation into essential metabolites.

We are still largely ignorant about the genetic basis of the different infection behaviour of mono- and dikaryotic rust mycelia. Many questions remain to be answered, such as: How are the mating type genes organized, how do they control mono- and dikaryon-specific gene expression, and how does this result in very different developmental programs? For instance, the penetration of monokaryotic hyphae through intact epidermal cells and into the vascular tissue gave rise to the speculation that they secrete larger amounts of cell wall degrading enzymes than dikaryotic hyphae (Larous and Loesel, 1993). Cloning of the corresponding genes and analysis of their expression in both nuclear phases could shed some light on the different kinds of parasitic growth by rust fungi.

Acknowledgements

I would like to thank my colleagues Christine Struck, Ralf Voegele and Stefan Wirsel for helpful comments on the manuscript. The author's work cited in this review was performed in the laboratory of Kurt Mendgen to whom I am grateful for his support.

References

Adhikar, K.N., and McIntosh, R.A. (1998) Susceptibility in oats to stem rust induced by co-infection with leaf rust, Plant Pathol. **47**, 420-426.

Aist, J.R., and Bushnell, W.R. (1991) Invasion of plants by powdery mildew fungi, and cellular mechanisms of resistance, in G. T. Cole and H. C. Hoch (eds.), *The Fungal Spore and Disease Initiation in Plants and Animals*, Plenum Press, New York, pp. 321-345.

Allen, E.A., Hazen, B.E., Hoch, H.C., Kwon, Y., Leinhos, G.M.E., Staples, R.C., Stumpf, M. A., and Terhune, B.T. (1991) Appressorium formation in response to topographical signals by 27 rust species, Phytopathology **81**, 323-331.

Alspaugh, J.A., Perfect, J.R., and Heitman, J. (1998) Signal transduction pathways regulating differentiation and pathogenicity of *Cryptococcus neoformans*, Fungal Genet. Biol. **25**, 1-14.

Altschul, S. F., Madden, T. L., Schaffer, A. A., Zhang, J., Zhang, Z., Miller, W., and Lipman, D. J. (1997) Gapped BLAST and PSI-BLAST: a new generation of protein database search programs, Nucl. Acids Res. **25**, 3389-402.

Bachem, U., and Mendgen, K. (1995) Endoplasmic reticulum subcompartments in a plant parasitic fungus and in baker's yeast: Differential distribution of lumenal proteins, Exp. Mycol. **19**, 137-152.

Baka, Z.A.M., Larous, L., and Loesel, D.M. (1995) Distribution of ATPase activity at the host-pathogen interfaces of rust infections, Physiol. Mol. Plant Pathol. **47**, 67-82.

Baka, Z.A.M., and Loesel, D.M. (1998) Ultrastructure and lectin-gold cytochemistry of the interaction between the rust fungus *Melampsora euphorbiae* and its host, *Euphorbia peplus*, Mycol. Res. **102**, 1387-1398.

Barja, F., Corrêa Jr., A., Staples, R.C., and Hoch, H.C. (1998) Microinjected antisense *Inf*24 oligonucleotides inhibit appressorium development in *Uromyces*, Mycol. Res. **102**, 1513-1518.

Beissmann, B., and Kogel, K.H. (1991) Identification and characterization of suppressors, in H. F. Linsken and J. F. Jackson (eds.), *Plant Toxin Analysis*, Springer, Berlin, pp. 259-275.

Berndt, R. (1995) The parasitic interaction of *Puccinia gouaniae* (Uredinales), Sydowia **47**, 129-137.

Berndt, R. (1997) Morphology of haustoria of *Ravenelia* and *Kernkampella* spp., Mycol. Res. **101**, 23-34.

Bhairi, S.M., and Staples, R.C. (1992) Transient expression of the ß-glucuronidase gene introduced into *Uromyces appendiculatus* uredospores by particle bombardment, Phytopathology **82**, 986-989.

Bhairi, S.M., Staples, R.C., Freve, P., and Yoder, O.C. (1989) Characterization of an infection structure-specific gene from the rust fungus *Uromyces appendiculatus*, Gene **81**, 237-243.

Bhattacharya, P. K., and Shaw, M. (1967) The physiology of host-parasite relations. 18. Distribution of tritium-labelled cytidine, uridine and leucine in wheat leaves infected with the stem rust fungus, Can. J. Bot. **45**, 555-563.

Biffen, R.H. (1905) Mendel's laws of inheritance and wheat breeding, J. Agric. Sci. **1**, 4-48.

Bohland, C.,T.,B., Loers, G., Feussner, I., and Grambow, H.J. (1997) Differential induction of lipoxygenase isoforms in wheat upon treatment with rust fungus elicitor, chitin oligosaccharides, chitosan, and methyl jasmonate, Plant Physiol. **114**, 679-685.

Broers, L.H.M., and López-Atilano, R.M. (1996) Effect of quantitative resistance in wheat on the development of *Puccinia striiformis* during early stages of infection, Plant Dis. **80**, 1265-1268.

Bushnell, W.R. (1984) Structural and physiological alterations in susceptible host tissue, in W. R. Bushnell and A. P. Roelfs (eds.), *The Cereal Rusts. Volume 1.* Academic Press, Orlando, pp. 477-507.

Cantrell, L.C., and Deverall, B.J. (1993) Isolation of haustoria from wheat leaves infected by the leaf rust fungus, Physiol. Mol. Plant Pathol. **42**, 337-343.

Chaubal, R., Wilmot, V.A., and Wynn, W.K. (1991) Visualization, adhesiveness and cytochemistry of the extracellular matrix produced by germ tubes of *Puccinia sorghi*, Can. J. Bot. **69**, 2044-2054.

Chauhan, E., Cowan, D.S., and Hall, J.L. (1991) Cytochemical localization of plasma membrane ATPase activity in plant cells. A comparison of lead and cerium-based methods, Protoplasma **165**, 27-36.

Chen, C.Y., and Heath, M.C. (1992) Effect of stage of development of the cowpea rust fungus on the release of a cultivar-specific elicitor of necrosis, Physiol. Mol. Plant Pathol. **40**, 23-30.

Chen, C.Y., and Heath, M.C. (1990) Cultivar-specific induction of necrosis by exudates from basidiospore germlings of the cowpea rust fungus, Physiol. Mol. Plant Pathol. **37**, 169-177.

Chen, C.Y., and Heath, M.C. (1994) Features of the rapid cell death induced in cowpea by the monokaryon of the cowpea rust fungus or the monokaryon-derived cultivar-specific elicitor of necrosis, Physiol. Mol. Plant Pathol. **44**, 157-170.

Clement, J.A., Butt, T.M., and Beckett, A. (1993a) Characterization of the extracellular matrix produced *in vitro* by urediniospores and sporelings of *Uromyces viciae-fabae*, Mycol. Res. **97**, 594-602.

Clement, J.A., Martin, S.G., Porter, R., Butt, T.M., and Beckett, A. (1993b) Germination and the role of extracellular matrix in adhesion of urediniospores of *Uromyces viciae-fabae* to synthetic surfaces, Mycol. Res. **97**, 585-593.

Clement, J.A., Porter, R., Butt, T.M., and Beckett, A. (1994) The role of hydrophobicity in attachment of urediniospores and sporelings of *Uromyces viciae-fabae*, Mycol. Res. **98**, 1217-1228.

Clement, J.A., Porter, R., Butt, T.M., and Beckett, A. (1997) Characteristics of adhesion pads formed during imbibition and germination of urediniospores of *Uromyces viciae-fabae*, Mycol. Res. **101**, 1445-1458.

Coffey, M.D., and Shaw, M. (1972) Nutritional studies with axenic cultures of the flax rust, *Melampsora lini*, Physiol. Plant Pathol. **2**, 37-46.

Collins, T.J., and Read, N.D. (1997) Appressorium induction by topographical signals in six cereal rusts, Physiol. Mol. Plant Pathol. **51**, 169-179.

Cook, D.R. (1999) Medicago truncatula – a model in the making!, Curr. Opin. Plant Biol. **2**, 301-304.

Corrêa, A.J., and Hoch, H.C. (1995) Identification of thigmoresponsive loci for cell differentiation in *Uromyces* germlings, Protoplasma **186**, 34-40.

Corrêa, J.A., Staples, R.C., and Hoch, H.C. (1996) Inhibition of thigmostimulated cell differentiation with RGD-peptides in *Uromyces* germlings, Protoplasma **194**, 91-102.

Craigie, J.H. (1927) Discovery of the function of the pycnia of the rust fungi, Nature **120**, 765-767.

D'Silva, I.D., and Heath, M.C. (1997) Purification and characterization of two novel hypersensitive response-inducing specific elicitors produced by the cowpea rust fungus, J. Biol. Chem. **272**, 3924-3927.

DeBary, A. (1865) Neue Untersuchungen über die Uredineen, insbesondere die Entwicklung der *Puccinia graminis* und den Zusammenhang derselben mit *Aecidium berberidis*, Monatsbericht Königl. Preuss. Akad. Wiss. Berlin, pp. 15-50.

Deising, H., Frittrang, A.K., Kunz, S., and Mendgen, K. (1995a) Regulation of pectin methylesterase and polygalacturonate lyase activity during differentiation of infection structures in *Uromyces viciae-fabae*, Microbiology **141**, 561-571.

Deising, H., Jungblut, P.R., and Mendgen, K. (1991) Differentiation-related proteins of the broad bean rust fungus *Uromyces viciae-fabae*, as revealed by high resolution two-dimensional polyacrylamide gel electrophoresis, Arch. Microbiol. **155**, 191-198.

Deising, H., Nicholson, R.L., Haug, M., Howard, R. J., and Mendgen, K. (1992) Adhesion pad formation and the involvement of cutinase and esterases in the attachment of uredospores to the host cuticle, Plant Cell **4**, 1101-1111.

Deising, H., Rauscher, M., Haug, M., and Heiler, S. (1995b) Differentiation and cell wall degrading enzymes in the obligately biotrophic rust fungus *Uromyces viciae-fabae*, Can. J. Bot. **73**, S624-S631.

Deising, H., and Siegrist, J. (1995) Chitin deacetylase activity of the rust *Uromyces viciae-fabae* is controlled by fungal morphogenesis, FEMS Microbiol. Lett. **127**, 207-212.

DeRisi, J.L., Iyer, V.R., and Brown, P.O. (1997) Exploring the metabolic and genetic control of gene expression on a genomic scale, Science **278**, 680-686.

Deverall, B.J., Saverimuttu, N., Cantrill, L.C., and McIntosh, R.A. (1994) Genetic control of responsiveness of wheat to elicitors in intercellular washing fluids from leaf rust-infected leaves, Physiol. Mol. Plant Pathol. **45**, 189-194.

Dunn-Colemann, N., and Prade, R. (1998) Toward a global filamentous fungus genome sequencing effort, Nature Biotechnol. 16: 5.

Durbin, R.D. (1984) Effects of rust on plant development in relation to the translocation of inorganic and organic solutes, in W. R. Bushnell and A. P. Roelfs (eds.), *The Cereal Rusts. Volume 1*, Academic Press, Orlando, pp. 509-528.

Ellis, J.G., Lawrence, G.J., Luck, J.E., and Dodds, P.N. (1999) Identification of regions in alleles of the flax rust resistance gene L that determine differences in gene-for-gene specificity, Plant Cell **11**, 495-506.

Epstein, L., Laccetti, L.B., Staples, R.C., and Hoch, H.C. (1987) Cell-substratum adhesive protein involved in surface contact responses of the bean rust fungus, Physiol. Mol. Plant Pathol. **30**, 373-388.

Farrar, J.F., and Lewis, D.H. (1987) Nutrient relations in biotrophic infections, in G. F. Pegg and P. G. Ayres (eds.), *Fungal Infection of Plants*, Cambridge University Press, pp. 92-132.

Fasters, M.K., Daniels, U., and Moerschbacher, B.M. (1993) A simple and reliable method for growing the wheat stem rust fungus, *Puccinia graminis* f. sp. *tritici*, in liquid culture, Physiol. Mol. Plant Pathol. **42**, 259-265.

Fernandez, M.R., and Heath, M.C. (1991) Interactions of the non-host French bean plant (*Phaseolus vulgaris*) with parasitic and saprophytic fungi. IV. Effect of preinoculation with the bean rust fungus on growth of parasitic fungi nonpathogenic on beans, Can. J. Bot. **69**, 1642-1646.

Feuillet, C., Schachermayr, G., and Keller, B. (1997) Molecular cloning of a new receptor-like kinase gene encoded at the Lr10 disease resistance locus of wheat, Plant J. **11**, 45-52.

Fink, W., Haug, M., Deising, H., and Mendgen, K. (1991) Early defence responses of cowpea (*Vigna sinensis* L.) induced by non-pathogenic rust fungi, Planta **185**, 246-254.

Fink, W., Liefland, M., and Mendgen, K. (1988) Chitinases and ß-1,3-glucanases in the apoplastic compartment of oat leaves (*Avena sativa* L.), Plant Physiol. **88**, 270-275.

Flor, H.H. (1956) The complementary gene systems in flax and flax rust, Adv. Genet. **8**, 29-54.

French, R.C. (1992) Volatile chemical germination stimulators of rust and other fungal spores, Mycologia **84**, 277 - 88.

French, R.C., Nester, S.E., and Stavely, J.R. (1993) Stimulation of germination of teliospores of *Uromyces appendiculatus* by volatile aroma compounds, J. Agricult. Food Chem. **41**, 1743-1747.

Freytag, S., Bruscaglioni, L., Gold, R.E., and Mendgen, K. (1988) Basidiospores of rust fungi (*Uromyces* species) differentiate infection structures *in vitro*, Exp. Mycol. **12**, 275-283.

Freytag, S., and Mendgen, K. (1991) Surface carbohydrates and cell wall structure of *in vitro* induced uredospore infection structures of *Uromyces viciae-fabae* before and after treatment with enzymes and alkali, Protoplasma **161**, 94-103.

Frittrang, A.K., Deising, H., and Mendgen, K. (1992) Characterization and partial purification of pectinesterase, a differentiation-specific enzyme of *Uromyces viciae-fabae*, J. Gen. Microbiol. **138**, 2213-2218.

Gil, F., and Gay, J.L. (1977) Ultrastructural and physiological properties of the host interfacial components of haustoria of *Erysiphe pisi* in vivo and in vitro, Physiol. Plant Pathol. **10**, 1-12.

Gilchrist, D. G. (1998). Programmed cell death in plant disease: the purpose and promise of cellular suicide, Annu. Rev. Phytopathol. **36**,393-414.

Godt, D.E., and Roitsch, T. (1997) Regulation and tissue-specific distribution of mRNAs for three extracellular invertase isoenzymes of tomato suggests an important function in establishing and maintaining sink metabolism, Plant Physiol. **115**, 273-282.

Gold, R.E., and Mendgen, K. (1983) Activation of teliospore germination in *Uromyces appendiculatus* var. *appendiculatus*. I. Ageing and temperature, Phytopathol. Z. **108**, 267-280.

Gold, R.E., and Mendgen, K. (1991) Rust basidiospore germlings and disease initiation, in G.T. Cole and H.C Hoch.(eds.), *The Fungal Spore and Disease Initiation in Plants and Animals*, Plenum Press, New York, London, pp. 67-99.

Grambow, J.-J. (1977) The influence of volatile leaf constitutents on the in vitro differentiation and growth of *Puccinia graminis* f. sp. *tritici*, Z. Pflanzenphysiol. **85**, 361-372.

Greenland, A.J., and Lewis, D.H. (1983) Changes in the activities and predominant molecular forms of acid invertase in oat leaves infected by crown rust, Physiol. Plant Pathol. **22**, 293-312.

Hahn, M., and Mendgen, K. (1992) Isolation by ConA binding of haustoria from different rust fungi and comparison of their surface qualities, Protoplasma **170**, 95-103.

Hahn, M., and Mendgen, K. (1997) Characterization of *in planta*-induced rust genes isolated from a haustorium-specific cDNA library, Mol. Plant-Microbe Interact. **10**, 427-437.

Hahn, M., Neef, U., Struck, C., Göttfert, M., and Mendgen, K. (1997) A putative amino-acid transporter is specifically expressed in haustoria of the rust fungus *Uromyces fabae*, Mol. Plant-Microbe Interact. **10**, 438-445.

Harder, D.E. (1984) Developmental ultrastructure of hyphae and spores, in W.E. Bushnell and A.P. Roelfs (eds.), *The Cereal Rusts. Volume 1*, Academic Press, INC, Orlando, pp. 333-373.

Harder, D.E. (1989) Rust fungal haustoria - past, present, future, Can. J. Plant Pathol. **11**, 91-99.

Harder, D.E., and Chong, J. (1984) Structure and physiology of haustoria, in W.R. Bushnell and A.P. Roelfs (eds.), *The Cereal Rusts, Volume 1*, Academic Press, New York, pp. 431-476.

Harder, D.E., and Chong, J. (1991) Rust Haustoria, in K. Mendgen and D.E. Lesemann (eds.), *Electron Microscopy of Plant Pathogens*, Springer Verlag, Berlin, pp. 235-250.

Harder, D.E., Chong, J., Rohringer, R., Mendgen, K., Schneider, A., Welter, K., and Knauf, G. (1989) Ultrastructure and cytochemistry of extramural substances associated with intercellular hyphae of several rust fungi, Can. J. Bot. **67**, 2043-2051.

Hardham, A.R., and Mitchell, H.J. (1998) Use of molecular cytology to study the structure and biology of phytopathogenic and mycorrhizal fungi, Fungal Genet. Biol. **24**, 252-284.

Heath, M. (1989) A comparison of fungal growth and plant responses in cowpea and bean cultivars inoculated with urediospores or basidiospores of the cowpea rust fungus, Physiol. Mol. Plant Pathol. **34**, 415-426.

Heath, M.C. (1981) The suppression of the development of silicon-containing deposits in French bean leaves by exudates of the bean rust fungus and extracts from bean rust-infected tissues, Physiol. Plant Pathol. **18**, 149-155.

Heath, M.C. (1990) Cytological and cytochemical features of the haustorium of the cowpea rust fungus, *Uromyces vignae*, formed in the absence of a living host cell. II. Electron microscopy, Can. J. Bot. **68**, 278-287.

Heath, M.C. (1997a) Evolution of plant resistance and susceptibility to fungal parasites, in G. C. Carroll and P. Tudzynski (eds.), *The Mycota, Volume V: Plant Relationships*, Springer Verlag, Berlin, pp. 257-276.

Heath, M.C. (1997b) Signaling between pathogenic rust fungi and resistant or susceptible host plants, Ann. Bot. **80**, 713-720.

Heath, M.C., Nimchuk, Z.L., and Xu, H. (1997) Plant nuclear migrations as indicators of critical interactions between resistant or susceptible cowpea epidermal cells and invasion hyphae of the cowpea rust fungus, New Phytol. **135**, 689-700.

Heath, M.C., and Perumalla, C.J. (1988) Haustorial mother cell development by *Uromyces vignae* on collodium membranes, Can. J. Bot. **66**, 736-741.

Heath, M.C., and Skalamera, D. (1997) Cellular interactions between plants and biotrophic fungal parasites, Adv. Bot. Res. **24**, 195-225.

Heiler, S., Mendgen, K., and Deising, H. (1993) Cellulolytic enzymes of the obligately biotrophic rust fungus *Uromyces viciae-fabae* are regulated differentiation-specifically, Mycol. Res. **97**, 77-85.

Heisterüber, D., Schulte, P., and Moerschbacher, B.M. (1994) Soluble carbohydrates and invertase activity in stem rust-infected, resistant and susceptible near-isogenic wheat leaves, Physiol. Mol. Plant Pathol. **44**, 111-123.

Heller, A. (1995) False-positive labeling of the haustorial cell wall of bean rust: a problem in immunogold labeling of thin sections, Micron **26**, 15-24.

Herbers, K., Meuwly, P., Frommer, W.B., Métraux, J.P., and Sonnewald, U. (1996) Systemic acquired resistance mediated by the ectopic expression of invertase: Possible hexose sensing in the secretory pathway, Plant Cell **8**, 793-803.

Hoch, H.C., and Staples, R.C. (1984) Evidence that cyclic AMP initiates nuclear division and infection structure formation in the bean rust fungus, *Uromyces phaseoli*, Exp. Physiol. **8**, 37-46.

Hoch, H.C., and Staples, R.C. (1987) Structural and chemical changes among the rust fungi during appressorium development, Annu. Rev. Phytopathol. **25**, 231-247.

Hoch, H.C., Staples, R.C., Whitehead, B., Comeau, J., and Wolf, E.D. (1987a) Signaling for growth orientation and cell differentiation by surface topography in *Uromyces*, Science **235**, 1659-1662.

Hoch, H.C., Tucker, B.E., and Staples, R.C. (1987b) An intact microtubule cytoskeleton is necessary for mediation of the signal for cell differentiation in *Uromyces*, Eur. J. Cell Biol. **45**, 209-218.

Hoffmann, J., and Mendgen, K. (1998) Endocytosis and membrane turnover in the germ tube of *Uromyces fabae*, Fungal Genet. Biol. **24**, 77-85.

Horsefall, J.G., and Dimond, A.E. (1957) Interactions of tissue sugar, growth substances, and disease susceptibility, Z. Pflanzenkr. Pflanzensch. **64**, 415-421.

Hu, G.G., and Rijkenberg, F.H.J. (1998) Subcellular localization of ß 1,3 glucanase in *Puccinia recondita* f. sp. *tritici* infected wheat leaves, Planta **204**, 324-334.

Ikegawa, T., Mayama, S., Nakayashiki, H., and Kato, H. (1996) Accumulation of diferulic acid during the hypersensitive response of oat leaves to *Puccinia coronata* f. sp. *avenae* and its role in the resistance of oat tissues to cell wall degrading enzymes, Physiol. Mol. Plant Pathol. **48**, 245 - 255.

Jäger, K., and Reisener, H.-J. (1969) Host-parasite metabolic relationship between *Puccinia graminis* var. *tritici* and *Triticum aestivum* (wheat). I. Uptake of amino acids from the host, Planta **85**, 57-72.

Jang, J.C., and Sheen, J. (1994) Sugar sensing in higher plants, Plant Cell **6**, 1665-79.

Jang, J.C., and Sheen, J. (1997) Sugar sensing in higher plants, Trends Plant Sci. **2**, 208-214.

Johal, G.S., Gray, J., Gruis, D., and Briggs, S. P. (1995) Convergent insights into mechanisms determining disease and resistance response in plant-fungal interactions, Can. J. Bot. **73**, S468-S474.

Katz, D.B., Sussman, M.R., Mierzwa, R.J., and Evert, R.F. (1988) Cytochemical localization of ATPase activity in oat roots localizes a plasma membrane-associated soluble phosphatase, not the proton pump, Plant Physiol. **86**, 841-847.

Kemp, G., Botha, A.-M., Kloppers, F. J., and Pretorius, Z. A. (1999) Disease development and ß-1,3-glucanase expression following leaf rust infection in resistant and susceptible near-isogenic wheat seedlings. Physiol. Mol. Plant Pathol. **55**, 45-52.

Kershaw, M. J., and Talbot, N. J. (1998) Hydrophobins and repellents: proteins with fundamental roles in fungal morphogenesis, Fungal Genet. Biol. **23**, 18-33.

Kilian, A., Chen, J., Han, F., Steffenson, B., and Kleinhofs, A. (1997) Towards map-based cloning of the barley stem rust resistance genes Rpg1 and Rpg4 using rice as an intergenomic cloning vehicle, Plant Mol. Biol. **35**, 187-95.

Knauf, G. M., Welter, K., Müller, M., and Mendgen, K. (1989) The haustorial host-parasite interface in rust-infected bean leaves after high-pressure freezing, Physiol. Mol. Plant Pathol. **34**, 519-530.

Kneale, J., and Farrar, J.F. (1985) The localization and frequency of haustoria in colonies of brown rust on barley leaves, New Phytol. **101**, 495-505.

Knogge, W. (1997) Elicitors and suppressors of the resistance response, in H. Hartleb, R. Heitefuss and H.-H. Hoppe (eds.), *Resistance of Crop Plants against Fungi*, Fischer Verlag, Jena, pp. 159-182.

Kogel, G., Beissmann, B., Reisener, H.-J., and Kogel, K.-H. (1988) A single glycoprotein from *Puccinia graminis* f. sp. *tritici* cell walls elicits the hypersensitive lignification response in wheat, Physiol. Mol. Plant Pathol. **33**, 173-185.

Kogel, G., Beissmann, B., Reisener, H.-J., and Kogel, K.-H. (1991) Specific binding of a hypersensitive lignification elicitor from *Puccinia graminis* f. sp. *tritici* to the plasmamembrane from wheat (*Triticum aestivum* L.), Planta **183**, 164-169.

Kogel, K.-H., Heck, B., Kogel, B., Moerschbacher, B., and Reisener, H.-J. (1985) A fungal elicitor of the resistance response in wheat, Z. Naturforsch. **40**, 743-744.

Kogel, K.-H., and Beissmann, B. (1992) Isolation and characterization of elicitors, in H. F. Linskens and J. F. Jackson (eds.), *Modern Methods of Plant Analysis*, Springer Verlag, Berlin, pp. 239-257.

Kramer, C.L., and Eversmeyer, M.G. (1992) Effect of temperature on germination and germ-tube development of *Puccinia recondita* and *P. graminis* urediniospores, Mycol. Res. **96**, 689-693.

Kronstad, J.,W. (1997) Virulence and cAMP in smuts, blasts and blights, Trends Plant Sci. **2**, 193-199.

Kwon, Y.H. , and Hoch, H. C. (1991) Temporal and spatial dynamics of appressorium formation in *Uromyces appendiculatus*, Exp. Mycol. **15**, 116 -131.

Kwon, Y.H., Hoch, H.C., and Aist, J.R. (1991a) Initiation of appressorium formation in *Uromyces appendiculatus*: organization of the apex, and the responses involving microtubules and apical vesicles, Can. J. Bot. **69**, 2560 - 2573.

Kwon, Y.H., Hoch, H.C., and Staples, R.C. (1991b) Cytoskeletal organization in *Uromyces* urediospore germling apices during appressorium formation, Protoplasma **165**, 37 - 50.

Lamboy, J.S., Staples, R.C., and Hoch, H.C. (1995) Superoxide dismutase: a differentiation protein expressed in *Uromyces* germlings during early appressorium development, Exp. Mycol. **19**, 284-296.

Langen, G., Beissmann, B., Reisener, H.J., and Kogel, K.-H (1992) A ß-1,3-D-endo-mannanase from culture filtrates of the hyperparasites *Verticillium lecanii* and *Aphanocladium album* that specifically lyses the germ pore plug from uredospores of *Puccinia graminis* f. sp. *tritici*, Can. J. Bot. **70**, 853-860.

Larous, L., and Loesel, D.M. (1993) Strategies of pathogenicity in monokaryotic and dikaryotic phases of rust fungi, with special reference to vascular infection, Mycol. Res. **97**, 415-420.

Laugé R, Joosten M.H.A.J., Haanstra J.P.W., Goodwin P.H., Lindhout P., and DeWit P.J.G.M. (1998) Successful search for a resistance gene in tomato targeted against a virulence factor of a fungal pathogen, Proc. Natl. Acad. Sci. USA **95**, 9014-9018.

Laugé, R., Joosten, M.H., Van den Ackerveken G.F., Van den Broek, H.W., and De Wit, P.J.G.M. (1997) The *in planta*-produced extracellular proteins ECP1 and ECP2 of *Cladosporium fulvum* are virulence factors., Mol. Plant-Microbe Interact. **10**, 725-734.

Li, A., Altosaar, I., Heath, M.C., and Horgen, P. (1993) Transient expression of the beta-glucuronidase gene delivered into urediniospores of *Uromyces appendiculatus* by particle bombardment, Can. J. Plant Pathol. **15**, 1-6.

Lin, K.C., Bushnell, W.R., Smith, A.G., and Szabo, L.J. (1998) Temporal accumulation patterns of defence response gene transcripts in relation to resistant reactions in oat inoculated with *Puccinia graminis*, Physiol. Mol. Plant Pathol. **52**, 95-114.

Lin, K.C., Bushnell, W.R., Szabo, L.J., and Smith, A.G. (1996) Isolation and expression of a host response gene family encoding thaumatin-like proteins in incompatible oat-stem rust fungus interactions, Mol Plant-Microbe Interact **9**, 511-22.

Littlefield, L.J. (1981). *Biology of the Plant Rusts. An Introduction.* Iowa State University Press, Ames.

Liu, Z., Szabo, L.J., and Bushnell, W.R. (1993) Molecular cloning and analysis of abundant and stage-specific mRNAs from *Puccinia graminis*, Mol. Plant-Microbe Interact. **6**, 84-91.

Lohaus, G., Winter,H., Riens,B., and Heldt, H.W. (1995) Further studies of the pholoem loading process in leaves of barley and spinach. The comparison of metabolite concentrations in the apoplastic compartment with those in the cytosolic compartment and in the sieve tubes, Bot. Acta **108**, 270-275.

Long, D.E., Fung, A.K., McGee, E.E.M., Cooke, R. C., and Lewis, D.H. (1975) The activity of invertase and its relevance to the accumulation of storage polysaccharides in leaves infected with biotrophic fungi, New Phytol. **74**, 173-182.

Luck, J.E., Lawrence, G.J., Finnegan, E.J., Jones, D.A., and Ellis, J.G. (1998) A flax transposon identified in two spontaneous mutant alleles of the L6 rust resistance gene, Plant J. **16**, 365-9.

Luke, H. H., Barnett, R.D., and Pfahler, P.L. (1987) Xenoparasite-nonhost reactions in *Puccinia*-gramineae pathosystems, Phytopathology **77**, 1488-1491.

Mackie, A.J., Roberts, A.M., Callow, J.A., and Green, J.R. (1991) Molecular differentiation in pea powdery-mildew haustoria. Identification of a 62-kDa N-linked glycoprotein unique to the haustorial plasma membrane, Planta **183**, 399-408.

Mackie, A.J., Roberts, A.M., Green, J.R., and Callow, J.A. (1993) Glycoproteins recognised by monoclonal antibodies UB7, UB8 and UB10 are expressed early in the development of pea powdery mildew haustoria, Physiol. Mol. Plant Pathol. **43**, 124-136.

Macko, V. (1981) Inhibitors and stimulants of spore germination and infection structure formation in fungi, in G. Turian and H.R. Hohl (eds.), *The Fungal Spore*, Academic Press, New York, pp. 565-584.

Maclean, D.J. (1982) Axenic culture of rust fungi, in K.J. Scott and A.K. Chakravorty (eds.), *The Rust Fungi*, Academic Press, London, pp. 37-120.

Manetti, A.G., Rosetto, M., and Maundrell, K.G. (1994) *nmt2* of fission yeast: a second thiamine-repressible gene co-ordinately regulated with *nmt1*, Yeast **10**, 1075-1082.

Manners, J.M., Maclean, D.J., and Scott, K.J. (1982) Pathways of glucose assimilation in *Puccinia graminis*, J. Gen. Microbiol. **128**, 2621-2630.

Manners, J.M., Maclean, D.J., and Scott, K.J. (1984) Hexitols as major intermediates of glucose assimilation by mycelium of *Puccinia graminis*, Arch. Microbiol. **139**, 158-161.

Manners, J.M., Maclean, D.J., and Scott, K.J. (1988) Metabolism of 2-deoxy-D-glucose by axenically grown mycelia of *Puccinia graminis*, Exp. Mycol. **12**, 350-356.

Marticke, K.H., Reisener, H.J., Fischer, R., and Hippe-Sanwald, S. (1998) In situ detection of a fungal glycoprotein-elicitor in stem rust- infected susceptible and resistant wheat using immunogold electron microscopy, Eur. J. Cell Biol. **76**, 265-73.

Mayama, S., Bordin, A P.A., Morikawa, T., Tanpo, H., and Kato, H. (1995) Association between avenalumin accumulation, infection hypha length and infection type in oat crosses segregating for resistance to *Puccinia coronata* f. sp. *avenae* race 226, Physiol. Mol. Plant Pathol. **46**, 263-274.

Mendgen, K. (1979) Microautoradiographic studies on host-parasite interactions. II The exchange of ^{3}H-lysine between *Uromyces phaseoli* and *Phaseolus vulgaris*, Arch. Microbiol. **123**, 129-135.

Mendgen, K. (1981) Nutrient uptake in rust fungi, Phytopathology **71**, 983-989.

Mendgen, K., Bachem, U., Stark-Urnau, M., and Xu, H. (1995) Secretion and endocytosis at the interface of plants and fungi, Can. J. Bot. **73**, S640-S648.

Mendgen, K., and Deising, H. (1993) Infection structures of fungal plant pathogens - a cytological and physiological evaluation, New Phytol. **124**, 193-213.

Mendgen, K., Hahn, M., and Deising, H. (1996) Morphogenesis and mechanisms of penetration by plant pathogenic fungi, Annu. Rev. Phytopathol. **34**, 367-386.

Mendgen, K., Welter, K., Scheffold, F., and Knauf-Beiter, G. (1991) High pressure freezing of rust infected plant leaves, in K. Mendgen and D. E. Lesemann (eds.), Electron Microscopy of Plant Pathogens, Springer Verlag, Berlin, pp. 31-42.

Mims, C.W. (1991) Using electron microscopy to study plant pathogenic fungi, Mycologia **83**, 1-19.

Moerschbacher, B.M., Noll, U.M., Flott, B.E., and Reisener, H.J. (1990a) Specific inhibition of lignification breaks hypersensitive resistance of wheat to stem rust, Plant Physiol. **93**, 465-470.

Moerschbacher, B.M., Schrenk, K., Graessner, B., Noll, U., and Reisener, H.-J. (1990b) A wheat cell wall fragment suppresses elicitor-induced resistance responses and disturbs fungal development, J. Plant Physiol. **136**, 761-764.

Moerschbacher, B.M., and Reisener, H.-J. (1997) The hypersensitive resistance reaction, in H. Hartleb, R. Heitefuss and H.-H. Hoppe (eds.), *Resistance of Crop Plants against Fungi*, Gustav Fischer, Jena, pp. 126-158.

Moloshok, T.D., and Leinhos, G. M.E. (1993) The autogenic extracellular environment of *Uromyces appendiculatus* urediospore germlings, Mycologia **85**, 392 - 400.

Moloshok, T.D., Terhune, B.T., Lamboy, J.S., and Hoch, H.C. (1994) Fractionation of extracellular matrix components from urediospore germlings of *Uromyces*, Mycologia **86**, 787-794.

Morin, L., Brown, J.F., and Auld, B.A. (1992) Teliospore germination, basidiospore formation and the infection process of *Puccinia xanthii* on *Xanthium occidentale*, Mycol. Res. **96**, 661-669.

Mould, M. J. J., and Heath, M. C. (1999) Ultrastructural evidence of differential changes in transcription, translation, and cortical microtubules during *in planta* penetration of cells resistant or susceptible to rust infection, Physiol. Mol. Plant Pathol. **55**, 225-236.

Münch-Garthoff, S., Neuhaus, J.M., Boller, T., Kemmerling, B., and Kogel, K.H. (1997) Expression of beta-1,3-glucanase and chitinase in healthy, stem-rust-affected and elicitor-treated near-isogenic wheat lines showing Sr5-or Sr24-specified race-specific rust resistance, Planta **201**, 235-44.

Murdoch, L.J., Kobayashi, I., and Hardham, A.R. (1998) Production and characterization of monoclonal antibodies to cell wall components of the flax rust fungus, Eur. J. Plant Pathol. **10**, 331-346.

Navarre, D.A., and Wolpert, T.J. (1999) Victorin induction of an apoptotic/senescence-like response in oats, Plant Cell **11**, 237-249.

Nicholson, R.L., Kunoh, H., Shiraishi, T., and Yamada, T. (1993) Initiation of the infection process by *Erysiphe graminis*: Conversion of the conidial surface from hydrophobicity to hydrophilicity and influence of the conidial exudate on the hydrophobicity of the barley leaf surface, Physiol. Mol. Plant Pathol. **43**, 307-318.

O'Connell, R.J., and Ride, J.P. (1990) Chemical detection and ultrastructural localization of chitin in cell walls of *Colletotrichum lindemuthianum*, Physiol. Mol. Plant Pathol. **37**, 39-53.

Owera, S.A.P., Farrar, J.F., and Whitebread, R. (1983) Translocation from leaves of barley infected with brown rust, New Phytol. **94**, 111-123.

Petersen, R. H. (1974) The rust fungus life cycle, Bot. Rev. 40: 453-513.

Peterson, R.B., and Aylor, D.E. (1995) Chlorophyll fluorescence induction in leaves of *Phaseolus vulgaris* infected with bean rust (*Uromyces appendiculatus*), Plant Physiol. **108**, 163-171.

Pfeiffer, E., Jäger, K., and Reisener, H.-J. (1969) Untersuchungen über Stoffwechselbeziehungen zwischen Parasit und Wirt am Beispiel von *Puccinia graminis* var. *tritici* auf Weizen. II. Aufnahme von Hexosen aus dem Wirtsgewebe., Planta **85**, 194-201.

Raguso, R.A., and Roy, B.A. (1998) 'Floral' scent production by *Puccinia* rust fungi that mimic flowers, Mol. Ecol. **7**, 1127-1136.

Rauscher, M., Mendgen, K., and Deising, H. (1995) Extracellular proteases of the rust fungus *Uromyces viciae-fabae*, Exp. Mycol. **19**, 26-34.

Rauscher, M., Adam, A. L., Wirtz, S., Guggenheim, R., Mendgen, K., and Deising, H. B. (1999) PR-1 protein inhibits the differentiation of rust infection hyphae in leaves of acquired resistant broad bean. Plant J. **19**, 625-633.

Razin, S., Yogev, D., and Naot, Y. (1998) Molecular biology and pathogenicity of mycoplasmas, Microbiol. Mol. Biol. Rev. **62**, 1094-156.

Read, N.D., Kellock, L.J., Collins, T.J., and Gundlach, A.M. (1997) Role of topography sensing for infection-structure differentiation in cereal rust fungi, Planta **202**, 163-70.

Read, N.D., Kellock, L.J., Knight, H., and Trewavas, A.J. (1992) Contact sensing during infection by fungal pathogens, in J. A. Callow and J. R. Green (eds.), *Perspectives in Plant Cell Recognition*, Cambridge University, Cambridge, pp. 137-172.

Reisener, H.-J., Ziegler, E., and Prinzing, A. (1970) Zum Stoffwechsel des Mycels von *Puccinia graminis* var. *tritici* auf der Weizenpflanze, Planta **92**, 355-357.

Richberg, M.H., Aviv, D.H., and Dangl, J.L. (1998) Dead cells do tell tales, Curr. Opin. Plant Biol. **1**, 480-485.

Roberts, A.M., Mackie, A.J., Hathaway, C., Callow, J.A., and Green, J.R. (1993) Molecular differentiation in the extrahaustorial membrane of pea powdery mildew haustoria at early and late stages of development, Physiol. Mol. Plant Pathol. **43**, 147-160.

Roberts, A.M., and Walters, D.R. (1988) Photosynthesis in discrete regions of leek infected with the rust, *Puccinia allii* Rud., New Phytol. **110**, 371-376.

Roberts, J.K., and Pryor, A. (1995) Isolation of a flax (*Linum usitatissimum*) gene induced during susceptible infection by flax rust (*Melampsora lini*), Plant J. **8**, 1-8.

Rubiales, D., and Niks, R.E. (1996) Avoidance of rust infection by some genotypes of *Hordeum chilense* due to their relative inability to induce the formation of appressoria, Physiol. Mol. Plant Pathol. **49**, 89-101.

Russo, V.M., and Bushnell, W.R. (1989) Responses of barley cells to puncture by microneedles and to attempted penetration by *Erysiphe graminis* f. sp. *hordei*, Can. J. Bot. **67**, 2912-2921.

Ryerson, D.E., and Heath, M.C. (1996) Cleavage of nuclear DNA into oligonucleosomal fragments during cell death induced by fungal infection or by abiotic treatments, Plant Cell **8**, 393-402.

Saverimuttu, N., and Deverall, B.J. (1998) A cytological assay reveals pathotype and resistance gene specific elicitors in leaf rust infections of wheat, Physiol. Mol. Plant Pathol. **52**, 25-34.

Scholes, J.D., and Farrar, J.F. (1985) Photosynthesis and chloroplast functioning within individual pustules of *Uromyces muscari* on bluebell leaves, Physiol. Plant Pathol. **27**, 387-400.

Scholes, J.D., and Farrar, J.F. (1986) Increased rates of photosynthesis in localised regions of a barley leaf infected with brown rust, New Phytol. **104**, 601-612.

Scholes, J.D., and Rolfe, S.A. (1996) Photosynthesis in localised regions of oat leaves infected with crown rust (*Puccinia coronata*): Quantitative imaging of chlorophyll fluorescence, Planta **199**, 573-582.

Serrano, R. (1989) Structure and function of plasma membrane ATPase, Annu. Rev. Plant Physiol. Plant Mol. Biol. **40**, 61-94.

Skalamera, D., and Heath, M.C. (1995) Changes in the plant endomembrane system associated with callose synthesis during the interaction between cowpea (*Vigna unguiculata*) and the cowpea rust fungus (*Uromyces vignae*), Can. J. Bot. **73**, 1731-1738.

Skalamera, D., and Heath, M.C. (1996) Cellular mechanisms of callose deposition in response to fungal infection or chemical damage, Can. J. Bot. **74**, 1236-1242.

Skalamera, D., Jibodh, S., and Heath, M.C. (1997) Callose deposition during the interaction between cowpea (*Vigna unguiculata*) and the monokaryotic stages of the cowpea rust fungus (*Uromyces vignae*), New Phytol. **136**, 511 - 524.

Spencer-Phillips, P.T.N., and Gay, J.L. (1981) Domains of ATPase in plasma membranes and transport through infected plant cells, New Phytol. **89**, 393-400.

Stakman, E.C. (1915) Relation between *Puccinia graminis* and plants highly resistant to its attack, J. Agricult.. Res. **4**, 193-199.

Stakman, E.C., and Piemeisal, F.J. (1917) Biological forms of *Puccinia graminis* on cereals and grasses, J. Agricult. Res. **10**, 429-495.

Staples, R.C., and Hoch, H.C. (1988) Preinfection changes in germlings of a rust fungus induced by host contact, in P.A. Hedin, J.J. Menn and R.M. Hollingwort (eds.), *Biotechnology for Crop Protection*, American Chemical Society, Washington D.C., pp. 82-93.

Stark-Urnau, M., and Mendgen, K. (1993) Differentiation of aecidiospore- and uredospore-derived infection structures on cowpea leaves and on artificial surfaces by *Uromyces vignae*, Can. J. Bot. **71**, 1236-1242.

Stark-Urnau, M., and Mendgen, K. (1995) Sequential deposition of plant glycoproteins and polysaccharides at the host-parasite interface of *Uromyces vignae* and *Vigna sinensis*, Protoplasma **186**, 1-11.

Stichter, L. Mauch-Mani, B. and Metraux, J.-P. (1997) Systemic acquired resistance, Annu. Rev. Phytopathol. **35**, 235-270.

Struck, C., Hahn, M., and Mendgen, K. (1996) Plasma membrane H^+-ATPase activity in spores, germ tubes and haustoria of the rust fungus *Uromyces viciae-fabae*, Fungal Genet. Biol. **20**, 30-35.

Struck, C., Siebels, C., Rommel, O., Wernitz, M., and Hahn, M. (1998) The plasma membrane H^+-ATPase from the biotrophic rust fungus *Uromyces fabae*: Molecular characterization of the gene (*PMA1*) and functional expression of the enzyme in yeast, Mol. Plant-Microbe Interact. **11**, 458-465.

Sutherland, M.W., Deverall, B.J., Moerschbacher, B.M., and Reisener, H.-J. (1989) Wheat cultivar and chromosomal selectivity of two types of eliciting preparations from rust pathogens, Physiol. Mol. Plant Pathol. **35**, 535-541.

Talbot, N. J., Kershaw, M. J., G.E., W., de Vries, O. M. H., Wessels, J. G., and Hamer, J. E. (1996) *MPG1* encodes a fungal hydrophobin involved in surface interactions during infection-related development of *Magnaporthe grisea*, Plant Cell **8**, 985-999.

Tang, G.Q., Luscher, M., and Sturm, A. (1999) Antisense repression of vacuolar and cell wall invertase in transgenic carrot alters early plant development and sucrose partitioning, Plant Cell **11**, 177-89.

Terhune, B.T., Bojko, R.J., and Hoch, H.C. (1993) Deformation of stomatal guard cell lips and microfabricated artificial topographies during appressorium formation by *Uromyces*, Exp. Mycol. **17**, 70-78.

Terhune, B. T., and Hoch, H.C. (1993) Substrate hydrophobicity and adhesion of *Uromyces* urediospores and germlings, Exp. Mycol. **17**, 241-252.

Tetlow, I.J., and Farrar, J. F. (1993) Apoplastic sugar concentration and pH in barley leaves infected with brown rust, J. Exp. Bot. **44**, 929-936.

Tiburzy, R., Martins, E.M.F.M., and Reisener, H.-J. (1992) Isolation of haustoria of *Puccinia graminis* f. sp. *tritici* from wheat leaves, Exp. Mycol. **16**, 324-328.

Tiburzy, R., and Reisener, H.-J. (1990) Resistance of wheat to *Puccinia graminis* f. sp. *tritici*. Association of the hypersensitive reaction with the cellular accumulation of lignin-like material and callose, Physiol. Mol. Plant Pathol. **36**, 109-120.

Tiburzy, R., Rogner, U.C., Fischer, R., Beissmann, B., Kreuzaler, F.M., and Reisener, H.-J. (1991) Detection of an elicitor on infection structures of *Puccinia graminis* using monoclonal antibodies, Eur. J. Cell Biol. **55**, 174-178.

Truernit, E., Schmid, J., Epple, P., Illig, J., and Sauer, N. (1996) The sink-specific and stress-regulated Arabidopsis *STP4* gene: enhanced expression of a gene encoding a monosaccharide transporter by wounding, elicitors and pathogen challenge, Plant Cell **8**, 2169-2182.

Vallavielle-Pope, C.D., Huber, M., Leconte, M., and Goyeau, H. (1995) Comparative effects of temperature and interrupted wet periods on germination, penetration, and infection of *Puccinia recondita* f. sp. *tritici* and *P. striiformis* on wheat seedlings., Phytopathology **85**, 409-415.

Villar, M., Lefevre, F., Bradshaw, H.D., Jr., and Teissier du Cros, E. (1996) Molecular genetics of rust resistance in poplars (*Melampsora larici- populina* Kleb/Populus sp.) by bulked segregant analysis in a 2 x 2 factorial mating design, Genetics **143**, 531-6.

Wachstein, M., and Meisel, E. (1957) Histochemistry of hepatic phosphatases at a physiologic pH. With special reference to the demonstration of bile canaliculi, Am. J. Clin. Pathol. **27**, 13-23.

Wada, M., Kato, H., Malik, K., Sriprasertsak, P., Ichinose, Y., Shiraishi, T., and Yamada, T. (1995) A supprescin from a phytopathogenic fungus deactivates transcription of a plant defense gene encoding phenylalanine ammonia-lyase, J. Mol. Biol. **249**, 513-519.

Weber, R.W.S., Webster, J., and Al-Gharabally, D.H. (1998) *Puccinia distincta*, cause of the current daisy rust epidemic in Britain, im comparison with other rusts recorded on daisies, *P. obscura* and *P. lagenophora*, Mycol. Res. **102**, 1227-1232.

Welter, K., Müller, M., and Mendgen, K. (1988) The hyphae of *Uromyces appendiculatus* within the leaf tissue after high pressure freezing and freeze substitution, Protoplasma **147**, 91-99.

Wilcox, P.L., Amerson, H.V., Kuhlman, E.G., Liu, B.H., O'Malley, D.M., and Sederoff, R.R. (1996) Detection of a major gene for resistance to fusiform rust disease in loblolly pine by genomic mapping, Proc. Natl. Acad. Sci. USA **93**, 3859-64.

Williams, P.G. (1984) Obligate parasitism and axenic culture, in W.R. Bushnell and A.P. Roelfs (eds.), *The Cereal Rusts. Volume 1*, Academic Press, Orlando, pp. 399-430.

Xu, H., and Heath, M.C. (1998) Role of calcium in signal transduction during the hypersensitive response by basidiospore-derived infection of the cowpea rust fungus, Plant Cell **10**, 585-597.

Xu, H., and Mendgen, K. (1991) Early events in living epidermal cells of cowpea and broad bean during infection with basidiospores of the cowpea rust fungus, Can. J. Bot. **69**, 2279-2285.

Xu, H., and Mendgen, K. (1994) Endocytosis of 1,3-ß-glucans by broad bean cells at the penetration site of the cowpea rust fungus (haploid stage), Planta **195**, 282-290.

Xu, H., and Mendgen, K. (1997) Targeted cell wall degradation at the penetration site of cowpea rust basidiosporelings, Mol. Plant-Microbe Interact. **10**, 87-94.

Xuei, X.L., Bhairi, S., Staples, R.C., and Yoder, O.C. (1992) Characterization of INF56, a gene expressed during infection structure development of *Uromyces appendiculatus*, Gene **110**, 49-55.

Yamamoto, H. (1995) Pathogenesis and host-parasite specificity in rusts, in K. Kohmoto (ed.), *Pathogenesis and Host Specificity in Plant Diseases*, Pergamon, Oxford, pp. 203-215.

Yamaoka, N., Toyoda, K., Kobayashi, I., and Kunoh, H. (1994) Induced accessibility and enhanced inaccessibility at the cellular level in barley coleoptiles. XIII. Significance of haustorium formation by the pathogen *Erysiphe graminis* for induced accessibility to the nonpathogen *E. pisi* as assessed by nutritional manipulations, Physiol. Mol. Plant Pathol. **44**, 217-225.

Zhou, X.-L., Stumpf, M.A., Hoch, H.C., and Kung, C. (1991) A mechanosensitive channel in whole cells and membrane patches of the fungus *Uromyces*, Science **253**, 1415-1417.

SYMBIOTIC PARASITES AND MUTUALISTIC PATHOGENS

Clavicipitaceous Symbionts of Grasses

C.L. SCHARDL
University of Kentucky
Department of Plant Pathology
S-305 Agricultural Sciences Bldg.-N
Lexington, Kentucky 40546-0091
U.S.A.

1. Introduction

In perhaps no other system are mutualists and antagonists known to be so closely related as in the fungal family, Clavicipitaceae. Its type species is the infamous ergot fungus, *Claviceps purpurea*, dreaded throughout history as a deadly contaminant of grains. This and related fungi produce ergot alkaloids, and in the hallucinogenic 1960's a modified ergot alkaloid became a popular symbol of the counter culture. Lysergic acid diethylamide (LSD) is a potent psychotropic agent, but its natural relatives, the ergopeptines, are also gruesome killers. *Claviceps purpurea* produces large amounts of ergotamine and other ergopeptines concentrated in its sclerotia, the resting structures that are also known as ergots (Tudzynski et al 1995). The fungus infects florets of grasses, including grain crops such as rye, and produces ergots as marvelous mimics of the seeds they replace. Historically the ergots were difficult to remove from the grain because of their similar size and density. In modern times ingenious methods are employed to remove the ergots automatically. Still, *C. purpurea* remains important as a source of ergot alkaloids for legitimate medical use (Rehácek 1991), and as the exemplar of a fascinating group of fungi. Other genera in the family Clavicipitaceae also produce ergot alkaloids and numerous other metabolites with potent biological properties (Porter 1994), but they do so continuously while hiding in plain sight. These are symbionts of grasses and sedges that live perennially in their hosts, give little or no outward sign of their presence, and often provide great benefits as agents of biological protection from insects, vertebrate herbivores, fungal pathogens, parasitic nematodes and even drought. The best studied for their beneficial effects on grasses are *Epichloë* species and their asexual derivatives, the *Neotyphodium* species. This chapter discusses the plant-associated Clavicipitaceae with emphasis on the mutualistic species and their closest relatives.

J. W. Kronstad (ed.), Fungal Pathology, 307–345.

The family Clavicipitaceae comprises mainly arthropod parasites, few fungicolous species and several genera of plant symbionts. Confirmed reports identify plant hosts from only two monocot families, the Poaceae (Graminae; grasses) and Cyperaceae (sedges). Hosts are know from all major grass subfamilies. *Claviceps purpurea,* the type species, causes 'ergot poisoning' of people who eat bread with grain contaminated with its resting structures (called sclerotia or ergots). Approximately 1% dry mass of the *C. purpurea* ergot is ergopeptine (i.e. ergot cyclol) alkaloids and related compounds ('ergot alkaloids') (Porter et al 1987). Hydrolysis of ergopeptines releases lysergic acid amide (and a simple chemical modification yields LSD). The ergopeptines cause vasoconstriction so extreme that limbs become anoxic, necrotic and ultimately gangrenous (Berde and Schild 1978). The condition can be fatal to the poison victim. For all its infamy, the ergot is just one stage in the peculiar life cycle of *Claviceps* species. These fungi infect only individual florets of grasses, take over the ovaries, and develop ergots in place of seeds that would otherwise develop (Tudzynski et al 1995). Thus, *Claviceps* is a replacement pathogen, substituting its own resting structure for the host seed that would normally be produced. There is very little cellular reaction of the host in *Claviceps* infections, and the only damage to the plant is the envelopment and mummification of the plant ovary within the ergots.

Most plant-associated Clavicipitaceae have much longer associations with individual host plants than do *Claviceps* species. Fungi of the genera *Balansia* (Diehl 1950), *Myriogenospora* (Luttrell and Bacon 1977, Rykard et al 1985, Rykard et al 1982, White and Glenn 1994), *Atkinsonella* (Clay and Jones 1984, Leuchtmann and Clay 1988, Morgan-Jones and White 1989), *Echinodothis*(White 1994b), *Parepichloë* (White and Reddy 1998), *Epichloë* (Sampson 1933), and *Neotyphodium* (asexual descendants of *Epichloë* species; Glenn et al 1996) remain in vegetative plant tissues throughout the life of the plant. These 'endophytes' tend to be concentrated in and around the shoot meristems at the crown of the plant, while also growing in leaves, leaf sheaths, stems and inflorescences (Herd et al 1997). Some species can invade the ovules where, unlike *Claviceps* species, they grow benignly while the seed develops normally (Philipson and Christey 1986). This allows these species to transmit vertically; that is, by infecting seed progeny of infected mother plants. The long-term associations of these Clavicipitaceae with plants seem to select for fungal characteristics that improve growth and survivability of the hosts. Plant gigantism and activities against insect and vertebrate herbivores are the most widely reported benefits (Clay 1990a, Clay et al 1989). However, in most cases the plants suffer reduced fecundity because the fruiting fungus surrounds and mummifies immature florets (Diehl 1950, White 1994c). In some cases the fungus fruits on leaves but still, by an unknown mechanism, suppresses host flowering; in still other cases the host flowers normally (Clay 1990a, Clay 1990b, Clay et al 1989, Clay et al 1993).

There is a continuum of vertical transmission among clavicipitaceous endophytes. Most species do not do so. Often any seed panicles that develop normally and set seeds have escaped the fungal infection altogether. However, seed transmission is a well studied characteristic of *Atkinsonella hypoxylon* and several *Epichloë* species, as well as those asexual (*Neotyphodium*) species that are evolutionarily derived from *Epichloë* species (Clay 1994, Philipson and Christey 1986, Schardl and Clay 1997). *Atkinsonella hypoxylon* infects *Danthonia* species, which typically produce both open pollinated (chastogamous) and self pollinated (cleistogamous) seeds (Clay 1983). The latter are efficiently colonized by the symbiont, whereas self pollinated seeds are rarely colonized (Kover and Clay 1998). *Epichloë* species and related *Neotyphodium* species inhabit grasses of subfamily Pooideae. Of these, most sexual and all asexual species transmit in open pollinated seeds at very high frequencies; nearly 100% of the seeds from a mother plant carry clones of the symbiont that inhabit that plant (Fig. 1) (Philipson and Christey 1986, Siegel et al 1984, White et al 1991). During the infection process a few hyphae invade meristematic tissues of the embryo, thus ensuring that the seedling will possess the symbiont (Philipson and Christey 1986). Asexual *Neotyphodium* species, lacking a contagious spore state, depend absolutely on seeds as their sole means of dissemination. Symbionts that rely on vertical transmission in host diaspores are expected to be under strong selection to benefit their hosts (Dawkins 1989). Results of many studies reviewed later in this chapter strongly indicate that vertically transmitted *Neotyphodium* species often benefit their hosts greatly.

The most extensively investigated plant-endophyte interaction is that of tall fescue (*Lolium arundinaceum* = *Festuca arundinacea*) with *Neotyphodium coenophialum* (formerly, *Acremonium coenophialum*). The reasons for intense interest in this system are the following: (1) As the most widely planted forage grass species, tall fescue now inhabits a very large portion of pasture and other grasslands in the United States (Ball et al 1993). (2) The endophyte is a significant contributor to survival of tall fescue in much of the area where it is used for pasture, soil conservation and turf (Schardl and Phillips 1997). (3) The endophyte produces alkaloids that have toxic effects on grazing livestock, including synergy with heat stress (Spiers et al 1995, Thompson and Stuedemann 1993).

In natural and pasture settings tall fescue is usually infected with *N. coenophialum* or other *Neotyphodium* species, suggesting strong selection favoring the symbiosis (Clay 1990b). Numerous benefits of *N. coenophialum* are reported in the literature. Of course, anti-vertebrate activity is beneficial to the grass in the wild, though toxicity to livestock poses significant economic problems (Hoveland 1993), and toxicity to wild mammals and birds is a concern (Madej and Clay 1991). Clearly beneficial are endophyte enhanced drought tolerance (Arechavaleta et al 1989), nematode resistance (Kimmons et al 1990, Schöberlein et al 1997), anti-insect activities (Clay et al 1993, Clement et al 1994, Popay and Rowan 1994, Rowan and Gaynor 1986, Rowan and Latch 1994, Siegel et al 1990), and resistance to seedling infections by *Rhizoctonia zeae* (Gwinn and Gavin 1992). The enhanced growth and competitiveness of tall fescue with endophyte can be beneficial, but can also thwart efforts to established multi-species pasture stands (Buta and Spaulding 1989, Peters and Zam 1981).

310   C. L. Schardl

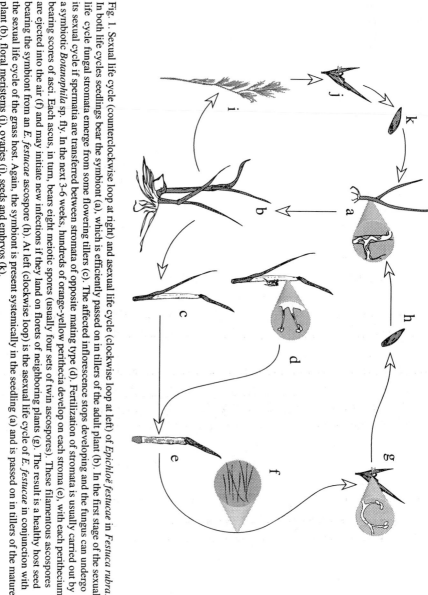

Fig. 1. Sexual life cycle (counterclockwise loop at right) and asexual life cycle (clockwise loop at left) of *Epichloë festucae* in *Festuca rubra*. In both life cycles seedlings bear the symbiont (a), which is efficiently passed on in tillers of the adult plant (b). In the first stage of the sexual life cycle fungal stromata emerge from some flowering tillers (c). The affected inflorescence stops developing and the fungus can undergo its sexual cycle if spermatia are transferred between stromata of opposite mating type (d). Fertilization of stromata is usually carried out by a symbiotic *Botanophila* sp. fly. In the next 3-6 weeks, hundreds of orange-yellow perithecia develop on each stroma (e), with each perithecium bearing scores of asci. Each ascus, in turn, bears eight meiotic spores (usually four sets of twin ascospores). These filamentous ascospores are ejected into the air (f) and may initiate new infections if they land on florets of neighboring plants (g). The result is a healthy host seed bearing the symbiont from an *E. festucae* ascospore (h). At left (clockwise loop) is the asexual life cycle of *E. festucae* in conjunction with the sexual life cycle of the grass host. Again, the symbiont is present systemically in the seedling (a) and is passed on in tillers of the mature plant (b), floral meristems (i), ovaries (j), seeds and embryos (k).

Livestock toxicosis is a major problem associated with alkaloids produced by *Claviceps* species and other plant-associated Clavicipitaceae, particularly several of the *Epichloë* and *Neotyphodium* species. Hoveland (1993) estimated that *N. coenophialum* in tall fescue caused a $600 million loss just to the U.S. beef industry in 1990, recognizing that substantial losses must also occur in other large animal industries.

Effects on horses can be devastating, including aborted foals and death of mares during delivery (Cross 1997). Likewise, ryegrass staggers is a major problem for sheep in New Zealand and Australia, and is clearly due to *Neotyphodium lolii* symbiotic with *Lolium perenne* (perennial ryegrass) (Gallagher et al 1984, Prestidge 1993). Breeders have removed endophyte from cultivars of both perennial ryegrass and tall fescue, but in neither case was this a generally acceptable solution. The effect in New Zealand was dramatic because the Argentine stem weevil (*Listronotus bonariensis*) devastated endophyte-free stands of ryegrass (Rowan and Latch 1994). The problems with endophyte-free tall fescue in the U.S. are not so clear cut, but there are persistent anecdotes that endophyte free stands are difficult to establish, difficult to maintain, and susceptible to overgrazing (Ball et al 1993).

In a sense the endophytes can be considered automatic pasture managers. In vegetative plant tissues, these fungi tend to be concentrated in the lower pseudostems (the whorl of leaf sheaths) and the crown where the shoot meristems reside (Rowan and Latch 1994). The anti-vertebrate alkaloids tend to be concentrated with the fungus in these critical growing points (Rottinghaus et al 1991, Siegel and Bush 1997). Animals on overgrazed pastures are forced to eat the crowns of the pasture grasses. Thus, although toxicoses are indeed suffered even on well managed pastures (and problems of fertility in cattle and stillbirths of foals are particularly common), severe toxicoses could be an indication of overgrazing.

Though known for many decades (Sampson 1937) clavicipitaceous endophytes were not intensively studied until they were implicated in major livestock toxicosis problems (Bacon et al 1977). In the past 22 years their image has gone from a problem requiring eradication, to a group of highly effective and important biological protection agents. In addition, they have become fascinating models for evolution of plant-fungus mutualisms (Schardl and Clay 1997, Wilkinson and Schardl 1997), while their diverse and often unique protective alkaloids have fascinated natural products chemists (Porter 1994).

2. The Symbiotic Continuum

2.1. COMING TO TERMS

The title of this chapter is meant to be provocative as well as informative. The terms 'symbiosis,' 'mutualism,' 'parasite,' and 'pathogen' are variously and often vaguely used. In order to discuss the nature of grass interactions with their best known endophytes, and in order to consider the selective forces driving evolution of these systems, it has been necessary to settle on precise definitions of such terms and, as I will explain, to coin another term. Symbiosis, to translate Anton deBary's original definition from German, is simply the "living together of differently named organisms" (cited in Lewis 1985). For most in the field this is taken to mean a physiological interaction between individuals of different species during a significant portion of their respective life cycles. Early after the term was coined it came to be used almost exclusively for interactions that benefited both partners, but the fundamental problems with this restricted usage are (1) the balance of relative costs and benefits would need to be known before the term is used, (2) where such balance is unknown or varies with time or environment there is no widely recognized alternative term, and (3) where the balance is mutually beneficial the obvious term to use is 'mutualism.' It is nontrivial to determine that benefits exceed costs for both partners. Many assume that if the host benefits the interaction is mutualistic, though it is conceivable that a symbiotic microbe is trapped and "parasitized" by a host that makes use of its unique metabolism. Also, as will be clear in this review, the manifestation of disease does not necessarily mean that the host is disadvantaged relative to is symbiont-free (therefore, non-diseased) counterpart.

The term 'parasitism' is problematic if used as an antonym for mutualism. This term suggests that nutrients are being acquired from a host, a situation very common in both pathogenic and mutualistic symbioses. The term 'antagonism' is better for a symbiosis where one partner benefits to the detriment of the other. If parasitism is restricted to the nutritional sense, endophytic symbionts of plants can all be called parasites regardless of whether they are mutualistic, commensal or antagonistic.

As will be discussed in this chapter, many endophytes simultaneously act as mutualists and as 'pathogens' on host individuals. The sexual endophytes, *Epichloë* species, cause disruption of normal inflorescence development, essentially replacing sexual structures of the plant with their own sexual structures (stromata) (White et al 1991). Host plants produce multiple inflorescences, and in many of these systems only some inflorescences have stromata while most develop normally (Fig. 1). This is not an escape phenomenon because the fungus is transmitted in seeds produced on the normal inflorescences. Furthermore, benefits of these interactions—such as greatly enhanced insect resistance (Clay 1990b)—are a basis for mutualism (the endophytes benefit as well, being provided the means and opportunity to grow, reproduce and disseminate).

Host-symbiont systems, whether mutualistic, antagonistic or otherwise, are more than the sum of their parts. Where the interactions can cause disease it must be remembered that the disease is not a characteristic of the pathogen *per se*, any more than of the host. Rather, disease development depends upon host genotype, pathogen genotype, and a slew of developmental and environmental factors. The host and pathogen genotypes taken together are the 'aegricorpus' genotype (Browder 1985, Loegering 1978), and the disease is a phenotype arising from interactions of aegricorpus genotype, development (e.g. host age, pathogen spore state, etc.) and environment. This concept can be extended to any symbiosis. For example, endophytes produce a number of alkaloids, but their alkaloid expression varies quantitatively or qualitatively with host genotype (different host species or genotypes within a species) (Siegel et al 1990). 'Aegricorpus' is not descriptive of such mutualistic interactions because it literally means 'disease body.' We have coined the term 'symbiotum' (pl., 'symbiota') to mean the interaction of a host individual and symbiont individual (Siegel and Schardl 1991). Thus, individual symbiotum genotypes give rise to certain phenotypes (e.g. alkaloid profiles) under certain environments and developmental stages. If the latter two factors are controlled and specified then the symbiotum can be said to produce certain alkaloids, exhibit partial, complete or no disease, and so forth. The term symbiotum can be applied to any interaction of an individual host and symbiont, whether pathogenic, mutualistic or otherwise.

In the field of plant pathology it is now common to refer to the phenotype of an aegricorpus as compatible or incompatible. Compatibility generally refers to disease, meaning the host is susceptible and the pathogen is virulent. Incompatibility is associated with disease resistance, though the situation can also arise from a lack of appropriate environmental or developmental factors for disease to progress. Grass-endophyte interactions do not fit these concepts of compatibility and incompatibility since there is no tissue damage in natural interactions of clavicipitaceous endophytes. However, when the spore states of these fungi occur the host is frequently inhibited wholly or partly from flowering. Thus, *Neotyphodium* species that do not sporulate on their hosts could be considered incompatible in this regard, but they are nevertheless compatible with vegetative plant parts in which they live, and their compatibility with embryos in seeds allows them to be vertically transmitted. Also, the expression of protective alkaloids or other benefits may be an indication of compatibility. Basically, one may consider compatibility to be that interaction that has been selected on the microbe side (in keeping with the term's usage for plant-pathogen interactions), and would thereby entail the predominant means of fungal dissemination either via spores, vertical transmission in host seeds, or a combination of both. Because survival of the endophyte also depends heavily on survival of the plant in which it lives year-round, compatible interactions should also be typified by healthy plants benefiting from the bioprotective characteristics of the symbiont.

2.2. RANGE OF INTERACTIONS

The continuum of associations between grasses and fungi of family Clavicipitaceae spans mutualism and antagonism (Schardl and Clay 1997). However, even the antagonisms are unusual among plant-pathogen interactions because little or no host tissue is directly damaged, and there is rarely anything resembling the classical defense responses that plants exhibit against many pathogenic fungi and bacteria (Koga et al 1993, Tudzynski et al 1995).

The *Claviceps* species cause ergot disease, a replacement of the ovary in the flower with fungal growth (termed sphacelium), which usually gives rise to a hard resting structure. This resting structure is formally known as a sclerotium and trivially as an ergot. In several *Claviceps* species, most famously *C. purpurea*, the ergots produce potent toxins that affect the central and peripheral nervous systems of insects and vertebrates (Lacey 1991, Raisbeck et al 1991). Thus, *Claviceps* species have negative effects on host fecundity (seed production) as well as on herbivores and granivores.

After overwintering the *C. purpurea* sclerotia germinate to produce stalked structures (i.e. stipate stromata) in which perithecia are imbedded. The perithecia, in turn, contain scores of asci, each with a coil of eight filamentous meiotic spores (ascospores). Once they mature ascospores are forcibly ejected, and those that land on host stigmata initiate new floret infections. In the ensuing sphacelial stage, asexual spores (conidia) are suspended in a sugary 'honeydew' that attracts insects, which in turn spread the infection to other florets. The *C. purpurea* life cycle is reviewed in detail by Tudzynski et al. (1995). The transient associations of *Claviceps* species with plant florets contrast with the systemic associations other members of the family have with plant hosts (Schardl and Clay 1997).

Other plant-associated Clavicipitaceae have such intimate and long-term interactions with plants that they are considered endophytes (though many technically live epiphytically). Systemic infections and perennial associations characterize interactions of grasses and sedges with Clavicipitaceae of genera *Epichloë*, *Atkinsonella*, *Balansia*, *Echinodothis*, *Myriogenospora*, and *Parepichloë* (Leuchtmann and Clay 1988, Leuchtmann and Clay 1989, Rykard et al 1985, Schardl et al 1997, White 1994b, White 1994c, White et al 1996, White and Glenn 1994, White et al 1991, White and Reddy 1998, White et al 1997, White et al 1995). In all such associations host plants are asymptomatic except at certain stages of development or on certain organs. Disease symptoms are strictly associated with fungal stroma or sclerotium development. Often, these structures encompass host florets or entire inflorescences and prevent the florets from developing and producing seeds. This is the case for all known *Atkinsonella* and *Epichloë* species, and for several *Balansia* and *Parepichloë* species (White 1994c, White and Reddy 1998). The other species of clavicipitaceous endophytes can produce stromata on leaves or nodes, yet in several cases their presence in the plant also inhibits flowering. The effects of the endophytes on host plants are always highly restricted in space and magnitude. Despite the developmental effects, there is little or no direct killing of plant tissues (Schardl and Clay 1997).

A key characteristic of most *Epichloë* species and the closely related *Neotyphodium* species is transmission in seeds (Philipson and Christey 1986, Schardl et al 1997, White 1988, White et al 1991). If a host plant is systemically infected with an *Epichloë* sp. endophyte, the seeds produced by that plant may be almost entirely infected with viable fungal hyphae, and the subsequent generation of plants will continue to harbor the endophyte. This process of vertical transmission is characteristic of *Epichloë festucae*, *E. brachyelytri* and *E. elymi*, and occurs in many but not all associations with *E. amarillans*, *E. bromicola*, and *E. sylvatica*. Vertical transmission is rare or nonexistent for *E. typhina*, *E. clarkii*, *E. baconii* and *E. glyceriae* (Leuchtmann and Schardl 1998, Schardl and Leuchtmann 1999).

Seed transmission is obligatory for the many asexual descendants of *Epichloë* species. The asexual endophytes are classified by taxonomic convention in a distinct genus, *Neotyphodium* (Glenn et al 1996; formerly *Acremonium* sect. *Albo-lanosa*), though they are phylogenetic congeners of *Epichloë* (Schardl et al 1994, Schardl et al 1991, Tsai et al 1994). The unusual and complex evolution of *Neotyphodium* species from *Epichloë* species is discussed later in this chapter. The fact that their propagation and proliferation is absolutely linked to host reproduction is important in their evolution because the success of such vertically transmitted symbionts depends absolutely on the success of their infected hosts. Thus, natural *Neotyphodium*-grass associations can be legitimately considered mutualistic. Nevertheless, some such associations may have arisen in hosts that suppress the *Epichloë* sexual state, such that the host has essentially turned the tables on the parasite (Schardl and Clay 1997).

All *Epichloë* species also transmit horizontally, a process mediated by ascospores as in most or all plant-associated Clavicipitaceae. The route of ascospore-mediated transmission shown in Figure 1 is based on a study in which ryegrass seeds became infected when the seed heads developed near mature, fertilized stromata of *E. typhina* (Chung and Schardl 1997a). In addition, recent evidence suggests that *E. sylvatica* ascospores can mediate transmission to adult plants of *Brachypodium sylvaticum* (Brem and Leuchtmann 1999).

2.3. FUNGAL TRANSMISSION, MUTUALISM AND ANTAGONISM

In certain *Epichloë*-grass symbiota flowering tillers are all or nearly all taken over by fungal stromata. The *Epichloë* spp. involved in such associations can be considered the most highly pathogenic of the genus. They clearly exact a cost on host plant fecundity by suppressing seed production. Growth enhancing effects of these endophytes can, however, promote vegetative propagation of the host. Thus, a reduction in fecundity may be partly offset by improved growth (Clay et al 1989). Also, the anti-herbivore effects of the symbionts further mitigate costs to the hosts. *Epichloë* species produce an array of alkaloids, some of which are toxic and some deterrent to vertebrate and invertebrate herbivores (Porter 1995, Rowan 1993, Siegel et al 1990). Especially profound effects against insects have been shown for *Glyceria striata-Epichloë glyceriae* symbiota (Clay et al 1993). Interestingly, when infected with this endophyte *G. striata* produces no seeds, though the grass abundantly proliferates by vegetative propagules (Hitchcock and Chase 1971, p. 81). For such a system it is very difficult to predict *a priori* whether there is a net benefit or detriment to the host. It would be interesting to observe competition between these symbiota and uninfected *G. striata* in natural settings, though such an experiment must control for possible contagious spread of the pathogen.

So far as is known the highly efficient transmission of endophytes in open-pollinated seeds is a unique feature of *Epichloë* and *Neotyphodium* species. Few even of the related fungi are capable of vertical transmission. An example of another vertically transmitted clavicipitaceous symbiont is *Balansia hypoxylon* (= *Atkinsonella hypoxylon*), which transmits in self-pollinated cleistogamous seeds of *Danthonia* spp. and, on rare occasions, can be transmitted via open pollinated seeds (Kover and Clay 1998). Usually open pollinated seeds are not produced in such symbiota because the fungal fruiting structure (stroma) overtakes the developing inflorescence. Also, genetic evidence casts doubt on the importance of vertical transmission in populations of *B. hypoxylon*, and suggests that horizontal transmission via sexually derived spores (ascospores) is prevalent (Kover and Clay 1998). Thus, the most intimate associations between clavicipitaceous endophytes and individual grass plants and lineages are those involving *Neotyphodium* and vertically transmissible *Epichloë* species.

The selfish-gene hypothesis (Dawkins 1989) predicts that reliance of a symbiont on vertical transmission selects for symbiont genotypes that enhance host fitness. Conversely, if the symbiont is incapable of causing net enhancement of host fitness, then the host should be under selection to resist vertical transmission. Whether this scenario is operative for grass-endophyte systems has never been tested, but it is interesting that the highest levels and broadest spectra of protective alkaloids are most often observed in symbiota where seed transmission prevails (Bush et al 1997).

3. Physiological Interactions of Mutualistic Endophytes in Grasses

3.1. HOW DOES THE ENDOPHYTE LIVE?

Cell-to-cell interactions of plants with clavicipitaceous endophytes are very unusual. These fungi can be loosely grouped into two categories based on where their hyphae are localized in the plant. In the classical interaction hyphae grow between plant cells, confined to the intercellular matrices and spaces (the plant apoplast) (Christensen et al 1997, Philipson and Christey 1986, Siegel et al 1987). Most *Neotyphodium* species interdigitate their cell walls with the host intercellular matrix.

Not all clavicipitaceous endophytes actually grow endophytically. *Myriogenospora, Atkinsonella, Echinodothis, Parepichloë* and several *Balansia* species grow epiphytically; specifically, they grow on leaf surfaces and between leaf primordia in the crown of the plant (Clay and Frentz 1993, Leuchtmann and Clay 1988, White 1994b, White and Glenn 1994, White and Reddy 1998). These epibionts may feed partly from the cuticle layer, while also establishing more intimate contact with epidermal cells by extending hyphae under the cuticle. Phylogenetic analysis of rDNA sequence indicates no simple phylogenetic grouping that would separate those species that grow endophytically (i.e. between cells) from those growing epibiotically (Kuldau et al 1997). Interestingly, when *Claviceps purpurea* infects florets it does so by a combination of endophytic and epiphytic growth (Tudzynski et al 1995).

Clavicipitaceae on plants never produce feeding structures resembling haustoria or arbuscules. Furthermore, there are no secondary hyphae growing through host cell walls except rarely in the embryo (Philipson and Christey 1986). It is possible that a bit of the host cell wall and intercellular matrix could be degraded for fungal nutrition. However, there is no report suggesting active invasion of host tissues. The distribution of *N. lolii* hyphae in perennial ryegrass (Herd et al 1997) seems consistent with growth of the fungus alongside plant cells as they divide and enlarge.

Of course, endophytes must obtain nutrition from their hosts. It is possible that leakage of some nutrients into the apoplast is normal, and this might explain the tendency of fungal hyphae to be abundant near nutrient sinks such as meristems. Nevertheless, endophytes almost certainly obtain some nutrition actively. *Epichloë* and *Neotyphodium* species produce invertase, a key enzyme for sucrose assimilation (Lam et al 1994, Lam et al 1995). In addition, they secrete a subtilisin-like proteinase (Reddy et al 1996). In certain *Neotyphodium* associations with *Poa* species the proteinase is extremely abundant, constituting approximately 1% soluble protein in symbiotum leaf sheathes (Lindstrom and Belanger 1994). A reasonable hypothesis is that this proteinase helps the endophyte acquire amino acids from extracellular proteins of the host. Low nutrient status induces expression of the proteinase, indicating that the endophyte in planta grows in a low nutrient environment. This observation is in keeping with the restriction of fungal hyphae to the apoplast. Nevertheless, *Epichloë* and *Neotyphodium* species also utilize simple amino acids (Kulkarni and Nielsen 1986) and inorganic nitrogen sources (Chung and Schardl 1997b) that are generally present in plant intercellular spaces (Pate 1973).

A clue to the intimacy and balance of these associations is the failure of many artificial symbiota where the endophyte is moved among even closely related hosts. For example, endophytes have been exchanged between three closely related species in *Lolium* and *Festuca*, resulting in various degrees and kinds of incompatibility. Often the endophyte failed to thrive in the plant (Koga et al 1993). In these cases ultrastructural examination indicated that adjacent plant cells accumulated densely staining material (possibly phenolics or phenolic polymers) in the cell walls. The endophyte cytoplasm had the appearance of entering a starvation phase, suggesting that the adjacent plant cells were blocking movement of nutrients (possibly also water) to the fungal hyphae. Other incompatible interactions result in plant crown necrosis or severe stunting (Christensen 1995). The latter may be due to the extraordinarily high level of endophyte around phloem companion cells in the vascular bundle, and it is speculated that such a situation may provide the endophyte so much access to nutrients and photosynthetic assimilates as to excessively deprive the host (Christensen et al 1997).

3.2. ENHANCED STRESS TOLERANCE

The most dramatic indication of mutualism in grass-endophyte interactions is that the endophyte helps protect the host from a variety of biotic and abiotic insults. Exactly which stresses are ameliorated by endophytes depends upon the particular symbiota. The best studied interaction is tall fescue with *N. coenophialum*. In these symbiota nearly all biotic or abiotic stress factors measured are reported to be partially abated by the endophyte. Anti-insect activities have been associated with the production of peramine and loline alkaloids (Siegel and Bush 1997). Anti-vertebrate activities are associated with the neurotropic ergot alkaloids, also products of *N. coenophialum* (Cunningham 1949, Lyons et al 1986). Of course, it is conceivable that unknown endophyte metabolites also contribute to anti-herbivore activities.

The physiological and biochemical basis for endophyte-enhanced resistance to the fungal pathogen, *Rhizoctonia zeae* (Gwinn and Gavin 1992), is unknown. This resistance to seedling disease can be very important to the establishment of tall fescue plants where *R. zeae* is a major soil microorganism. Also unknown are the bases for the very dramatic effects of *N. coenophialum* against two important nematodes, *Pratylenchus scribneri* and *Meloidogyne marylandi* (Kimmons et al 1990). Such effects may be due to metabolites produced by the fungus and moved into the roots. It is also possible that the endophyte induces physiological changes in the plant that are inimical to nematode reproduction.

Even abiotic stresses on tall fescue can be ameliorated by *N. coenophialum*. Several studies have indicated enhanced drought tolerance (Arechavaleta et al 1989, West 1994), an effect also of *N. uncinatum* in meadow fescue (Malinowski et al 1997b). Drought tolerance is enhanced even in the absence of biotic stress factors, though anti-nematode activity would obviously reduce potential drought stress. There are conflicting reports on the role of osmotic adjustment in endophyte-enhanced drought tolerance (Hill et al 1996, Richardson et al 1992, West 1994). It seems likely that several mechanisms are involved, working together or separately under different environmental and developmental conditions. Drought tolerance effects also depend on both host and fungal genotype.

Although drought tolerance effects have not been directly attributed to *N. lolii* in perennial ryegrass (Gleason et al 1990), the distribution of perennial ryegrass symbiota in Europe suggests that regions of more frequent water deficit (i.e. the Mediterranean region) select for greater incidences of endophyte symbiosis (Lewis et al 1997).

Recent studies indicate that *N. coenophialum* can enhance phosphorus uptake under low phosphorus availability (Malinowski and Belesky 1999, Malinowski et al 1998). As in drought tolerance, the mechanism for enhanced phosphorus uptake is unknown, depends upon host and fungal genotype and soil characteristics, and is not fully attributable to increased root growth. Though the endophyte is mainly associated with aerial plant tissues, *N. coenophialum* has also been detected in host roots (Azevedo and Welty 1995). However, in contrast to mycorrhizae the *Neotyphodium* and *Epichloë* species are not reported to grow into the soil, so this endophyte effect is probably indirect. The authors speculate that changes in root exudates may be involved in enhanced phosphorus utilization (Malinowski and Belesky 1999, Malinowski et al 1998).

If the endophyte effects are so profound, why don't all grasses have them? One explanation is that the benefit to the grass of maintaining endophyte undoubtedly depends on the particular stresses to which the grass is subjected (Lewis et al 1997). Since these symbioses are perennial and vertically transmitted, cumulative effects of small selective pressures are sure to contribute significantly to patterns observed today. Thus, even slight increases in host fitness may be very important in increasing frequencies of symbiosis. By the same token, even the slight cost of maintaining endophyte could eventually select for its elimination if the endophyte does not fully compensate its host over the long term. This is especially true if there is no option for contagious spread and, as a result, uninfected maternal lineages necessarily remain uninfected. Given that endophyte viability in seed is more easily lost (for example, by heat and humidity) than is viability of the seed itself (Rolston et al 1986), endophytes would expect to go extinct in host lineages on a regular basis. The competitive advantage of endophyte-containing conspecifics (Hill et al 1991) must also be great enough to counteract this effect if the symbiotic systems are to persist.

Some grass species may be so coadapted to their endophytes that the endophyte-free conspecifics are, in essence, ecological abnormalities. Clearly, though endophyte symbiosis is selected in stressful environments; the symbiosis has some cost and the grass must adapt physiologically to the endophyte's presence and activities. For example, endophytes produce compounds (particularly auxin) that regulate plant growth and development (Porter 1994). Gradually, the host genotypes would be selected that best accommodate the endophyte. This may explain, at least in part, why growth and fecundity are apparently improved by endophytes in fescues and ryegrasses (Hill et al 1990, Latch et al 1985, Malinowski et al 1997a, West et al 1988). Basically, removing endophyte from these species may establish the less normal physiological state and thereby significantly reduce growth. This proposition would be difficult to test, but fundamentally speaks to the strategy behind breeding of grasses intended to be used in their symbiotic state. For this, Pedersen and Sleper (1993) advise using symbiotic breeding stock to simultaneously select host, fungus and interactive effects (conversely, if endophyte-free cultivars are to be developed, the initial breeding stock should be endophyte free).

In summary, the high frequency of endophyte symbiosis in tall fescue and certain other species, and the multifarious benefits and growth enhancements due to these endophytes, strongly suggest a long history of coadaptation.

4. Endophyte-Host Coevolution

4.1. HOST SPECIALIZATION IN RELATIONSHIP TO *EPICHLOË* SPECIATION

Until the 1990's the binomial *Epichloë typhina* referred to all sexual clavicipitaceous symbionts that persist benignly in vegetative grasses and produce uniform stromata surrounding the flag leaf sheath of the immature inflorescence; the stromata containing embedded yellow, orange or tan perithecia. Hosts of *Epichloë* (and of the *Neotyphodium* species derived from them) include grasses classified in most recognized tribes of the subfamily Pooideae (Catalán et al 1997, Soreng and Davis 1998). Thus, hosts have been identified from tribes Brachyelytreae (whose inclusion in the Pooideae is still considered uncertain), Brachypodieae, Meliceae, Stipeae and the "crown" tribes, Poeae, Aveneae, Triticeae and Bromeae (Table 1).

Table 1. *Epichloë* species and their host interactions.

Species	Mating population	Host tribe	Transmission strategy	Symbiotic character
E. typhina	I	Aveneae	Horizontal	Antagonistic
		Poeae	Horizontal	Antagonistic
		Brachypodieae	Horizontal	Antagonistic
E. clarkii	I	Poeae	Horizontal	Antagonistic
E. festucae	II	Poeae	Mixed	Balanced
E. elymi	III	Triticeae	Mixed	Balanced
E. amarillans	IV	Aveneae	Mixed	Balanced
E. baconii	V	Aveneae	Horizontal	Antagonistic
E. bromicola	VI	Bromeae	Horizontal or	Antagonistic
		Vertical	or Benign	
E. sylvatica	VII	Brachypodieae	Mixed	Balanced
E. glyceriae	VIII	Meliceae	Horizontal	Antagonistic
E. brachyelytri	IX	Brachyelytreae	Mixed	Balanced

White and Bultman (1987) demonstrated bipolar heterothallism in an *Epichloë* species; that stromata of one mating type must be fertilized with spermatia from the opposite mating type to initiate perithecium development and production of meiotic spores (ascospores). Schardl and Tsai (1992) subsequently demonstrated that asexual spores (conidia) from *Epichloë* cultures were effective as spermatia, and that their mating types were discernible after transfer to stromata of appropriate tester strains. The initial reactions between opposite mating types were similar whether or not the mating gave rise to asci and ascospores, but the development of viable ascospores indicated whether mated strains belonged to the same interfertility group (i.e. mating population). White (1993) then identified distinct interfertility groups in Britain, described the new species *Epichloë baconii* and *E. clarkii*, and emended the description of *E. typhina* based on morphological characteristics of Persoon's type specimen. Since then, seven additional *Epichloë* species have been described in studies of interfertility, morphology, genetic variation, phylogenetics and host specificity (Leuchtmann and Schardl 1998, Leuchtmann et al 1994, Schardl and Leuchtmann 1999, White 1994a).

In White's (1993) description of *E. baconii* and *E. clarkii* he also restricted the definition of *E. typhina* based on morphology of the type specimen on the host, *Dactylis glomerata*. Leuchtmann and Schardl (1998) subsequently clarified the relationships between these species. Fertility tests and morphological evaluations indicated that *E. baconii* was associated with several related species in two genera (*Agrostis* and *Calamagrostis*) of the grass tribe Aveneae. The situation with *E. typhina* and *E. clarkii* is much more complex. Strains that are interfertile with *E. typhina* on *D. glomerata* have been found on diverse grass species within subfamily Pooideae. Furthermore, Leuchtmann and Schardl (1998) found that *E. clarkii* isolates from *Holcus lanatus* (the only host of this species so far identified) were actually interfertile with *E. typhina*. Thus, *E. typhina* and *E. clarkii* were grouped together in mating population I. Nevertheless, a morphological feature of the ascospores and low genetic diversity within the population from *H. lanatus* support the validity of *E. clarkii* as a separate species. A possible explanation is that *E. typhina* X *E. clarkii* progeny, should they occur in nature, have no ecological niche. That is, the progeny may not be able to survive in either parental host or any other host.

It is possible that *E. typhina* as presently defined actually includes several populations isolated by such host specialization. A study by Chung et al. (1997) supports this possibility. Matings between strains from orchardgrass (*Dactylis glomerata*) and perennial ryegrass usually had very low compatibility with both parental hosts. Interestingly, one of 84 progeny tested was able to complete its life cycle on orchardgrass (and this low frequency is likely due to experimental limitations). In fact, the population associated with orchardgrass in nature is much more diverse than populations associated with perennial ryegrass and several other host species (Schardl et al 1997). Thus, the emerging picture is that mating population I comprises subpopulations in restrictive hosts (such as perennial ryegrass and *H. lanatus*) harboring genetically related strains, plus populations in reservoir hosts (such as orchardgrass) that harbor a broader genetic diversity. The reservoir hosts may link most or all of the populations and thereby prevent the evolution of intersterility barriers even between populations on restrictive hosts.

For most other *Epichloë* species there is a clear relationship between host specificity, phylogenetic distinctness, and interfertility. Thus, *E. festucae* comprises phylogenetically related and interfertile strains symbiotic with *Festuca* and some *Lolium* species (Leuchtmann et al 1994; A. Leuchtmann and C.L. Schardl, unpubl. data). Host specificity is also indicated for *E. bromicola* from *Bromus* species (Leuchtmann and Schardl 1998), *E. brachyelytri* from *Brachyelytrum erectum*, *E. elymi* from *Elymus* species, and *E. glyceriae* from *Glyceria striata* (Schardl and Leuchtmann 1999). *Epichloë amarillans* is associated with North American species of *Agrostis, Sphenopholis* and *Calamagrostis*, hosts that form a related group within the Aveneae (Hitchcock and Chase 1971, Stebbins 1976). Although *E. baconii* also occurs on *Agrostis* and *Calamagrostis* species, it is a European species. *Epichloë sylvatica* corresponds to an interfertility group apparently restricted to *Brachypodium sylvaticum*. However phylogenetic analysis suggests that *E. sylvatica* arose from within the broad host range and genetically diverse species, *E. typhina* (Leuchtmann and Schardl 1998).

4.2. EVIDENCE OF COEVOLUTION

The observation that most biological species of *Epichloë* have very restricted host ranges, and that the host ranges of different species are distinct from each other, make reasonable an hypothesis that *Epichloë* species have coevolved with their hosts (Schardl et al 1997). Specifically, this hypothesis would hold that as plant species, genera or tribes diverged from each other, new *Epichloë* species arose in association with them and diverged from each other. Thus, an ancestral *Epichloë* species associated with an ancestral plant taxon would have given rise to daughter species associated with the daughter plant taxa as they emerged. A simple prediction of this kind of coevolution is that phylogenies of hosts and symbionts may track each other [this is actually an overly simplistic prediction because it assumes that the symbiont never speciates except while tracking the divergence of its host (Mitter and Brooks 1983)]. Thus, phylogenetic relationships among *Epichloë* species can be compared to phylogenetic relationships among host tribes to query for evidence of coevolution (Fig. 2).

The coevolution hypothesis can be eliminated for mating population I (*E. typhina* plus *E. clarkii*) because it has not specialized for any distinct host taxon. In fact, the host range of *E. typhina* spans the phylogenetic range of hosts for all other *Epichloë* species, including the sister tribes Poeae and Aveneae, plus the *deeply* branching tribe Brachypodieae (Leuchtmann and Schardl 1998).

The crown group of pooid tribes can serve as a starting point for inspecting the corresponding *Epichloë* and grass phylogenies (Fig. 2). Thus, *E. festucae* is associated with host tribe Poeae (which includes *Festuca*), while the related *E. amarillans* and *E. baconii* are associated with the sister tribe Aveneae. Likewise, *E. elymi* and *E. bromicola* are closely related, as are their host tribes Triticeae and Bromeae, respectively. However, these two *Epichloë* species are also related to *E. glyceriae*, with the specific branching order between them being poorly resolved. The host of *E. glyceriae* is in a tribe (Meliceae) that is deeply rooted relative to the pooid crown group (Catalán et al 1997, Davis and Soreng 1993). Therefore, the phylogenetic position of *E. glyceriae* (Fig. 2) argues against its coevolutionary origin. The positions of Brachyelytreae and Brachypodieae relative to the crown group are consistent with phylogenetic positions of *E. brachyelytri* and *E. sylvatica*, except that an earlier branching to *E. brachyelytri* would have been expected. However, support for the particular branching order in both cases — fungus and host — is not firmly established. Thus, there is good but not absolute correspondence of host and fungus phylogenies, with the position of *E. glyceriae* being the only strong case against phylogenetic tracking, and *E. typhina* / *E. clarkii* representing a case where such tracking is not expected (Schardl et al 1997).

A possible pattern of coevolution emerges when host interactions of *Epichloë* species are taken into account (Schardl et al 1997). Most *Epichloë species* are capable of both vertical and horizontal transmission. In contrast, *E. glyceriae*, *E. typhina*, *E. clarkii* and *E. baconii* are so effective at eliminating host flowering that they preclude vertical transmission (actually, on rare occasions *E. typhina* is transmitted in seeds of *Poa nemoralis*; Leuchtmann and Schardl 1998). Efficient vertical transmission is common in *E. amarillans*, *E. brachyelytri*, *E. sylvatica*, *E. bromicola*, *E. elymi*, and *E. festucae*. The phylogenetic relationships among these six species plus *E. baconii* is a close (but, as stated above, nonidentical) reflection of host phylogenetics (Schardl et al 1997)(Fig. 2).

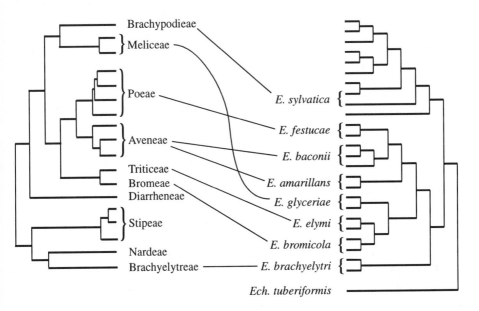

Fig. 2. Comparison of cladograms relating tribes of Pooideae (left) and species of *Epichloë* specialized to those tribes (right). Unlabelled termini are members of *Epichloë* mating population I (*E. typhina* and *E. clarkii*), which, taken as a group, exhibit a lack of host specialization. The outgroup to root the Epichloë tree was *Echinodothis tuberiformis*. Note the relationship of four host tribes in what is called the crown group for purposes of this discussion: sister tribes Poeae and Aveneae, plus sister tribes Triticeae and Bromeae. These relationships are reflected in the relationships of *E. festucae* (on Poeae), *E. baconii, E. amarillans* (both on Aveneae), *E. elymi* (on Triticeae), and *E. bromicola* (on Bromeae), but with *E. glyceriae* out of place. A past colonization of a new host tribe (Meliceae) is therefore hypothesized in the evolution of *E. glyceriae*. Relative to the crown group symbionts, *E. sylvatica* and *E. brachyelytri* are deeply rooted, as are their respective host tribes Brachyelytreae and Brachypodieae. The exact branching order of these tribes is not as firmly established as for the crown group, and in this analysis there is not a precise fit with the early *Epichloë* branches. In fact, the evolution of the *E. typhina/E. clarkii/E. sylvatica* clade appears more complex than can be explained by cospeciation and host jumps. Data for these analyses were compiled from Davis and Soreng (1993) and Schardl et al. (Schardl et al 1997).

Although preliminary, the suggestion of *Epichloë*-grass coevolution is intriguing; but why is this pattern largely restricted to the vertically transmissible *Epichloë* species? A simple answer could be that vertical transmission has entrained the evolutionary pathway to track that of its host. Arguing against this is a very surprising observation: Phylogenies of asexual *Neotyphodium* species (addressed later in this chapter) fail to mirror host phylogenies even though these fungi rely entirely on vertical transmission (Schardl et al 1991). An alternative hypothesis is that the balance between vertical transmission and horizontal transmission is best maintained by coevolution (Schardl et al 1997). The argument is that these interactions maximize benefits to both host and symbiont, providing protection for the former, a habitat and food for the latter, and both clonal and sexual reproduction for both. Extremes where sexual reproduction of either partner is completely suppressed may represent maladaptations and, as such, may be the likely outcome when a symbiont colonizes a new host with which it lacks a history of coevolution.

Interactions between *Epichloë/Neotyphodium* species and host grasses, like many host-symbiont interactions, require exquisite developmental coordination between the partners. In the case of mutualistic interactions this coordination must be particularly finely tuned such that the costs to each partner do not outweigh the mutual benefits. The fungus can exact a cost either by the normal pathogenic cycle, whereby host flowers are prevented from maturing (Schardl and Clay 1997), or by various negative effects on growth (Christensen 1995, Christensen et al 1997). Although it is not widely considered, the plant might also exact net fitness costs on the fungus. For example, it seems likely that certain hosts suppress the sexual state of certain *Epichloë* species, either by failing to trigger or support stroma production, or by actively suppressing this fungal stage. Although this hypothesis has not yet been tested directly, it is noteworthy that *E. festucae* and closely related endophytes are common in *Festuca* and *Lolium* species, but are only known to fruit on *Festuca* subgenus *Festuca* (Leuchtmann et al 1994).

What are the fates, then, of *Epichloë* genotypes that find themselves in new hosts? The most likely is that they fail to establish a compatible interaction and the plant escapes infection. In most, but not all experiments a high level of host specificity dictated this outcome (Leuchtmann and Clay 1993, Leuchtmann and Clay 1997). In those rare occasions of stable infections, incompatibility may still preclude either the sexual state of the fungus or its vertical transmission. If the fungus cannot be transmitted vertically, selection may favor increased spore production, thus enhancing the opportunities for the fungus to spread (Yamamura 1993). Alternatively, if the fungus cannot sporulate it will become an efficient vertically transmitted symbiont intimately tied to its host maternal lineage. Another possibility is, if the host plant is already infected with an endophyte (*Epichloë* or *Neotyphodium* species), then the new arrival and old resident may fuse hyphae. An unusual characteristic of *Epichloë* is that hyphae of different species can fuse and remain viable (Chung and Schardl 1997b). In fact, evidence presented in the next section indicates that such events have occurred numerous times in the past, with the result that new interspecific hybrids have arisen and spread throughout host populations.

4.3. HYBRID ORIGINS OF ENDOPHYTES

Evolutionary pathways of fungi in the *Epichloë/Neotyphodium* complex appear related to the interaction of each species with its host. A dramatic example of this is the tendency of the asexual endophytes (*Neotyphodium* species) to be hybrids. Evidence of hybrid origins includes large genome sizes relative to haploid *Epichloë* species (Kuldau et al 1999), and the occurrence of multiple copies in *Neotyphodium* species of genes that are single copy in the sexual relatives. For example, *N. coenophialum* possesses three β-tubulin genes (*tub2*) and two genes for translation elongation factor 1-α (*tef1*), whereas only a single copy of each gene has been found in each of 43 isolates from 10 *Epichloë* species (2-11 isolates each species) (Schardl et al 1997; T. Hsiau, K.D. Craven and C.L. Schardl, unpublished, Tsai et al 1994). Wherever multiple copies have been found, each copy is much more closely related to the single copies of different *Epichloë* species than to each other. Thus, for the three copies in *N. coenophialum*, one is identical to that of *N. uncinatum* and an *E. typhina* isolate, another to that of *E. festucae*, and a third is related to those of *E. baconii* (Tsai et al 1994; K.D. Craven and C.L. Schardl, unpubl. data)(Fig. 3).

Another clear example of a hybrid endophyte is a rare symbiont of *L. perenne*, exemplified by isolate Lp1 (Christensen et al 1993). This and a similar isolate were obtained in a population in southern France. Also found in southern France is an *E. typhina* genotype associated with *L. perenne*. Much more common than either of these as a symbiont of *L. perenne* is *Neotyphodium lolii*. Although *N. lolii* is genetically very similar to *E. festucae*, extensive tests with isolates from several populations have suggested that *N. lolii* cannot mate with *E. festucae* (pers. obs.). The *Epichloë* and *Neotyphodium* species from *L. perenne* are among the most thoroughly characterized genetically. Molecular phylogenetic analysis of *tub2* and the orotidine-5'-phosphate decarboxylase gene (*pyr4*) both indicate that Lp1 is derived from a hybrid of either *N. lolii* or *E. festucae* with the *L. perenne*-associated *E. typhina* genotype (Collett et al 1995, Schardl et al 1994). The latter also contributed rDNA, but the mitochondrial DNA profile of Lp1 appears similar to that of *N. lolii* rather than *E. festucae* or *E. typhina* (Schardl et al 1994). An RFLP analysis of the 4-(dimethylallyl)tryptophan synthase gene (an enzyme of ergot alkaloid biosynthesis) also supports *N. lolii*, rather than *E. festucae*, as the species that hybridized to *E. typhina* to produce the Lp1 genotype (J. Wang and C.L. Schardl, unpubl. data). Thus, the Lp1 genotype is likely to have originated by hybridization of an asexual endophyte, *N. lolii*, with an *E. typhina* genotype compatible with the same host species. To date, hybrid origins have been indicated for seven out of eight *Neotyphodium* species (Table 2).

Table 2. *Neotyphodium* species and their relationships to *Epichloë* species. [a]

Neotyphodium species	Host species	Closest rDNA relationships	Closest *tub2* relationships	Closest *tefl* relationships
Neotyphodium huerfanum	*Festuca arizonica*	*E. festucae*	*E. typhina, E. festucae*	nd [b]
N. coenophialum	*Lolium arundinaceum*	*E. festucae*	*N. uncinatum, E. baconii, E. festucae*	*E. baconii, E. festucae*
Neotyphodium sp. FaTG-2	*L. arundinaceum*	*E. festucae*	*E. festucae, E. baconii*	nd
Neotyphodium sp. FaTG-3	*L. arundinaceum*	*E. typhina*	*E. typhina, E. baconii*	nd
N. lolii	*L. perenne*	*E. festucae*	*E. festucae*	*E. festucae*
Neotyphodium sp. Lp1	*L. perenne*	*E. typhina*	*E. typhina, E. festucae*	*E. typhina, E. festucae*
N. uncinatum	*L. pratense*	*E. bromicola*	*E. typhina*	*E. bromicola*
Neotyphodium sp. e915	*L. pratense*	nd	*E. typhina*	*E. bromicola, E. festucae*
Neotyphodium sp. e187	*Poa ampla*	*E. typhina*	*E. typhina, E. amarillans*	nd

[a] compiled from Collett et al 1995, Schardl et al 1994, Schardl et al 1996, Tsai et al 1994; C.L. Schardl, C.D. Moon, K.D. Craven unpublished.
[b] nd = not determined

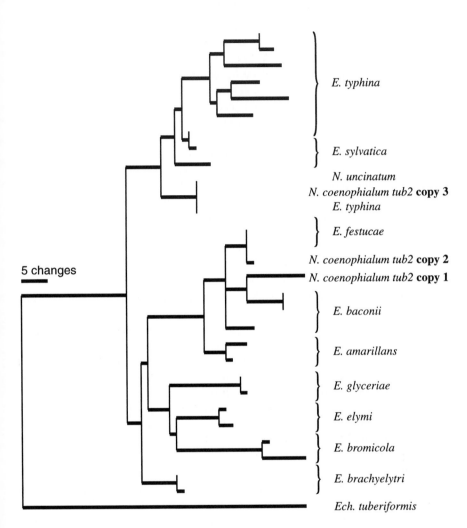

Fig. 3. Phylogenetic relationships of the three β-tubulin gene (*tub2*) copies in *Neotyphodium coenophialum* with the single copies in the *Epichloë* species. The tree shown is one of the three most parsimonious, which differ from each other in placement of the outgroup, *Ech. tuberiformis*. Each substitution and each unambiguously aligned indel (insertion or deletion of one or more nucleotides) is equally weighted. The relationships indicate that each copy is derived from a different ancestral *Epichloë* species, identifying the closest extant relatives of those ancestors as *E. typhina* (or *N. uncinatum*), *E. baconii* and *E. festucae*. Data for this analysis were compiled from Tsai et al. (1994) and Schardl et al. (1997).

What are the cause and effect relationships of hybrid origins and asexual life histories? One possibility is that hybridizations rendered the fungal endophytes incapable of sexual reproduction, while an alternative but not mutually exclusive possibility is that hybrids have strong selective advantage over nonhybrid asexual endophytes. While the former scenario is likely to be true in some instances, it is unlikely to explain all *Neotyphodium* hybrids. For example, if one assumes that hybrids of two *Epichloë* species would be indeed be asexual, then hybrids of three *Epichloë* species would likely have arisen by fusion of a hybrid ancestor with another *Epichloë* species. This would explain the three *tub2* genes of *N. coenophialum* (Tsai et al 1994). The alternative explanation, a single three-way hybridization event, should be very unlikely.

The propensity of *Neotyphodium* species to be hybrids suggests that hybrids have a selective advantage over nonhybrid asexual endophytes. A reasonable hypothesis is that mutational load eventually reduces fitness of the asexual endophytes relative to their sexual ancestors. Muller (1964) originally proposed that strictly clonal lineages inexorably accumulate marginally deleterious mutations, a process now known as Muller's ratchet. Sexual (or parasexual) recombination would counteract the ratchet by regenerating relatively unmutated genotypes.

Three additional considerations are needed in considering the effect of Muller's ratchet on *Neotyphodium* species: the effective population size, genetic bottlenecks, and the mutual importance of host and symbiont on each other's fitness. A large effective population size should slow the ratchet (Felsenstein 1974). Many asexual fungi produce huge numbers of propagules, but the *Neotyphodium* spp. population sizes are probably much more restricted because of the link to host seed production. On the other hand, bottlenecks in which genotypes must compete could help eliminate the more heavily mutated genotypes and are predicted to slow (not stop) Muller's ratchet (Bergstrom and Pritchard 1998). Such bottlenecks might well occur since very few hyphae typically invade each new tiller (Schmid et al 1999) and ovule, as well as each embryo of developing seeds (Philipson and Christey 1986, White et al 1991). It has been suggested (J. Schmid, pers. com.) that such bottlenecks counter Muller's ratchet and enhance long term fitness of the endophytes. This is a particularly intriguing idea in systems were the endophyte plays a major role in host fitness. Considerable selection should act on the symbiotum as a unified system. Frequent bottlenecking of the endophyte might be an important mechanism in maintaining symbiotum fitness.

Coupling genetic bottlenecking and strong selection on symbiota might slow Muller's ratchet, but some form of genetic exchange still seems necessary to actually counteract it. Interspecific hybridization between resident endophytes and co-infecting *Epichloë* species should do the trick. The *Epichloë* genome involved would most likely be derived from meiosis, since ascospores mediate horizontal transmission (Brem and Leuchtmann 1999, Chung and Schardl 1997a). Thus, the new genome would be from a lineage with a recent history of sex counteracting the ratchet. Also, the entire *Epichloë* genome can be absorbed by the endophyte, as apparently occurred in the origin of the Lp1 genotype in perennial ryegrass (Schardl et al 1994). Thus, any and all deleterious mutations in genes for life functions and symbiotic competence could be complemented by the incoming genome. Therefore, assuming Muller's ratchet has a significant effect on asexual endophytes it provides a very logical explanation why most such endophytes are hybrids.

5. Bioprotective Alkaloids

The clavicipitaceous endophytes produce a variety of beneficial metabolites of which four alkaloid classes are best known: lolines (saturated 1-aminopyrrolizidines with an oxygen bridge), peramine (a pyrrolopyrazine likely derived from cyclic prolylarginine), ergot alkaloids, and indolediterpenes. The lolines are insecticidal (Bush et al 1993), and peramine deters herbivory by insects (Rowan and Gaynor 1986). Peramine is thought to be very important in control of the Argentine stem weevil, an extremely damaging herbivore of perennial ryegrass accidentally introduced into New Zealand (Rowan 1993). Ergot alkaloids and indolediterpenes are well known for their anti-vertebrate activities.

5.1. ERGOT ALKALOIDS

The ergot fungus, *Claviceps purpurea*, synthesizes ergot alkaloids very similar to those of clavicipitaceous endophytes. A well-known hazard to human health is poisoning by ergots contaminating rye and other grains (Tudzynski et al 1995). The ergots contain very high levels of the ergopeptines, ergotamine and ergocristine (Porter et al 1987, Reháček 1991). Some major historic events are thought to be associated with ergotism. The problem is believed to be a major trigger of the first crusade (Billings 1996), and centuries later to have halted the march on the Ottoman Empire by Peter the Great of Russia (Kavaler 1965). Beyond dispute is the contribution of an ergot alkaloid derivative, lysergic acid diethylamide (LSD), to the drug culture of the 1960's. Symptoms of ergotism include neurological disorders as well as vascular constriction so severe as to cause gangrene of the limbs and death. Chronic exposure to ergopeptines can cause major fertility problems and difficulties in delivery (though low doses of ergotamine are sometimes used to induce labor). Similar effects on livestock (Thompson and Stuedemann 1993) and the strong association with grazing of tall fescue led to a search for relatives of *C. purpurea* as the cause of fescue toxicosis (Bacon et al 1975). Soon afterward, a tall fescue endophyte now known as *N. coenophialum* was implicated as the source of ergot alkaloids in tall fescue (Bacon et al 1977, Lyons et al 1986). Particularly abundant in tall fescue-*N. coenophialum* symbiota is ergovaline, which closely resembles ergotamine in structure and activity (Dyer 1993, Larson et al 1995, Strickland et al 1994, Zhang et al 1994).

Ergot alkaloids are present in some but not all grasses possessing symbiotic *Neotyphodium* or *Epichloë* species (Siegel et al 1990). Very high levels of ergonovine and lysergic acid amide are found in *Achnatherum inebrians* (=*Stipa inebrians*) from the central Asian steppe (Miles et al 1996). This grass is largely protected from grazing (Gwinn et al 1998), probably because of these alkaloids. In the southwestern United States a related grass, *Stipa robusta*, typically harbors a *Neotyphodium* sp. endophyte, has high levels of ergot alkaloids, and is known as sleepy grass for its effects on horses (Petroski et al 1992). As an example of 'horse sense,' animals that experience the effects of sleepy grass avoid it thereafter (understanding why humans often have the opposite response to LSD could address an important social problem).

A particularly interesting case is the attempted use in New Zealand of an endophyte isolate, Lp1, from a wild population of the common pasture grass, *Lolium perenne* (perennial ryegrass). This isolate was identified in a search of *L. perenne* endophytes in Europe, the grass's center of origin, and was determined by extensive genetic analysis to be a hybrid of *E. typhina* and *N. lolii* (Collett et al 1995, Schardl et al 1994). The agronomic importance of Lp1 was that it produced little or no indolediterpene alkaloid (Christensen et al 1993), known to be a toxicity problem to sheep grazed on *L. perenne* pastures (discussed below). Isolate Lp1 was introduced into *L. perenne* and *Lolium* hybrids as breeding lines for cultivar development. However, as trials began in the summer of 1992-93, most Lp1-containing cultivars were withdrawn due to very high levels of ergovaline and preliminary indications of ergotism in the grazing sheep (Anonymous 1992).

The biosynthetic pathway to lysergic acid and ergotamine in *C. purpurea* is well studied (Shibuya et al 1990). Assuming the same pathway in *Neotyphodium* species, an appropriate strategy to address tall fescue toxicosis is to genetically engineer *N. coenophialum* and knock out a key step. Toward this end, recent efforts are directed toward the first determinant and rate limiting step, dimethylallyltryptophan synthase (Fig. 4). This enzyme was purified to homogeneity from a clavine alkaloid producing culture of *Claviceps fusiformis* ATCC 26245 (Gebler and Poulter 1992), peptide fragments were isolated, and their N-terminal sequences used in cloning the gene (Tsai et al 1995). The identity of the gene, designated *dmaW*, was confirmed by heterologous expression in the yeast, *Saccharomyces cerevisiae*, and authentication of the enzyme product as 4-(γ,γ-dimethylallyl)tryptophan. The homologous gene has apparently been cloned from *C. purpurea* (Tudzynski et al 1999) and two *Neotyphodium* species (J. Wang and C.L. Schardl, unpubl. data), though these have not yet been authenticated by enzyme activity assays. There are two transcribed homologues in *N. coenophialum*. If both are found to encode active enzyme, they will both need to be disrupted to knock out the ergot alkaloid biosynthetic pathway. Once this is accomplished, the modified endophyte can be reintroduced into tall fescue lines to generate breeding stock to be tested for forage.

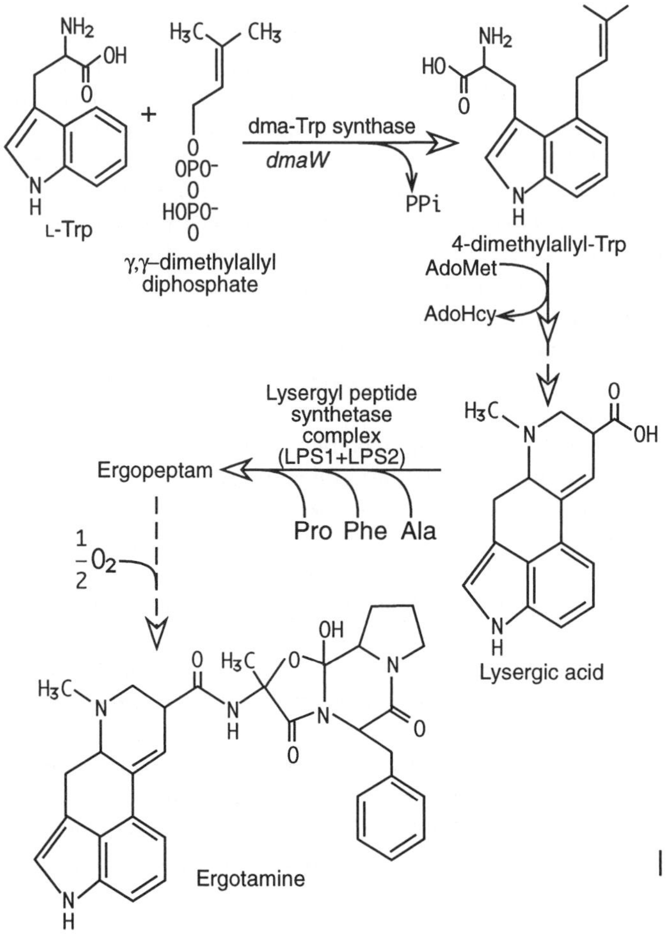

Fig. 4. Summary of the biosynthetic pathway for lysergic acid and the ergopeptine derivative, ergotamine (Shibuya et al 1990), in *Claviceps purpurea*. Standard amino acid abbreviations are used, and dashed arrows indicate multiple steps. It is likely that ergovaline synthesis in the *Neotyphodium* species is similar except that the penultimate step (lysergyl peptide synthetase; LPS) would link lysergic acid with Ala, Val and Pro. Logical targets for genetic disruption to address livestock toxicosis are the gene (*dmaW*) for the first determinant step, and the LPS genes.

5.2. LOLITREMS

Indolediterpenes are indole alkaloids, but otherwise have structures and functions very different from those of ergot alkaloids. The likely biosynthetic pathways of common indolediterpenes, such as paxilline and the lolitrems, have largely been inferred by isolation of likely intermediates and some precursor feeding studies (Gatenby et al 1999, Mantle and Weedon 1994, Miles et al 1992) (Fig. 5). It is likely that the indolediterpenes play a role in protection both from insects and vertebrate herbivores (Rowan 1993). Toxicity to livestock is associated with the lolitrems, which are prenylated derivatives of the indolediterpene, paxilline. These alkaloids have the sobriquet, tremorgens, referring to associated symptoms suffered by grazing livestock. Indolediterpene tremorgens are known from a variety of fungal sources, including *C. paspali* (Cole et al 1977) and more distantly related ascomycetes. An abundant source of paxilline in culture is *Penicillium paxilli*, which, however, is not known to elaborate this compound into lolitrems (Mantle and Weedon 1994, Penn and Mantle 1994).

In a survey of grass species with *Neotyphodium* and *Epichloë* spp. endophytes, Siegel et al. (1990) found lolitrems infrequently. However, reports of staggers associated with *Neotyphodium*-containing grasses are common, particularly in the southern hemisphere (Cabral et al 1999, Coetzer et al 1985, Miles et al 1998). Also, endophytes not associated with lolitrems in planta have been reported to synthesize these compounds in culture (reviewed in Siegel and Bush 1997). Thus, it is worth considering whether these compounds can be produced at low levels, whether under certain circumstances their levels can reach physiologically active concentrations, and whether even at low levels they may act in synergy with other metabolites produced by the fungus or plant.

Good information about the physiological activity of lolitrems has emerged only recently. These alkaloids inhibit high-conductance Ca^{2+}-activated K^+ (maxi-K) channels (Knaus et al 1994). They probably cause tremors by cholinergic activity in smooth and skeletal muscle, possibly as an indirect consequence of maxi-K channel inhibition (McLeay et al 1999).

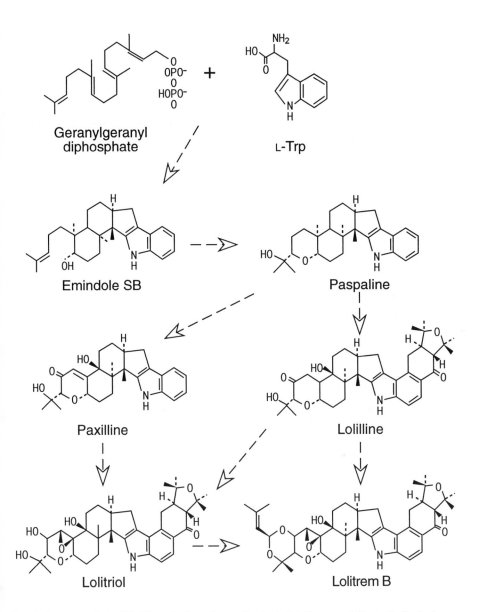

Fig. 5. Summary of possible biosynthetic pathways for the indolediterpene, lolitrem B. Dashed arrows indicate multiple steps are likely. Each of the metabolites shown has been identified in indolediterpene-producing fungal cultures. This scheme is based on those proposed by Gatenby et al. (1999) and Mantle and Weedon (1994).

5.3. LOLINES AND PERAMINE

Anti-insect effects of endophytes are well known and established for a wide variety of grasses (Clay 1990b, Funk et al 1994). It is likely that both known and unknown metabolites and physiological effects are involved in protection from insect herbivores (Popay and Rowan 1994). All four of the known alkaloid classes from *Neotyphodium* species have anti-insect activity. Lolines (saturated aminopyrrolizidines) and peramine (a pyrrolopyrazine) are especially active and specific against insects.

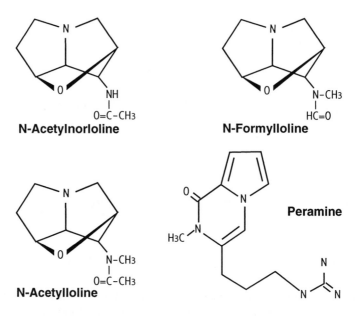

Fig. 6. The structures of three loline alkaloids and peramine, all of which are implicated in anti-insect activities.

Loline alkaloids (Fig. 6) are toxic to insects when applied in acetone solution to the cuticle or when added to nutrient medium (Siegel and Bush 1997, Yates et al 1989). Lolines and peramine are associated with activity against aphids on natural plant-endophyte symbiota (Siegel et al 1990). Such correlative information is intriguing, but a more definitive demonstration of the role is preferable. For this, a Mendelian analysis was conducted of *E. festucae* segregating for loline alkaloid production and effects on the greenbug aphid (*Schizaphis graminum*) (Fig. 7) and bird-cherry oat aphid (H.H. Wilkinson and C. L. Schardl, unpubl. data). In these studies, the statistically significant effect against aphids was entirely attributable to loline alkaloid expression. Since aphids are phloem feeders this finding suggests that lolines move into phloem, a scenario supported by the observation that lolines are highly mobile in plants where they move away from the main concentration of endophytes into leaf blades and roots (Bush et al 1997, Justus et al 1997). In addition to reducing herbivory, lolines may be principally responsible for an additional host benefit, a decrease in infection by the aphid transmitted barley yellow dwarf virus (Mahmood et al 1993).

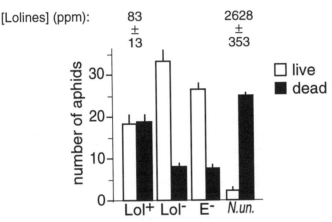

Fig. 7. Response of *Schizaphis graminum* (greenbug aphid) to meadow fescue plants symbiotic with Lol⁺ vs. Lol⁻ full-sibling progeny from an *E. festucae* cross, endophyte-free (E–) meadow fescue, and meadow fescue with its common endophyte *Neotyphodium uncinatum* (*N.un.*). Error bars are standard error of the mean (SE). Loline levels ± SE are given for plants with Lol⁺ progeny and *N. uncinatum*. Data are from Heather H. Wilkinson, M. R. Siegel and C. L. Schardl, unpublished observations.

Though the biosynthesis of loline alkaloids is by an unknown pathway, feeding studies with radiolabeled putrescine and spermidine, together with known loline structures, suggest that they are most likely derived from a polyamine (Bush et al 1997). Plant pyrrolizidines are derived from homospermidine (Graser and Hartmann 1997), but spermidine or spermine is the likely precursor for lolines. Other differences from common plant pyrrolizidines are that lolines have a saturated necine ring structure and only simple decorations on the pyrrolizidine ring. In contrast, plant pyrrolizidines tend to have a monounsaturated ring and a large aliphatic chain or ring, and often are

N-oxygenated. All of these features of plant pyrrolizidines are associated with hepatotoxicity in mammals (Hartmann 1999, Hincks et al 1991), whereas lolines have no reported cytotoxicity.

Peramine (Fig. 6) has been characterized as a feeding deterrent to insect herbivores, and is best known for protecting perennial ryegrass from the Argentine stem weevil (Prestidge et al 1985). Such protection is essential for survival of the grass in much of New Zealand, where this South American insect would otherwise be a major pest on this European grass (Rowan and Latch 1994). Of course, this is an artificial association of plant and insect, but the frequent occurrence of peramine in a wide variety of symbiota suggests its expression has also been selected in many natural circumstances. Like lolines, peramine is associated with activity against greenbug aphid, but unlike lolines it has no significant effect on bird-cherry oat aphid (Siegel et al 1990). The biosynthesis of peramine is unstudied. However, its simple structure suggests cyclic prolylarginine as a precursor. From there, two oxidative desaturations and a methylation would yield the alkaloid (Porter 1994).

6. Conclusions

For centuries Clavicipitaceae received steady attention by mycologists and secondary product biochemists because of their importance in human pharmaceuticals and as threats to the food supply. Then, there ensued a real explosion of research following the 1977 identification of the tall fescue endophyte and realization of its impact on the U.S. beef industry. In the brief time since then there have been extensive toxicological, ecological, taxonomic, biochemical and evolutionary studies focussing on how clavicipitaceous endophytes interaction with plants, insects, mammals, birds, nematodes, other fungi and abiotic stress factors. Four major classes of alkaloids have become closely associated with protective effects against certain insects and mammals. Many of their evolutionary relationships, particularly among *Epichloë* and *Neotyphodium* species, have been well illuminated. Modification of alkaloid profiles by genetic engineering is quickly approaching reality.

As the field matures and we feel some intimate understanding of these systems, this is an excellent time to consider the many things remaining to be learned. Their widespread occurrence in natural grasses and diversity of chemical profiles (sometimes lacking the known alkaloids) begs the question, what are the selective forces that maintain and propagate these fungal symbionts in diverse ecosystems? Drought tolerance effects and protection from other pathogens and parasites are obvious possibilities, but the biochemical and physiological mechanisms for these benefits remain mysterious. Also, what happens on evolutionary time scales as grasses colonize new habitats, some where endophytes are important for survival, and others where they may provide little benefit and be too costly to maintain? What dictates the balance of mutualistic and antagonistic effects of endophytes? Why are many grasses commonly symbiotic with *Neotyphodium* or *Epichloë* species, but many of their close relatives are not? How much of an effect do the endophytes have on grass evolution? Can renewed investigations reveal novel metabolites responsible for increasing reports of livestock toxicoses or insect protection not readily explained by the known alkaloids? Can the mechanisms of host protection be somehow borrowed for use in grain crops or other crops lacking these endophytes?

7. Acknowledgments

Research by the author and coworkers receives the following support: U.S. National Science Foundation grants DEB-9707427 on endophyte evolution and IBN-9808554 on loline alkaloid biosynthesis, and U.S. Department of Agriculture National Research Initiative grant 98-35303-6663 on ergot alkaloid biosynthesis.

8. References

Anonymous (1992) New ryegrass withdrawn after side-effects found. Evening Standard, Palmerston North, New Zealand, October 16, 1992, pp 31.

Arechavaleta M, Bacon CW, Hoveland CS, Radcliffe DE (1989) Effect of the tall fescue endophyte on plant response to environmental stress. Agron J **81**, 83-90.

Azevedo MD, Welty RE (1995) A study of the fungal endophyte *Acremonium coenophialum* in the roots of tall fescue seedlings. Mycologia **87**, 289-297.

Bacon CW, Porter JK, Robbins JD (1975) Toxicity and occurrence of *Balansia* on grasses from toxic fescue pastures. Appl Microbiology **29**, 553-556.

Bacon CW, Porter JK, Robbins JD, Luttrell ES (1977) *Epichloe typhina* from toxic tall fescue grasses. Appl Environ Microbiol **34**, 576-581.

Ball DM, Pedersen JF, Lacefield GD (1993) The tall-fescue endophyte. American Scientist 81, 370-379.

Berde B, Schild HO (1978) Ergot alkaloids and related compunds. Handbook of Experimental Pharmacology 49. New York: Springer-Verlag.

Bergstrom CT, Pritchard J (1998) Germline bottlenecks and the evolutionary maintenance of mitochondrial genomes. Genetics **149**, 2135-2146.

Billings M (1996) The Crusades: five centuries of holy wars. New York: Sterling Publishing Company.

Brem D, Leuchtmann A (1999) High prevalence of horizontal transmission of the fungal endophyte *Epichloë sylvatica*. Bulletin of the Geobotanical Institute ETH **65**, 3-12.

Browder LE (1985) Parasite:host:environment specificity in the cereal rusts. Ann Rev Phytopathol **23**, 201-222.

Bush LP, Fannin FF, Siegel MR, Dahlman DL, Burton HR (1993) Chemistry, occurrence and biological effects of saturated pyrrolizidine alkaloids associated with endophyte-grass interactions. Agric Ecosyst Environ **44**, 81-102.

Bush LP, Wilkinson HH, Schardl CL (1997) Bioprotective alkaloids of grass-fungal endophyte symbioses. Plant Physiology **114**, 1-7.

Buta JG, Spaulding DW (1989) Allelochemicals in tall fescue—abscisic and phenolic acids. J Chem Ecol **15**, 1629-1636.

Cabral D, Cafaro MJ, Saidman B, Lugo M, Reddy PV, White JF, Jr (1999) Evidence supporting the occurrence of a new species of endophyte in some South American grasses. Mycologia **91**, 315-325.

Catalán P, Kellogg EA, Olmstead RG (1997) Phylogeny of Poaceae subfamily Pooideae based on chloroplast *ndhF* gene sequences. Mol Phylogenet Evol **8**, 150-166.

Christensen MJ (1995) Variation in the ability of *Acremonium* endophytes of perennial rye-grass (*Lolium perenne*), tall fescue (*Festuca arundinacea*) and meadow fescue (*F. pratensis*) to form compatible associations in three grasses. Mycol Res **99**, 466-470.

Christensen MJ, Ball OJ-P, Bennett R, Schardl CL (1997) Fungal and host genotype effects on compatibility and vascular colonisation by *Epichloë festucae*. Mycol Res **101**, 493-501.

Christensen MJ, Leuchtmann A, Rowan DD, Tapper BA (1993) Taxonomy of *Acremonium* endophytes of tall fescue (*Festuca arundinacea*), meadow fescue (*F. pratensis*), and perennial rye-grass (*Lolium perenne*). Mycol Res **97**, 1083-1092.

Chung K-R, Hollin W, Siegel MR, Schardl CL (1997) Genetics of host specificity in *Epichloë typhina*. Phytopathology **87**, 599-605.

Chung K-R, Schardl CL (1997a) Sexual cycle and horizontal transmission of the grass symbiont, *Epichloë typhina*. Mycol Res **101**, 295-301.

Chung K-R, Schardl CL (1997b) Vegetative compatibility between and within *Epichloë* species. Mycologia **89**, 558-565.

Clay K (1983) The differential establishment of seedlings from chasmogamous and cleistogamous flowers in natural populations of the grass *Danthonia spicata* (L.) Beauv. Oecologia **57**, 183-188.

Clay K (1990a) Comparative demography of three graminoids infected by systemic, clavicipitaceous fungi. Ecology **7**, 558-570.

Clay K (1990b) Fungal endophytes of grasses. Annu Rev Ecol System **21**, 275-295.

Clay K (1994) Hereditary symbiosis in the grass genus *Danthonia*. New Phytol **126**, 223-231.

Clay K, Cheplick GP, Marks S (1989) Impact of the fungus *Balansia henningsiana* on *Panicum agrostoides*: frequency of infection, plant growth and reproduction and resistance to pests. Oecologia **80**, 374-380.

Clay K, Frentz IC (1993) *Balansia pilulaeformis*, an epiphytic species. Mycologia **85**, 527-534.

Clay K, Jones JP (1984) Transmission of the fungus *Atkinsonella hypoxylon* (Clavicipitaceae) by cleistogamous seed of *Danthonia spicata* (Gramineae). Can J Bot **62**, 2893-2898.

Clay K, Marks S, Cheplick GP (1993) Effects of insect herbivory and fungal endophyte infection on competitive interactions among grasses. Ecology **74**, 1767-1777.

Clement SL, Kaiser WJ, Eichenseer H (1994) Acremonium endophytes in germplasms of major grasses and their utilization for insect resistance. In: Bacon CW, White JF, Jr. eds. Biotechnology of Endophytic Fungi of Grasses. Boca Raton, Florida: CRC Press. pp 185-199.

Coetzer JAW, Kellerman TS, Naude TW (1985) Neurotoxicoses of livestock caused by plants and fungi in southern Africa. Onderstepoort, South Africa: Technical Communication, Department of Agriculture, South Africa. 38 pp.

Collett MA, Bradshaw RE, Scott DB (1995) A mutualistic fungal symbiont of perennial ryegrass contains two different *pyr4* genes, both expressing orotidine-5'-monophosphate decarboxylase. Gene **158**, 31-39.

Cross DL (1997) Fescue toxicosis in horses. In: Hill NS, Bacon CW eds. Neotyphodium/grass interactions. New York: Plenum Press. pp 289-309.

Cunningham IJ (1949) A note on the cause of tall fescue lameness in cattle. Australian Vet J **25**, 27-28.

Davis JI, Soreng RJ (1993) Phylogenetic structure in the grass family (Poaceae) as inferred from chloroplast DNA restriction site variation. Am J Bot **80**, 1444-1454.

Dawkins R (1989) The Selfish Gene. Oxford: Oxford University Press. 352 pp.

Diehl WW (1950) Balansia and the Balansiae in America. Agricultural Monograph 4. Washington, D.C.: United States Department of Agriculture. Agricultural Monograph. 82 pp.

Dyer DC (1993) Evidence that ergovaline acts on serotonin receptors. Life Sci **53**, L223-PL228.

Felsenstein J (1974) The evolutionary advantage of recombination. Genetics **78**, 737-756.

Funk CR, Belanger FC, Murphy JA (1994) Role of endophytes in grasses used for turf and soil conservation. In: Bacon CW, White JF, Jr. eds. Biotechnology of Endophytic Fungi of Grasses. Boca Raton, Florida: CRC Press. pp 201-209.

Gallagher RT, Hawkes AD, Steyn PS, Vleggaar R (1984) Tremorgenic neurotoxins from perennial ryegrass causing ryegrass staggers disorder of livestock: structure elucidation of lolitrem B. J Chem Soc Chem Commun **1984**, 614-616.

Gatenby WA, Munday-Finch SC, Wilkins AL, Miles CO (1999) Terpendole M, a novel indole-diterpenoid isolated from *Lolium perenne* infected with the endophytic fungus *Neotyphodium lolii*. J Agric Food Chem **47**, 1092-1097.

Gebler JC, Poulter CD (1992) Purification and characterization of dimethylallyltryptophan synthase from *Claviceps purpurea*. Archives of Biochemistry and Biophysics **296**, 308-313.

Gleason ML, Christians NE, Agnew M (1990) Effect of endophyte infection of perennial ryegrass on growth under drought stress. Phytopathology **80**, 1031 (Abstr.).

Glenn AE, Bacon CW, Price R, Hanlin RT (1996) Molecular phylogeny of *Acremonium* and its taxonomic implications. Mycologia **88**, 369-383.

Graser G, Hartmann T (1997) Biosynthetic incorporation of the aminobutyl group of spermidine into pyrrolizidine alkaloids. Phytochemistry **45**, 1591-1595.

Gwinn KD, Fribourg HA, Waller JC, Saxton AM, Smith MC (1998) Changes in *Neotyphodium coenophialum* infestation levels in tall fescue pastures due to different grazing pressures. Crop Sci **38**, 201-204.

Gwinn KD, Gavin AM (1992) Relationship between endophyte infestation level of tall fescue seed lots and *Rhizoctonia zeae* seedling disease. Plant Dis 76, 911-914.

Hartmann T (1999) Chemical ecology of pyrrolizidine alkaloids. Planta **207**, 483-495.

Herd S, Christensen MJ, Saunders K, Scott DB, Schmid J (1997) Quantitative assessment of in planta distribution of metabolic activity and gene expression of an endophytic fungus. Microbiology **143**, 267-275.

Hill NS, Belesky DP, Stringer WC (1991) Competitiveness of tall fescue (*Festuca arundinacea* Schreb.) as influenced by endophyte (*Acremonium coenophialum* Morgan-Jones and Gams). Crop Sci **31**, 185-190.

Hill NS, Pachon JG, Bacon CW (1996) *Acremonium coenophialum*-mediated short- and long-term drought acclimation in tall fescue. Crop Sci **36**, 665-672.

Hill NS, Stringer WC, Rottinghaus GE, Belesky DP, Parrot WA, Pope DD (1990) Growth, morphological, and chemical component responses of tall fescue to *Acremonium coenophialum*. Crop Sci **30**, 156-161.

Hincks JR, Kim HY, Segall HJ, Molyneux RJ, Stermitz FR, Coulombe RA (1991) DNA cross-linking in mammalian cells by pyrrolizidine alkaloids: structure-activity relationships. Toxicol Appl Pharmacol **111**, 90-98.

Hitchcock AS, Chase A (1971) Manual of the Grasses of the United States. New York: Dover Publications, Inc. 1051 pp.

Hoveland C (1993) Importance and economic significance of the *Acremonium* endophytes to performance of animals and grass plants. Agric Ecosyst Environ **44**, 3-12.

Justus M, Witte L, Hartmann T (1997) Levels and tissue distribution of loline alkaloids in endophyte-infected *Festuca pratensis*. Phytochemistry **44**, 51-57.

Kavaler L (1965) Mushrooms, Molds, and Miracles: The Strange Realm of Fungi. New York: The John Day Company, Ltd. 318 pp.

Kimmons CA, Gwinn KD, Bernard EC (1990) Nematode reproduction on endophyte-infected and endophyte-free tall fescue. Plant Dis **74**, 757-761.

Knaus HG, McManus OB, Lee SH, Schmalhofer WA, Garcia-Calvo M, Helms L, Sanchez M, Giangiacomo K, Reuben JP, Smith A, III, Kaczorowski GJ, Garcia ML (1994) Tremorgenic indole alkaloids potently inhibit smooth muscle high-conductance calcium-activated potassium channels. Biochem **33**, 5819-5828.

Koga H, Christensen MJ, Bennett RJ (1993) Cellular interactions of some grass/*Acremonium* endophyte associations. Mycol Res **97**, 1237-1244.

Kover PX, Clay K (1998) Trade-off between virulence and vertical transmission and the maintenance of a virulent plant pathogen. Am Nat **152**, 165-175.

Kuldau GA, Liu J-S, White JF, Jr., Siegel MR, Schardl CL (1997) Molecular systematics of Clavicipitaceae supporting monophyly of genus *Epichloë* and form genus *Ephelis*. Mycologia **89**, 431-441.

Kuldau GA, Tsai H-F, Schardl CL (1999) Genome sizes of *Epichloë* species and anamorphic hybrids. Mycologia **91**, 776-782.

Kulkarni RK, Nielsen BD (1986) Nutritional requirements for growth of a fungal endophyte of tall fescue grass. Mycologia **78**, 781-786.

Lacey J (1991) Natural occurrence of mycotoxins in growing and conserved forage crops. In: Smith JE, Henderson RS eds. Mycotoxins and Animal Foods. Boca Raton, Florida: CRC Press. pp 363-414.

Lam CK, Belanger FC, White JF, Jr., Daie J (1994) Mechanism and rate of sugar uptake by *Acremonium typhinum*, an endophytic fungus infecting *Festuca rubra*: Evidence for presence of a cell wall invertase in endophytic fungi. Mycologia **86**, 311-460.

Lam CK, Belanger FC, White JF, Jr., Daie J (1995) Invertase activity in *Epichloë/Acremonium* fungal endophytes and its possible role in choke disease. Mycol Res **99**, 867-873.

Larson BT, Samford MD, Camden JM, Piper EL, Kerley MS, Paterson JA, Turner JT (1995) Ergovaline binding and activation of D2 dopamine receptors in GH4ZR7 cells. J Animal Sci **73**, 1396-1400.

Latch GCM, Hunt WF, Musgrave DR (1985) Endophytic fungi affect growth of perennial ryegrass. N Z J Agric Res **28**, 165-168.

Leuchtmann A, Clay K (1988) *Atkinsonella hypoxylon* and Balansia cyperi, epiphytic members of the Balansiae. Mycologia **80**, 192-199.

Leuchtmann A, Clay K (1989) Morphological, cultural and mating studies in *Atkinsonella, including A. texensis*. Mycologia **81**, 692-701.

Leuchtmann A, Clay K (1993) Nonreciprocal compatibility between *Epichloë* and four host grasses. Mycologia **85**, 157-163.

Leuchtmann A, Clay K (1997) Population biology of grass endophytes. In: Carroll GC, Tudzynski P eds. The Mycota V: Plant Relationships, part B, Berlin: Springer-Verlag. pp 185-202.

Leuchtmann A, Schardl CL (1998) Mating compatibility and phylogenetic relationships among two new species of *Epichloë* and other congeneric European species. Mycol Res **102**, 1169-1182.

Leuchtmann A, Schardl CL, Siegel MR (1994) Sexual compatibility and taxonomy of a new species of *Epichloë* symbiotic with fine fescue grasses. Mycologia **86**, 802-812.

Lewis DH (1985) Symbiosis and mutualism: crisp concepts and soggy semantics. In: Boucher DH eds. The Biology of Mutualism. New York: Oxford University Press. pp 29-39.

Lewis GC, Ravel C, Naffaa W, Astier C, Charmet G (1997) Occurrence of *Acremonium* endophytes in wild populations of *Lolium* spp. in European countries and a relationship between level of infection and climate in France. Ann Appl Biol **130**, 227-238.

Lindstrom JT, Belanger FC (1994) Purification and characterization of an endophytic fungal proteinase that is abundantly expressed in the infected host grass. Plant Physiol **106**, 7-16.

Loegering WQ (1978) Current concepts in interorganismal genetics. Ann Rev Phytopathol **16**, 309-320.

Luttrell ES, Bacon CW (1977) Classification of *Myriogenospora* in the Clavicipitaceae. Can J Bot **55**, 2090-2097.

Lyons PC, Plattner RD, Bacon CW (1986) Occurrence of peptide and clavine ergot alkaloids in tall fescue grass. Science **232**, 487-489.

Madej CW, Clay K (1991) Avian seed preference and weight loss experiments: the role of fungal endophyte-infected tall fescue seeds. Oecologia **88**, 296-302.

Mahmood T, Gergerich RC, Milus EA, West CP, D'arcy CJ (1993) Barley yellow dwarf viruses in wheat, endophyte-infected and endophyte-free tall fescue, and other hosts in Arkansas. Plant Dis **77**, 225-228.

Malinowski D, Leuchtmann A, Schmidt D, Nösberger J (1997a) Growth and water status in meadow fescue (Festuca pratensis) is affected by *Neotyphodium* and *Phialophora* endophytes. Agron J **89**, 673-678.

Malinowski D, Leuchtmann A, Schmidt D, Nösberger J (1997b) Symbiosis with *Neotyphodium uncinatum* endophyte may increase the competitive ability of meadow fescue. Agron J **89**, 833-839.

Malinowski DP, Belesky DP (1999) *Neotyphodium coenophialum*-endophyte infection affects the ability of tall fescue to use sparingly available phosphorus. J Plant Nutrition **22**, 835-853.

Malinowski DP, Belesky DP, Hill NS, Baligar VC, Fedders JM (1998) Influence of phosphorus on the growth and ergot alkaloid content of *Neotyphodium coenophialum*-infected tall fescue (*Festuca arundinacea* Schreb.). Plant Soil **198**, 53-61.

Mantle PG, Weedon CM (1994) Biosynthesis and transformation of tremorgenic indole-diterpenoids by *Penicillium paxilli* and *Acremonium lolii*. Phytochemistry **36**, 1209-1217.

McLeay LM, Smith BL, Munday-Finch SC (1999) Tremorgenic mycotoxins paxilline, penitrem and lolitrem B, the non-tremorgenic 31-epilolitrem B and electromyographic activity of the reticulum and rumen of sheep. Res Vet Sci **66**, 119-127.

Miles CO, Di Menna ME, Jacobs SWL, Garthwaite I, Lane GA, Prestidge RA, Marshall SL, Wilkinson HH, Schardl CL, Ball OJP, Latch GCM (1998) Endophytic fungi in indigenous Australasian grasses associated with toxicity to livestock. Appl Environ Microbiol **64**, 601-606.

Miles CO, Lane GA, Di Menna ME, Garthwaite I, Piper EL, Ball OJP, Latch GCM, Allen JM, Hunt MB, Bush LP, Min FK, Fletcher I, Harris PS (1996) High levels of ergonovine and lysergic acid amide in toxic *Achnatherum inebrians* accompany infection by an *Acremonium*-like endophytic fungus. J Agric Food Chem **44**, 1285-1290.

Miles CO, Wilkins AL, Gallagher RT, Hawkes AD, Munday SC, Towers NR (1992) Synthesis and tremorgenicity of paxitriols and lolitriol: possible biosynthetic precursors of lolitrem-B. J Agric Food Chem **40**, 234-238.

Mitter C, Brooks DR (1983) Phylogenetic aspects of coevolution. In: Futuyma DJ, Slatkin M eds. Coevolution. Sunderland, Massachusetts: Sinauer Associates. pp 65-98.

Morgan-Jones G, White JF, Jr (1989) Concerning *Atkinsonella texensis*, a pathogen of the grass *Stipa leucotrichia*: Developmental morphology and mating system. Mycotaxon **35**, 455-467.

Muller HJ (1964) The relation of recombination to mutational advance. Mutation Res 1, 2-9.

Pate JS (1973) Uptake, assimilation and transport of nitrogen compunds by plants. Soil Biol Biochem 5, 109-119.

Pedersen JF, Sleper DA (1993) Genetic manipulation of tall fescue. Agric Ecosyst Environ 44, 187-193.

Penn J, Mantle PG (1994) Biosynthetic intermediates of indole-diterpenoid mycotoxins from selected transformations at C-10 of paxilline. Phytochemistry 35, 921-926.

Peters EJ, Zam AHBM (1981) Allelopathic effects of tall fescue genotypes. Agron J 73, 56-58.

Petroski R, Powell RG, Clay K (1992) Alkaloids of Stipa robusta (sleepygrass) infected with an *Acremonium* endophyte. Nat Toxins 1, 84-88.

Philipson MN, Christey MC (1986) The relationship of host and endophyte during flowering, seed formation, and germination of *Lolium perenne*. N Z J Bot 24, 125-134.

Popay AJ, Rowan DD (1994) Endophytic fungi as mediators of plant insect interactions. In: Bernays EA eds. Insect-Plant Interactions V. Boca Raton, Florida: CRC Press. pp 83-103.

Porter JK (1994) Chemical constituents of grass endophytes. In: Bacon CW, White JF, Jr. eds. Biotechnology of Endophytic Fungi of Grasses. Boca Raton, Florida: CRC Press. pp 103-123.

Porter JK (1995) Analysis of endophyte toxins: fescue and other grasses toxic to livestock. J Animal Sci 73, 871-880.

Porter JK, Bacon CW, Plattner RD, Arrendale RF (1987) Ergot peptide alkaloid spectra of *Claviceps*-infected tall fescue, wheat and barley. J Agric Food Chem 35, 359-361.

Prestidge RA (1993) Causes and control of perennial ryegrass staggers in New Zealand. Agric Ecosyst Environ 44, 283-300.

Prestidge RA, Lauren DR, van der Zujpp SG, di Menna ME (1985) Isolation of feeding deterrents to Argentine stem weevil in cultures of endophytes of perennial ryegrass and tall fescue. N Z J Agric Res 28, 87-92.

Raisbeck MF, Rottinghaus GE, Kendall JD (1991) Effects of naturally occurring mycotoxins on ruminants. In: Smith JE, Henderson RS eds. Mycotoxins and Animal Foods. Boca Raton, Florida: CRC Press. pp 647-677.

Reddy PV, Lam CK, Belanger FC (1996) Mutualistic fungal endophytes express a proteinase that is homologous to proteases suspected to be important in fungal pathogenicity. Plant Physiology 111, 1209-1218.

Rehácek Z (1991) Physiological controls and regulation of ergot alkaloid formation. Folia Microbiologica 36, 323-342.

Richardson MD, Chapman GW, Jr., Hoveland CS, Bacon CW (1992) Sugar alcohols in endophyte-infected tall fescue under drought. Crop Sci 32, 1060-1061.

Rolston MP, Hare MD, Moore KK, Christensen MJ (1986) Viability of *Lolium* endophyte fungus in seed stored at different moisture contents and temperatures. N Z J Agric Res 14, 297-300.

Rottinghaus GE, Garner GB, Cornell CN, Ellis JL (1991) HPLC method for quantitating ergovaline in endophyte-infested tall fescue: seasonal variation of ergovaline levels in stems with leaf sheaths, leaf blades, and seed heads. J Agric Food Chem 39, 112-115.

Rowan DD (1993) Lolitrems, paxilline and peramine: mycotoxins of the ryegrass/endophyte interaction. Agric Ecosyst Environ 44, 103-122.

Rowan DD, Gaynor DL (1986) Isolation of feeding deterrents against Argentine stem weevil from ryegrass infected with the endophyte *Acremonium loliae*. J Chem Ecol 12, 647-658.

Rowan DD, Latch GCM (1994) Utilization of endophyte-infected perennial ryegrasses for increased insect resistance. In: Bacon CW, White JF, Jr. eds. Biotechnology of Endophytic Fungi of Grasses. Boca Raton, Florida: CRC Press. pp 169-183.

Rykard DM, Bacon CW, Luttrell ES (1985) Host relations of *Myriogenospora atrementosa* and *Balansia epichloe* (Clavicipitaceae). Phytopathology **75**, 950-956.

Rykard DM, Luttrell ES, Bacon CW (1982) Development of the conidial state of *Myriogenospora atramentosa*. Mycologia **74**, 648-654.

Sampson K (1933) The systemic infection of grasses by *Epichloe typhina* (Pers.) Tul. Trans Brit Mycol Soc **18**, 30-47.

Sampson K (1937) Further observations on the systemic infection of *Lolium*. Trans Brit Mycol Soc **21**, 84-97.

Schardl CL, Clay K (1997) Evolution of mutualistic endophytes from plant pathogens. In: Carroll GC, Tudzynski P eds. The Mycota V: Plant Relationships, part B: Springer-Verlag. pp 221-238.

Schardl CL, Leuchtmann A (1999) Three new species of *Epichloë* symbiotic with North American grasses. Mycologia **91**, 95-107.

Schardl CL, Leuchtmann A, Chung K-R, Penny D, Siegel MR (1997) Coevolution by common descent of fungal symbionts (*Epichloë* spp.) and grass hosts. Mol Biol Evol **14**, 133-143.

Schardl CL, Leuchtmann A, Tsai H-F, Collett MA, Watt DM, Scott DB (1994) Origin of a fungal symbiont of perennial ryegrass by interspecific hybridization of a mutualist with the ryegrass choke pathogen, *Epichloë typhina*. Genetics **136**, 1307-1317.

Schardl CL, Liu J-S, White JF, Finkel RA, An Z, Siegel MR (1991) Molecular phylogenetic relationships of nonpathogenic grass mycosymbionts and clavicipitaceous plant pathogens. Plant Systemat Evol **178**, 27-41.

Schardl CL, Phillips TD (1997) Protective grass endophytes: Where are they from and where are they going? Plant Dis **81**, 430-437.

Schardl CL, Tsai H-F (1992) Molecular biology and evolution of the grass endophytes. Nat Toxins **1**, 171-184.

Schardl CL, Tsai H-F, Chung K-R, Leuchtmann A, Siegel MR (1996) Evolution of *Epichloë* species symbioses with grasses. In: Stacey G, Mullin B, Gresshoff PM eds. Biology of Plant-Microbe Interactions: Proceedings of the 8th International Symposium on Molecular Plant-Microbe Interactions. St. Paul, Minnesota, U.S.A.: International Society for Molecular Plant-Microbe Interactions. pp 541-546.

Schmid J, Spiering MJ, Christensen MJ (1999) Metabolic activity, distribution and propagation of grass endophytes in planta: investigations using the GUS system. In: White JF, Jr., Bacon CW eds. Evolution of endophytes of land plants. New York: Marcel Dekker. (in press).

Schöberlein W, Eggestein S, Pfannmöller M, Szabová M (1997) Investigation of interactions between Acremonium uncinatum in *Festuca pratensis* and various nematode species in the soil. In: Hill NS, Bacon CW eds. *Neotyphodium*/grass interactions. New York: Plenum Press. pp 201-203.

Shibuya M, Chou H-M, Fountoulakis M, Hassam S, Kim S-U, Kobayashi K, Otsuka H, Rogalska E, Cassady JM, Floss HG (1990) Stereochemistry of the isoprenylation of tryptophan catalyzed by 4-(γ,γ-dimethylallyl)tryptophan synthase from *Claviceps*, the 1st pathway-specific enzyme in ergot alkaloid biosynthesis. J Am Chem Soc **112**, 297-304.

Siegel MR, Bush LP (1997) Toxin production in grass/endophyte associations. In: Carroll GC, Tudzynski P eds. The Mycota V: Plant Relationships, part A The Mycota V: Plant Relationships, part A. Berlin: Springer-Verlag. pp 185-207.

Siegel MR, Jarlfors U, Latch GCM, Johnson MC (1987) Ultrastructure of *Acremonium coenophialum*, Acremonium lolii, and *Epichloe typhina* endophytes in host and nonhost Festuca and Lolium species of grasses. Can J Bot **65**, 2357-2367.

Siegel MR, Johnson MC, Varney DR, Nesmith WC, Buckner RC, Bush LP, Burrus PB, II., Jones TA, Boling JA (1984) A fungal endophyte in tall fescue: incidence and dissemination. Phytopathology **74**, 932-937.

Siegel MR, Latch GCM, Bush LP, Fannin FF, Rowan DD, Tapper BA, Bacon CW, Johnson MC (1990) Fungal endophyte-infected grasses: alkaloid accumulation and aphid response. J Chem Ecol **16**, 3301-3315.

Siegel MR, Schardl CL (1991) Fungal endophytes of grasses: detrimental and beneficial associations. In: Andrew JH, Hirano SS eds. Microbial Ecology of Leaves. Berlin: Springer Verlag. pp 198-221.

Soreng RJ, Davis JI (1998) Phylogenetics and character evolution in the grass family (Poaceae): Simultaneous analysis of morphological and chloroplast DNA restriction site character sets. Bot Rev **64**, 1-85.

Spiers DE, Zhang Q, Eichen PA, Rottinghaus GE, Garner GB, Ellersieck MR (1995) Temperature-dependent responses of rats to ergovaline derived from endophyte-infected tall fescue. J Animal Sci **73**, 1954-1961.

Stebbins GL (1971) Chromosomal Evolution in Higher Plants. London: Edward Arnold (Publishers) Ltd. 216 pp.

Strickland JR, Cross DL, Birrenkott GP, Grimes LW (1994) Effect of ergovaline, loline, and dopamine antagonists on rat pituitary cell prolactin release in vitro. Am J Vet Res **55**, 716-721.

Thompson FN, Stuedemann JA (1993) Pathophysiology of fecue toxicosis. Agric Ecosyst Environ **44**, 263-281.

Tsai H-F, Liu J-S, Staben C, Christensen MJ, Latch GCM, Siegel MR, Schardl CL (1994) Evolutionary diversification of fungal endophytes of tall fescue grass by hybridization with *Epichloë* species. Proc Nat Acad Sci USA **91**, 2542-2546.

Tsai H-F, Wang H, Gebler JC, Poulter CD, Schardl CL (1995) The *Claviceps purpurea* gene encoding dimethylallyltryptophan synthase, the committed step for ergot alkaloid biosynthesis. Biochem Biophys Res Commun **216**, 119-125.

Tudzynski P, Hölter K, Correia T, Arntz C, Grammel N, Keller U (1999) Evidence for an ergot alkaloid gene cluster in *Claviceps purpurea*. Mol Gen Genet **261**, 133-141.

Tudzynski P, Tenberge KB, Oeser B (1995) *Claviceps purpurea*. In: Kohmoto K, Singh US, Singh RP eds. Pathogenesis and host specificity in plant diseases: Histopathological, biochemical, genetic and molecular bases II: Eukaryotes. Tarrytown, New York: Elsevier Science Ltd. pp 161-187.

West CP (1994) Physiology and drought tolerance of endophyte-infected grasses. In: Bacon CW, White JF, Jr. eds. Biotechnology of Endophytic Fungi of Grasses. Boca Raton, Florida: CRC Press. pp 87-99.

West CP, Izekor E, Oosterhuis DM, Robbins RT (1988) The effect of *Acremonium coenophialum* on the growth and nematode infestation of tall fescue. Plant Soil **112**, 3-6.

White JF (1988) Endophyte-host associations in forage grasses. XI. A proposal concerning origin and evolution. Mycologia **80**, 442-446.

White JF, Jr (1993) Endophyte-host associations in grasses. XIX. A systematic study of some sympatric species of *Epichloë* in England. Mycologia **85**, 444-455.

White JF, Jr (1994a) Endophyte-host associations in grasses. XX. Structural and reproductive studies of Epichloë amarillans sp. nov. and comparisons to *E. typhina*. Mycologia **86**, 571-580.

White JF, Jr (1994b) Structure and mating system of the graminicolous fungal epiphyte *Echinodothis tuberiformis* (Clavicipitales). Am J Bot **80**, 1465-1471.

White JF, Jr (1994c) Taxonomic relationships among the members of the Balansieae (Clavicipitales). In: Bacon CW, White JF, Jr. eds. Biotechnology of Endophytic Fungi of Grasses. Boca Raton, Florida: CRC Press. pp 3-20.

White JF, Bultman TL (1987) Endophyte-host associations in forage grasses. VIII. heterothallism in *Epichloë typhina*. Am J Bot **74**, 1716-1721.

White JF, Jr., Drake TE, Martin TI (1996) Endophyte-host associations in grasses .23. A study of two species of Balansia that form stromata on nodes of grasses. Mycologia **88**, 89-97.

White JF, Jr., Glenn AE (1994) A study of two fungal epibionts of grasses: structural features, host relationships, and classification in genus *Myriogenospora* Atk. (Clavicipitales). Am J Bot **81**, 216-223.

White JF, Jr., Morrow AC, Morgan-Jones G, Chambless DA (1991) Endophyte-host associations in forage grasses. XIV. Primary stromata formation and seed transmission in *Epichloe typhina*: Developmental and regulatory aspects. Mycologia **83**, 72-81.

White JF, Jr., Reddy PV (1998) Examination of structure and molecular phylogenetic relationships of some graminicolous symbionts in genera *Epichloe* and *Parepichloe*. Mycologia **90**, 226-234.

White JF, Jr., Reddy PV, Glenn AE, Bacon CW (1997) An examination of structural features and relationships in Balansia subgenus *Dothichloe*. Mycologia **89**, 408-419.

White JF, Jr., Sharp LT, Martin TI, Glenn AE (1995) Endophyte-host associations in grasses. XXI. studies on the structure and development of *Balansia obtecta*. Mycologia **87**, 172-181.

Wilkinson HH, Schardl CL (1997) The evolution of mutualism in grass-endophyte associations. In: Bacon CW, Hill N eds. *Neotyphodium*/grass interactions. New York: Plenum Press. pp 13-25.

Yamamura N (1993) Vertical transmission and evolution of mutualism from parasitism. Theor Pop Biol **44**, 95-109.

Yates SG, Fenster JC, Bartelt RJ (1989) Assay of tall fescue seed extracts, fractions, and alkaloids using the large milkweed bug. J Agric Food Chem **37**, 354-357.

Zhang Q, Spiers DE, Rottinghaus GE, Garner GB (1994) Thermoregulatory effects of ergovaline isolated from endophyte-infected tall fescue seed on rats. J Agric Food Chem **42**, 954-958.

USTILAGO MAYDIS, THE CAUSATIVE AGENT OF CORN SMUT DISEASE

REGINE KAHMANN, GERO STEINBERG, CHRISTOPH BASSE, MICHAEL FELDBRÜGGE AND JÖRG KÄMPER
Institut für Genetik und Mikrobiologie der Universität München, Maria-Ward-Str.1a, D-80638 München, Germany

1. Introduction

Smut diseases are caused by Basidiomycetes of the order Ustilaginales. Their occurrence is worldwide and cereals and grasses serve as most common hosts. Among the estimated 1100 species of smut fungi the most prevalent causing significant losses in yield are of the genus *Ustilago*, causing loose smut of oats (*U. avenae*), barley (*U. nuda*) and wheat (*U. tritici*), covered smut of barley and oats (*U. hordei*), causing corn (*U. maydis*) and sugarcane smut (*U. scitaminea*), respectively; of the genus Tilletia inducing covered smut or bunt of wheat (*T. caries* and *T. foetida*), dwarf bunt of wheat (*T. contraversa*); and of the genus *Sporisorium* causing smut diseases in sorghum as well as head smut in sorghum and corn (see Agrios, 1988). In general, the host range of most smut fungi is rather narrow and only closely related plant species are infected. Many smut fungi develop within the grain kernels and replace them with masses of dark teliospores resembling smut or soot, giving the name to this disease. As a result considerable losses in yield are encountered and healthy seed is contaminated with spores during harvest. While some smuts infect germinating seedlings and grow internally without causing symptoms until flowering occurs others are able to infect all aerial parts of the plant and cause local disease symptoms around the site of infection. The most prominent disease symptoms are inflorescences which are completely smutted and in which all individual kernels are replaced by masses of dark teliospores. Initially the developing spores are surrounded by a membrane that breaks when teliospore development is completed and sets free the massive amounts of spores. In other instances, like infections with *U. maydis*, spore development takes place within plant tumors that are induced by the fungus and appear to provide the ideal environment for fungal proliferation and teliospore production. Teliospores developing within the infected tissue are invariably diploid. Upon germination meiosis takes place and the haploid form, the so-called sporidium is generated. These haploid forms can be propagated on artificial media in the laboratory, however, this form is unable to cause disease when applied to host plants in pure culture. Prerequisite for generating the infectious stage is the mating of two compatible, haploid sporidia and the generation of the dikaryon. In contrast to the haploid form this stage cannot be propagated outside the host plant. Once in contact with the plant characteristic infection structures are produced that allow penetration and subsequent proliferation within the infected plant. This leads to the characteristic disease symptoms already described and culminates with the

J. W. Kronstad (ed.), Fungal Pathology, 347–371.

formation of teliospores which represent the resting stage able to survive harsh environmental conditions like drought and winter.

It is not a coincidence that one of these smut fungi, *U. maydis*, has received particular attention by geneticists and plant pathologists alike. Contrary to most other smuts, disease development after infection with *U. maydis* can occur on the leaf, stem, tassel and ear of the infected maize plant. The infection of germinated seedlings results in the appearance of plant tumors only about 7 days after inoculation under controlled conditions in the green house, i.e., contrary to most of the other smuts, disease symptoms can be scored without having to wait for the appearance of flowers. In addition, the genetic identification of the two mating-type loci termed *a* and *b* and the finding that these loci control discrete steps of fungal development has made this system attractive from a molecular as well as a plant pathologist´s view (Rowel and DeVay, 1954; Rowell, 1955; Holliday, 1961; Puhalla, 1970; Day *et al.*, 1971; Banuett and Herskowitz, 1989). Through the same set of studies it was shown that the generation of the infective dikaryon required two mating partners differing at the *a* and *b* mating-type loci. The analysis of stable diploid lines demonstrated, furthermore, that the *b* locus is the central control locus for pathogenic development and heterozygosity at this locus is the sole prerequisite for initiating disease. In retrospect it appears as a lucky coincidence that the molecular techniques already established for other systems could be transferred to *U. maydis* without major problems in the late eighties. These studies were mainly carried out by the group of S. Leong at Madison, who established the transformation system relying on ectopic integration, adopted pulse field analysis for karyotyping *U. maydis* chromosomes and demonstrated the feasibility for generating gene knock-outs by homologous recombination (Kinscherf and Leong, 1988; Wang *et al.*, 1988; Holden *et al.*, 1989). These tools were extended by the development of autonomously replicating plasmid vectors allowing high frequency of transformation for complementation analyses (Tsukuda *et al.*, 1988). With these tools at hand *U. maydis* has become one of the few models to study fungal disease as well as the interaction with the host plant. In this chapter we give an overview of its basic biology and the current status of approaches aimed at understanding the molecular concepts of disease.

2. Development and differentiation

Infection of the plant starts with recognition of two fungal sporidia of opposite mating type, and is followed by the formation of long, often coiled conjugation tubes (Figure 1, stage 1) (Bowmann, 1946; Snetselaar and Mims, 1992; Snetselaar, 1993). These structures usually appear at one pole of each cell and grow towards each other, probably following a pheromone gradient (Snetselaar, 1993; Snetselaar *et al.*, 1996). This reaction occurs on the surface of the host plant (Snetselaar and Mims, 1992; Snetselaar *et al.*, 1996) as well as on solid agar medium supplemented with plant extract (Rowell, 1955) or activated charcoal (Day and Anagnostakis, 1971; Banuett and Herskowitz, 1994b). After fusion of the progenitor cells (plasmogamy), a straight dikaryotic hypha emerges (Rowell, 1955), which contains the two nuclei and the cytoplasm of the mating partners (Figure 1, stage 2) (Sleumer, 1932; Fischer and Holton, 1957). The tip cell of

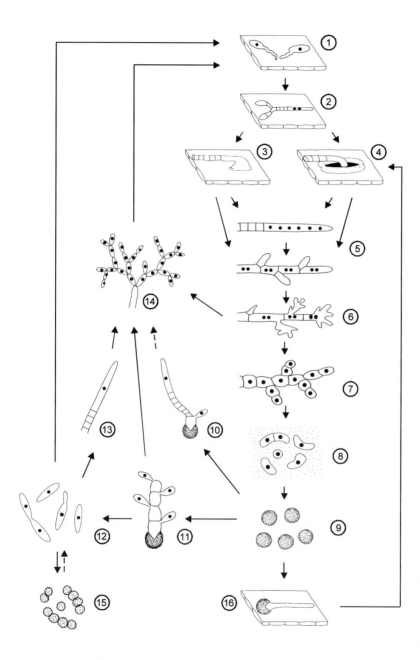

Figure 1. Life cycle of *U. maydis*. Nuclei are marked by filled circles. Numbers refer to developmental stages discussed in the text. Broken arrows indicate assumed, yet unverified pathways.

this hypha is cell cycle arrested and grows by apical cell expansion. On the leaf surface hyphal tip growth extends to the site of penetration, leaving behind the original sporidial wall as well as empty cell compartments (Brefeld, 1883; Christensen, 1963). Entry into the host can be mediated by appressoria-like structures (Figure 1, stage 3) (Snetselaar and Mims, 1992; 1993), probably supported by secreted hydrolytic enzymes (see below), or via host stomata (Figure 1, stage 4) (Banuett and Herskowitz, 1996). All kinds of green maize tissue can be invaded, however, for reasons unknown hyphal penetration is restricted to young plant cells (Christensen, 1963; Brefeld, 1895). Initial growth occurs intracellularly (Snetselaar and Mims, 1992), with the plasma membrane of the plant cell remaining intact. On the ultrastructural level no obvious defense reaction of the plant cell is observed (Snetselaar and Mims, 1993), however, 2-3 days after infection anthocyanin formation (Banuett and Herskowitz, 1996) and limited necrosis at areas of initial penetration can be detected. Within the first two days inside the host, the fungal hypha continues to grow as an unbranched often multinucleated tip cell leaving highly vacuolated portions behind, which often collapse as the tip cell passes through several host cells (Figure 1, stage 5) (Brefeld, 1895; Snetselaar and Mims, 1992; Banuett and Herskowitz, 1996). About five days after infection, the fungus induces formation of plant tumors which consist of an increased number of enlarged host cells (Snetselaar and Mims, 1992; 1994) and some fungal hyphae. At this stage hyphae tend to grow intercellularly and form clamp-like structures and branches (Figure 1, stage 6) (Hanna, 1929; Snetselaar and Mims, 1992). The latter often reach into adjoining plant cells and resemble finger-like projections, which might represent haustoria-like feeding structures (Snetselaar and Mims, 1994) observed for example in infections by the related rust fungi. About seven days after infection hyphae proliferate in the intercellular space (Snetselaar and Mims, 1992) or within host cells of the plant tumor (Banuett and Herskowitz, 1996). At this stage most segments of the multi-cellular hyphae contain pairs of nuclei (Sleumer, 1932; Christensen, 1963), while other segments appear to consist of mono-nucleated cells (Snetselaar and Mims, 1994). At the onset of spore formation, convoluted and highly branched sporogenous hyphae fill the tumor (Figure 1, stage 7) (Snetselaar and Mims, 1994). Most of these cells contain a single nucleus suggesting that nuclear fusion (karyogamy) occurred. Subsequently, the sporogenous hyphae undergo fragmentation and give rise to short fragments and single cells (Banuett and Herskowitz, 1996), which can be regarded as immature spores (Figure1, stage 8) (Snetselaar and Mims, 1994). These structures are embedded in a gelatinous matrix, which is believed to derive from the former hyphal wall (Fisher and Holton, 1957). However, this material appears to be digestible by cellulases (Fletcher, 1981), indicating that it might be composed of cellulose provided by the host. After maturation teliospores are surrounded by a structured and pigmented spore wall and are able to survive for several years (Figure 1, stage 9) (Brefeld, 1883). It has been estimated that a plant tumor of medium size contains 200 billion of these teliospores or brandspores that are spread by wind and rain (Christensen, 1963). Under appropriate conditions these spores are able to germinate (Figure 1, stage 10), and give rise to structures which are reminiscent of aerial hyphae (see below) (Brefeld, 1883). Teliospore germination is supported by nutrients and it is believed that in eutrophic soil most spores develop into multi-cellular promycelia, in which meiosis occurs (Figure 1, stage 11) (Christensen, 1963). The resulting haploid nuclei migrate into elongated single cells, named sporidia or basidiospores (Figure 1, stage 12) (Fischer and Holton, 1957; Christensen, 1963). These haploid cells grow saprophytically and can be propagated indefinitely under laboratory conditions. They grow yeast-like by polar budding and are accessible to molecular genetics and cytological methods (Banuett, 1992). However, their relevance in nature is not known (Christensen, 1963). In stationary cultures sporidia are able to form aerial hyphae (Figure 1, stage 13) and subsequently aerial conidia develop (Figure 1, stage 14) (Brefeld, 1895), which are believed to represent the inoculum for plant infection. Under certain conditions sporidia can differentiate into haploid chlamydospores (Figure 1, stage 15) (Kusch and Schauz, 1989). Moreover, it has been shown that teliospores can directly infect host tissue after germination (Figure 1, stage 16) (Walter, 1934; Millis and Kotze, 1981).

Depending on environmental conditions the life cycle outlined may be diverted. For example under high humidity hyphae growing within the plant tissue have been described to emerge as aerial hyphae on the plant surface (Christensen, 1963). In addition, depending on nutrient conditions, the germination of teliospores can lead either to aerial conidia or to a promycelium from which haploid sporidia are budded off (Brefeld, 1883, 1895). Therefore, it appears that morphological and developmental stages of *U. maydis* are highly flexible, which might reflect the necessity to cope with changes in environment both during growth in the soil as well as when inside the plant host.

3. Cellular organization

3.1. POLAR GROWTH, CELL WALL AND THE CYTOSKELETON

Haploid sporidia are 10-20 μm long and 2-3 μm in diameter. The cells have a doubling time of approximately 120 min in complete medium and in a growing culture about half of the cells are unbudded (Snetselaar and McCann, 1997). During budding, cells are in G2, followed by mitosis, which occurs within less than 5 minutes (G. Steinberg, unpublished). Septation starts with the formation of a primary septum and an adjoining secondary septum, followed by vacuolization of the enclosed cytoplasm and cell separation (O´Donell and McLaughlin, 1984b). The new bud emerges at the opposite cell pole using the former or an associated bud site (Jacobs *et al.*, 1994). However, environmental parameters like nutrient condition can influence this growth pattern and can lead to septation without cell separation and to bipolar growth (Snetselaar and McCann, 1997). Thus, sporidia are able to escape from the budding program and grow in a hyphal-like fashion, which could lead to cell chains with multiple growth sites. These structures are reminiscent of the fungal hyphae formed during biotrophic growth and this growth mode can be induced by environmental parameters like pH alterations (Ruiz-Herrera *et al.*, 1995). However, defects in polar secretion were shown to lead to similar multicellular structures. In particular, such a phenotype was reported for mutants defective in γ-adaptin (Keon *et al.*, 1995), a protein that participates in coating of Golgi-derived vesicles, as well as in a vesicle fusion protein required for endocytotic vesicle transport (Wedlich-Söldner *et al.*, 2000). Moreover, mutations in a protein kinase from *U. maydis*, Ukc1, which is a homologue of *cot-1* from *Neurospora crassa* also lead to multi-cellular chains of haploid sporidia (Dürrenberger and Kronstad, 1999). In *N. crassa* it was shown that cytoplasmic dynein, a microtubule dependent motor molecule responsible for certain aspects of intracellular transport, is a multi-copy suppressor of *cot-1* (Plaman et al., 1994) suggesting that Cot-1 like kinases are involved in regulation of microtubule dependent vesicle transport. The formation of multi-cellular chains by haploid sporidia is therefore likely to reflect the ability of *U. maydis* to grow as multi-cellular hyphae within the plant. However, it remains to be demonstrated to which extent these multi-cellular structures resemble the fungal hyphae found during growth within the plant tissue.

The shape of a fungal cell is determined by the cell wall, and polar delivery of wall components and associated enzymes along the cytoskeleton are of crucial importance for polar growth (Steinberg, 1998). The cell wall of *U. maydis* consists mainly of proteins, neutral polysaccharides and chitin (Ruiz-Herrera *et al.*, 1996). The latter is synthesized by chitin synthase, and representatives of six distinct classes of chitin synthase have been identified in *U. maydis* (Bowen *et al.*, 1992; Gold and Kronstad, 1994; Xoconostle-Cazares *et al.*, 1996; 1997). Surprisingly, none of the investigated genes were found to be essential, but in certain deletion mutants the chitin content of the wall was altered and growth defects were visible (Xoconostle-Cazares *et al.*, 1996; 1997; Gold and Kronstad, 1994).

Mating of haploid sporidia occurs on the epidermis of the host (see above) and it is believed that sporidia and dikaryotic infection hyphae are attached to the plant surface by an adhesive material. Ultrastructural studies revealed fibrous strands between hyphae and the plant epidermis (Snetselaar and Mims, 1993). Fibrous surface structures, so called fimbriae, were identified in several fungi including *U. maydis* (Poon and Day, 1974, 1975; Xu and Day, 1992). These structures consist of distinct proteins that assemble into fine strands several micrometers in length. However, although the cellular role of fimbriae is not completely understood, evidence exists for a role in mating (Day and Poon, 1975), rather than in surface adhesion.

Growth of infectious aerial hyphae requires protection against water loss. In fungi this is achieved by specialized proteins named hydrophobins that assemble into a hydrophobic layer on the cell surface (Wessels, 1996). While *U. maydis* contains at least one canonical hydrophobin gene (Bohlmann, 1996), surface hydrophobicity of hyphae was shown to be conferred primarily by a new class of amphipatic peptides encoded by *rep1* (Wösten *et al.*, 1996). These peptides are not related to hydrophobins in primary structure but may serve a similar function in enabling the hypha to maintain a positive water balance. In addition to limited water supplies on the plant surface the infection hypha, while searching for the site of entry, has to cope with low nutrient supplies. It is conceivable that smut fungi have adapted to this situation by synthesizing new cell wall material without extending the volume of the cytoplasm and without dividing. The accumulation of all cytoplasm in the single tip cell can conceivably allow rapid growth over considerable distances on stored resources while only an empty tube of cell wall is left behind (Fischer and Holton, 1957). This growth mode is achieved by the formation of a basal vacuole which appears to expand at the same rate as the tip growth and is left behind while a new septum cuts off the dead portion of the hypha (Steinberg *et al.*, 1998). Recently it has been shown that microtubules and an associated motor molecule of the kinesin family, Kin2 (Lehmler *et al.*, 1997), are involved in polar growth and the formation of this basal vacuole (Steinberg *et al.*, 1998). Mutants disrupted in *kin2* showed impaired mating and growth, but were able to form filaments. However, due to impaired vacuole organization these infection hyphae failed to leave empty cell portions behind. As a likely consequence pathogenicity was significantly reduced (Lehmler *et al.*, 1997). After the dikaryotic hypha reaches the infection site, penetration itself may be supported by polar secretion of hydrolytic enzymes. A cellulase gene, *egl1*, was identified, which is strongly induced in filaments, but not expressed in sporidia (Schauwecker *et al.*, 1995). However, *egl1* disruption mutants showed no attenuation in pathogenicity suggesting the existence of a battery of such enzymes which can substitute for one another as has been shown in other plant pathogenic fungi (Apel-Birkhold and Walton, 1996). In summary, it becomes obvious from these and other results (G. Steinberg, unpublished) that defects in cellular organization and secretion can drastically reduce pathogenicity of *U. maydis*. This confirms a central role of the cytoskeleton and associated transport processes for the biotrophic stage of the life cycle.

3.2. NUCLEAR MIGRATION AND DIVISION

Sporidia contain a haploid nucleus which is positioned in the center of the cell until the growing bud has reached a size of about 50% of the mother cell. While the cell still enlarges the nucleus migrates into the bud and undergoes mitosis in the mother-daughter constriction or within the bud. Finally, nuclear division is followed by positioning of the new nuclei in the center of mother and daughter cell (Holliday, 1965; O´Donnel and McLaughlin, 1984b). Detailed ultrastructural studies ascribed a central role of the spindle pole body and microtubules to nuclear migration, meiosis and mitosis of *U. maydis* (O´Donnell and McLaughlin, 1984b; 1984a; O´Donnell, 1992). During interphase microtubules are in close proximity with the nuclear envelope, however, they appear to have no contact with the spindle pole body, which is associated with the nuclear envelope and contains two globular regions. While the nucleus migrates into the bud, microtubules increase in number and have contact with the leading edge of the nucleus. At the onset of mitosis the globular regions of the spindle pole body separate and both spindle pole bodies nucleate astral (= cytoplasmic) as well as spindle microtubules. During prophase/metaphase the nuclear envelope partially breaks down and a spindle of 1-2.5 µm is formed. Segregation of chromosomes is supported by a rapid anaphase. During telophase both daughter nuclei gain a nuclear envelope and migrate towards their final position within their cells while still being connected by numerous microtubules.

After mating of two compatible sporidia, their nuclei migrate into the infection hyphae (Sleumer, 1932; Snetselaar and Mims, 1992). These nuclei are positioned in defined proximity in the middle of the hyphal tip cell (Snetselaar and Mims, 1992; Steinberg *et al.*, 1998) which does not divide and hence must be arrested in cell cycle. After invasion of the host, the cell cycle arrest is lifted resulting in hyphal cells that contain either multiple numbers of nuclei (Lutman, 1910; Snetselaar and Mims, 1993) or paired nuclei (Ehrlich, 1958; Banuett and Herkowitz, 1994b; Snetselaar and Mims, 1994). At present it cannot be excluded that the multi-nucleate type might be converted into the dikaryotic type at a later stage (Snetselaar and Mims, 1993). Although some contradictory reports exist (Christensen, 1963; Snetselaar and Mims, 1994) it is most likely that karyogamy occurs in sporogenous hyphae prior to sporogenesis (Rawitscher, 1912; Sleumer, 1932; Ehrlich, 1958).

4. The mating-type loci

The most critical event within the life cycle of *U. maydis,* the switch from saprophytic to parasitic growth, is genetically controlled by two distinct mating-type loci termed *a* and *b*. While the *a* locus exists in two different alleles, *a1* and *a2,* the *b* locus is multiallelic and 33 different alleles have been isolated from nature (DeVay, quoted in Wong and Wells, 1985). For a compatible interaction that leads to cell fusion and establishment of the infectious dikaryon, the mating partners have to carry different alleles of both the *a* and the *b* loci (for example, *a1b1* and *a2b2*). For the formation of conjugation tubes and for the cell fusion event, it is sufficient when the cells are different at the *a* locus (Rowell, 1955; Puhalla, 1969; Trueheart and Herskowitz, 1992; Banuett and Herskowitz, 1994b; Snetselaar *et al.,* 1996). However, after cell fusion the presence of different *b* loci is required in addition to establish a stable, filamentous dikaryon (Holliday, 1961; Puhalla, 1968; Banuett and Herskowitz, 1989). The feasibility to generate stable diploid cells allowed the study of the contribution of the two loci to filamentous growth and pathogenicity. Diploid cells with different alleles of *b,* but the same allele of *a* (*a1a1b1b2*) grow yeast-like but are able to infect corn plants and to induce tumors, whereas diploid cells heterozygous at *a* but homozygous at *b* (*a1a2b1b1*) are yeast-like and nonpathogenic. Thus, for pathogenic development, diploid strains require different alleles at *b,* but not at *a,* which defines the *b* locus as central pathogenicity locus. For filamentous growth on artificial media, on the other hand, different alleles of both *a* and *b* are required (Banuett and Herskowitz, 1989).

4.1. REQUIREMENTS FOR CELL FUSION

The two alleles of the *a* mating-type locus, *a1* and *a2,* comprise 4.5 and 8 kb of nonhomologous DNA, respectively (Figure 2). Both alleles have in common that they encode a precursor for a lipopeptide mating factor (Mfa1/2) and a receptor for the pheromone secreted by the opposite mating type (Pra1/2, Bölker *et al.,* 1992; see Figure 3). Thus, in contrast to *S. cerevisiae* the primary determinants of cell/cell recognition are directly provided by the *a* mating-type locus. The two *a* alleles differ with respect to two additional genes *lga2* and *rga2* present in the *a2* locus (Figure 2).

Besides their role in cellular recognition the gene products of the *a* locus are also needed in concert with the *b* gene products to induce filamentous growth of the fungus, a postfusion event occurring naturally in the dikaryon. Since treatment of diploids carrying identical *a* but different *b* alleles with purified or synthetic pheromone also induces filamentous growth, an active pheromone signalling pathway appears necessary to induce the filamentous growth program under these conditions (Spellig *et al.,* 1994).

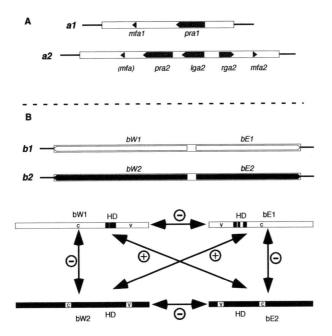

Figure 2. Organization of the *a* and *b* mating-type loci. A) The *a1* and *a2* alleles are indicated by an open bar with gray arrows marking the respective genes and their direction of transcription. Identical flanking sequences are indicated by the solid line. (*mfa*) marks a presumed pheromone pseudogene in the *a2* locus. Data were adopted from Urban *et al.*, 1996b. B) In the upper part the gene organization is shown for both the *b1* and *b2* locus. The lower part illustrates which polypeptides are able to interact (+) and which combinations unable to form a heterodimer (-). c marks the constant part, v the variable portion and HD the homeodomain of the respective proteins.

Haploid cells that receive a pheromone signal respond with the formation of conjugation tubes that grow towards the source of pheromone ensuring that the two tubes meet and fuse (Snetselaar *et al.*, 1996). On the molecular level, all genes present at the mating-type loci are transcriptionally activated upon mating. This is achieved through the action of pheromone response factor 1 (Prf1), a transcriptional activator that recognizes pheromone response elements (PREs) via its DNA-binding domain of the HMG box type (high mobility group) (see Figure 3). The PREs are located in the regulatory regions of pheromone-induced genes and are sufficient to confer pheromone-dependent expression of reporter genes (Hartmann *et al.*, 1996; Urban *et al.*, 1996a). The analysis of *prf1* deletion strains revealed that they are sterile and nonpathogenic. Their sterile phenotype reflects that such strains neither show basal nor pheromone-induced expression of the pheromone and receptor genes. The lack of pathogenicity was even observed in solopathogenic strains, which do not need to fuse in order to cause tumors (Bölker *et al.*, 1995a). However, in such strains pathogenicity could be restored by providing for constitutive expression of a pair of active b proteins (see below). This indicates that Prf1 is the key link between pheromone signalling, filamentous growth and pathogenic development (Hartmann *et al.*, 1996) (see Figure 3). With this in mind it is now possible to understand why diploid strains carrying identical *a* but different *b* alleles are unable to grow filamentous on plates: in these strains Prf1 activity is not high enough to increase the expression of the *b* genes. Since Prf1 plays such a central role it is not surprising that its activity is regulated in a complex manner, both on the transcriptional as well as on the posttranscriptional level. *prf1* transcription is activated by carbon source, such as glucose and fructose, through an upstream activating sequence. The same promoter element provides for negative control of *prf1* gene transcription at high cAMP levels (Hartmann *et al.*, 1999). Strains in which *prf1* expression is driven by a constitutively active promoter still respond to pheromone and cAMP treatment with an elevated level of *mfa1* expression. Thus, there exist at least two signalling processes that regulate Prf1 activity on the posttranscriptional level (Hartmann *et al.*, 1999). Current research in this area is devoted to unravel the signalling events that lead from the activated pheromone-bound receptor at the membrane to the activation of the transcription factor Prf1 that exerts its function in the nucleus. Recently, it was shown that the MAP kinase Kpp2 appears to be involved in this process. First hints for the regulation of Prf1 activity by a MAP kinase were provided by the observation that mutations of six putative MAP kinase phosphorylation sites in combination with a putative MAP kinase interaction domain present in Prf1 interfered with the function of Prf1 during mating (Müller *et al.*, 1999). Subsequently, a PCR approach led to the identification of the MAP kinase Kpp2. The same MAP kinase gene (termed *ubc3*) has also been identified in a suppressor screen for *uac1* mutants lacking adenylate cyclase (Mayorga and Gold, 1999). A strain carrying a deletion in this MAP kinase gene exhibited a reduced mating capability because it neither produced nor responded to pheromone (Müller *et al.*, 1999). Therefore it is likely that comparable to the situation in yeast, a MAP kinase cascade consisting of a MAP kinase kinase kinase, a MAP kinase kinase, and the MAP kinase Kpp2/Ubc3 is transmitting the pheromone signal (Figure 3). A possible candidate for the MAPK kinase, designated Fuz7, was already identified some years ago. Based on the fact that Δ*fuz7* strains are unable to form conjugation tubes this kinase was assigned to the pheromone MAP kinase module (Banuett and Herskowitz, 1994a). This assignment has gained additional support by the recent observation, that *fuz7* mutants, as *ubc3* mutants, act as suppressors for the filamentous growth of *uac1* strains (Mayorga and Gold, 1999). However, the pheromone-induced *mfa1* gene expression is not disturbed in Δ*fuz7* strains (Regenfelder *et al.*, 1997). Thus, it appears unlikely that Fuz7 transmits the signal to Kpp2/Ubc3 during pheromone signalling and therefore a second MAP kinase module has been postulated to exist as part of the pheromone signalling network (Figure 3; Kahmann *et al.*, 1999). An additional level of complexity is provided through cross talk between pheromone and cAMP signalling. Kronstad and coworkers discovered the close connection between cAMP signalling and morphology in *U. maydis*: based on a genetic screen they showed that mutations in the adenylate cyclase gene *uac1* lead to a filamentous growth phenotype that is independent of pheromone signalling and of heterozygosity at the *b* locus (Gold *et al.*, 1994) (Figure 3). Subsequently, additional components of the cAMP signalling pathway that is evolutionarily highly conserved in other fungi were identified (Kronstad and Staben, 1997). Upon activation of a heterotrimeric G protein, the α subunit Gpa3 (Krüger *et al.*, 1998; Regenfelder *et al.*,

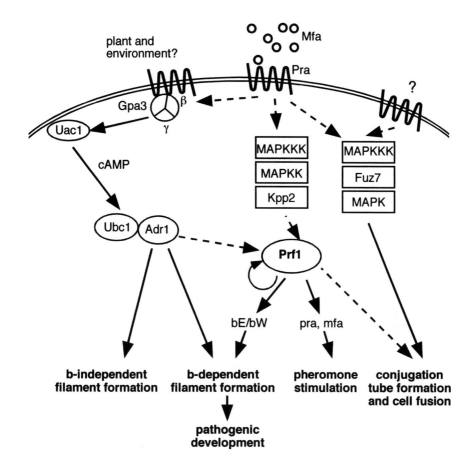

Figure 3. The regulatory network controlling cAMP and pheromone signalling. On the left the known components of the cAMP signalling cascade, Gpa3, Uac1, Ubc1 and Adr1 are shown. The center displays the proposed MAP kinase module transmitting the pheromone signal. Kpp2/Ubc3 is the proposed MAP kinase of this module. A second MAP kinase module shown at the right is predicted to contain Fuz7 and is proposed to act in parallel to control conjugation tube formation. Broken arrows or question marks indicate missing components or unknown inputs, respectively. Arrows indicate activation, broken arrows indicate that the mechanism of activation has not been established. Details are described in the text.

1997) activates the adenylate cyclase Uac1 (Gold *et al.*, 1994). The increase in cAMP concentration results in a dissociation of the regulatory subunit Ubc1 (Gold *et al.*, 1997) from the catalytic subunit (Adr1) of a cAMP dependent protein kinase (Dürrenberger *et al.*, 1998), which in turn phosphorylates various target proteins not yet identified (Figure 3). Mutations in genes that are expected to lower the intracellular cAMP level cause filamentous growth of haploid cells (Gold *et al.*, 1994; Dürrenberger *et al.*, 1998). Conversely, a disruption of the *ubc1* gene, which should reflect a high intracellular cAMP level, exhibits a phenotype termed multiple budding (Gold *et al.*, 1997). The *gpa3* gene encoding the Gα subunit was originally placed in the pheromone signalling pathway consistent with its cell fusion defect and its failure to react to pheromone (Regenfelder *et al.*, 1997). However, the filamentous phenotype and the mating defect of compatible *gpa3* mutants could be suppressed by exogeneous cAMP indicating a possible connection with cAMP signalling. This assumption was firmly established by a detailed genetic analysis revealing that Gpa3 functions epistatic to the adenylate cyclase Uac1 (Krüger *et al.*, 1998). Furthermore, *ubc1* deletion mutants and strains expressing a constitutively active form of Gpa3 both show elevated levels of *mfa1* gene expression (Krüger *et al.*, 1998). According to a current working model Gpa3 elicits an elevated cAMP level resulting in an activation of Adr1 (Figure 3). A putative target of Adr1 is Prf1. This is based on the observation that strains carrying a Prf1 allele, in which PKA phosphorylation sites are eliminated by mutation, are severely compromised in mating (M. Feldbrügge and R. Kahmann, unpublished observation). Thus, we hypothesize that Prf1 is the central integrator between cAMP and pheromone signalling (Figure 3).

The complex signalling network used by *U. maydis* to orchestrate the mating event between two haploid cells may reflect an adaptation to its plant host. Fungal cells could sense surface and/or nutritional environment of the plant with a receptor coupled to Gpa3 (see Figure 3). Under favourable conditions, an elevated level of mating pheromone and receptor are produced via induction of the cAMP signalling cascade. If a mating partner is present the cells are able to fuse after activation of the MAP kinase cascade, switch to filamentous growth and form infection structures.

4.2. THE *B* LOCUS AS MASTER REGULATORY LOCUS FOR PATHOGENICITY

Cloning of the *b* locus revealed that it contains a pair of divergently transcribed genes termed *bE* and *bW* (Figure 2B). The predicted proteins of these genes comprise 473 and 645 amino acids, respectively. They do not share significant sequence identities, except for a homeodomain box, a motif highly conserved in eukaryotes and known to permit sequence specific binding to DNA. Comparison of different alleles revealed that in both proteins the region carboxy-terminal to the homeodomain is highly conserved, and that allelic differences cluster within the region amino-terminal to both homeodomains (Kronstad and Leong, 1990; Schulz *et al.*, 1990; Gillissen *et al.*, 1992; see Figure 2B). Thus, bE and bW proteins share a similar organization. The analysis of mutants deleted either for bE or bW revealed that *b* locus function, i.e. pathogenic development is seen only when *bE* and *bW* genes originating from different alleles are combined (Gillissen *et al.*, 1992). Therefore, in a cross of two haploid wild type strains, two active bE/bW combinations must exist that are probably redundant in their regulatory potential (Figure 2B). This creates the fascinating scenario that with 33 different alleles existing in nature, there are predicted to be 1056 active and only 33 inactive bE/bW combinations.

The need for pairwise action of the proteins became understood with the finding that bE and bW proteins can form heterodimeric complexes when they are derived from different alleles, but not when the proteins originate from the same allele. Therefore, haploid cells are nonpathogenic because no bE/bW heterodimer can be formed. However, haploid strains engineered to express bE and bW proteins from different alleles become pathogenic (Bölker *et al.*, 1995a) illustrating that it is indeed only the inability to form heterodimers that maintains the noninfectious state. As predicted, it was found that the interacting domains of the bE and bW proteins map to the variable amino-terminal ends of bE as well as bW. These regions of about 100 amino acids in bE and 150 amino acids in bW, respectively, have the potential to discriminate self from nonself in all possible allelic combinations (Kämper *et al.*, 1995). To accommodate this situation the existence of a cohesive surface for promiscuous interactions within the variable regions has been proposed. It is thought that this is mainly governed by hydrophobic and/or polar interactions, a situation reminiscent of proteins that engage in combinatorial interaction through coiled-coil motifs. To accommodate the dimerization properties of deletion variants in the variable domains, bE and bW are divided into subdomains thought to have matching subdomains in the corresponding dimerization partner. A combination of a bE and a bW subdomain can result either in attraction or repulsion. The multitude of bE/bW interactions is explained by the assertion that in any given combination with bE and bW originating from different alleles the sum of these forces decides for or against dimerization. In a bE/bW combination originating from the same allele the repulsing forces would surpass the attracting forces and prevent dimerization (Kämper *et al.*, 1995). The model is supported by the finding that single amino acid substitutions in bE2 lead to dimerization with bW2. Further support comes from an elegant genetic approach in which sets of chimeric bE and bW proteins each consisting of a segment of *b1* and *b2* of various length were generated. Crosses between strains carrying such chimeric bE and bW alleles identified two short blocks of amino acids within the variable domain of the proteins that influence specificity and might correspond to one of the proposed subdomains (Yee and Kronstad, 1993; Yee and Kronstad, 1998). However, for the definitive resolution of the self/nonself recognition mechanism one has to await solution of the three dimensional structure of these domains.

The native bE/bW heterodimer can be substituted by an artificial translational fusion protein consisting of bE tethered to bW via a flexible linker region (Romeis *et al.*, 1997). Haploid strains expressing only this fusion protein were shown to be pathogenic on corn. Interestingly, both dimerization domains of bE and bW could be deleted without effect on the biological function of the fusion protein, indicating that these domains do not have additional functions besides dimerization (Romeis *et al.*, 1997).

In contrast to the tetrapolar mating system of *U. maydis* the barley pathogen *U. hordei* has a bipolar system consisting of only one locus with two alleles, *MAT-1* and *MAT-2*. Interestingly, it was found that in the *U. hordei MAT* locus genes with functions analogous to the *mfa*, *pra*, *bE* and *bW* genes are linked so that they propagate as one locus (Bakkeren and Kronstad, 1993; Bakkeren and Kronstad, 1994; Bakkeren and Kronstad, 1996). The *U. hordei bE* and *bW* genes are also divergently transcribed and show an organization and structure similar to the *b* genes from *U. maydis*. Interestingly, the *U. hordei b* genes are able to initiate the pathogenic program when a nonallelic pair is introduced in *U. maydis* by transformation, indicating that the genes are true homologues to the *U. maydis b* genes (Bakkeren and Kronstad, 1993). Since sequences homologies to the *b* mating-type locus can be detected in a variety of other smut fungi by heterologous hybridization, it is likely the self/nonself discrimination of homeodomain proteins is a conserved mechanism in this group of fungi (Bakkeren *et al.*, 1992).

The presence of the homeodomain suggests that the bE/bW heterodimer plays its crucial role in pathogenic development by acting as transcriptional regulator of pathogenicity genes (Figure 4). Both homeodomains are essential for the function of a bE1/bW2 heterodimer and amino acid substitutions or deletion of amino acids known to be crucial for function of homeodomains render bE as well as bW proteins inactive (Schlesinger *et al.*, 1997). This is in contrast to the situation found in the basidiomycetes *Coprinus cinnereus* and *Schizophyllum commune* and also to the ascomycete *Saccharomyces cerevisiae* where only one of the homeodomains of the heterodimeric homeodomain protein complex is essential for function (Asante Owusu *et al.*, 1996; Luo *et al.*, 1994). Similar to *U. maydis*, in these fungi activation of development is regulated via heterodimerization of two unrelated homeodomain proteins (reviewed in Casselton and Kües, 1994). The role of the bE/bW heterodimer as a transcriptional regulator was validated by the identification of the *lga2* gene (Figures 2 and 4) as the first direct target gene of the bE/bW complex. *lga2* is a gene with so far unknown function located in the *a2* mating-type locus that is transcriptionally activated in a *b* dependent manner (Urban *et al.*, 1996a). The heterodimer was found to bind to a short defined region in the *lga2* gene promoter mediating the *b* dependent activation of the gene (T. Romeis, A. Brachmann, R. Kahmann and J. Kämper, unpublished results). In the case of *lga2* the bE/bW heterodimer acts as a transcriptional activator. However, this does not rule out that in a different promoter context the protein could serve as a transcriptional repressor, as it has been described for different systems in animals, plants and fungi (Miner and Yamamoto, 1991; Hoecker *et al.*, 1995; Tsukiyama and Wu, 1997). In the latter case, the central function of the bE/bW heterodimer during pathogenic development could lie in the repression of a repressor for pathogenicity genes as has been hypothesized by Banuett (1992).

In general, two different types of genes regulated by *b* have to be distinguished, the class 1 genes that are directly regulated by binding of the bE/bW heterodimer to *cis* regulatory sequences, and the class 2 genes that are regulated indirectly via a *b* dependent signal cascade (Figure 4). The various genes involved in establishment of the pathogenic stage could be directly regulated by bE/bW. This situation would be reminiscent to the situation in *Drosophila melanogaster* where the simultaneous regulation of hundreds of different genes is achieved by direct binding of the Even-skipped and Fushi tarazu homeodomain proteins to promoter elements (Liang and Biggin, 1998; Lichter and Mills, 1997). On the other hand, if pathogenicity genes are regulated via a *b* dependent signal cascade (class 2 genes, Figure 4) only a limited number of genes with central regulatory function will turn out to be directly controlled by the bE/bW heterodimer (class 1 genes, Figure 4). Up to now it is unclear which of these two different, but not exclusive scenarios describe the central role of the bE/bW heterodimer for pathogenic development. The latter scenario is supported by a yet uncharacterized *U. maydis* mutant (*rtf1*) which bypasses the bE/bW heterodimer requirement for pathogenic development. Rtf1 has consequently been defined a central repressor of pathogenicity (Banuett, 1991). However, all attempts to recreate such a mutant have failed so far (M. Reichmann and R. Kahmann, unpublished).

A number of genes have been isolated that are likely to be regulated via a *b* dependent signal cascade (class 2 genes, Figure 4). *egl1* codes for an endoglucanase (Schauwecker *et al.*, 1995), *rep1* and *hum2* for a repellent and a hydrophobin, respectively, both hydrophobic surface proteins (Wösten *et al.*, 1996; Bohlmann, 1996), and the genes *dik1* and *dik6* which do not show homologies to proteins in databases

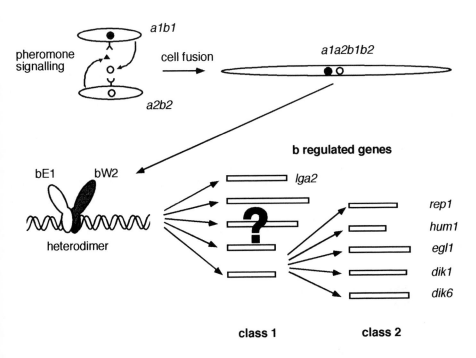

Figure 4.   Pheromone-mediated cell fusion and the proposed action of the bE/bW heterodimer as transcriptional regulator. In the upper part the process of cell recognition through pheromones and their cognate receptors is shown. Cell fusion leads to the generation of the filamentous dikaryon through the action of the bE/bW complex. In the lower part the heterodimer has recognized a particular site on DNA and the respective gene is up or down regulated (not shown). Class 1 genes do represent direct targets of the bE/bW complex. Class 2 genes are indirectly regulated by the bE/bW complex through additional regulator(s) present among the class 1 genes.

(Bohlmann *et al.*, 1994). All of these genes have in common that they are strongly induced in cells with an active bE/bW complex. In addition, the *pra* and *mfa* genes located in the *a* mating-type locus are repressed in cells with an active bE/bW heterodimer (Urban *et al.*, 1996a). Since expression of these genes is a prerequisite for cell fusion, the *b*-mediated repression explains the observation that cell fusion of such strains is significantly attenuated (Laity *et al.*, 1994). However, all these *b*-regulated genes differ significantly in the onset of regulation by the bE/bW heterodimer and individual mutations in these genes do not affect pathogenic development (Bohlmann, 1996; Regenfelder *et al.*, 1997). This was taken to indicate that *b* is not a direct regulator of these genes. The identification of regulatory proteins necessary for their stage-specific induction may, however, provides a valuable tool to fill in the missing links that connect the expression of these presumed class 2 genes to the bE/bW heterodimer (Figure 4).

5. *U. maydis*/maize interaction

The interaction with maize initiates with the attachment of haploid sporidia or teliospores on aerial organs of the plant. Compatible haploid sporidia fuse on the leaf surface and generate the dikaryon which can form extended filaments on the plant surface. Penetration through the plant cell wall can be accomplished by the formation of an appressorium-like structure. Hyphal growth within the plant is a dynamic process and requires several stages of differentiation, which culminate in the formation of sexual, diploid teliospores. The molecular triggers for these developmental processes are unknown. Because these events require the host plant it has been hypothesized that specific plant compounds are perceived by the fungus and result in the activation or repression of specific sets of genes. Therefore, the identification of genes that affect pathogenic development as well as the isolation of genes which are stage-specifically expressed is likely to provide insight into the molecular events during the biotrophic life cycle of *U. maydis*.

5.1. PATHOGENICITY GENES

The REMI (restriction enzyme mediated integration) method (Schiestl and Petes, 1991) has proven successful in the identification of pathogenicity genes in a number of phytopathogenic fungi including *U. maydis* (see Kahmann and Basse, 1999). In this method, linearized plasmid DNA is transformed into the organism of choice in the presence of a restriction enzyme. As a result, plasmid integration often occurs as single copy event at a corresponding restriction site in the genome and thus provides for easy identification of genes through the tag introduced. The integrated plasmid can subsequently be excised from genomic DNA together with flanking sequences, which can be sequenced using primers complementary to the vector backbone. In *U. maydis* the application of this methodology became feasible when solopathogenic haploid strains that express an active bE/bW heterodimer became available (Bölker *et al.*, 1995a). Approximately 50 % of the REMI transformants in such a strain contained single copy integrations of the transformed plasmid (Bölker *et al.* 1995b). From a comprehensive screen comprising 6000 *U. maydis* REMI transformants, pathogenicity genes with functions in amino acid metabolism, signalling, gene regulation, as well as enzymatic functions have been isolated. In addition, about 50 % of all genes identified do not have homologues in available databases (see Kahmann and Basse, 1999; H. Böhnert and R. Kahmann, unpublished). To learn about the function of these genes the establishment of *GFP* (green fluorescent protein, Cubitt *et al.*, 1995) as a reporter gene in *U. maydis* has been instrumental (Spellig *et al.* 1996). The *GFP* gene can be used to study developmental fates of pathogenicity mutants or to investigate temporal and spatial expression patterns of pathogenicity genes during development of the fungus within plant tissue.

Besides these pathogenicity genes identified by an unbiased screen of mutants, a respectable number of genes identified by other means proved to affect pathogenicity as well. This is particularly evident for mutants in cAMP signalling (*gpa3*, *uac1*, *ubc1*, *adr1*) as well as for mutations in MAP kinase cascade components hypothesized to transmit the pheromone signal (*fuz7*, *kpp2/ubc3*). As the pathogenicity defect of mutants in cAMP signalling cannot be bypassed by expressing an active bE/bW heterodimer from constitutive promoters (Regenfelder *et al.*, 1997), Adr1 must have additional targets besides its hypothesized target Prf1. It is expected that their identification will shed considerable light on the entire process of pathogenic development. For the components of the MAP kinase module that attenuate pathogenic development it is interesting that the more upstream signalling components like pheromone and receptor genes can be deleted without adverse effects on pathogenic development while more downstream components cannot. This implies that signalling components may not be restricted in their function to a single module as has been exemplified in *S. cerevisiae* (see Madhani and Fink, 1998)

5.2. PLANT-INDUCED GENES IN *U. MAYDIS*

The identification of genes whose expression is confined to the biotrophic growth phase can follow a number of different strategies. In *U. maydis* REMI has been successfully combined with an enhancer trap method using *GFP* as a reporter. For this purpose a plasmid containing *GFP* fused to a basal *U. maydis* promoter was constructed and transformed by the REMI method. In resulting transformants the *GFP* gene is either not transcribed, constitutively expressed, or stage specifically expressed depending on its genomic context (C. Aichinger and R. Kahmann, unpublished). In addition, the method of differential display (Liang and Pardee, 1992) has been used to identify *U. maydis* genes which are strongly expressed during the interaction with maize. RNA extracted from leaf tumor tissue was compared to corresponding RNA isolated from non-infected leaf tissue. From a comprehensive screen eight maize genes that are upregulated between 3 and 60-fold in tumor tissue, two maize genes that are downregulated upon infection, and four fungal genes that are strongly upregulated in tumor tissue were identified. Among these fungal genes strongly expressed during the tumor stage, *mig1* and *mig2* displayed a similar regulatory profile (Basse *et al.*, 2000). The *mig1* gene of *U. maydis* represents the first gene identified in this organism whose expression is tightly coupled to the biotrophic phase. The *mig1* gene encodes a small, highly charged, secreted protein of unknown function. Expression of *mig1* mRNA was not detectable in haploid strains and was weak in a diploid strain heterozygous for *a* and *b*. This indicates that *mig1* gene expression responds to some extent to the *a* and *b* mating-type loci, which are known to regulate a large number of genes prior to infection (Hartmann *et al.*, 1996). Based on *GFP* reporter gene activity the expression of *mig1* is undetectable during hyphal growth on the leaf surface and formation of infection structures but is immediately switched on after penetration, and remains high during fungal colonization between 4 and 11 d post infection. Subsequently, the expression appears downregulated in proliferating sporogenic hyphae and becomes virtually undetectable in mature teliospores (Basse *et al.*, 2000). For the *Cladosporium fulvum Avr9* and a number of other plant induced genes in fungi it has been demonstrated that expression can also be increased under conditions of nitrogen and/or carbon deprivation (Lauge and deWit, 1998). In contrast, the *mig1* gene is not induced under comparable starvation conditions. Expression of the *mig1* gene is subjected to negative as well as positive regulation as assessed from promoter deletion studies. In particular, a 0.5 kb region upstream of the translational start in the *mig1* promoter contains several sequence elements specifically required for high levels of induction during fungal growth *in planta* (Basse *et al.*, 2000). The availability of functional *mig1* reporter gene fusions will now allow one to address signalling between *U. maydis* and its host genetically. Through the isolation of *U. maydis* mutants derepressed in *mig1* gene expression and subsequent cloning of the affected gene(s) by complementation, information will be gained on the numbers of genes involved, as well as their specific function in repressing the *mig1* gene promoter. Once such mutants are available, they will also provide important tools for addressing the positive regulation acting on the *mig1* promoter. It is conceivable that relief from negative regulation may be prerequisite for the strong induction observed during fungal growth *in planta*. For this reason the availability of *mig1* repressor mutants may also allow the identification of the inducing maize compounds directly or allow a screen to identify the respective activator gene from an expression library (Basse *et al.*, 2000).

5.3. TUMOR INDUCTION BY *U. MAYDIS*

Among different *Ustilago* species, *U. maydis* is the only one that is able to cause large tumors in its host plant. A particularly intriguing question is how the massive proliferation of plant tissue is induced after infection. Presently, the molecular inducers for this morphologic transition are unknown, however, it is conceivable that plant hormones participate in this process. Biochemical analyses of fungal culture supernatants and of plant tumors have indicated that *U. maydis* is capable of producing plant growth promoting substances. *U. maydis* was shown to produce indole-3-acetic acid (IAA) when grown in medium containing tryptophan (Wolf, 1952). In addition, in tumor tissue 5 to 20-fold elevated levels of auxin have been demonstrated (Wolf, 1952; Turian and Hamilton, 1960). The presence of elevated IAA levels in leaf tumor tissue was recently confirmed by mass spectroscopic analysis (C. Basse, unpublished). IAA has profound effects on biological processes such as cell expansion and cell division (see Klee and Estelle, 1991). Since tumor development in plants infected by *U. maydis* is characterized by increased mitotic divisions and an enlargement of cells (Christensen, 1963; Callow and Ling, 1973) it is conceivable that IAA produced by *U. maydis* is involved in this process.

   In *U. maydis* two enzymes Iad1 and Iad2 converting indole-3-acetaldehyde to IAA were characterized. Indole-3-acetaldehyde is the final intermediate in a possible IAA pathway starting from tryptophan. The *iad1* gene was cloned and corresponding null-mutant strains were inable to convert indole-3-acetaldehyde to IAA in the presence of glucose. However, in the presence of arabinose this ability was regained due to the activity of Iad2. Since *iad1* mutant strains caused tumors indistinguishable from wild-type strains it is possible that *iad2* and *iad1* have redundant functions. Therefore, an assessment of their involvement in tumor induction will have to await the cloning of the *iad2* gene and the construction of double mutants (Basse *et al.*, 1996).

   Additional biosynthetic pathways for plant hormones have not yet been described in *U. maydis*, although previous studies report on the formation of cytokinins and gibberellins in *U. maydis* during axenic growth (Mills and van Staden, 1978; Sokolovskaya and Kuznetsov, 1984). Gibberellins are implicated in cell elongation and could conceivably be implicated in tumor formation. The giberellin biosynthesis pathway has been analyzed in detail in *Gibberella fujikuroi* (Tudzynski, 1999) and the gene encoding copalyl pyrophosphate synthase required for the conversion of geranylgeranyl pyrophosphate to copalyl pyrophosphate, a key intermediate in gibberellin biosynthesis, has been cloned from this organism (Tudzynski *et al.*, 1998). However, attempts to isolate a related gene from *U. maydis* have so far not been successful (C. Basse and R. Kahmann, unpublished). Besides the possibility that increased hormone levels in tumor tissue are directly caused by phytohormone production of *U. maydis* the alternative possibility of altering the hormone balance in the host plant through a fungal compound also has to be considered.

5.4. THE NUTRITIONAL SITUATION IN THE HOST

During biotrophic growth *U. maydis* has to adapt to a novel nutritional environment within the host plant. It is evident that an efficient supply of nutrients from the host must occur to provide for the massive fungal proliferation seen in tumor tissue. Presently, it is unclear which carbon sources are used by *U. maydis* during growth within the plant and how they accumulate in tumor tissue. Previous investigations have pointed to a redirection of carbohydrates from photosynthetically active organs to infected tissue where imported assimilates are preferentially localized in tumor tissue (Billett and Burnett, 1978). It will be a challenging task to elucidate how these processes are triggered by *U. maydis*. In this respect, the characterization of maize genes differentially expressed during tumor development may shed light on the underlying molecular events.

An additional limitation during biotrophic growth may be the uptake of iron as is the case for some human, animal and plant pathogens (see Leong and Expert, 1989). This aspect was also addressed in *U. maydis* by the isolation of a class of *U. maydis* mutants defective in the formation of the ferric iron chelating peptides (siderophores) ferrichrome and ferrichrome A. Such mutants were unable to grow under conditions of iron deprivation. However, compatible strains both carrying the respective mutation were not affected in pathogenicity. This could indicate that iron uptake is either not critical during pathogenic development or functional gene redundancy could exist (Mei *et al.*, 1993).

6. Genome variability

Karyotype analysis by orthogonal field-alternation gel electrophoresis (OFAGE) or clamped homogeneous electric fields (CHEF) electrophoresis of numerous *U. maydis* laboratory strains and freshly collected field isolates revealed that considerable chromosome length polymorphisms exists in this species. Although general features of the banding pattern of chromosome-sized DNAs were conserved between strains, all strains tested had different karyotypes (Kinscherf and Leong, 1988; J. Kämper, unpublished data). Different strains contain up to 20 chromosomes with sizes ranging from 300 kb to about 3 Mb in size, giving the haploid genome of *U. maydis* a complexity of about 20 Mb (Kinscherf and Leong, 1988; J. Kämper, unpublished data). The variation in chromosome size is most probably due to frequent translocations, deletions or insertions as it has been also noticed for *U. hordei* (McCluskey *et al.*, 1994; Gaudet *et al.*, 1998). The observed genome flexibility could serve as a mechanism of pathogenic microorganisms that allows the rapid adaptation to the host which, in the case of biotrophic microorganisms, could be a prerequisite for the survival of the species.

7. The *U.maydis* genome project

The *Ustilago maydis* genome project launched by the Bayer AG (Leverkusen, Germany) will allow another rationale to identify genes involved in pathogenicity. The comparison of genomes of different pathogenic and nonpathogenic fungi (*U. maydis* and *Magnaporthe grisea* in comparison with *S. cerevisiae* and *Schizosaccharomyces pombe*, for example) will allow the identification of genes that are unique to the pathogenic species. The availability of the genomic *U. maydis* DNA sequence will also allow the establishment of microarrays for transcriptional profiling. For this technique, specific probes, oligonucleotides or DNA fragments, corresponding to each single gene or open reading frame, is attached to a solid matrix. These microarrays are then probed with labelled cDNA obtained from RNA representing two different cell states, and from differences in signal intensity the mRNA abundance in the two stages can be calculated (see Duggan et al, 1999 and other reviews in the same issue). This technique should allow the evaluation of genome-wide changes in expression, for example after controlled induction of the master regulator for pathogenic development, the bE/bW heterodimer. In all cases, candidate genes must subsequently be evaluated by gene disruption experiments. Gene replacement could be accomplished individually for each gene; however this labor-intensive strategy is only applicable to a limited number. For a systematic approach it is instrumental to use mutant libraries for rapid isolation of the desired mutations. Such libraries could be made either by insertional mutagenesis using REMI or through transposon tagging strategies, a technique not yet established for *U maydis*.

8. Outlook

The design of the required tools has allowed the identification of some of the genes required at various stages of pathogenic development in this system. The initial studies already reveal the enormous level of complexity to be expected. Nevertheless, with the advent of functional genomics, global regulatory connections are anticipated to emerge which should allow the concentration on key regulators of pathogenicity. It will be challenging to use this model system to investigate gene function of obligate parasites for which no knockout strategies have been developed by performing mutational analysis of the homologous genes in *U. maydis*. Another fascinating prospect is the identification of molecules on which the communication between plant and fungus is built.

9. References

Agrios, G.N. (1988) *Plant Pathology*. Academic Press, Inc., San Diego.

Apel-Birkhold, P.C. and Walton, J.D. (1996) Cloning, disruption, and expression of two endo-beta 1, 4-xylanase genes, XYL2 and XYL3, from *Cochliobolus carbonum*. *Appl. Envirom. Microbiol.*, **62**, 4129-4135.

Asante Owusu, R.N., Banham, A.H., Bohnert, H.U., Mellor, E.J. and Casselton, L.A. (1996) Heterodimerization between two classes of homeodomain proteins in the mushroom *Coprinus cinereus* brings together potential DNA-binding and activation domains. *Gene*, **172**, 25-31.

Bakkeren, G. and Kronstad, J.W. (1993) Conservation of the b mating-type gene complex among bipolar and tetrapolar smut fungi. *Plant Cell*, **5**, 123-136.

Bakkeren, G. and Kronstad, J.W. (1994) Linkage of mating-type loci distinguishes bipolar from tetrapolar mating in basidiomycetous smut fungi. *Proc. Natl. Acad. Sci USA*, **91**, 7085-7089.

Bakkeren, G. and Kronstad, J.W. (1996) The pheromone cell signaling components of the *Ustilago a* mating-type loci determine intercompatibility between species. *Genetics*, **143**, 1601-1613.

Bakkeren, G., Gibbard, B., Yee, A., Froeliger, E., Leong, S. and Kronstad, J. (1992) The a and b loci of *Ustilago maydis* hybridize with DNA sequences from other smut fungi. *Mol. Plant. Microbe. Interact.*, **5**, 347-355.

Banuett, F. (1991) Identification of genes governing filamentous growth and tumor induction by the plant pathogen *Ustilago maydis*. *Proc. Natl. Acad. Sci. USA*, **88**, 3922-3926.

Banuett, F. (1992) *Ustilago maydis*, the delightful blight. *Trends Genet.*, **8**, 174-180.

Banuett, F. and Herskowitz, I. (1989) Different a alleles are necessary for maintenance of filamentous growth but not for meiosis. *Proc. Natl. Acad. Sci. USA*, **86**, 5878-5882.

Banuett, F. and Herskowitz, I. (1994a) Identification of fuz7, a *Ustilago maydis* MEK/MAPKK homolog required for a-locus-dependent and -independent steps in the fungal life cycle. *Genes Dev*, **8**, 1367-1378.

Banuett, F. and Herskowitz, I. (1994b) Morphological transitions in the life cycle of *Ustilago maydis* and their genetic control by the a and b loci. *Exp. Mycol.*, **18**, 247-266.

Banuett, F. and Herskowitz, I. (1996) Discrete developmental stages during teliospore formation in the corn smut fungus, *Ustilago maydis*. *Development*, **122**, 2965-2976.

Basse, C.W., Lottspeich, F., Steglich, W. and Kahmann, R. (1996) Two potential indole-3-acetaldehyde dehydrogenases in the phytopathogenic fungus *Ustilago maydis*. *Eur. J Biochem*, **242**, 648-656

Basse, C.W., Stumpferl, S. and Kahmann, R. (2000) Characterization of a *Ustilago maydis* gene specifically induced during the biotrophic phase: evidence for negative as well as positive regulation. *Mol. Cell. Biol.*, **20**, 329-339.

Billett, E.E. and Burnett, J.H. (1978) The host-parasite physiology of the maize smut fungus, *Ustilago maydis*. II. Translocation of ^{14}C-labelled assimilates in smutted maize plants. *Physiol. Plant Pathol.*, **12**, 103-112.

Bohlmann, R. (1996) Isolierung und Charakterisierung von filamentspezifisch exprimierten Genen aus *Ustilago maydis*. *Fakultät für Biologie*. Ludwig-Maximilians-Universität, München.

Bohlmann, R., Schauwecker, F., Basse, C. and Kahmann, R. (1994) Genetic regulation of mating and dimorphism in *Ustilago maydis*. In Daniels, M.J. (ed.) *Advances in Molecular Genetics of Plant-Microbe Interactions*. Kluwer Acad. Publ., Dordrecht, Vol. 3, pp. 239-245.

Bölker, M., Genin, S., Lehmler, C. and Kahmann, R. (1995a) Genetic regulation of mating, and dimorphism in *Ustilago maydis*. *Can. J. Bot.*, **73**, 320-325.

Bölker, M., Böhnert, H.U., Braun, K.H., Görl, J. and Kahmann, R. (1995b) Tagging pathogenicity genes in *Ustilago maydis* by restriction enzyme-mediated integration (REMI). *Mol. Gen. Genet.*, **248**, 547-552.

Bölker, M., Urban, M. and Kahmann, R. (1992) The a mating type locus of *U. maydis* specifies cell signaling components. *Cell*, **68**, 441-450.

Bowen, A.R., Chen-Wu, J.L., Momany, M., Young, R., Szaniszlo, P.J. and Robbins, P.W. (1992) Classification of fungal chitin synthases. *Proc. Natl. Acad. Sci. USA*, **89**, 519-523.

Bowman, D.H. (1946) Sporidial fusion in *Ustilago maydis*. *J. Agric. Res.*, **72**, 233-243.

Brefeld, O. (1883) Untersuchungen aus dem Gesammtgebiet der Mykologie. , **Heft 5**, 67-75.

Brefeld, O. (1895) Untersuchungen aus dem Gesammtgebiet der Mykologie. , **Heft 11**, 52-92.

Callow, J.A. and Ling, I.T. (1973) Histology of neoplasms and chlorotic lesions in maize seedlings following the infection of sporidia of *Ustilago maydis* (DC) Corda. *Physiol. Plant Pathol.*, **3**, 489-494.

Casselton, L.A. and Kües, U. (1994) Mating type genes in homobasidiomycetes. In Wessels, J.G.H. and Meinhardt, F. (eds.), *The Mycota I: Growth, Differentiation and Sexuality*. Springer-Verlag, Heidelberg, pp. 307-321.

Christensen, J.J. (1963) Corn smut induced by *Ustilago maydis*. *Amer. Phytopathol. Soc. Monogr.*, **2**.

Cubitt, A.B., Heim, R., Adams, S.R., Boyd, A.E., Gross, L.A. and Tsien, R.Y. (1995) Understanding, improving and using green fluorescent proteins. *Trends Biochem. Sci.*, **20**, 448-455.

Day, A.W. and Poon, N.H. (1975) Fungal fimbriae. II. Their role in conjugation in *Ustilago violacea*. *Can. J. Microbiol.*, **21**, 547-57.

Day, P.R. and Anagnostakis, S.L. (1971) Corn smut dikaryon in culture. *Nature New Biol.*, **231**, 19-20.

Day, P.R., Anagnostakis, S.L. and Puhalla, J.E. (1971) Pathogenicity resulting from mutation at the *b* locus of *Ustilago maydis*. *Proc. Natl. Acad. Sci. USA*, **68**, 533-535.

Dürrenberger, F. and Kronstad, J. (1999) The ukc1 gene encodes a protein kinase involved in morphogenesis, pathogenicity and pigment formation in *Ustilago maydis*. *Mol. Gen. Genet.*, **261**, 281-289.

Dürrenberger, F., Wong, K. and Kronstad, J.W. (1998) Identification of a cAMP-dependent protein kinase catalytic subunit required for virulence and morphogenesis in *Ustilago maydis*. *Proc. Natl. Acad. Sci. USA*, **95**, 5684-5689.

Duggan, D.J., Bittner, M., Chen, Y., Meltzer, P. and Trent, J.M. (1999) Expression profiling using cDNA microarrays. *Nat. Genet.*, **21**, 10-14.

Ehrlich, H.G. (1958) Nuclear behavior in mycelium of a solopathogenic line and in a cross of two haploid lines of *Ustilgo maydis* (DC.) Cda. *Mycologia*, **50**, 622-627.

Fischer, G.W. and Holton, C.S. (1957) Biology and control of the smut fungi. Ronald Press Co., New York.

Fletcher, H.L. (1981) A search for synaptonemal complexes in *Ustilago maydis*. *J. Cell Sci.*, **50**, 171-180.

Gaudet, D.A., Gusse, J. and Laroche, A. (1998) Origins and inheritance of chromosome-length polymorphisms in the barley covered smut fungus, *Ustilago hordei*. *Curr. Genet.*, **33**, 216-224.

Gillissen, B., Bergemann, J., Sandmann, C., Schröer, B., Bölker, M. and Kahmann, R. (1992) A two-component regulatory system for self/non-self recognition in *Ustilago maydis*. *Cell*, **68**, 647-657.

Gold, S.E., Duncan, G., Barrett, K. and Kronstad, J. (1994) cAMP regulates morphogenesis in the fungal pathogen *Ustilago maydis*. *Genes Dev*, **8**, 2805-2816.

Gold, S.E. and Kronstad, J.W. (1994) Disruption of two genes for chitin synthase in the phytopathogenic fungus *Ustilago maydis*. *Mol. Microbiol.*, **11**, 897-902.

Gold, S.E., Brogdon, S.M., Mayorga, M.E. and Kronstad, J.W. (1997) The *Ustilago maydis* regulatory subunit of a cAMP-dependent protein kinase is required for gall formation in maize. *Plant Cell*, **9**, 1585-1594.

Hanna, W.F. (1929) Studies in the physiology and cytology of *Ustilago zeae* and *Sorosporium reilianum*. *Phytopathology*, **19**, 415-443.

Hartmann, H.A., Kahmann, R. and Bölker, M. (1996) The pheromone response factor coordinates filamentous growth and pathogenicity in *Ustilago maydis*. *EMBO J.*, **15**, 1632-1641.

Hartmann, H.A., Krüger, J., Lottspeich, F. and Kahmann, R. (1999) Environmental signals controlling sexual development of the corn Smut fungus *Ustilago maydis* through the transcriptional regulator Prf1. *Plant Cell*, **11**, 1293-1306.

Hoecker, U., Vasil, I.K. and McCarty, D.R. (1995) Integrated control of seed maturation and germination programs by activator and repressor functions of *Viviparous-1* of maize. *Genes Dev.*, **9**, 2459-2469.

Holden D.W., Kronstad J.W. and Leong, S.A. (1989) Mutation in a heat-regulated hsp70 gene of *Ustilago maydis*. *EMBO J* , **8**, 1927-1934.

Holliday, R. (1961) The genetics of *Ustilago maydis*. *Genet. Res. Camb.*, **2**, 204-230.

Holliday, R. (1965) Induced mitotic crossing-over in relation to genetic replication in synchronously dividing cells of *Ustilago maydis*. *Genet. Res. Camb.*, **6**, 104-120.

Jacobs, C.W., Mattichak, S.J. and Knowles, J.F. (1994) Budding patterns during cell cycle of the maize smut pathogen *Ustilago maydis*. *Can. J. Bot.*, **72**, 1675-1680.

Kahmann, R. and Basse, C. (1999) REMI (Restriction Enzyme Mediated Integration) and its impact on the isolation of pathogenicity genes in fungi attacking plants. *Eur. J. Plant Pathol.*, **105**, 221-229.

Kahmann, R., Basse, C. and Feldbrügge, M. (1999) Fungal-plant signalling in the *Ustilago maydis*-maize pathosystem. *Curr. Opin. Microbiol.*, **2**, 647-650.

Kämper, J., Reichmann, M., Romeis, T., Bölker, M. and Kahmann, R. (1995) Multiallelic recognition: nonself-dependent dimerization of the bE and bW homeodomain proteins in *Ustilago maydis*. *Cell*, **81**, 73-83.

Keon, J.P., Jewitt, S. and Hargreaves, J.A. (1995) A gene encoding _-adaptin is required for apical extension growth in *Ustilago maydis*. *Gene*, **162**, 141-145.

Kinscherf, T.G. and Leong, S.A. (1988) Molecular analysis of the karyotype of *Ustilago maydis*. *Chromosoma*, **96**, 427-433.

Klee, H. and Estelle, M. (1991) Molecular genetic approaches to plant hormone biology. *Annu. Rev. Plant Physiol. Plant Mol. Biol.*, **42**, 529-551.

Kronstad, J.W. and Leong, S.A. (1990) The *b* mating-type locus of *Ustilago maydis* contains variable and constant regions. *Genes Dev.*, **4**, 1384-1395.

Kronstad, J.W. and Staben, C. (1997) Mating type in filamentous fungi. *Annu. Rev. Genet.*, **31**, 245-276.

Krüger, J., Loubradou, G., Regenfelder, E., Hartmann, A. and Kahmann, R. (1998) Crosstalk between cAMP and pheromone signalling pathways in *Ustilago maydis*. *Mol Gen Genet*, **260**, 193-198.

Kusch, G. and Schauz, K. (1989) Light and electron microscopic studies of chlamydospore development in *Ustilago maydis* (Ustilaginales, Basidiomycetes). *Crypt. Bot.*, **1**, 230-235.

Laity, C., Giasson, L., Campbell, R. and Kronstad, J. (1995) Heterozygosity at the b mating-type locus attenuates fusion in *Ustilago maydis*. *Curr. Genet.*, **27**, 451-459.

Lauge, R. and De Wit, P.J. (1998) Fungal avirulence genes: structure and possible functions. *Fungal Genet. Biol.*, **24**, 285-297.

Lehmler, C., Steinberg, G., Snetselaar, K. M., Schliwa, M., Kahmann, R. and Bölker, M. (1997) Identification of a motor protein required for filamentous growth in *Ustilago maydis*. *EMBO J.*, **16**, 3464-3473.

Leong, S.A. and Expert, D. (1989) In Nester, E. and Kosuge, T. (eds.), *Plant-Microbe interactions-a molecular genetic perspective*, McGraw-Hill, New York, NY, pp. 62-83.

Liang, P. and Pardee, A.B. (1992) Differential display of eukaryotic messenger RNA by means of the polymerase chain reaction. *Science*, **257**, 967-971.

Liang, Z. and Biggin, M.D. (1998) Eve and ftz regulate a wide array of genes in blastoderm embryos: the selector homeoproteins directly or indirectly regulate most genes in *Drosophila*. *Development*, **125**, 4471-4482.

Lichter, A. and Mills, D. (1997) Fil1, a G-protein alpha-subunit that acts upstream of cAMP and is essential for dimorphic switching in haploid cells of *Ustilago hordei*. *Mol Gen Genet*, **256**, 426-435.

Luo, Y., Ullrich, R.C. and Novotny, C.P. (1994) Only one of the paired *Schizophyllum commune* A alpha mating-type, putative homeobox genes encodes a homeodomain essential for A alpha-regulated development. *Mol. Gen. Genet.*, **244**, 318-324.

Lutman, B.F. (1910) Some contributions to the life history and cytology of the smuts. *Trans. Wisconsin Acad. Sci.*, **16**, 1191-1244.

Madhani, H.D. and Fink, G.R. (1998) The control of filamentous differentiation and virulence in fungi. *Trends Cell Biol.*, **8**, 348-353.

Mayorga, M.E. and Gold, S.E. (1999) A MAP kinase encoded by the ubc3 gene of *Ustilago maydis* is required for filamentous growth and full virulence. *Mol. Microbiol.*, **34**, 485-497.

McCluskey, K., Agnan, J. and Mills, D. (1994) Characterization of genome plasticity in *Ustilago hordei*. *Curr. Genet.*, **26**, 486-493.

Mei, B., Budde, A.D. and Leong, S.A. (1993) sid1, a gene initiating siderophore biosynthesis in *Ustilago maydis*: molecular characterization, regulation by iron, and role in phytopathogenicity. *Proc. Natl. Acad. Sci. U S A*, **90**, 903-907.

Millis, L.J. and Kotze, J.M. (1981) Scanning electron microscopy of the germination, growth and infection of *Ustilago maydis* on maize. *Phytopth. Z.*, **102**, 21-27.

Mills, L.J. and Van Staden, J. (1978) Extraction of cytokinins from maize, smut tumors of maize and *Ustilago maydis* cultures. *Physiol. Plant Pathol.*, **13**, 73-80.

Miner, J.N. and Yamamoto, K.R. (1991) Regulatory crosstalk at composite response elements. *Trends Biochem. Sci.*, **16**, 423-426.

Müller, P., Aichinger, C., Feldbrügge, M. and Kahmann, R. (1999) The MAP kinase Kpp2 regulates mating and pathogenic development in *Ustilago maydis*. *Molecular Microbiology*, **34**, 1007-1017.

O'Donell, K. (1992) Ultrastructure of meiosis and the spindle pole body cycle in freeze-substituted basidia of the smut fungi *Ustilago maydis* and *Ustilago avenae*. *Can. J. Bot.*, **70**, 629-638.

O'Donell, K.L. and McLaughlin, D.J. (1984a) Ultrastructure of meiosis in *Ustilago maydis*. *Mycologia*, **76**, 468-485.

O'Donell, K.L. and McLaughlin, D.J. (1984b) Postemeiotic mitosis, basidiospore development, and septation in *Ustilago maydis*. *Mycologia*, **76**, 486-502.

Plamann, M., Minke, P.F., Tinsley, J.H. and Bruno, K.S. (1994) Cytoplasmic dynein and actin-related protein Arp1 are required for normal nuclear distribution in filamentous fungi. *J. Cell Biol.*, **127**, 139-149.

Poon, H. and Day, A.W. (1974) 'Fimbriae' in the fungus *Ustilago violacea*. *Nature*, **250**, 648-649.

Poon, N.H. and Day, A.W. (1975) Fungal fimbriae. I. Structure, origin, and synthesis. *Can. J. Bot.*, **21**, 537-546.

Puhalla, J.E. (1968) Compatibility reactions on solid medium and interstrain inhibition in *Ustilago maydis*. *Genetics*, **60**, 461-474.

Puhalla, J.E. (1969) The formation of diploids of *Ustilago maydis* on agar medium. *Phytopathology*, **59**, 1771-1772.

Puhalla, J.E. (1970) Genetic studies on the *b* incompatibility locus of *Ustilago maydis*. *Genet. Res. Camb.*, **16**, 229-232.

Rawitscher, F. (1912) Beiträge zur Kenntnis der Ustilagineen I. *Ztschr. Bot.*, **4**, 673-706.

Regenfelder, E., Spellig, T., Hartmann, A., Lauenstein, S., Bölker, M. and Kahmann, R. (1997) G proteins in *Ustilago maydis*: Transmission of multiple signals? *EMBO J.*, **16**, 1934-1942.

Romeis, T., Kämper, J. and Kahmann, R. (1997) Single-chain fusions of two unrelated homeodomain proteins trigger pathogenicity in *Ustilago maydis*. *Proc. Natl. Acad. Sci. USA*, **94**, 1230-1234.

Rowell, J.B. (1955) Functional role of compatibility factors and an in vitro test for sexual compatibility with haploid lines of *Ustilag zeae*. *Phytopathology*, **45**, 370-374.

Rowell, J.B. and DeVay, J.E. (1954) Genetics of *Ustilago zea* in relation to basic problems of its pathogenicity. *Phytopathology*, **44**, 356-362.

Ruiz-Herrera, J., Leon Claudia, G., Guevara-Olvera, L. and Carabez-Trejo, A. (1995) Yeast-mycelial dimorphism of haploid and diploid strains of *Ustilago maydis*. *Microbiology*, **141**, 695-703.

Ruiz-Herrera, J., Leon, C.G., Carabez-Trejo, A. and Reyes-Salinas, E. (1996) Structure and chemical composition of the cell walls from the haploid yeast and mycelial forms of *Ustilago maydis*. *Fungal Genet. Biol.*, **20**, 133-142.

Schauwecker, F., Wanner, G. and Kahmann, R. (1995) Filament-specific expression of a cellulase gene in the dimorphic fungus *Ustilago maydis*. *Biol. Chem. Hoppe-Seyler*, **376**, 617-625.

Schiestl, R.H. and Petes, T.D. (1991) Integration of DNA fragments by illegitimate recombination in *Saccharomyces cerevisiae*. *Proc. Natl. Acad. Sci. U.S.A*, **88**, 7585-7589.

Schlesinger, R., Kahmann, R. and Kämper, J. (1997) The homeodomains of the heterodimeric bE and bW proteins of *Ustilago maydis* are both critical for function. *Mol. Gen. Genet.*, **254**, 514-519.

Schulz, B., Banuett, F., Dahl, M., Schlesinger, R., Schäfer, W., Martin, T., Herskowitz, I. and Kahmann, R. (1990) The *b* alleles of *U. maydis*, whose combinations program pathogenic development, code for polypeptides containing a homeodomain-related motif. *Cell*, **60**, 295-306.

Sleumer, H.O. (1932) Über Sexualität und Zytologie von *Ustilago zeae* (Beckm.) Unger. *Z. Botan.*, **25**, 209-263.

Snetselaar, K.M. (1993) Microscopic observation of *Ustilago maydis* mating interactions. *Exp. Mycol.*, **17**, 345-355.

Snetselaar, K.M. and Mims, C.W. (1992) Sporidial fusion and infection of maize seedlings by the smut fungus *Ustilago maydis*. *Mycologia*, **84**, 193-203.

Snetselaar, K.M. and Mims, C.W. (1993) Infection of maize stigmas by *Ustilago maydis*: Light and electron microscopy. *Phytopathology*, **83**, 843-850.

Snetselaar, K.M. and Mims, C.W. (1994) Light and electron microscopy of *Ustilago maydis* hyphae in maize. *Mycol. Res.*, **98**, 347-355.

Snetselaar, K.M. and McCann, M.P. (1997) Using microdensitometry to correlate cell morphology with the nuclear cycle in *Ustilago maydis*. *Mycologia*, **89**, 689-697.

Snetselaar, K.M., Boelker, M. and Kahmann, R. (1996) *Ustilago maydis* mating hyphae orient their growth toward pheromone sources. *Fungal Genet. Biol.*, **20**, 299-312.

Sokolovskaya, I.V. and Kuznetsov, L.V. (1984) Gibberellin-like substances in the mycelium of haploid and diploid strains of the smut fungus *Ustilago-zeae*. *Appl. Biochem. Microbiol.*, **20**, 397-401.

Spellig, T., Boelker, M., Lottspeich, F., Frank, R.W. and Kahmann, R. (1994) Pheromones trigger filamentous growth in *Ustilago maydis*. *EMBO J.*, **13**, 1620-1627.

Spellig, T., Bottin, A. and Kahmann, R. (1996) Green fluorescent protein (GFP) as a new vital marker in the phytopathogenic fungus *Ustilago maydis*. *Mol. Gen. Genet.*, **252**, 503-509.

Steinberg, G. (1998) Organelle transport and molecular motors in fungi. *Fungal Genet. Biol.*, **24**, 161-177.

Steinberg, G., Schliwa, M., Lehmler, C., Boelker, M., Kahmann, R. and McIntosh, J.R. (1998) Kinesin from the plant pathogenic fungus *Ustilago maydis* is involved in vacuole formation and cytoplasmic migration. *J. Cell Sci.*, **111**, 2235-2246.

Trueheart, J. and Herskowitz, I. (1992) The a locus governs cytoduction in *Ustilago maydis*. *J. Bacteriol.*, **174**, 7831-7833.

Tsukiyama, T. and Wu, C. (1997) Chromatin remodeling and transcription. *Curr. Opin. Genet. Dev.*, **7**, 182-191.

Tsukuda, T., Carleton, S., Fotheringham, S. and Holloman, W.K. (1988) Isolation and characterization of an autonomously replicating sequence from *Ustilago maydis*. *Mol. Cell. Biol.*, **8**, 3703-3709.

Tudzynski, B. (1999) Biosynthesis of gibberellins in *Gibberella fujikuroi*: biomolecular aspects. *Appl. Microbiol. Biotechnol.*, **52**, 298-310.

Tudzynski, B., Kawaide, H. and Kamiya, Y. (1998) Gibberellin biosynthesis in *Gibberella fujikuroi*: cloning and characterization of the copalyl diphosphate synthase gene. *Curr. Genet.*, **34**, 234-240.

Turian, G. and Hamilton, R.H. (1960) Chemical detection of 3-indolylacetic acid in *Ustilago zeae* tumors. *Biochem. Biophys. Acta*, **41**, 148-150.

Urban, M., Kahmann, R. and Bölker, M. (1996a) Identification of the pheromone response element in *Ustilago maydis*. *Mol. Gen. Genet.*, **251**, 31-37.

Urban, M., Kahmann, R. and Bölker, M. (1996b) The biallelic *a* mating type locus of *Ustilago maydis*: remnants of an additional pheromone gene indicate evolution from a multiallelic ancestor. *Mol. Gen. Genet.*, **250**, 414-420.

Walter, J.M. (1934) The mode of entrance of *Ustilago zeae* into corn. *Phytopathology*, **24**, 1012-1020.

Wang, J., Holden, D.W. and Leong, S.A. (1988) Gene transfer system for the phytopathogenic fungus *Ustilago maydis*. *Proc. Natl. Acad. Sci. USA*, **85**, 865-869.

Wedlich-Söldner, R., Bölker, M., Kahmann, R. and Steinberg, G. (2000) *EMBO J., submitted.*

Wessels, J.G.H. (1996) Fungal hydrophobins: proteins that function at an interface. *Trends Plant Sci.*, **1**, 9-15.

Wolf, F.T. (1952) The production of indole acetic acid by *Ustilago zeae*, and its possible significance in tumor formation. *Proc. Natl. Acad. Sci. USA*, **38**, 106-111.

Wong, G.J. and Wells, K. (1985) Modified bifactorial incompatibility in *Tremella mesenterica*. *Trans. Br. Mycol. Soc.*, **84**, 95-109.

Wösten, H.A., Bohlmann, R., Eckerskorn, C., Lottspeich, F., Bölker, M. and Kahmann, R. (1996) A novel class of small amphipathic peptides affect aerial hyphal growth and surface hydrophobicity in *Ustilago maydis*. *EMBO J.*, **15**, 4274-4281.

Xoconostle-Cazares, B., Leon-Ramirez, C. and Ruiz-Herrera, J. (1996) Two chitin synthase genes from *Ustilago maydis*. *Microbiology*, **142**, 377-387.

Xoconostle-Cazares, B., Specht Charles, A., Robbins Phillips, W., Liu, Y., Leon, C. and Ruiz-Herrera, J. (1997) Umchs5, a gene coding for class IV chitin synthase in *Ustilago maydis*. *Fungal Genet. Biol.*, **22**, 199-208.

Xu, J. and Day Alan, W. (1992) Multiple forms of fimbriae on the sporidia of corn smut, *Ustilago maydis*. *Int. J. Plant Sci.*, **153**, 531-540.

Yee, A.R. and Kronstad, J.W. (1993) Construction of chimeric alleles with altered specificity at the *b* incompatibility locus of Ustilago maydis. *Proc. Natl. Acad. Sci. USA*, **90**, 664-668.

Yee, A.R. and Kronstad, J.W. (1998) Dual sets of chimeric alleles identify specificity sequences for the *bE* and *bW* mating and pathogenicity genes of *Ustilago maydis*. *Mol. Cell. Biol.*, **18**, 221-232.

Index

Index